Big Data Analytics in Chemoinformatics and Bioinformatics

Big Data Analytics in Chemoinformatics and Bioinformatics

With Applications to Computer-Aided Drug Design, Cancer Biology, Emerging Pathogens and Computational Toxicology

Edited by

Subhash C. Basak
Department of Chemistry and Biochemistry, University of Minnesota, Duluth, MN, United States

Marjan Vračko
Theory Department, Kemijski inštitut/National Institute of Chemistry, Ljubljana, Slovenia

ELSEVIER

Elsevier
Radarweg 29, PO Box 211, 1000 AE Amsterdam, Netherlands
The Boulevard, Langford Lane, Kidlington, Oxford OX5 1GB, United Kingdom
50 Hampshire Street, 5th Floor, Cambridge, MA 02139, United States

ISBN: 978-0-323-85713-0

For Information on all Elsevier publications
visit our website at https://www.elsevier.com/books-and-journals

Publisher: Susan Dennis
Acquisitions Editor: Charlotte Rowley
Editorial Project Manager: Kyle Gravel
Production Project Manager: Sujatha Thirugnana Sambandam
Cover Designer: Greg Harris

Typeset by MPS Limited, Chennai, India

Contents

Section 3 Bioinformatics and computatioanl toxicology section

List of contributors

Anshika Agarwal In silico Research Laboratory, Eminent Biosciences, Indore, Madhya Pradesh, India

Sarah Albogami Department of Biotechnology, College of Science, Taif University, Taif, Saudi Arabia

Nandadulal Bairagi Department of Mathematics, Centre for Mathematical Biology and Ecology, Jadavpur University, Kolkata, West Bengal, India

Krishnan Balasubramanian School of Molecular Sciences, Arizona State University, Tempe, AZ, United States

Subhash C. Basak Department of Chemistry and Biochemistry, University of Minnesota, Duluth, MN, United States

Emilio Benfenati Laboratory of Environmental Chemistry and Toxicology, Istituto di Ricerche Farmacologiche Mario Negri IRCCS, Milano, Italy

Apurba K. Bhattacharjee Department of Microbiology and Immunology, Biomedical Graduate Research Organization, School of Medicine, Georgetown University, Washington, DC, United States

Anushka Bhrdwaj In silico Research Laboratory, Eminent Biosciences, Indore, Madhya Pradesh, India; Department of Bioinformatics, Computer Aided Drug Designing and Molecular Modeling Lab, Alagappa University, Karaikudi, Tamil Nadu, India

Suman K. Chakravarti MultiCASE Inc., Beachwood, OH, United States

Pratim Kumar Chattaraj Department of Chemistry, Indian Institute of Technology Kharagpur, Kharagpur, West Bengal, India

Samrat Chatterjee Complex Analysis Group, Translational Health Science and Technology Institute, NCR Biotech Science Cluster, Faridabad, Haryana, India

Ramana V. Davuluri Department of Preventive Medicine, Division of Health and Biomedical Informatics, Northwestern University Feinberg School of Medicine, Chicago, IL, United States

Tathagata Dey Centre for Interdisciplinary Research and Education, Kolkata, West Bengal, India; Department of Computer Science & Engineering, Indian Institute of Technology Bombay, Mumbai, Maharashtra, India

Abhik Ghosh Indian Statistical Institute, Kolkata, West Bengal, India

Indira Ghosh School of Computational & Integrative Sciences, Jawaharlal Nehru University, New Delhi, Delhi, India

Giuseppina Gini Politecnico di Milano, DEIB, Piazza Leonardo da Vinci, Milano, Italy

Lima Hazarika In silico Research Laboratory, Eminent Biosciences, Indore, Madhya Pradesh, India

Guang Hu Department of Bioinformatics, Center for Systems Biology, School of Biology and Basic Medical Sciences, Soochow University, Suzhou, P.R. China

Chiakang Hung Politecnico di Milano, DEIB, Piazza Leonardo da Vinci, Milano, Italy

Tajamul Hussain Biochemistry Department, College of Science, King Saud University, Riyadh, Saudi Arabia; Center of Excellence in Biotechnology Research, College of Science, King Saud University, Riyadh, Saudi Arabia

Yanrong Ji Department of Preventive Medicine, Division of Health and Biomedical Informatics, Northwestern University Feinberg School of Medicine, Chicago, IL, United States

Isha Joshi In silico Research Laboratory, Eminent Biosciences, Indore, Madhya Pradesh, India

Taushif Khan Immunology and Systems Biology Department, OPC-Sidra Medicine, Ar-Rayyan, Doha, Qatar

Ravina Khandelwal In silico Research Laboratory, Eminent Biosciences, Indore, Madhya Pradesh, India

Pawan Kumar National Institute of Immunology, Aruna Asaf Ali Marg, New Delhi, Delhi, India

Shivam Kumar Complex Analysis Group, Translational Health Science and Technology Institute, NCR Biotech Science Cluster, Faridabad, Haryana, India

Min Li Department of Bioinformatics, Center for Systems Biology, School of Biology and Basic Medical Sciences, Soochow University, Suzhou, P.R. China

Jie Liao Department of Pathology, Northwestern University Feinberg School of Medicine, Chicago, IL, United States

Claudiu N. Lungu Department of Chemistry, Faculty of Chemistry and Chemical Engineering, Babes-Bolyai University, Cluj, Romania; Department of Surgery, Faculty of Medicine and Pharmacy, University of Galati, Galati, Romania

Subhabrata Majumdar AI Vulnerability Database, Seattle, WA, USA; Bias Buccaneers, Seattle, WA, USA

Rama K. Mishra Department of Biochemistry and Molecular Genetics, Feinberg School of Medicine, Northwestern University, Chicago, IL, United States

Manju Mohan In silico Research Laboratory, Eminent Biosciences, Indore, Madhya Pradesh, India

Ashesh Nandy Centre for Interdisciplinary Research and Education, Kolkata, West Bengal, India

Anuraj Nayarisseri In silico Research Laboratory, Eminent Biosciences, Indore, Madhya Pradesh, India; Department of Bioinformatics, Computer Aided Drug Designing and Molecular Modeling Lab, Alagappa University, Karaikudi, Tamil Nadu, India; Biochemistry Department, College of Science, King Saud University, Riyadh, Saudi Arabia; Bioinformatics Research Laboratory, LeGene Biosciences Pvt Ltd, Indore, Madhya Pradesh, India

Shahul H. Nilar Global Blood Therapeutics, San Francisco, CA, United States

Ranita Pal Advanced Technology Development Centre, Indian Institute of Technology Kharagpur, Kharagpur, West Bengal, India

Aditi Pande In silico Research Laboratory, Eminent Biosciences, Indore, Madhya Pradesh, India

Guillermo Restrepo Max Planck Institute for Mathematics in the Sciences, Leipzig, Germany; Interdisciplinary Center for Bioinformatics, Leipzig University, Leipzig, Germany

Dipanka Tanu Sarmah Complex Analysis Group, Translational Health Science and Technology Institute, NCR Biotech Science Cluster, Faridabad, Haryana, India

Dwaipayan Sen Centre for Interdisciplinary Research and Education, Kolkata, West Bengal, India

Sanjeev Kumar Singh Department of Bioinformatics, Computer Aided Drug Designing and Molecular Modeling Lab, Alagappa University, Karaikudi, Tamil Nadu, India

Chillamcherla Dhanalakshmi Srija In silico Research Laboratory, Eminent Biosciences, Indore, Madhya Pradesh, India

Revathy Arya Suresh In silico Research Laboratory, Eminent Biosciences, Indore, Madhya Pradesh, India

Muyun Tang Department of Bioinformatics, Center for Systems Biology, School of Biology and Basic Medical Sciences, Soochow University, Suzhou, P.R. China

Garima Thakur In silico Research Laboratory, Eminent Biosciences, Indore, Madhya Pradesh, India

Xin Tong Department of Pathology, Northwestern University Feinberg School of Medicine, Chicago, IL, United States

Marjan Vracko Theory Department, Kemijski inštitut/National Institute of Chemistry Ljubljana, Slovenia

Marjan Vračko National Institute of Chemistry, Hajdrihova 19, Ljubljana, Slovenia; Theory Department, Kemijski inštitut/National Institute of Chemistry, Ljubljana, Slovenia

Ze Wang Department of Pharmaceutical Sciences, Zunyi Medical University at Zhuhai Campus, Zhuhai, P.R. China

DanDan Xu Department of Pathology, Northwestern University Feinberg School of Medicine, Chicago, IL, United States

Guang-Yu Yang Department of Pathology, Northwestern University Feinberg School of Medicine, Chicago, IL, United States

Preface

"We adore chaos because we love to produce order."

—M.C. Escher

"...shall we stay our upward course? In that blessed region of Four Dimensions, shall we linger at the threshold of the Fifth, and not enter therein? Ah, no! Let us rather resolve that our ambition shall soar with our corporal ascent. Then, yielding to our intellectual onset, the gates of the Six Dimension shall fly open; after that a Seventh, and then an Eighth..."

—Edwin Abbott, In: Flatland

"I'm tired of sailing my little boat

Far inside of the harbor bar;

I want to be out where the big ships float—

Out on the deep where the Great Ones are!...

—Daisy Rinehart

In science there is and will remain a Platonic element which could not be taken away without ruining it. Among the infinite diversity of singular phenomena science can only look for invariants.

—Jacques Monod

We are currently living in an age when many spheres of science and life are flushed with the explosion of big data. We are familiar with the term "data is the new oil," but often hear about information overload or data deluge. We need to systematically manage, model, interpret, visualize, and use such data in diverse decision-support systems in basic research, technology, health care, and business, to name just a few.

If we look at the main focus of this book—applications of big data analytics in chemoinformatics, bioinformatics, new drug discovery, and hazard assessment of environmental pollutants—, it is evident that data in all these fields are exploding.

Regarding the size of chemical space, the GDB-17 database contains 166.4 billion molecules containing up to 17 atoms of C, N, O, S, and halogens which fall within the size range containing many drugs and are typical for druggable lead compounds. The sequence data on DNA, RNA, and proteins are increasing each day by new depositions by researchers worldwide. A simple combinatorial exercise of sequence possibility for a 100-residue long protein suggests 20^{100} different possible sequences (considering 20 frequently occurring natural amino acids). Modern computer software can calculate many hundreds, sometimes thousands of descriptors for a molecule or a macromolecular sequence. The Vs of big data, viz., validity, vulnerability, volatility, visualization, volume, value, velocity, variety, veracity, and variability, increase the complexity of big data analytics immensely.

Here, we come face to face with the stark reality of the curse of dimensionality in the big data space of chemistry and biology. Following the parsimony principle, we need to be careful in feature selection and use of robust validation techniques in model building. Finally, analysis and visualization of models to understand their meaning and derive actionable knowledge from the vast information space for practical implementation in the decision-support systems of science and society are of paramount importance.

The first section, General Section, of the book has three chapters. Chapter 1 briefly traces the history of the development of chemodescriptors and biodescriptors spanning three centuries—from the eighteenth century to the present. It is observed by the author that the initial characterization of structures, both chemical and biological, were qualitative which was gradually followed by the development of quantitative chemodescriptors and biodescriptors. The author concluded that in the socially and economically important areas of new drug discovery and hazard assessment of chemicals use of a combined set of chemodescriptors and biodescriptors for model building using big data would be a useful and practical paradigm. Chapter 2 deals with the problem of robust model building from noisy high-dimensional data, focusing primarily on the robustness aspects against data contamination. The author also demonstrates the utility of his method in the prediction of salmonella mutagenicity of a set of amines, a class priority pollutants. Chapter 3 delves into the ethical issues associated with the landscape of desirable qualities such as fairness, transparency, privacy, and robustness of currently used machine learning (ML) methods of big data analysis.

The second section, Chemistry and Chemoinformatics Section, of the book has nine chapters. Chapter 4 discusses the use of big data in the characterization of adverse outcome pathways (AOPs), a novel paradigm in toxicology. The author integrated "big data"—the–omics and high-throughput (HT) screening data—to derive AOPs for chemical carcinogens. Chapter 5 discusses the latest progress in the use of ML and DL (deep learning) methods in creating systems that automatically mine patterns and learn from data. The author also discuss the challenges and usefulness of DL for quantitative structure–activity relationship (QSAR) modeling. Chapter 6 describes retrosynthetic planning and analysis of organic compounds in the synthetic space using big data sets and in silico algorithms. Chapter 7 discusses that the vast amount of historical chemical information is not only a rich source of data, but also a useful tool for studying the evolution of chemistry, chemoinformatics, and bioinformatics through a computational approach to the history of chemistry. The author exemplifies that by a case study of recent results on the computational analysis of the evolution of the chemical space. Chapter 8 gives a detailed description of combinatorial techniques useful in studying large data sets with hypercubes and halocarbons as the main focus. Quantum chemical techniques discussed here can generate electronic parameters that have potential for use in QSAR for toxicity prediction of big data sets. Chapter 9 deals with the use of computed high-level quantum chemical descriptors derived from the density functional theory in the prediction of property/toxicity of chemicals. Chapter 10 covers the important area of the use of computed pharmacophores in practical drug design from analysis of large databases. Chapter 11 uses ML based classification methods for the detection of hot spots in protein–protein interactions and prediction of new hotspots. Chapter 12 discusses applications of decision tree methods like recursive partitioning, phylogenetic-like trees, multidomain classification, and fuzzy clustering within the context of small molecule drug discovery from analysis of large databases.

The third section, Bioinformatics and Computatioanl Toxicology Section, of the book has seven chapters. Chapter 13 discusses their contributions in the emerging area of mathematical proteomics approach in developing biodescriptors for the characterization of bioactivity and toxicity of drugs and pollutants. Chapter 14 discusses the important role of efficient computational frameworks developed to catalog and navigate the protein space to help the drug discovery process. Chapter 15 discusses applications of ML and DL approaches to HT sequencing data in the development of precision medicine using single-nucleotide polymorphisms as a tool of reference. Chapter 16 discusses the development and use of a new class of sequence comparison methods based on alignment-free sequence descriptors in the characterization of emerging global pathogens like the Zika virus and coronaviruses (SARS, MERS, and SARS-CoV-2). Chapter 17 discusses the important and emerging issue of different ways of building QSARs from large and diverse data sets that can be continuously updated and expanded over time. The importance of modularity in scalable QSAR system development is also discussed. Chapter 18 deals with the applications of network analysis and big data to study interactions of drugs with their targets in the biological systems. The authors point out that a paradigm shift integrating big data and complex network is needed to understand the expanding universe of drug molecules, targets, and their interactions. Finally, Chapter 19 reports the use of ML approaches consisting of supervised and unsupervised techniques in the analysis of RNA sequence data of breast cancer to derive important biological insights. They were able to pinpoint some disease-related genes and proteins in the breast cancer network.

Finally, we would like to specially mention that in drug research and toxicology, we are witnessing an explosion of data, which are expressed by four principal Vs—volume, velocity, variety, and veracity. However, the data per se is useless, the real challenge is the transition to the last two steps on the three-step path to knowledge: data – information – knowledge. When we talk about big data in drug research and toxicology, we often think of omics data and in vitro data derived from HT screening. On the other hand, a pool of high-quality "small" data exists, which has been collected in the past. Under the label "small data" we have the standard toxicological data based on well-defined toxic effects. A future challenge for us is to integrate both data platforms—big and small—into a new and integrated knowledge extraction system.

Subhash C. Basak
Marjan Vračko

Section 1

General section

Chemoinformatics and bioinformatics by discrete mathematics and numbers: an adventure from small data to the realm of emerging big data

1

Subhash C. Basak
Department of Chemistry and Biochemistry, University of Minnesota Duluth, Duluth, MN, United States

1.1 Introduction

"Oh, the thirst to know

how many!

The hunger

to know

how many

stars in the sky!

—

We spent

our childhood counting

stones and plants, fingers and

toes, grains of sand, and teeth,

our youth was past counting

petals and comets' tails.

We counted

colors, years,

lives, and kisses;

in the country,

Big Data Analytics in Chemoinformatics and Bioinformatics. DOI: https://doi.org/10.1016/B978-0-323-85713-0.00028-1

oxen; by the sea,

the waves. Ships

became proliferating ciphers.

Numbers multiplied."

Pablo Neruda, In: Ode to numbers

A currently emerging trend in many scientific disciplines is their tendency of being gradually transformed/evolved into some form of information science (Basak et al., 2015; Dehmer and Basak, 2012; Kerber et al., 2014). In the realm of chemoinformatics and bioinformatics, in particular, methods of discrete mathematics like graph theory, network theory, information theory etc. are gaining momentum as useful tools in the representation, characterization, and comparison of molecular and biological systems and their structures as well as in the prediction of property/ bioactivity/ toxicity of chemicals for new drug discovery and environmental protection (Basak, 1987, 2010, 2013a, 2014; Basak et al., 1988b, 2015; Bayda et al., 2019; Bielinska-Waz et al., 2007; Braga et al., 2018; Chakravarti, 2021; Ciallella and Zhu, 2019; Diudea et al., 2018; Gini et al., 2013; Guo et al., 2001; Kerber et al., 2014; Khan et al., 2018; Nandy, 2015; Kier and Hall, 1986, 1999; Nandy et al., 2006; Osolodkin et al., 2015; Randic et al., 2000, 2001, 2004, 2011; Restrepo and Villaveces, 2013; Rouvray, 1991; Sabirov et al., 2021; Toropov and Toropova, 2021; Vračko et al., 2018, 2021a,b; Wang et al., 2021; Winkler et al., 2014).

The impetus for the development of chemoinformatics and bioinformatics tools/ methods has come from different directions. In new drug design, thousands of derivatives of the initially discovered "lead" compound have to be synthesized and tested in order to find one useful drug. This journey of the lead from the chemist's desk to the bedside of the patient involves a span of about 10 years and an expenditure of over US$2 billion (DiMasi et al., 2016). Synthesis and testing of all possible chemical derivatives of the identified lead compound is prohibitively costly. Under such circumstances in silico approaches of chemoinformatics can give us fast and cost-effective estimation of properties of promising derivatives of the lead chemicals necessary for the prediction of the most probable pharmacological and toxicological profiles (Table 1.1). Thus, chemoinformatics tools can assist the drug designer as a *decision support system*. It has been noted that currently no drug is developed without the prior evaluation by quantitative structure–activity relationship (QSAR) methods (Santos-Filho et al., 2009).

The Toxic Substances Control Act (TSCA, 2021) Inventory, maintained by the United States Environmental Protection Agency (USEPA), currently has more than 86,000 chemicals. Most of the TSCA chemicals have very little or no experimental data required for their toxicity estimation. Detailed laboratory testing of all these chemicals and their possible metabolites produced in the exposed organisms including humans would be prohibitively costly. In the face of this lack of available data, two approaches are used by the regulatory agencies: (a) class-specific QSAR models and (b) quantitative molecular similarity analysis (QMSA)-based modeling of

Table 1.1 A partial list of important physical, pharmacological, and toxicological properties prerequisite to the evaluation of chemicals for new drug discovery and environmental protection.

Physicochemical	Pharmacological/toxicological
Molar volume	*Macromolecule level*
Boiling point	Receptor binding (K_D)
Melting point	Michaelis constant (Km)
Vapor pressure	Inhibitor constant (Ki)
Water solubility	DNA alkylation
Dissociation constant (pKa)	Unscheduled DNA synthesis
Partition coefficient	*Cell level*
Octanol-water (log P)	Salmonella mutagenicity
Air-water	Mammalian cell transformation
Sediment-water	*Organism level (acute)*
Reactivity (electrophile)	Algae
	Invertebrates
	Fish
	Birds
	Mammals
	Organism level (chronic)
	Bioconcentration
	Carcinogenicity
	Reproductive toxicity
	Delayed neurotoxicity
	Biodegradation

properties using structural analogs (Auer et al., 1990). The situation becomes more numerous and complex if one considers the biotransformation and pharmacokinetic data of the chemicals (TSCA metabolism and pharmacokinetics, 2021). A similar situation exists in the European Union with the chemicals in commerce (European Chemicals Agency, 2021) list showing more than 100,000 chemicals registered with the system.

Table 1.1 provides a partial list of physicochemical, pharmacological, and toxicological properties that drug designers and risk assessors of chemicals frequently use in evaluating their beneficial and deleterious effects (Basak et al., 1990).

1.2 Chemobioinformatics—a confluence of disciplines?

"At quite uncertain times and places,

The atoms left their heavenly path,

And by fortuitous embraces,

Engendered all that being hath.

And though they seem to cling together,

And form "associations" here,

Yet, soon or late, they burst their tether,

And through the depths of space career."

—*James Clerk Maxwell*

The current QSAR paradigm did not arise out of one or a few "***aha***" moments, but it emerged through the confluence of a diverse set of ideas originated by quite a few researchers of different disciplines over the past couple of centuries. For a recent review, please see Basak (2021a). Some seminal aspects of the developments of modern chemoinformatics are discussed as follows.

1.2.1 Physical property: colligative versus constitutive

"In order to describe an aspect of holistic reality we have to ignore certain factors such that the remainder separates into facts. Inevitably, such a description is true only within the adopted partition of the world, that is, within the chosen context."
—*Hans Primas (1981), Chemistry, Quantum Mechanics and Reductionism*

In physical chemistry, a *colligative property*, for example, lowering of vapor pressure, elevation of boiling point, depression of freezing point, and osmotic pressure, of solutions is a property that depends solely upon the concentration of solute molecules or ions, being independent of the constitution or identity of the solute. *Constitutive property*, on the other hand, depends on the constitution or structure of the substance. The American Heritage Dictionary of the English Language, 5th Edition, states the following regarding the word constitutive:

"In physical chemistry, a term introduced by Ostwald to denote those properties of a compound which depend on the constitution of the molecule, or on the mode of union and arrangement of the atoms in the molecule."

1.2.2 Early biochemical observations on the relationship between chemical structure and bioactivity of molecules

For almost a century, various researchers in biochemistry and pharmacology generated data on the relation between the structure of molecules and their bioactivities. Most probably one of the earliest was the 1928 finding of Quastel and Wooldridge (1928) that malonic acid competitively inhibited the activity of the Krebs cycle enzyme succinic dehydrogenase. Although the substrate succinic acid and the inhibitor malonic acid differed by one methylene ($-CH_2$) group, the catalytic site of the enzyme still recognized malonic acid. This seminal observation may be looked upon as the rational basis for the synthesis of analogs of nucleic acid bases for cancer chemotherapy (Hitchings and Elion, 1954) and the more modern chemoinformatics approach to computer-aided drug design using the concept of pharmacophore (Bhattacharjee, 2015). The antibiotic penicillin inhibits cell wall biosynthesis in bacteria by interfering with the transpeptidation reaction responsible for the crosslinking of mucopeptide chains in the cell wall polymer. This is attributed to its putative structural similarity to the D-alanyl-D-alanine portion of the peptide chain (Goodman and Gilman, 1990).

1.2.3 Linear free energy relationship: the multiparameter Hansch approach to quantitative structure–activity relationship

As described by Hansch and Leo (1979), in the early 1900s the English school of organic chemists (Ingold, 1953) became interested in the mechanisms of reactions of organic molecules. One approach was to make a set of structural modifications in a parent molecule and then observe the effects of the substitutions on the rates or equilibria of a reaction with a reactant under standard conditions. One could draw conclusions about the electronic and steric requirements of a given reaction from the analysis of the perturbations of the reaction center by the substituents. The problems of applications of these concepts, as indicated by Hansch and Leo (1979), were:

> *"The difficulty with these early and important ideas was that no numerical scales were available that could be used to quantify each of these effects that could operate singly or in concert.* **Even when such scales had been devised, it was difficult to make progress in the separation of substituent effects before high-speed computers became generally available** *(approximately 1960)."*

One important breakthrough in the field of mechanistic organic chemistry came when Hammett (1937) proposed the now well-known Hammett equation. He defined the parameter σ as follows:

$$\sigma = \log K_x - \log K_H \tag{1.1}$$

where K_H is the ionization constant for benzoic acid in water at 25°C and K_x is the ionization constant for its meta or para derivative under the same experimental conditions. Positive values of σ indicate electron withdrawal by the substituent from the aromatic ring and negative values represent electron release from the substituent to the ring.

In the second half of the 20th century, Taft (1952) formulated the linear free energy-related steric descriptor E_s.

The multiparameter linear free energy relationship (LFER) approach, popularly known as the "Hansch Analysis," to quantitative structure–property–activity relationship (QSPR/QSAR), derived from physical organic chemistry, attempted to predict property/bioactivity of molecules using a combination of their electronic, steric, and hydrophobic parameters (Hansch and Leo, 1995):

$$Log\,BA = a \log P + b\sigma + c\,Es + \text{constant (linear)} \tag{1.2}$$

$$\log BA = a \log P + b\,(\log P)^2 + c\sigma + d\,Es + \text{constant} \tag{1.3}$$

In Eq. (1.2), BA stands for biological activity, log P stands for the logarithm of the partition coefficient (experimentally determined or calculated from structure) of the chemical, σ usually represents Hammett's (1937) electronic descriptor, and E_s usually symbolizes Taft's steric parameter (Taft, 1952). A perusal of LEFR-based QSAR models would indicate that different varieties of hydrophobic, steric, and electronic parameters have been developed and used in numerous correlation studies (Hansch and Leo, 1995). A short description of the historical timeline for the evolution of the LFER approach is depicted in Fig. 1.1.

The LFER approach gives good predictive models for congeneric sets of molecules. As discussed above, both for drug design and hazard assessment of chemicals

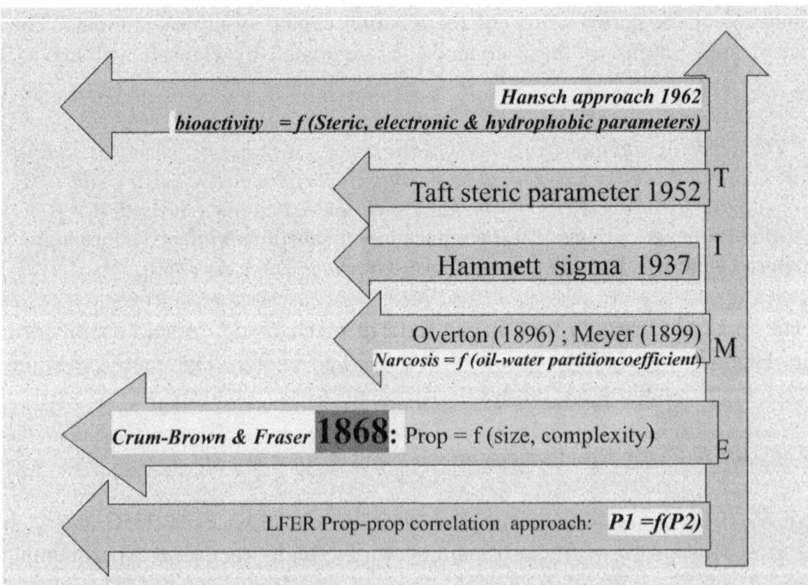

Figure 1.1 A short history (1868-date) of the development of linear free energy relationship approach for quantitative structure–activity relationship modeling based on physical properties and substituent constants derived from physical organic chemistry. For more information please see: Basak (2013a, 2021a) and Hansch and Leo (1995). In this approach, a property (P_1) of a molecule is estimated from another available property (P_2) or a combination of other properties.

we need to estimate the properties and bioactivities of chemicals which are structurally diverse (Basak and Majumdar, 2016). Sometimes, one could wish to estimate pharmacological and toxicological profiles of chemicals not yet synthesized. Models based on the LFER-type experimental data are of little utility in such cases. Furthermore, for applications in chemical engineering and technological processes we need to know the values of many properties of substances (Drefahl and Reinhard, 1998; Lyman et al., 1990). The use of good quality experimental property values are always desirable, but such data are often unavailable. The use of QSAR-predicted properties utilizing computed descriptors as the independent variables is generally the practical alternative (Katritzky et al., 1995, 2001). More recently, many large and diverse databases of properties needed for drug design and predictive toxicology are becoming available in the public domain. These are resources available for the development of broad-based models for property/bioactivity estimation (Gadaleta et al., 2019; Mansouri et al., 2018; Meng et al., 2021).

During the second half of the 20th century and the first quarter of this century, various chemoinformatics approaches have given us molecular descriptors which can be computed directly from the molecular structure without the input of any other experimental data. Such descriptors are finding side spread applications in the formulation of useful QSAR models (Basak, 2021a, 2012b, 2013a, 2014; Drefahl and Reinhard, 1998; Katritzky et al., 1995, 2001; Kier and Hall, 1986, 1999).

1.2.4 Chemical graph theory and quantum chemistry as the source of chemodescriptors

"By convention sweet and by convention bitter, by convention hot, by convention cold, by convention color; but in reality atoms and void."

—*Democritus*

"The fundamental laws necessary for the mathematical treatment of a large part of physics and the whole of chemistry are thus completely known, and the difficulty lies only in the fact that application of these laws leads to equations that are too complex to be solved."

—*Paul Dirac*

1.2.4.1 Topological indices—graph theoretic definitions and calculation methods

A graph, G, is defined as an ordered pair consisting of two sets V and R, $G=[V(G), R]$, where $V(G)$ represents a finite nonempty set of points, and R is a binary relation defined on the set $V(G)$. The elements of V are called vertices and the elements of R, also symbolized by $E(G)$ or E, are called edges. Such an abstract graph is commonly visualized by representing elements of $V(G)$ as points and by connecting each pair (u, v) of elements of $V(G)$ with a line if and only if $(u, v) \in R$. The vertex, v, and edge, e, are incident with each other, as are u and e. Two vertices u and

v in G are called adjacent if $(u, v)\epsilon R$, that is, they are connected by an edge. A walk of a graph is a sequence beginning and ending with vertices in which vertices and edges alternate and each edge is incident with vertices immediately preceding and following it. A walk of the form $v_0, e_1, v_1, e_2, \ldots, v_n$ joins vertices v_0 and v_n. The length of a walk is the number of edges in the walk. A walk is closed if $v_0 = v_n$, otherwise it is open. A closed walk with n points is a cycle if all its points are distinct and $n \geq 3$. A path is an open walk in which all vertices are distinct. A graph G is connected if every pair of its vertices is connected by a path. A graph G is a multigraph if it contains more than one edge between at least one pair of adjacent vertices, otherwise, G is a simple graph. The distance $d\ (u, v)$ between vertices u and v in G is the length of the shortest path connecting u and v.

Because of the general nature of graph-theoretic (GT) methods in the representation of objects this method has been used in such diverse areas as theoretical physics, chemistry, biological and social sciences, engineering, computer science and linguistics (Harary, 1986). For example, GT has been used in the representation and comparison of proteins, characterization of the nucleotide sequence topology in DNA and RNA sequences (Nandy, 2015; Nandy et al., 2006; Randic et al., 2000, 2011), representation of protein spots of proteomics maps (Randić et al., 2001), folding patterns in protein structures (Khan et al., 2018; Liu et al., 2006), structural characterization of nanosubstances (Toropov and Toropova, 2021), to name just a few.

For chemical graph theory research and applications (Basak, 2013a; Basak et al., 2011; Janezic et al., 2015), a molecular graph represents molecular topology where V represents the set of atoms and E usually symbolizes the set of covalent bonds present in the molecule. It should be noted, however, that the set E should not be limited to covalent bonds only. In fact, elements of E may symbolize any type of bond, viz., covalent, ionic, or hydrogen bonds, etc. It was emphasized by Basak et al. (1988a) that weighted pseudographs constitute a very versatile model for the representation of a wide range of chemical species. Fig. 1.2 depicts the chemical structure, labeled hydrogen-filled graph and labeled hydrogen-suppressed graph of the molecule acetamide. It may be mentioned here that a large number of molecules

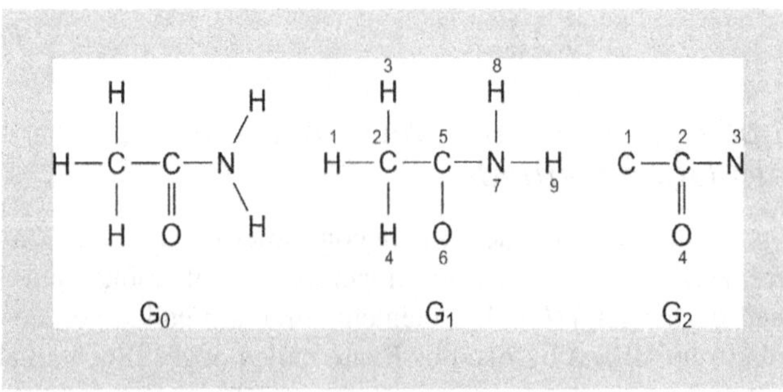

Figure 1.2 Structural formula (G_0), labeled hydrogen-filled graph (G_1), and labeled hydrogen-suppressed graph (G_2) of acetamide.

can be represented by planar graphs as defined by the absence of the subgraphs K_5 and $K_{3,3}$ it its architecture (Harary, 1969; Kuratowski, 1930)

Graph invariants may be used for the characterization of molecular graphs (Harary, 1969; Janezic et al., 2015). Numerical graph invariants that quantitatively characterize molecular structure are called topological indices (Hosoya, 1971). Many topological indices (TIs) can be conveniently derived from various matrices including the adjacency matrix $A(G)$ and the distance matrix $D(G)$ of a chemical graph G (Devillers and Balaban, 1999; Harary, 1969; Janezic et al., 2015). These matrices are usually constructed from labeled graphs of hydrogen-suppressed molecular skeletons. For such a graph G with vertex set $\{v_1, v_2, \ldots, v_n\}$, the adjacency matrix $A(G)$ is defined to be the $n \times n$ matrix (a_{ij}), where a_{ij} may have only two different values as follows:

$$
\begin{aligned}
a_{ij} &= 1, \quad \text{if vertices } v_i \text{ and } v_j \text{ are adjacent in } G,\\
a_{ij} &= 0, \quad \text{otherwise.}
\end{aligned}
$$

The distance matrix $D(G)$ of a nondirected graph G with n vertices is a symmetric $n \times n$ matrix (d_{ij}), where d_{ij} is equal to the distance between vertices v_i and v_j in G. Each diagonal element d_{ij} of $D(G)$ is equal to zero. Since topological distance in a graph is not related to the physical property or weight attached to each edge (chemical bond), $D(G)$ does not represent valence bond structures of molecules or attributes of multigraphs containing more than one covalent bond between adjacent vertices (atoms).

The adjacency matrix $A(G_2)$ and the distance matrix $D(G_2)$ for the labeled graph G_2 in Fig. 1.1 may be written as follows:

$$
A(G_2) = \begin{array}{c} \\ 1 \\ 2 \\ 3 \\ 4 \end{array}
\begin{array}{c} \begin{array}{cccc} (1) & (2) & (3) & (4) \end{array} \\
\begin{bmatrix} 0 & 1 & 0 & 0 \\ 1 & 0 & 1 & 1 \\ 0 & 1 & 0 & 0 \\ 0 & 1 & 0 & 0 \end{bmatrix} \end{array}
$$

$$
D(G_2) = \begin{array}{c} \\ 1 \\ 2 \\ 3 \\ 4 \end{array}
\begin{array}{c} \begin{array}{cccc} (1) & (2) & (3) & (4) \end{array} \\
\begin{bmatrix} 0 & 1 & 2 & 2 \\ 1 & 0 & 1 & 1 \\ 2 & 1 & 0 & 2 \\ 2 & 1 & 2 & 0 \end{bmatrix} \end{array}
$$

Wiener (1947) was the first to put forward the idea of a structural index (topological index) for the estimation of properties of molecules from their structure using topological indices. This index is popularly known as the Wiener index, W. It can be calculated from the distance matrix $D(G)$ of a hydrogen-suppressed graph G as the sum of entries in the upper triangular submatrix (Hosoya, 1971):

$$
W = 1/2 \sum_{ij} d_{ij} = \sum_{h} h \cdot g_h \tag{1.4}
$$

where g_h represents the number of unordered pairs of vertices whose distance is h.

From the adjacency matrix of a graph with n vertices, it is possible to calculate δ_i, the degree or valence of the ith vertex, as the sum of all entries in the ith row:

$$\delta_i = \sum_{j=1}^{n} a_{ij} \tag{1.5}$$

The zero-order connectivity index, $^0\chi$, is defined as (Kier and Hall, 1976)

$$^0\chi = \sum_i (\delta_i)^{-1/2} \tag{1.6}$$

Randić's connectivity index or first-order connectivity $^1\chi$, is defined as (Randic, 1975)

$$^1\chi = \sum_{\text{all edges}} (\delta_i \delta_j)^{-1/2} \tag{1.7}$$

Based on these two indices, Kier and Hall (1976) developed a generalized connectivity index ^{h} considering paths of type v_0, v_1, … v_h of length h in the molecular graph:

$$^h\chi = \sum (\delta_{v0}, \delta_{v1}, \ldots, \delta_{vh})^{-1/2} \tag{1.8}$$

where the summation is taken over all paths of length h.

Kier and Hall (1976, 1986, 1999) extended the molecular graph connectivity approach to developing the valence connectivity indices, electrotopological indices as well as a cluster, path-cluster, and cyclic types of connectivity indices.

Another class of graph theoretic indices, the information-theoretic topological indices, are calculated by applying information theory to chemical graphs. An appropriate set A of n elements is derived from a molecular graph G depending upon certain structural characteristics. On the basis of an equivalence relation defined on A, the set A is partitioned into disjoint subsets A_i of order n_i

$$n_i(i = 1, 2, \ldots, h);$$

$$\sum_i n_i = n$$

A probability distribution is then assigned to the set of equivalence classes:

$$\begin{pmatrix} A_1, A_2, \ldots, A_h \\ p_1, p_2, \ldots, p_h \end{pmatrix}$$

where $pi = n_i/n$ is the probability that a randomly selected element of A will occur in the ith subset.

The mean information content, IC, of an element A is defined by Shannon's relation (Shannon and Weaver, 1949):

$$IC = -\sum_{i=1}^{h} p_i \log_2 p_i \tag{1.9}$$

The logarithm is taken at base 2 for measuring the information content in bits. The total information content (TIC) of the set A is then n times IC.

Another information-theoretic measure, structural information content (SIC_r) for a particular order r of topological neighborhood was defined as (Basak et al., 1980)

$$\text{SIC}_r = \text{IC}_r/\log_2 n \tag{1.10}$$

where IC_r is calculated as in Eq. (1.9) and n is the total number of vertices of the molecular graph.

Another information-theoretic invariant, complementary information content (CIC_r) was defined as (Basak and Magnusson, 1983)

$$\text{CIC}_r = \log_2 n - \text{IC}_r \tag{1.11}$$

CIC_r represents the difference between maximum possible complexity of a graph (where each vertex is in a separate equivalence class) and the realized topological information of a chemical species as defined by IC_r. It has been pointed out by Bonchev (1983) that, in many cases the equivalent vertices in the neighborhood symmetry formalism belong to the same orbit of the automorphism group of the graph.

It is noteworthy that information content of graphs of molecules or biomolecules is not uniquely defined. It depends on the manner in which set A is derived from graph G as well as on the equivalence relation used by the researcher in partitioning A into disjoint subsets A_i.

To account for the chemical nature of atoms (vertices) as well as their bonding pattern, in the last quarter of the 20th century Basak and coworkers (Basak, 1987; Basak et al., 1980; Roy et al., 1984) calculated information content of molecular graphs on the basis of an equivalence relation in which two atoms of the same chemical element are considered equivalent if they possess an identical first-order topological neighborhood. Since properties of atoms or reaction centers in molecules are often modulated by the inductive effects of immediately bonded and distant neighbors (Hammett, 1937; Hansch and Leo, 1979; Morrison and Boyd, 1987), that is, neighbors of immediately bonded neighbors, it was deemed essential to extend this approach to account for higher-order neighbors of atoms. This can be accomplished by defining open spheres for all vertices of a chemical graph. If r is any nonnegative real number and v is a vertex of the graph G, the open sphere $S(v, r)$ is defined as the set consisting of all vertices v_j in G such that $d(v, vj) < r$. Obviously, $S(v,0) = \emptyset$, $S(v, r) = v$ for $0 < r < 1$, and $S(v, r)$ is the set consisting of v and all vertices v_j of G situated at unit distance from v, if $1 < r < 2$. For the formulation of information indices of different orders of neighborhoods of atoms, please see Basak (1987), Basak and Magnusson (1983), Basak et al. (1980), and Roy et al. (1984). Bonchev and Trinajstic (1977) and Raychaudhury et al. (1984) applied information theory to the distance matrix $D(G)$ of molecular graphs to develop different types of information-theoretic indices.

A short history of the development of topological indices in the modern era starting from the work of Wiener (1947) preceded by seminal contributions of Euler (1736) in the first development of graph theory and the idea of Sylvester (1878) that molecular structures are basically graphs is given in Fig. 1.3.

Table 1.2 gives a list of molecular descriptors frequently used by Basak et al. (1980) in their QSAR/QSTR studies and calculated by Basak et al. (1988), Triplet (Basak et al., 1993), and MolconnZ, Version 4.05 (2003) software as well as quantum chemical routines like MOPAC (Stewart, 1990).

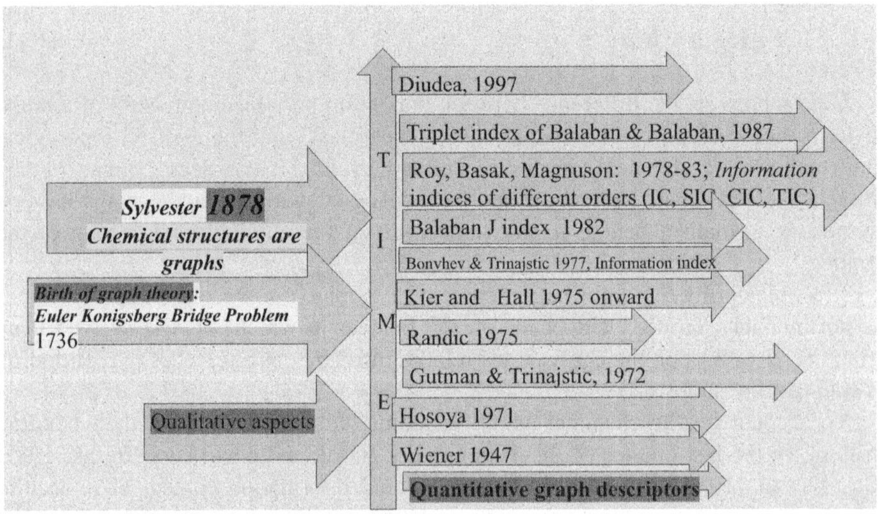

Figure 1.3 A brief history of the discovery of graph theory by Euler and the development of topological indices and their use. Euler discovered graph theory in 1736. Sylvester (1878) recognized that the conventional chemical structures are graphs, and this was a qualitative concept. Since 1947 formulation of quantitative descriptors of molecular topology (structure) was done by Wiener (1947); Hosoya (1971); Gutman and Trinajstic (1972); Randic (1975); Kier and Hall (1976 onward); Bonchev and Trinajstic (1977), Basak and coworkers (1980 onward), Balaban (1982, 1987). Diudea (1997). For a detailed discussion on the contents and scientific basis of Fig. 1.3, please see Basak (2021a).

Table 1.2 Symbols, definitions and classification of structural molecular descriptors.

Topostructural (TS)	
$I^{W}{}_{D}$	Information index for the magnitudes of distances between all possible pairs of vertices of a graph
$\overline{I_D^W}$	Mean information index for the magnitude of distance
W	Wiener index = half-sum of the off-diagonal elements of the distance matrix of a graph
I^D	Degree complexity
H^V	Graph vertex complexity
H^D	Graph distance complexity
$\overline{\mathrm{IC}}$	Information content of the distance matrix partitioned by frequency of occurrences of distance h
M_1	A Zagreb group parameter = sum of square of degree over all vertices
M_2	A Zagreb group parameter = sum of cross-product of degrees over all neighboring (connected) vertices
${}^{h}\chi$	Path connectivity index of order $h = 0-10$
${}^{h}\chi_C$	Cluster connectivity index of order $h = 3-6$
${}^{h}\chi_{PC}$	Path-cluster connectivity index of order $h = 4-6$
${}^{h}\chi_{Ch}$	Chain connectivity index of order $h = 3-10$
P_h	Number of paths of length $h = 0-10$
J	Balaban's J index based on topological distance
nrings	Number of rings in a graph
ncirc	Number of circuits in a graph
DN^2S_y	Triplet index from distance matrix, square of graph order, and distance sum; operation $y = 1-5$
DN^21_y	Triplet index from distance matrix, square of graph order, and number 1; operation $y = 1-5$
$AS1_y$	Triplet index from adjacency matrix, distance sum, and number 1; operation $y = 1-5$
$DS1_y$	Triplet index from distance matrix, distance sum, and number 1; operation $y = 1-5$
ASN_y	Triplet index from adjacency matrix, distance sum, and graph order; operation $y = 1-5$
DSN_y	Triplet index from distance matrix, distance sum, and graph order; operation $y = 1-5$
DN^2N_y	Triplet index from distance matrix, square of graph order, and graph order; operation $y = 1-5$
ANS_y	Triplet index from adjacency matrix, graph order, and distance sum; operation $y = 1-5$
$AN1_y$	Triplet index from adjacency matrix, graph order, and number 1; operation $y = 1-5$
ANN_y	Triplet index from adjacency matrix, graph order, and graph order again; operation $y = 1-5$
ASV_y	Triplet index from adjacency matrix, distance sum, and vertex degree; operation $y = 1-5$
DSV_y	Triplet index from distance matrix, distance sum, and vertex degree; operation $y = 1-5$

(*Continued*)

Table 1.2 (Continued)

Topostructural (TS)	
ANV_y	Triplet index from adjacency matrix, graph order, and vertex degree; operation $y = 1-5$
Topochemical (TC)	
O	Order of neighborhood when IC_r reaches its maximum value for the hydrogen-filled graph
O_{orb}	Order of neighborhood when IC_r reaches its maximum value for the hydrogen-suppressed graph
I_{ORB}	Information content or complexity of the hydrogen-suppressed graph at its maximum neighborhood of vertices
IC_r	Mean information content or complexity of a graph based on the rth ($r = 0-6$) order neighborhood of vertices in a hydrogen-filled graph
SIC_r	Structural information content for r^{th} ($r = 0-6$) order neighborhood of vertices in a hydrogen-filled graph
CIC_r	Complementary information content for rth ($r = 0-6$) order neighborhood of vertices in a hydrogen-filled graph
${}^h\chi^b$	Bond path connectivity index of order $h = 0-6$
${}^h\chi^b_C$	Bond cluster connectivity index of order $h = 3-6$
${}^h\chi^b_{Ch}$	Bond chain connectivity index of order $h = 3-6$
${}^h\chi^b_{PC}$	Bond path-cluster connectivity index of order $h = 4-6$
${}^h\chi^v$	Valence path connectivity index of order $h = 0-6$
${}^h\chi^v_C$	Valence cluster connectivity index of order $h = 3-6$
${}^h\chi^v_{Ch}$	Valence chain connectivity index of order $h = 3-6$
${}^h\chi^v_{PC}$	Valence path-cluster connectivity index of order $h = 4-6$
J^B	Balaban's J index based on bond types
J^X	Balaban's J index based on relative electronegativities
J^Y	Balaban's J index based on relative covalent radii
AZV_y	Triplet index from adjacency matrix, atomic number, and vertex degree; operation $y = 1-5$
AZS_y	Triplet index from adjacency matrix, atomic number, and distance sum; operation $y = 1-5$
ASZ_y	Triplet index from adjacency matrix, distance sum, and atomic number; operation $y = 1-5$
AZN_y	Triplet index from adjacency matrix, atomic number, and graph order; operation $y = 1-5$
ANZ_y	Triplet index from adjacency matrix, graph order, and atomic number; operation $y = 1-5$
DSZ_y	Triplet index from distance matrix, distance sum, and atomic number; operation $y = 1-5$
DN^2Z_y	Triplet index from distance matrix, square of graph order, and atomic number; operation $y = 1-5$
nvx	Number of non-hydrogen atoms in a molecule
$nelem$	Number of elements in a molecule
fw	Molecular weight
${}^h\chi^v$	Valence path connectivity index of order $h = 7-10$
${}^h\chi^v_{Ch}$	Valence chain connectivity index of order $h = 7-10$

(*Continued*)

Table 1.2 (Continued)

Topochemical (TC)	
si	Shannon information index
totop	Total Topological Index *t*
sumI	Sum of the intrinsic state values *I*
sumdelI	Sum of delta-*I* values
tets2	Total topological state index based on electrotopological state indices
phia	Flexibility index (kp_1* kp_2/nvx)
Idcbar	Bonchev–Trinajsti information index
IdC	Bonchev–Trinajsti information index
Wp	Wienerp
Pf	Plattf
Wt	Total Wiener number
knotp	Difference of chi-cluster-3 and path/cluster-4
knotpv	Valence difference of chi-cluster-3 and path/cluster-4
nclass	Number of classes of topologically (symmetry) equivalent graph vertices
NumHBd	Number of hydrogen bond donors
NumHBa	Number of hydrogen bond acceptors
SHCsats	E-State of C sp^3 bonded to other saturated C atoms
SHCsatu	E-State of C sp^3 bonded to unsaturated C atoms
SHvin	E-State of C atoms in the vinyl group, = CH–
SHtvin	E-State of C atoms in the terminal vinyl group, = CH_2
SHavin	E-State of C atoms in the vinyl group, = CH–, bonded to an aromatic C
SHarom	E-State of C sp^2 which are part of an aromatic system
SHHBd	Hydrogen bond donor index, sum of Hydrogen E-State values for –OH, = NH, –NH_2, –NH–, –SH, and #CH
SHwHBd	Weak hydrogen bond donor index, sum of C–H Hydrogen E-State values for hydrogen atoms on a C to which a F and/or Cl are also bonded
SHHBa	Hydrogen bond acceptor index, sum of the E-State values for –OH, = NH, –NH_2, –NH–, >N–, –O–, –S–, along with –F and –Cl
Qv	General polarity descriptor
NHBint$_y$	Count of potential internal hydrogen bonders ($y = 2–10$)
SHBint$_y$	E-State descriptors of potential internal hydrogen bond strength ($y = 2–10$)
	Electrotopological State index values for atoms types:*SHsOH, SHdNH, SHsSH, SHsNH2, SHssNH, SHtCH, SHother, SHCHnX, Hmax Gmax, Hmin, Gmin, Hmaxpos, Hminneg, SsLi, SssBe, Sssss, Bem, SssBH, SsssB, SssssBm, SsCH3, SdCH2, SssCH2, StCH, SdsCH, SaaCH, SsssCH, SddC, StsC, SdssC, SaasC, SaaaC, SssssC, SsNH3p, SsNH2, SssNH2p, SdNH, SssNH, SaaNH, StN, SsssNHp, SdsN, SaaN, SsssN, SddsN, SaasN, SssssNp, SsOH, SdO, SssO, SaaO, SsF, SsSiH3, SssSiH2, SsssSiH, SssssSi, SsPH2, SssPH, SsssP, SdsssP, SsssssP, SsSH, SdS, SssS, SaaS, SdssS, SddssS, SssssssS, SsCl, SsGeH3, SssGeH2, SsssGeH, SssssGe, SsAsH2, SssAsH, SsssAs, SdsssAs, SsssssAs, SsSeH, SdSe, SssSe, SaaSe, SdssSe, SddssSe, SsBr, SsSnH3, SssSnH2, SsssSnH, SssssSn, SsI, SsPbH3, SssPbH2, SsssPbH, SssssPb*
Geometrical (3D)/shape	
kp_0	Kappa zero
kp_1-kp_3	Kappa simple indices

(*Continued*)

Table 1.2 (Continued)

Geometrical (3D)/shape	
ka_1-ka_3	Kappa alpha indices
V_W	Van der Waals volume
^{3D}W	3D Wiener number based on the hydrogen-suppressed geometric distance matrix
$^{3D}W_H$	3D Wiener number based on the hydrogen-filled geometric distance matrix
Quantum chemical (QC)	
E_{HOMO}	Energy of the highest occupied molecular orbital
E_{HOMO-1}	Energy of the second highest occupied molecular
E_{LUMO}	Energy of the lowest unoccupied molecular orbital
E_{LUMO+1}	Energy of the second lowest unoccupied molecular orbital
ΔHf	Heat of formation
μ	Dipole moment

1.2.4.2 What do the topological indices represent about molecular structure?

Each topological index quantifies certain aspects of molecular structure as represented by the starting *model object*, that is, the particular type of molecular graph used. A quantitative descriptor which is then derived from such a model object may be considered as a theoretical model (Bunge, 1973). As conceptualized by Basak (2013b) topological indices quantify intuitive/qualitative chemical concepts like branching, complexity, cyclicity, etc. For example, Randic (1975), Bonchev and Trinajstic (1977) and Raychaudhury et al. (1984) developed new types of topological indices to discriminate among differently branched isomers in the collections of congeneric sets of molecules. But when Basak et al. (1988b) calculated 90 TIs for a structurally diverse set of 3692 molecules and carried out principal components analysis (PCA) it was found that the first principal component (PC_1) was strongly correlated with the connectivity class of branching index $^1\chi$ and also with the size of the molecular graph. Therefore, the crucial "take home message" from the PCA results was that the ***structural meaning*** of TIs is related to the molecular landscape of the data set that is being analyzed. Intuitive concepts like branching and their quantification by TIs may not survive when we go from a congeneric and small set to a diverse set of molecules (Basak and Majumdar, 2016). As pointed out by the philosopher Bertrand Russell (1950) regarding human intuition:

> *"Intuition, in fact, is an aspect and development of instinct, and, like all instincts, is admirable in those customary surroundings which have moulded the habits of the animal in question, but totally incompetent as soon as the surroundings are changed in a way which demands some non-habitual mode of action."*

The multidimensional space derived from PCA provided an unexpected insight into the structural meaning of topological indices. It may be mentioned here that each TI usually maps the structure of the molecular graph into a set of real numbers (Johnson et al., 1988b; Trinajstić, 1992). Although many TIs with different discriminating power to distinguish among structures of congeneric sets of molecular graphs have been devised (Balaban, 1982; Raychaudhury et al., 1984), no decent set of TIs is known which can completely characterize molecular graphs up to isomorphism (Read and Corneil, 1977). ***Graph theoretical approach*** (Natarajan et al., 2007) ***has also been used to develop quantitative chirality index to quantify relative chirality of diastereomers.***

1.3 Bioifnormatics: quantitative inforamtics in the age of big biology

> *"And indeed this theme has been at the centre of all my research since 1943, both because of its intrinsic fascination and my conviction that a knowledge of sequences could contribute much to our understanding of living matter."*
>
> *—Frederick Sanger*

> *"As soon as you go into any biological process in any real detail, you discover it's open-ended in terms of what needs to be found out about it."*
>
> *—Joshua Lederberg*

> *"Life is a partial, continuous, progressive, multi-form, and conditionally interactive self-realization of the potentialities of atomic electron states"*
>
> *—John Desmond Bernal*

The 20th century and the first part of this century (Fig. 1.4) witnessed a lot of important developments in our understanding of the storage and processing of the genetic information of organisms. The work of Avery et al. (1944), Crick (1968) and Hershey and Chase (1952) showed that nucleic acids contain the genetic information. Watson and Crick's (1953) work showed how genetic information is stored at the molecular level. The central dogma that genetic information stored in DNA is transcribed into RNA which, in turn, is translated into proteins that are the workhorses of the biological system was an important advancement in molecular biology. Deciphering that a triplet of nucleic acid bases codes for one amino acid in a protein was proved by Nirenberg and Leder (1964). Sanger et al. (1977) development of nucleic acid sequencing method speeded up the sequencing process. The completion of the Human Genome Project (HGP, 2003) gave us the ability to read nature's complete genetic blueprint for building up a human being. The efficient DNA sequencing method of Sanger et al. (1977) and subsequent next-generation sequencing methods speeded up the process of accumulation of nucleic acid

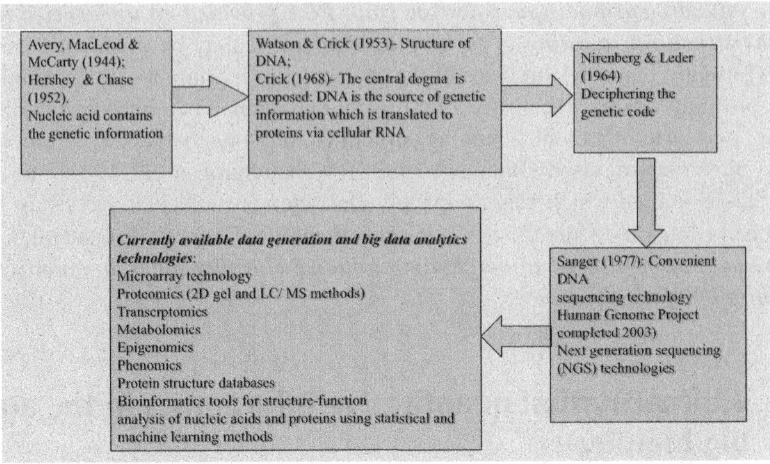

Figure 1.4 A brief history of the recognition of DNA as the genetic material, development of sequencing technologies and the emergence of bioinformatics.

sequence data in various publicly available databases. Subsequent development of microarray, proteomics, and large-scale protein sequencing technologies and formulation of computer-assisted bioinformatics tools catapulted us into the modern era of bioinformatics (Heather and Chain, 2016).

The information coded in the nucleic acid sequences in DNA and RNA can be quantified using graph theoretical approaches. The process consists of first converting the sequences into graphs based on some rules and then extracting invariants from such graphs (Nandy, 2015; Nandy et al., 2006). For details on the development and use of alignment-free sequence descriptors (AFSDs) and their applications see Chapter 16.

Proteins are structurally complex and highly organized macromolecules and are the workhorses of the cellular and metabolic processes. The structure–activity relationship of proteins was pioneered by structural chemists, as described in a recent article (Gauthier et al., 2019):

> *"Although much of the pre-1960s research in biochemistry focused on the mechanistic modeling of enzymes, Emile Zuckerkandl and Linus Pauling departed from this paradigm by investigating biomolecular sequences as "carriers of information." Just as words are strings of letters whose specific arrangement convey meaning, the molecular function (i.e. meaning) of a protein results from how its amino acids are arranged to form a 'word."*

Early researchers like Dayhoff and Ledley (1962) and Pauling and Corey (1951) contributed immensely to the development of bioinformatics.

These days various protein and nucleic acid databases are available online and are updated regularly. The enormous speed by which the sequence data is increasing is evident from a look at the GenBank and WGS Statistics. W***hereas on 3 December 1982 the database had 680,338 bases in 606 sequences, in October***

2021 the corresponding numbers were 1,014,763,752,113 bases of 233,642,893 sequences (https://www.ncbi.nlm.nih.gov/genbank/statistics/).

According to PDB Statistics (https://www.rcsb.org/stats/growth/growth-released-structures) on the overall growth of available protein structures, ***in 2012 this database had 86,194 entries whereas in 2021 that number increased to 184,700.***

Very recently, with the help of the deep learning algorithm developed by Google, its offshoot DeepMind has made a gigantic step in predicting the protein three-dimensional (3D) structure with near to the experimental accuracy (https://www.nature.com/articles/d41586-020-03348-4) using their AlphaFold 2 algorithm.

Descriptors derived from biological systems like omics (DNA/RNA sequences, proteomics patterns, etc.) have been called "biodescriptors" by Basak group of researchers (Basak, 2010, 2013a).

Data derived from proteomics technologies can be used to predict the bioactivity of drugs, toxicants, and nanosubstances. Basak and coworkers (Basak, 2010; Vračko et al., 2018) formulated different approaches for the quantification of proteomics patterns using diverse mathematical methods. Such proteomics-based biodescriptors can be used in combination with molecular chemodescriptors in assessing bioactivity and toxicity of chemicals as shown in Fig. 1.5.

1.4 Major pillars of model building

"In God we trust; all others bring data."

—*W. Edwards Deming*

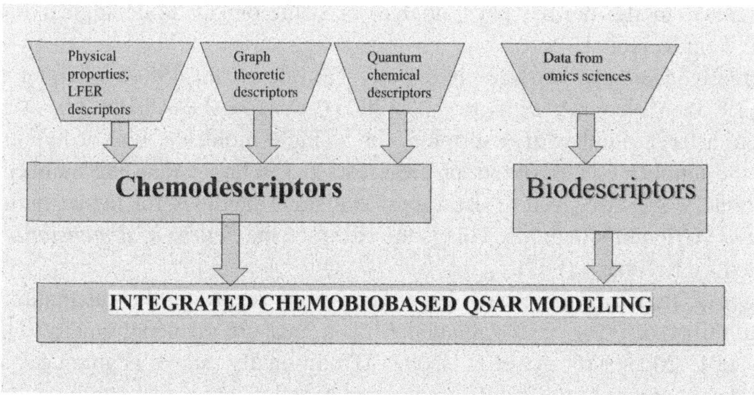

Figure 1.5 Confluence of descriptors for predictive model building: There are two main sources of descriptors for assessing physicochemical and biological properties. For physical property estimation chemodescriptors consisting of calculated topological indices, substructures, three-dimensional or geometrical descriptors; and quantum chemical descriptors may be used. For complex biological properties a combination of chemodescriptors and omics-derived biodescriptors may be used as intendent variables in predictive models Basak, 2021a,b).

"Tho' kingdoms fall,

This pillar never shall

Decline or waste at all;

But stand for ever by his own

Firm and well-fixed foundation."

Robert Herrick, The Pillar of Fame

"To call in the statistician after the experiment is done may be no more than asking him to perform a post-mortem examination: he may be able to say what the experiment died of"

—*Ronald Fisher*

The major pillars of high-quality QSAR model developments are (a) sufficient number of high-quality data (dependent variable), (b) descriptors or features (independent variables) that quantify relevant aspects of molecular or biomolecular structures related to the property under investigation, and (c) correct applications of robust statistical or machine learning methods for the construction and validation of models (Basak, 2013a). For biodescriptors derived from DNA/RNA sequences (Nandy, 2015) or omics technologies like proteomics (Basak, 2010; Vračko et al., 2018) the selected independent variables should be linked to the biochemical pathways related to the normal physiological or pathological/ toxicological functions that we want to model.

Currently available software like Polly (Basak et al., 1988b), Triplet (Basak et al., 1993), MolconnZ, Version 4.05 (2003), Dragon (DRAGON 7.0, 2021) can calculate a large number of descriptors for a single molecule, real or hypothetical. When the number of descriptors pr predictors (p) is larger than the number (n) of data points, $n > p$, we need to use robust statistical methods for model building in such ***rank deficient*** situations. This is the realm of the ***"curse of dimensionality"*** as indicated by Bellman (1961).

In QSAR literature there have been discussions regarding the calculation methods and utility of q^2 or cross-validated R^2 as a measure of model quality (Hawkins et al., 2001, 2003; Tropsha et al., 2003). Traditionally, at the beginning stages of QSAR development in the 1960s, researchers faced situations where the number (n) of chemicals was much larger than the number of descriptors or predictors (p). Currently, as discussed above, with the advent of software capable of calculating many molecular descriptors for a single molecule, we often face the rank-deficient situation.

Modeling data that is high-dimensional and/or rank deficient often contains a screening step where either a smaller number of important predictors are selected from the much larger, full set of predictors, or some transformation like principal component analysis (PCA) is applied on the descriptor data to reduce the intrinsic dimension of the effective predictor space (Basak et al., 1988a; Basak and Majumdar, 2015; Majumdar and Basak, 2016).

Researchers should take special care when performing validation on a QSAR model that incorporates such a screening step. Although it is more intuitive to first select the important predictors or reduce the dimension of predictor space, and then perform the validation steps, this is not the statistically correct approach. The correct method is two-deep cross-validation which gives the correct or true q^2 instead of the wrong or naïve q_2 (Fig. 1.6). Instead of performing the first screening step only once on the full data, this validation technique performs it for each training-test split used in the validation procedure (see Fig. 1.6). Although this results in longer computation times, owing to the speed of the present-day computing architecture that is not an issue.

One very important area to consider while validating QSAR models is the method of validation used. QSAR literature has seen three types of cross-validation: Leave one out (LOO), External validation and k-fold cross-validation. For sample

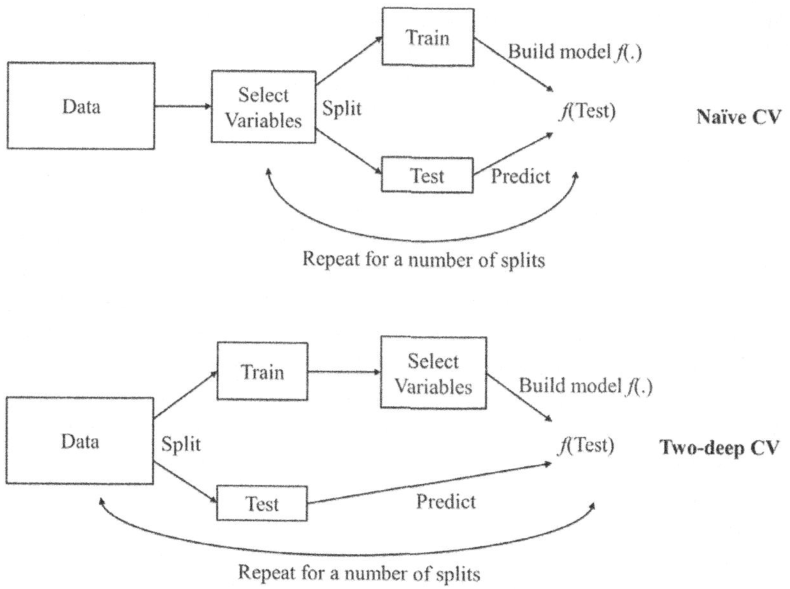

Figure 1.6 A schematic representation of two-deep cross-validation schemes. The first one leads to the calculation of naïve or wrong q^2 whereas the second one gives the statistically correct or true q^2.

size n, the LOO procedure builds n models leaving each of the samples out from the training set and predicts the activity of a sample using its corresponding leave-out model. For high values of n this needs prohibitively high computer time. A way to mitigate this is to split the data into k disjoint partitions and get predictions for compounds in each partition from the model trained on all compounds not belonging to that partition. This is the k-fold cross-validation.

As seen discussed above, both in LOO and k-fold cross-validation, all chemicals in the data set are used in turn to check the comparison between experimental and QSAR-predicted values of activity/property. In the external validation method, only one subset of the n compounds is held back for checking model quality while the rest is used in model building. It is a widely held belief that the use of a hold-out test set is always the best method of model validation (Tropsha et al., 2003). However, theoretical argument and empirical study (Hawkins et al., 2001, 2003) show that the LOO cross-validation approach is preferred to the use of only one hold-out test set unless the data set to be modeled is very large. The drawbacks of holding out a test set include (a) structural features of the held-out chemicals are not included in the modeling process resulting in a loss of information, (b) predictions are made on only a subset of the available compounds, whereas LOO method predicts the activity values for all compounds, (c) there is no scientific tool that can guarantee the similarity between the training and test sets, and (d) personal bias can easily affect the selection of the external test set, which in turn influences the prediction performance because of the small size of the test set. The reader is referred to Hawkins et al. (2001, 2003) for further discussion of proper model validation techniques.

According to the Organization for Economic Cooperation and Development (OECD) QSAR guidelines, evaluation of out-of-sample performance is essential in determining the plausibility of a QSAR model to be applicable in situations that involve real-world decision-making. The cross-validated q^2 metric is an effective way of quantifying the model performance, and we hope the above discussion will encourage researchers in mathematical chemistry, QSAR, drug discovery, and predictive toxicology to navigate several subtle issues in the calculation of q^2 values that reflect the actual predictive capability of a modeling technique.

A properly validated QSAR model based on computed molecular descriptors can be used in the design of new drugs and specialty chemicals as part of the decision support system. As stated by Professor Dennis Rouvray (1991):

> *"Assigning numbers to molecular structures allow us to predict the behavior of a wide arrary of chemical substances. It also offers a new approach to the design of plastics, drugs and even tastier beer."*

1.5 Discussion

> *"A good decision is based on knowledge and not on numbers."*
>
> —*Plato*

"The endless cycle of idea and action,

Endless invention, endless experiment,

Brings knowledge of motion, but not of stillness;

Knowledge of speech, but not of silence;

Knowledge of words, and ignorance of the Word.

Where is the wisdom we have lost in knowledge?

Where is the knowledge we have lost in information?"

—T.S. Eliot, In: The Rock

"No human inquiry can be a science unless it pursues its path through mathematical exposition and demonstration."

—Leonardo da Vinci

Understanding the structural basis of properties of molecules and biomolecules has been a sustained interest in chemistry and biology. The brief history of the development of the LFER approach and numerical graph invariant or topological indices (Basak, 2021a; Hansch and Leo, 1979) indicate that initially qualitative aspects of small and congeneric sets of molecules were investigated in these approaches. Subsequently, quantitative and mathematically based methods were developed. Each approach has its own model object and specific method of developing quantitative descriptors or theoretical models (Basak et al., 1990; Bunge, 1973).

In the realm of molecular graph invariants, Basak (2013b) noted that *such chemodescriptors quantify different qualitative concepts of molecular structure, viz., branching, complexity, cyclitic*, etc. Whereas from the middle of the 20th century until the early 1980s topological indices were used mainly to develop QSARs of small and congeneric sets of molecules, in the subsequent time period larger sets of graph theoretic descriptors were used to develop models of large and diverse data sets of chemicals. This was possible principally by (a) availability of progressively higher computer power as envisaged by Moore's law (2021), (b) availability of large and diverse databases for properties (Basak et al., 1988a; Gadaleta et al., 2019; Meng et al., 2021; Mansouri, et al., 2018), (c) need for property estimation of molecules, existent or not synthesized, for new drug discovery and predictive toxicology (Drefahl and Reinhard, 1998; Lyman et al., 1990), and (d) easy access to statistical and machine learning software for analysis of such large data sets using descriptors which could be computed fast.

Topological indices and substructures of molecular/ biomolecular graphs have been used for many purposes some of which are as follows:

(a) Characterization of structure and discrimination of closely related structures like isospectral graphs (Balasubramanian and Basak, 1998) and structural isomers with the same empirical formula (Balaban, 1982; Bonchev and Trinajstic, 1977; Raychaudhury et al., 1984).
(b) Development of robust and high-quality regression models for the assessment of physicochemical, pharmacological and toxicological properties of large and diverse sets of chemicals (Basak, 2013a; Basak and Majumdar, 2016; Chakravarti, 2018, 2021; Majumdar et al., 2019).
(c) Discrimination among different modes of action (MOAs) of priority pollutants (Basak et al., 1998).
(d) Clustering of large chemical databases into smaller subsets using topological descriptors as to support practical decision support systems. Topological indices were used to cluster a virtual library of 248,832 psoralen derivatives (Basak et al., 2010), JP_8 jet fuel constituents (Basak, et al., 2000) and a large set of proprietary chemicals (Lajiness, 1990) for pharmaceutical drug design.
(e) Measuring structural similarity/dissimilarity of chemicals for drug design and hazard assessment of pollutants (Basak, 2014; Lajiness, 1990).
(f) Characterization of the structural basis of emerging drug resistance using differential QSAR (DiffQSAR) approach (Basak, et al., 2011).
(g) Design of novel peptide vaccines for emerging global pathogens (Dey et al., 2017).
(h) Discrimination and classification of pathogens like Zika virus (Vračko et al., 2021a), severe acute respiratory syndrome (SARS) virus, Middle East respiratory syndrome (MERS) virus and severe acute respiratory syndrome Coronavirus 2 or SARS-CoV-2 or COVID-19 (Vračko et al., 2021b) using alignment-free sequence descriptors (AFSDs) calculated from the RNA sequences of such viruses.
(i) Computational characterization of emerging mutants/ variants of the global viral pathogen like SARS-CoV-2 (Dey et al., 2021).
(j) Computer-assisted rational design of immunosuppressive compounds (Grassy et al., 1998).
(k) Development of QSARS of priority pollutants using TIs and proteomics-based biodescriptors (Basak, 2010; Chakravarti, 2021).

Basak and coworkers (Basak, 2013a; Basak et al., 2003) used topological, geometrical or 3D, and different levels of quantum chemical (QC) descriptors in a hierarchical or graduated manner in the formulation of hierarchical QSAR (HiQSAR) models. Results of such studies showed that in many cases the addition of 3D and QC descriptors to the set of independent variables made very little or no improvement in model quality. Further studies are needed to validate this finding. The development of high-quality predictive models using graph invariants which can be calculated fast using currently available software as compared to the high computational resource-intensive quantum chemical routines will have many useful applications in the emerging big data analytics area. Because complex biological properties of chemicals cannot be predicted only by chemodescriptors or from their physicochemical properties alone (Arcos, 1987; Basak, 2010, 2013a, 2021a), we will have to use a combination of physiochemical properties, computed chemodescriptors and omics-derived biodescriptors to predict complex endpoints as indicated in Fig. 1.5.

1.6 Conclusion

"There are two possible outcomes: if the result confirms the hypothesis, then you've made a measurement. If the result is contrary to the hypothesis, then you've made a discovery."

—Enrico Fermi

"Good tests kill flawed theories; we remain alive to guess again."

—Karl Popper

"I have become my own version of an optimist. If I can't make it through one door, I'll go through another door - or I'll make a door. Something terrific will come no matter how dark the present."

—Rabindranath Tagore

"Nature shows us only the tail of the lion. But there is no doubt in my mind that the lion belongs with it even if he cannot reveal himself to the eye all at once because of his huge dimension. We see him only the way a louse sitting upon him would."

—Albert Einstein

The fields of chemoinformatics and bioinformatics are witnessing the advent of big data. Progressively larger data sets of properties needed for new drug discovery and environmental protection are being available in the public domain. In the realm of computational chemistry, currently available combination software is capable of calculating thousands of descriptors for a single molecule. This is like the invention of the microscope by Antonie van Leeuwenhoek which gave us a magnified view of the microbial world not seen by anyone before that. In the realm of bioinformatics and computational biology, in the postgenomic era sequence and structural data on proteins and nucleic acids are providing a vast amount of data for modeling and analysis (Cartwright et al., 2016; Gadaleta et al., 2019; Mansouri et al., 2018; Meng et al., 2021). When we venture into the realm of analysis of these data sets with robust statistical and machine-learning methods we will probably find many new relationships (Basak et al., 2015, 2021a). This is particularly highly expected for biological processes/ domains because of the phenomenon of emergent properties of complex systems (Basak et al., 1990). Also, when we go from congeneric sets to large and structurally diverse sets of chemicals, the structural meaning of the descriptors may change quite significantly (Basak, et al., 1988a).

When stable relationships are found, we will have to develop explanation underlying those empirically discovered relationships.

When divergent fields come together the dividing lines among them may fall apart. That is what is happening right now in the chemobioinformatics area (Basak, 2021a,b). Different fields like the LFER approach, computational chemistry including graph theoretic formalisms, quantum chemical methods, and omics techniques had their own specific process of development following their unique "***connect the dots***" lanes in their respective domain. It is tempting to speculate that for big data analytics we may need a bold and convergent paradigm of "***connecting the connectors***" that will bring together a grand fusion of data, methodology and expertise from diverse disciplines for the solution of practical problems of science and society. An illustration of such a need comes from the letter of Professor Corwin Hansch, a pioneer of LFER based QSAR modeling, written to Subhash C. Basak in response to his dedication of one of his papers to honor Professor Hansch:

> *"Chemistry Department*
>
> *Pomona College*
>
> *June 13, 1994*
>
> *Dr. Subhash C. Basak*
>
> *Center for Water and the Environment 5013 Miller Trunk Highway*
>
> *Duluth, Minnesota 55811 Dear Dr. Basak:*
>
> *It was quite a pleasant surprise to get your paper dedicated to me!! I'd forgotten about your mentioning it in Atlanta.*
>
> *Looking over your papers on the Graph Theoretic approach I'm impressed by all that you are doing. However, I'm not up on this area to the point that I can say that I really understand all that you are attempting to do. Your approach is quite abstract.*
>
> *Hope our paths will soon cross again. Sincerely,*
>
> *e:*
>
> *Corwin Hansch*
>
> *CH/pa"*

Big data analytics will demand from us integrative approaches derived from a combination of elements from a diverse range of disciplines like mathematics, physics, chemistry, biology, toxicology, statistics, artificial intelligence, genomics, proteomics and many others yet to evolve and come to the forefront of science and technology.

Acknowledgment

I have immensely benefitted for more than four decades from collaborations with a large number of colleagues spread over four continents. I would like to specifically mention Kanika Basak, Douglas Hawkins, Gregory Grunwald, Ashesh Nandy, Milan Randic, Alexandru T. Balaban, Nenad Trinajstic, Dilip K. Sinha, Smarajit Manna, Subhabrata Majumdar, Claudiu Lungu, Amiya B. Roy, Guillermo Restrepo, and Suman Chakravarti for their helpful collaborations in my research.

References

American Heritage Dictionary <https://www.wordnik.com/words/constitutive#: ~ :text = Having %20power%20to%20enact%20or,which%20see%2C%20unite%20regulative> (accessed 16.11.22).

Arcos, J.C., 1987. Structure-activity relationships: criteria for predicting the carcinogenic activity of chemical compounds. Environ. Sci. Technol. 21, 743–745.

Auer, C.M., Nabholz, J.V., Baetcke, K.P., 1990. Mode of action and the assessment of chemical hazards in the presence of limited data: use of structure-activity relationships (SAR) under TSCA, Section 5. Environ. Health Perspect. 87, 183–197.

Avery, O.T., MacLeod, C.M., McCarty, M., 1944. Studies on the chemical nature of the substance inducing transformation of pneumococcal types. J. Exp. Med. 79, 137–158.

Balaban, A.T., 1982. Highly discriminating distance-based topological index'. Chem. Phys. Lett. 80, 399–404.

Balasubramanian, K., Basak, S.C., 1998. Characterization of isospectral graphs using graph invariants and derived orthogonal parameters. J. Chem. Inf. Comput. Sci. 38, 367–373.

Basak, S.C., 1987. Use of molecular complexity indices in predictive pharmacology and toxicology: a QSAR approach. Med. Sci. Res. 15, 605–609.

Basak, S.C., 2010. Role of mathematical chemodescriptors and proteomics-based biodescriptors in drug discovery. Drug. Dev. Res. 72, 1–9.

Basak, S.C., 2013a. Mathematical descriptors for the prediction of property, bioactivity, and toxicity of chemicals from their structure: a chemical-cum-biochemical approach. Curr. Comput. Aided Drug. Des. 9, 449–462.

Basak, S.C., 2013b. Philosophy of mathematical chemistry: a personal perspective. HYLE 19, 3–17.

Basak, S.C., 2014. Molecular similarity and hazard assessment of chemicals: a comparative study arbitrary and tailored similarity spaces. J. Eng. Sci. Manage. Educ. 7 (III), 178–184.

Basak, S.C., 2021a. My tortuous pathway through mathematical chemistry and QSAR research with memories of some personal interactions and collaborations with Professors Milan Randic and Mircea Diudea. Croat. Chem. Acta 93 (4), 247–258.

Basak, S.C., 2021b. Some comments on the three-pronged chemobiodescriptor approach to QSAR—a historical view of the emerging integration. Curr. Comput. Aided Drug Des. 2022, in press.

Basak, S.C., Balasubramanian, K., Gute, B.D., Mills, D., Gorczynska, A., Roszak, S., 2003. Prediction of cellular toxicity of halocarbons from computed chemodescriptors: a hierarchical QSAR. Approach. J. Chem. Inf. Comput. Sci. 43, 1103–1109.

Basak, S.C., Grunwald, G.D., Balaban, A.T., 1993. TRIPLET, Copyright of the Regents of the University of Minnesota.

Basak, S.C., Grunwald, G.D., Host, G.E., Niemi, G.J., Bradbury, S.P., 1998. A comparative study of molecular similarity, statistical and neural network methods for predicting toxic modes of action of chemicals. Environ. Toxicol. Chem. 17, 1056–1064.

Basak, S.C., Grunwald, G.D., Gute, B.D., Mills, D., 2000, Clustering of JP-8 chemicals using property spaces and structure spaces: a novel tool for hazard assessment, Second Indo-US Workshop on Mathematical Chemistry (with applications to Drug Discovery, Environmental toxicology, Chemoinformatics and Bioinformatics), University of Minnesota Duluth, Duluth, MN, USA, Volume: 1 <https://www.researchgate.net/publication/271830175_Clustering_of_JP-8_chemicals_using_property_spaces_and_structure_spaces_A_novel_tool_for_hazard_assessment>.

Basak, S.C., Harriss, D.K., Magnuson, V.R., 1988b. POLLY v. 2.3: Copyright of the University of Minnesota, USA.

Basak, S.C., Magnusson, V.R., 1983. Molecular topology and narcosis: a quantitative structure-activity relationship (QSAR) study of alcohols using complementary information content (CIC). Arzneim. Forsch. Drug. Res. 33, 501–503.

Basak, S.C., Majumdar, S., 2015. Prediction of mutagenicity of chemicals from their calculated molecular descriptors: a case study with structurally homogeneous versus diverse datasets. Curr. Comput. Aided Drug. Des. 11, 117–123.

Basak, S.C., Magnuson, V.R., Niemi, G.J., Regal, R.R., 1988a. Determining structural similarity of chemicals using graph-theoretic indices. Discret. Appl. Math. 19, 17–44.

Basak, S.C., Majumdar, S., 2016. Exploring two QSAR paradigms-congenericity principle versus diversity begets diversity principle analyzed using computed mathematical chemodescriptors of homogeneous and diverse sets of chemical mutagens. Curr. Comput. Aided Drug. Des. 12, 1–3.

Basak, S.C., Mills, D., Gute, B.D., Balaban, A.T., Basak, K., Grunwald, G.D., 2010. Use of mathematical structural invariants in analyzing combinatorial libraries: a case study with psoralen derivatives. Curr. Comput. Aided Drug. Des. 6, 240–251.

Basak, S.C., Mills, D., Hawkins, D.M., 2011. Characterization of dihydrofolate reductases from multiple strains of Plasmodium falciparum using mathematical descriptors of their inhibitors. Chem. Biodivers. 8, 440–453.

Basak, S.C., Niemi, G.J., Veith, G.D., 1990. Optimal characterization of structure for prediction of properties. J. Math. Chem. 4, 185–205 (1990).

Basak, S.C., Roy, A.B., Ghosh, J.J., 1980. Study of the structure–function relationship of pharmacological and toxicological agents using information theory. In: Avula, X.J.R., Bellman, R., Luke, Y.L., Rigler, A.K. (Eds.), Proceeding of the 2nd International Conference on Mathematical Modelling, vol. 2. University of Missouri–Rolla, Rolla, Missouri, USA, pp. 851–856.

Basak, S.C., Villaveces, J.L., Restrepo, G. (Eds.), 2015. Advances in Mathematical Chemistry and Applications, Volume 1 & 2. Elsevier & Bentham Science Publishers, Amsterdam & Boston.

Bayda, S., Adeel, M., Tuccinardi, T., Cordani, M., Rizzolio, F., 2019. The history of nanoscience and nanotechnology: from chemical-physical applications to nanomedicine. Molecules (Basel, Switz.) 25 (1), 112. Available from: https://doi.org/10.3390/molecules25010112.

Bellman, R.E., 1961. Adaptive control processes. A Guided Tour. Princeton University Press, Princeton, NJ.

Bhattacharjee, A.K., 2015. Role of in silico stereoelectronic properties and pharmacophores in aid of discovery of novel antimalarials, antileishmanials, and insect Repellents. In: Basak, S.C., Restrepo, G., Villaveces, J.L. (Eds.), Advances in Mathematical Chemistry and Applications, Elsevier & Bentham Science. Publishers, Amsterdam & Boston, pp. 273–305.

Bielinska-Waz, D., Waz, P., Nowak, W., Nandy, A., Basak, S.C., 2007. Similarity and dissimilarity of DNA/RNA sequences. In: Simos, T.E., Maroulis, G. (Eds.), Computation in Modern Science and Engineering, Proceedings of the International Conference on Computational Methods in Science and Engineering 2007 (ICCMSE 2007). American Institute of Physics, Melville, New York, pp. 28–30.

Bonchev, D., 1983. Information Theoretic Indices For Characterization of Chemical Structures. Research Studies Press, Letchworth, Hertfordshire, U.K.

Bonchev, D., Trinajstic, N., 1977. Information theory, distance matrix, and molecular branching. J. Chem. Phys. 38, 4517–4533.

Braga, R.C., Melo-Filho, C.C., Moreira-Filho, J.T., Muratov, E.N., Andrade, C.H., 2018. QSAR-based virtual screening: advances and applications in drug discovery. Front. Pharmacol. 9, 1275. Available from: https://doi.org/10.3389/fphar.2018.01275.

Bunge, M., 1973. Method, Model and Matter. Reidel, Dordrecht, Springer, The Netherlands.

Cartwright, J.H.E., Giannerini, S., González, D.L., 2016. DNA as information: at the crossroads between biology, mathematics, physics and chemistry. Phil. Trans. R. Soc. A 374, 20150071. Available from: http://doi.org/10.1098/rsta.2015.0071.

Ciallella, H.L., Zhu, H., 2019. Advancing computational toxicology in the big data era by artificial intelligence: data-driven and mechanism-driven modeling for chemical toxicity. Chem. Res. Toxicol. 32 (4), 536–547.

Chakravarti, S.K., 2018. Distributed representation of chemical fragments. ACS Omega. 31 (3), 2825–2836. 3.

Chakravarti. S.K., 2021, Scalable QSAR systems for predictive toxicology, Chapter 17, in this book.

Crick, F.H., 1968. The origin of the genetic code. J. Mol. Biol. 38, 367–379.

Dayhoff, M.O., Ledley, R.S., 1962. Comprotein: a computer program to aid primary protein structure determination. Proceedings of the December 4–6, 1962, Fall Joint Computer Conference. ACM, New York, NY, pp. 262–274.

Dehmer, M., Basak, S.C. (Eds.), 2012. Statistical and Machine Learning Approaches for Network Analysis. Wiley, Hoboken, New Jersey, USA.

Devillers, J., Balaban, A.T. (Eds.), 1999. Topological Indices and Related Descriptors in QSAR and QSPR; Gordon and Breach. Amsterdam, The Netherlands.

Dey, S., Nandy, A., Basak, S.C., Nandy, P., Das, S., 2017. A Bioinformatics approach to designing a Zika virus vaccine. Comput. Biol. Chem. 68, 143–152.

Dey, T., Chatterjee, S., Manna, S., Nandy, A., Basak, S, C., 2021. Identification and computational analysis of mutations in SARS-CoV-2. Comput Biol Med. 129:104166. Available from: https://doi.org/10.1016/j.compbiomed.2020.104166. Epub 2020 Dec 28. PMID: 33383528; PMCID: PMC7837166.

DiMasi, J.A., Grabowski, H.G., Hansen, R.W., 2016. Innovation in the pharmaceutical industry: new estimates of R&D costs. J. Health Econ. 47, 20–33.

Diudea, M.V., Lungu, C.N., Nagy, C.L., 2018. Cube-rhombellane related structures: a drug perspective. Molecules 23 (10), 2533. Available from: https://doi.org/10.3390/molecules23102533.

Diudea, M.V., 1997. Indices of reciprocal property or harary indices. J. Chem. Inf. Comput. Sci. 37, 292–299.

DRAGON 7.0 <https://chm.kode-solutions.net/pf/dragon-7-0/> (accessed 04.12.21).

Drefahl, A., Reinhard, M., 1998. Handbook for Estimating Physicochemical Properties of Organic Compounds, 1st edition Wiley-Interscience.

Euler, I., 1736. Solutio problematis ad geometriam situs pertinentis. Comment. Acad. Sci. U. Petrop. 8, 128–140.

European Chemicals Agency (ECHA) <https://echa.europa.eu/sl/registration-statistics> (accessed 11.11.21).

Gadaleta, D., Vuković, K., Toma, C., et al., 2019. SAR and QSAR modeling of a large collection of LD50 rat acute oral toxicity data. J. Cheminform 11, 58. Available from: https://doi.org/10.1186/s13321-019-0383-2.

Gauthier, J., Vincent, A.T., Charette, S.J., Derome, N., 2019. A brief history of bioinformatics. Brief. Bioinform 20 (6), 1981–1996. Available from: https://doi.org/10.1093/bib/bby063. PMID: 30084940.

GenBank and WGS Statistics <https://www.ncbi.nlm.nih.gov/genbank/statistics/> (accessed 03.12.21).

Gini, G., Ferrari, T., Cattaneo, D., Bakhtyari, N.G., Manganaro, A., Benfenati, E., 2013. Automatic knowledge extraction from chemical structures: the case of mutagenicity prediction. SAR. QSAR Env. Res. 24 (5), 365–383.

Goodman and Gilman, 1990. The Pharmacological Basis of Therapeutics, Eighth Edition, Pergamon Press, New York.

Grassy, G., Calas, B., Yasri, A., Lahana, R., Woo, J., Iyer, S., et al., 1998. Nat. Biotechnol. 16, 748–752.

Guo, X.F., Randić, M., Basak, S.C., 2001. A novel 2-D graphical representation of DNA sequences of low degeneracy. Chem. Phys. Lett. 350, 106–112.

Gutman, I., Trinajstic, N., 1972. Graph theory and molecular orbitals. Total φ-electron energy of alternant hydrocarbons. Chem. Phys. Lett. 17, 535–538.

Hammett, L.P., 1937. The effect of structure upon the reactions of organic compounds. benzene derivatives. J. Am. Chem. Soc. 59, 96–103.

Hansch, C., Leo, A., 1979. Substituent Constants For Correlation Analysis in Chemistry and Biology. Wiley.

Hansch, C., Leo, A., 1995. Exploring QSAR: fundamentals and applications in, 1st edition Chemistry and Biology, Volume 1. American Chemical Society, Washington, D. C.

Harary, F., 1969. Graph Theory, 2nd ed. Reading, MA, Addison-Wesley.

Harary, F., 1986. Graph theory as applied mathematics. J. Graph. Theory 10, iii–iv.

Hawkins, D.M., Basak, S.C., Shi, X., 2001. QSAR with few compounds and many features. J. Chem. Inf. Comput. Sci. 41, 663–670.

Hawkins, D.M., Basak, S.C., Mills, D., 2003. Assessing model fit by cross-validation. J. Chem. Inf. Comput. Sci. 43, 579–586.

Heather, J.M., Chain, B., 2016. The sequence of sequencers: the history of sequencing DNA. Genomics 107 (1), 1–8.

Hitchings, G.H., Elion, G.B., 1954. The chemistry and biochemistry of purine analogs. Ann. NY. Acad. Sci. 60, 195–199.

Hershey, A.D., Chase, M., 1952. Independent functions of viral protein and nucleic acid in growth of bacteriophage. J. Gen. Physiol. 36, 39–56.

Hosoya, H., 1971. Topological index. A newly proposed quantity characterizing the topological nature of structural isomers of saturated hydrocarbons. Bull. Chem. Soc. Jpn. 44, 2332–2339.

Human Genome Project (HGP), 2003 <https://www.genome.gov/human-genome-project>.

Ingold, C.K., 1953. Structure And Mechanism in Organic Chemistry. Cornell Univ. Press, Ithaca, N. Y.

Janezic, D., Milicevic, A., Nikolic, S., Trinajstic, N., 2015. Graph-Theoretical Matrices in Chemistry, 1st edition CRC Press, Boca Raton, FL.

Johnson, M., Basak, S.C., Maggiora, G., 1988b. A characterization of molecular similarity methods for property prediction. Mathl. Comput. Model. 11, 630–634.

Katritzky, A.R., Lobanov, V.S., Karelson, M., 1995. QSPR: the correlation and quantitative prediction of chemical and physical properties from structure. Chem. Soc. Rev. 24, 279–287.

Katritzky, A.R., Putrukhin, R., Tatham, D., Basak, S.C., Benfenati, E., Karelson, M., et al., 2001. Interpretation of quantitative structure-property and -activity relationships. J. Chem. Inf. Comput. Sci. 41, 679–685.

Kerber, A., Laue, R., Meringer, M., Rücker, C., Schymanski, E., 2014. Mathematical Chemistry and Chemoinformatics: Structure Generation, Elucidation and Quantitative Structure-Property Relationships. De Gruyter, Berlin, Germany.

Khan, T., Panday, S.K., Ghosh, I., 2018. ProLego: tool for extracting and visualizing topological modules in protein structures. BMC Bioinforma. 19, 167.

Kier, L.B., Hall, L.H., 1976. Molecular Connectivity in Chemistry and Drug Research. Academic Press, New York.

Kier, L.B., Hall, L.H., 1986. Molecular Connectivity in Structure Activity Analysis. Wiley, London.

Kier, L.B., Hall, L.H., 1999. Molecular Structure Description: The Electrotopological State. Academic Press, San Diego, CA.

Kuratowski, C., 1930. Sur les probl'emes des courbes gauches en Topologie. Fund. Math. 15, 271–283.

Lajiness, M.S., 1990. Molecular similarity-based methods for selecting compounds for screening. In: Rouvray, D.H. (Ed.), Computational Chemical Graph Theory. Nova, New York, pp. 299–316.

Liu, Y., Carbonell, J., Weigele, P., Gopalakrishnan, V., 2006. Protein fold recognition using segmentation conditional random fields (SCRFs). J. Comput. Biol. 13, 394–406.

Lyman, W.J., Reehl, W.F., Rosenblatt, D.H., 1990. Handbook of Chemical Property Estimation Methods. American Chemical Society, Washington D. C.

Majumdar, S., Basak, S.C., 2016. Exploring intrinsic dimensionality of chemical spaces for robust QSAR model development: a comparison of several statistical approaches. Curr. Comput. Aided Drug. Des. 12, 294–301.

Majumdar, S., Basak, S.C., Lungu, C.N., Diudea, M.V., Grunwald, G.D., 2019. Finding needles in a haystack: determining key molecular descriptors associated with the blood-brain barrier entry of chemical compounds using machine learning. Mol. Inform. 38 (8–9), e1800164. Available from: https://doi.org/10.1002/minf.201800164. Epub 2019 Jul 19. PMID: 31322827.

Mansouri, K., Grulke, C.M., Judson, R.S., Williams, A.J., 2018. OPERA models for predicting physicochemical properties and environmental fate endpoints. J. Cheminform. 10 (1), 10. Available from: https://doi.org/10.1186/s13321-018-0263-1. PMID: 29520515; PMCID: PMC5843579.

Meng, F., Xi, Y., Huang, J., et al., 2021. A curated diverse molecular database of blood-brain barrier permeability with chemical descriptors. Sci. Data 8, 289. Available from: https://doi.org/10.1038/s41597-021-01069-5.

MolconnZ, Version 4.05, 2003. Hall Ass. Consult.; Quincy, MA.

Moore's law <https://www.investopedia.com/terms/m/mooreslaw.asp> (accessed 11.12.21).

Morrison, R.T., Boyd, Robert N., 1987. Organic Chemistry, 5th Edition Allyn and Bacon, Boston.

Nandy, A., 2015. The GRANCH techniques for analysis of DNA, RNA and protein sequences. In: Basak, S.C., Villaveces, J.L., Restrepo, G. (Eds.), Advances in Mathematical Chemistry and Applications, Volume 2. Elsevier & Bentham Science Publishers, Amsterdam & Boston, pp. 96–124. , Advances in Mathematical Chemistry and Applications.

Nandy, A., Harle, M., Basak, S.C., 2006. Mathematical descriptors of DNA sequences: development and application. ARKIVOC 9, 211–238.

Natarajan, R., Basak, S.C., Neumann, T.S., 2007. Novel approach for the numerical characterization of molecular chirality. J. Chem. Inf. Model. 47, 771–775.

Nirenberg, M., Leder, P., 1964. RNA codewords and protein synthesis. The effect of trinucleotides upon the binding of sRNA to ribosomes. Science 145, 1399–1407.

Osolodkin, D.I., Radchenko, E.V., Orlov, A.A., Voronkov, A.E., Palyulin, V.A., Zefirov, N. S., 2015. Progress in visual representations of chemical space. Expert. Opin. Drug. Discov. 10 (9), 959–973.

Pauling, L., Corey, R.B., 1951. Configurations of polypeptide chains with favored orientations around single bonds. Proc. Natl Acad. Sci. USA 37, 729–740.

Primas, H., 1981. Chemistry, Quantum Mechanics and Reductionism. Springer-Verlag, Berlin.

Quastel, J.H., Wooldridge, W.R., 1928. Some properties of the dehydrogenating enzymes of bacteria. Biochem. J. 22 (3), 689–702.

Randić, M., Vračko, M., Nandy, A., Basak, S.C., 2000. On 3-D graphical representation of DNA primary sequences and their numerical characterization. J. Chem. Inf. Comput. Sci. 40, 1235–1244.

Randic, M., 1975. Characterization of molecular branching. J. Am. Chem. Soc. 97, 6609–6615.

Randic, M., Zupan, J., Balaban, A.T., Vikić-Topić, D., Dejan Plavsic, D., 2011. Graphical representation of proteins. Chem. Rev. 111, 790–862.

Randić, M., Witzmann, F., Vračko, M., Basak, S.C., 2001. On characterization of proteomics maps and chemically induced changes in proteomes using matrix invariants: application to peroxisome proliferators. Med. Chem. Res. 10, 456–479.

Randić, M., Lerš, N., Plavšić, D., Basak, S.C., 2004. On invariants of a 2-D proteome map derived from neighborhood graphs. J. Proteome Res. 3, 778–785.

Raychaudhury, C., Ray, S.K., Ghosh, J.J., Roy, A.B., Basak, S.C., 1984. Discrimination of isomeric structures using information theoretic topological indices. J. Comput. Chem. 5, 581–588.

Read, R.C., Corneil, D.G., 1977. The graph isomorphism disease. J. Graph. Theory 1 (4), 339–363.

Restrepo, G., Villaveces, J.L., 2013. Discrete mathematical chemistry: social aspects of its emergence and reception. HYLE – Int. J. Philosophy Chem. 19 (1), 19–33.

Rouvray, D.H., 1991. Making molecules by numbers, New Scientist, 20 March <https://www.newscientist.com/article/mg12917625-800>.

Roy, A.B., Basak, S.C., Harriss, D.K., Magnuson, V.R., 1984. Neighborhood complexities and symmetry of chemical graphs and their biological applications. In: Avula, X.J.R., Kalman, R.E., Liapis, A.I., Rodin, E.Y. (Eds.), Math. Model. Sci. Technol. Pergamum Press, pp. 745–750.

Russell, B., 1950. Mysticism and Logic. George Allen & Unwin, Ltd, London.

Sabirov, D.S., Tukhbatullina, A.A., Shepelevich, I.S., 2021. Molecular size and molecular structure: discriminating their changes upon chemical reactions in terms of information entropy. J. Mol. Graph. Model. 110 (2022), 108052Volume. Available from: https://www.sciencedirect.com/science/article/abs/pii/S1093326321002230.

Sanger, F., Nicklen, S., Coulson, A.R., 1977. DNA sequencing with chain-terminating inhibitors. Proc. Natl Acad. Sci. USA 74, 5463–5467.

Santos-Filho, O.A., Hopfinger, A.J., Cherkasov, A., de Alencastro, R.B., 2009. The receptor dependent QSAR paradigm: an overview of the current state of the art. Med. Chem. (Shariqah) 5, 359–366.

Shannon, C.E., Weaver, W., 1949. The Mathematical Theory of Communication. The University of Illinois Press, Urbana, Urbana, IL.

Stewart, J., 1990. MOPAC 6.00, QCPE #455, Frank J. Seiler Research Laboratory, US Air Force Academy, CO <http://scistore.cambridgesoft.com/mopac>.

Sylvester, J.J., 1878. On an application of the new atomic theory to the graphical representation of the invariants and covariants of binary quantics. Amer. J. Math. 1, 64–125.

Taft, R.W., 1952. Linear free energy relationships from rates of esterification and hydrolysis of aliphatic and ortho-substituted benzoate esters. J. Am. Chem. Soc. 74, 2729–2732.

Toropov, A.A., Toropova, A.P., 2021. The system of self-consistent models for the uptake of nanoparticles in PaCa2 cancer cells. Nanotoxicology 15, 995–1004.

Trinajstić, N., 1992. Chemical Graph Theory. CRC Press, Boca Raton, FL, USA.

Tropsha, A., Gramatica, P., Gombar, V.K., 2003. The importance of being earnest: validation is the absolute essential for successful application and interpretation of QSPR models. Mol. Inform. 22, 69–77.

TSCA Chemical Substance Inventory <https://www.epa.gov/tsca-inventory/about-tsca-chemical-substance-inventory> (accessed 08.11.21).

TSCA Metabolism and Pharmacokinetics <https://www.law.cornell.edu/cfr/text/40/799.9748> (accessed 08.11.21).

Vračko, M., Basak, S.C., Sen, D., Nandy, A., 2021a. Clustering of zika viruses originating from different geographical regions using computational sequence descriptors. Curr. Comput. Aided Drug. Des. 17 (2), 314–322.

Vračko, M., Basak, S.C., Dey, T., Nandy, A., 2021b. Cluster analysis of coronavirus sequences using computational sequence descriptors: with applications to SARS, MERS and SARS-CoV-2 (CoVID-19). Curr. Comput. Aided Drug. Des. 2021 (17), 936–945.

Vračko, M., Basak, S.C., Witzmann, F., 2018. A possible chemo-biodescriptor framework for the prediction of toxicity of nanosubstances: an integrated computational approach. Curr. Comput. Aided Drug. Des. 14 (1), 2–4.

Wang, F., Han, S., Yang, J., Yan, W., Hu, G., 2021. Knowledge-guided "community network" analysis reveals the functional modules and candidate targets in non-small-cell lung cancer. Cells 10 (2), 402. Available from: https://doi.org/10.3390/cells10020402.

Watson, J.D., Crick, F.H.C., 1953. Molecular structure of nucleic acids: a structure for deoxyribose nucleic acid. Nature 171, 737–738.

Wiener, H., 1947. Structural determination of paraffin boiling points. J. Am. Chem. Soc. 69, 17–20.

Winkler, D.A., Burden, F.R., Yan, B., Weissleder, R., Tassa, C., Shaw, S., et al., 2014. Modelling and predicting the biological effects of nanomaterials. SAR. QSAR Env. Res. 25, 161–172.

Robustness concerns in high-dimensional data analyses and potential solutions

2

Abhik Ghosh
Indian Statistical Institute, Kolkata, West Bengal, India

2.1 Introduction

> *The twenty-first century is a digital book. Zola taught Hydra how to read it. Your bank records, medical histories, voting patterns, e-mails, phone calls, your damn SAT scores. Zola's algorithm evaluates people's past to predict their future.*[1]

Although we are at a long distance from developing *Zola's algorithm*, the initial *small baby steps* have already been taken globally to extract decisive information from the widespread data around us for greater goods of the community, through the process of big data analytics. It involves the development of advanced algorithms to efficiently store, analyze, and interpret the large amount of data being generated everyday, from social media to planned research experiments with advanced technologies. Resulting insights are then used for further innovations in different domains of science and technology, as well as governmental policy formulations. These applications include the field of biology and chemistry which can together lead to significant discoveries and innovations in medical sciences to overcome the ever increasing global threats of new pathogens and associated drug requirements. This led Steve Jobs to comment, "*I think the biggest innovations of the twenty-first century will be the intersection of biology and technology. A new era is beginning*"! And, in this new era, appropriate statistical procedures and computer algorithms are essential tools for the progress of big data analytics in its wider applications, including chemo- and bio-informatics.

For modern big datasets, both the number of observations (say n) as well as the number of features (say p) can be extremely large. When we have enough observation, more than the number of features, the challenge is mostly algorithmic to efficiently store and analyze such large amount of data; but the standard statistical methodologies are still applicable. However, when we have more features than the number of observations, quite frequent in medical sciences (e.g., Omics data), all classical statistical methods fail and we need to develop advanced inference procedures to analyze such data (see, e.g., Fan and Li, 2006), in addition to the

[1] From the movie *Captain America: The Winter Soldier.*

Big Data Analytics in Chemoinformatics and Bioinformatics. DOI: https://doi.org/10.1016/B978-0-323-85713-0.00032-3

algorithmic challenges. Such data, having $p \gg n$, are commonly referred to as the high-dimensional data and, in this chapter, we focus on the specific statistical procedures required for their analyses.

In order to obtain legitimate statistical inference with high-dimensional data, one needs to invoke the concept of "sparsity," which says that only a small number (say $s \ll n$) of features, among the vast pool of p ($\gg n$) of them, are indeed relevant for the inference; although the challenge remains in that we do not know which s are important. The objectives are then twofold—correctly identify these s important variables and appropriately use them for our targeted inference. The famous LASSO (least absolute shrinkage and selection operator) of Tibshirani (1996) is the first statistical method developed to achieve both these objectives simultaneously under a standard linear regression set-up; it was later extended to cover several aspects of its performance and applications (see Section 2.2). However, such early developments ignored another hidden challenge of noises (contamination) in the observed datasets. As we are having more and more data everyday, they are also becoming more prone to different sorts of data contamination (e.g., outliers) which are difficult to identify separately given the large number of features available within such data. So, one should use appropriate robust procedures to unveil the correct research insights in the presence of data contamination. These robustness concerns under the high-dimensional data started getting more attention very lately, almost a decade after the development of the LASSO.

The present chapter is devoted to discuss this extremely important issue of contamination in high-dimensional data and different robust statistical procedures as its potential solution, along with a real-life illustration. It is important to mention that here we are talking about parametric robustness which is different from the nonparametric one. If we model the contaminated data by a mixture of the original distribution (majority) and a contaminated part (minority), then a nonparametric procedure tries to estimate the full mixture without any parametric assumption. But a parametric robust procedure estimates the major original component of the data, via an appropriate parametric distribution, suppressing the effects of the noisy contaminated part. Among others, a major advantage of the parametric robust procedures is their significantly high efficiency compared to the nonparametric methods; see Hampel et al. (1986) for related discussions and illustrations under the classical low-dimensional ($n < p$) set-ups.

Here, for brevity, we restrict our focus only on the important regression set-ups, the building block of supervised learning procedures, with high-dimensional covariates. For the sake of completeness, we start with a brief review of the initially developed (nonrobust) procedures for the high-dimensional linear and generalized linear models (GLMs) in Section 2.2. Then, we discuss their lack of robustness concerns in the presence of data contamination in Section 2.3 and two popular approaches for generating robust inference in Sections 2.4 and 2.5, respectively. Finally, in Section 2.6, we illustrate these robust and sparse procedures through a real data application. The chapter ends with some concluding remarks and future research directions in Section 2.7.

For a wider spectrum of readers, throughout this chapter, we have focused mostly on the concepts and nontechnical descriptions of the methods, avoiding mathematical technicalities, except for Section 2.5.2 that contains new theoretical results on the minimum penalized density power divergence estimators (MPDPDEs). However, appropriate references are provided continuously for advanced readers interested on the theoretical details. A list of few important R-packages, which would be useful for any real-life application of the high-dimensional statistical procedures discussed in this chapter, is provided in Appendix.

2.2 Sparse estimation in high-dimensional regression models

2.2.1 *Starting of the era: the least absolute shrinkage and selection operator*

Let us start with the simplest and most popular linear regression model (LRM) where a continuous response variable Y is assumed to depend linearly on p-variate explanatory variable $\mathbb{X}\in\mathbb{R}^p$. Suppose that we have n independent observations $(y_1, x_1), \ldots, (y_n, x_n)$ from $(Y, \mathbb{X})$. Then, we can write the corresponding LRM, in a matrix form, as

$$\boldsymbol{y} = \boldsymbol{X}\beta + \varepsilon, \tag{2.1}$$

where $\boldsymbol{y} = (y_1, \ldots, y_n)^t$, $\boldsymbol{X} = (x_1, \ldots, x_n)^t$, $\boldsymbol{\beta} = \left(\beta_1, \ldots, \beta_p\right)^t$ is the unknown regression coefficient vector, and $\varepsilon = (\varepsilon_1, \ldots, \varepsilon_n)^t$ is the vector of random errors having mean zero and variance $\sigma^2\boldsymbol{I}_n$, with $\boldsymbol{I}_n$ being the identity matrix of order n. When $p \gg n$, the classical least-squares or other estimation methods no longer provide a unique legitimate estimate of $\boldsymbol{\beta}$. To develop an appropriate estimate of $\boldsymbol{\beta}$ under such high-dimensional set-up, invoking "*sparsity*," we assume that there are only s ($\ll n$) nonzero components in the true value, say $\boldsymbol{\beta}_0$, of the regression coefficient $\boldsymbol{\beta}$, although we do not know their exact positions. For this purpose, Tibshirani (1996) proposed the LASSO estimate of $\boldsymbol{\beta}$ defined as

$$\hat{\beta}_{LASSO}(\lambda) = \arg\min_{\beta} \left(\frac{1}{n}\left\|\boldsymbol{y}-\boldsymbol{X}\beta\right\|_2^2 + \lambda\left\|\beta\right\|_1\right), \tag{2.2}$$

where $\|\cdot\|_2$ and $\|\cdot\|_1$ denote the l_2 and l_1 norms, respectively, and $\lambda > 0$ is a regularization parameter controlling the strength of the penalty. Note that, $\lambda = 0$ leads to the (nonpenalized) ordinary least-squares estimate which is not uniquely defined for $p \gg n$. But, for $\lambda > 0$, some coefficients in $\hat{\beta}_{LASSO}$ are shrunk exactly to zero, simultaneously selecting important variables (with nonzero estimated coefficients); this striking property makes the LASSO useful even for the high-dimensional set-ups. As the value of λ increases, the penalty function gets more weight in the combined objective function and, hence, more components of $\hat{\beta}_{LASSO}$ become zero leading to the selection of fewer covariates. But, $\hat{\beta}_{LASSO}$ does not have a closed form; it needs to be computed numerically using appropriately developed algorithms (see, e.g., Hastie et al., 2015).

After the start of this new area of high-dimensional statistics with the development of LASSO, three main themes of evaluating such procedures prevail—estimation accuracy for the regression coefficient, variable selection accuracy in terms of correct discovery of the true active set (the set of variable indices having nonzero regression coefficients), and the prediction accuracy of the resulting regression model. Several results have been derived for evaluating the LASSO estimator (2.2) in terms of these three criteria; see, among others, Zhao and Yu (2006), Bunea et al. (2007), Zhang and Huang (2008), Bickel et al. (2009), van de Geer and Bühlmann (2009), Wainwright (2009), and El Karoui et al. (2013). In summary, LASSO has a slow convergence rate for prediction accuracy, but a fast convergence rate can be obtained for estimation error bound under appropriate conditions. Additionally, the LASSO asymptotically selects the true active set of important variable with probability tending to one but, in practice, it often yields too many false positives (nonzero estimated coefficients for irrelevant variables). We refer to Bühlmann and Van de Geer (2011) and Wainwright (2019) for an extensive treatment of LASSO.

To address the problem of high false positives in LASSO, Zou (2006) developed a two-stage generalization of LASSO that replaces the l_1-penalty in (2.2) by a reweighted version $\lambda \sum_{j=1}^{p} \frac{|\beta_j|}{|\hat{\beta}_{init,j}|}$, with $\hat{\beta}_{init,j}$ being an initial estimator of β_j for all j; the resulting estimator, referred to as the adaptive LASSO, yields consistent estimate of the active set under much weaker conditions (Zou, 2006; Huang et al., 2008). Some other extensions of LASSO, improving its performance under different criteria, include the relaxed LASSO (Meinshausen, 2007), the group LASSO (Yuan and Lin, 2006), the multistep adaptive LASSO (Bühlmann and Meier, 2008), and the fused LASSO (Tibshirani et al., 2005). Another popular related procedure, namely the Dantzig selector introduced by Candes and Tao (2007), has similar statistical properties as the LASSO (Bickel et al., 2009).

2.2.2 Likelihood-based extensions of the LASSO

While extending the LASSO procedure for other high-dimensional regression setups beyond linear regression, it appears that the squared-error loss in (2.2) may not always be useful and, hence, the closely related likelihood-based loss function has been explored. Let us consider the class of GLMs, where the distribution of the response variable, given a covariate value x, is assumed to have a density, from an exponential family, as give by

$$f(y; x^t\beta) = \exp\{y\theta - b(\theta) + c(y)\}, \quad \text{with} \quad E[Y|x] = b'(\theta) = g^{-1}(x^t\beta), \tag{2.3}$$

where $b(\cdot)$ and $c(\cdot)$ are some appropriate known functions, g is a known monotone differentiable link function, and $\beta = (\beta_1, \ldots, \beta_p)^t \in \mathbb{R}^p$ is the vector of unknown regression coefficients which characterizes the canonical parameter θ via the linear predictor $\eta = x^t\beta$ [second equation in (2.3)]. The corresponding log-likelihood, based on n independent observations (y_i, x_i), $i = 1,\ldots,n$, has the form $\sum_{i=1}^n \log f(y_i; x_i^t\beta)$. In low-dimensional GLMs (with $p < n$), the maximum likelihood (ML) estimates of β are obtained by maximizing this log-likelihood. For extending this idea to the high-dimensional set-ups with $p \gg n$, we use the negative log-likelihood to define a loss function as

$$L_n(\beta) = L_n(\beta|\mathbf{y}, \boldsymbol{X}) = \frac{1}{n}\sum_{i=1}^{n} \rho_\beta(x_i, y_i), \tag{2.4}$$

with $\rho_\beta(x, y) = -\log f(y; x^t\beta)$. Note that, this loss function coincides with the squared-error loss under the LRM (1) if the errors are assumed to be normal with unit variance (known). The LASSO estimator of β under the high-dimensional GLM (2.3) is then defined, by replacing the squared-error loss in (2.2) with the likelihood-based loss (2.4), as

$$\hat{\beta}_{LASSO}(\lambda) = \arg\min_{\beta} \left(L_n(\beta) + \lambda\|\beta\|_1\right). \tag{2.5}$$

As in the case of LRM, the two-stage adaptive LASSO estimator can also be defined for the GLMs just by replacing the l_1 penalty in (2.5) by the adaptive LASSO penalty. The properties of the LASSO or adaptive LASSO in GLMs are very similar to the LRM case; see Chapters 3 and 6 of Bühlmann and Van De Geer (2011) for detailed theory and examples.

The properties of the LASSO estimator have been further generalized for any general convex loss $L_n(\beta)$, defined as in (2.4) but with a general $\rho_\beta(x, y)$, by Loubes and van de Geer (2002) and van de Geer (2008). The LASSO estimator has also been extended to a few more complex models involving nonconvex likelihood-based loss (2.4); these include the works of Städler et al. (2010) for the mixture of high-dimensional regression models and Schelldorfer et al. (2011, 2014) for the high-dimensional linear and generalized linear mixed models.

2.2.3 Search for a better penalty function

Despite the popularity of l_1-penalty (also referred to as the LASSO penalty) in high-dimensional statistical procedures, two major drawbacks remain in terms of slower convergence for prediction accuracy and high false positive rates. These motivated scientists to look for the possibility of some better penalty functions. For simplicity of presentation, let the penalty function to be the same for all coefficients, denoted by $p_\lambda(|.|)$, given a regularization parameter $\lambda > 0$. In this situation, a general penalized estimator under the high-dimensional GLM (2.3) is defined as

$$\hat{\beta}(\lambda) = \arg\min_{\beta} \left(L_n(\beta) + \sum_{j=1}^{p} p_\lambda\left(|\beta_j|\right) \right). \tag{2.6}$$

One simple example of extended penalty function is the l_q-penalty, defined as $p_\lambda(|\beta_j|) = \lambda|\beta_j|^q$, for $q > 0$. In particular $q = 2$ leads to the popular ridge regression estimate which has high prediction power but, being nonsparse, cannot be used in the high-dimensional context. Combining this with the LASSO penalty, we get the elastic-net penalty $p_\lambda(|\beta_j|) = \lambda((1-\alpha)|\beta_j| + \alpha|\beta_j|^q)$ with $\alpha \in [0, 1]$; the associated estimator is sparse under the high-dimensional set-ups and also has higher prediction accuracy compared to the LASSO estimate.

To solve the second issue of higher false positives, as mentioned previously, there is multistage advancement of LASSO (e.g., adaptive LASSO). However, due to the complexity already present in the large high-dimensional datasets, a one-step approach leading to consistent variable selection is of extreme importance. In a pioneer paper, Fan and Li (2001) characterized the desired behaviors of a penalty function for (one-step) sparse estimation with consistent variable selection. According to them, a "good" penalty function p_λ should satisfy the following three properties.

1. unbiasedness, which holds if $p'_\lambda(|s|) = 0$ for large s;
2. sparsity, which holds if $\min(|s| + p'_\lambda(|s|)) > 0$;
3. continuity, for which $\min(|s| + p'_\lambda(|s|))$ is attained at $s = 0$.

Interestingly, none of the l_q-penalties satisfies all three conditions simultaneously; in particular, a l_q penalty does not satisfy the sparsity, unbiasedness, or the continuity condition, respectively, if $q > 1$, $q = 1$, or $0 < q < 1$. Fan (1997) and Fan and Li (2001) described a popular smoothly clipped absolute deviation (SCAD) penalty which satisfies all the desired properties from (1) to (3) mentioned above. It is defined based on two fixed parameters $a > 0$ and $\lambda > 0$ and is given by

$$p_\lambda(|s|) = \begin{cases} \lambda|s| & \text{if} \quad |s| \le \lambda \\ \dfrac{2a\lambda|s| - |s|^2 - \lambda^2}{2(a-1)} & \text{if} \quad \lambda < |s| \le a\lambda \\ \dfrac{(a+1)\lambda^2}{2} & \text{if} \quad |s| > a\lambda \end{cases}. \tag{2.7}$$

Later, Zhang (2010) developed another useful penalty function, known as the minimax concave penalty (MCP), which satisfies properties (1) and (2), but not (3), and is given by

$$p_\lambda(|s|) = \begin{cases} \lambda|s| - \dfrac{|s|^2}{2a} & \text{if} \quad |s| \le a\lambda \\ \dfrac{a\lambda^2}{2} & \text{if} \quad |s| > a\lambda \end{cases} \quad (a > 1, \lambda > 0). \tag{2.8}$$

Extending these ideas, the variable selection consistency of the general penalized estimators is proved to hold, with the likelihood loss under the high-dimensional

GLM (2.3), for a general class of nonconcave penalty functions satisfying the following assumption (see, e.g., Fan and Li, 2001, 2006; Kim et al., 2008; Fan and Lv, 2011).

Assumption (P): $p_\lambda(s)$ is continuously differentiable, increasing, and concave in $s \in [0, \infty)$. Also $p_\lambda'(s)/\lambda$ is increasing in λ and $\rho(p_\lambda) := p_\lambda'(0+)/\lambda > 0$ is independent of λ.

The SCAD and the MCP penalties satisfy Assumption (P) and are the commonly used ones, after the LASSO penalty. Recently, the general penalized estimators, involving a penalty satisfying (P), are studied by Ghosh and Thoresen (2018) for the more general nonconvex loss functions, with an application to the high-dimensional linear mixed-effect models.

2.3 Robustness concerns for the penalized likelihood methods

The issue of robustness has been largely overlooked in the sparse learning literature, while this aspect is extremely critical in dealing with high-dimensional noisy data. Most of the initial statistical procedures developed under the high-dimensional LRM or GLMs, discussed in the previous section, use the squared-error loss or the likelihood-based loss in their objective functions. However, although they satisfy asymptotic optimality properties (e.g., oracle consistency) under appropriate conditions, these likelihood-based (and least-squares) estimators lack resilience to outliers or other types of data contamination. To see this for a general penalized procedure as in (2.6), note that, the loss function is the only part that depends on the observed data and the penalty does not in most cases (except, e.g., adaptive penalties). So, the effect of any data contamination comes to the resulting estimator (or inference) only from the loss function that is being minimized, along with a penalty, to obtain the regularized estimators. And, it is already known, in the literature of classical regression models, that the squared-error and the likelihood-based loss functions yield highly nonrobust inference in the presence of data contamination. The estimators obtained by minimizing either of these two loss functions have a (finite sample) breakdown point of $1/n$, that is, even if we move only one observation (either of response or covariate), out of n sample observations, away from the center of the data cloud (tends to infinity), the absolute bias of the resulting estimator increases continuously (tends to infinity). The same effect continue to hold for the high-dimensional regularized procedures using these two loss functions, unless the regularization parameter is chosen to be extremely high (downweighting the effect of the loss function and, hence, overall effect of the observed data, compared to the penalty). In practice, however, we must chose the regularization parameter appropriately to give enough weight to the sample data and the resulting estimators of regression coefficients get highly affected in the presence of data contamination.

This need of generalizing the loss function in account of their robustness against data contamination has been noted much later but, from then, several robust procedures have been developed. Many such studies have confirmed, both theoretically and numerically, that the outliers and/or heavy-tailed noise can severely influence the variable selection accuracy of existing sparse learning methods. The remedial robust statistical procedures, developed under the high-dimensional parametric regression models, can be divided broadly into two classes. The first class is the regularized extension of the popular M-estimation method from the classical robust statistics, and the second one is the penalized extension of the minimum distance approach. We discuss them, respectively, in the next two sections.

2.4 Penalized M-estimation for robust high-dimensional analyses

As the need of robust procedures becomes evident for high-dimensional data analysis, scientist have started exploring different (robust) loss functions in place of the (nonrobust) squared-error or the likelihood-based losses. Wang et al. (2007) considered sparse estimation of the regression coefficients under a high-dimensional LRM by using the absolute-error (l_1) loss along with the LASSO (l_1) penalty; the resulting procedure is referred to as the LAD-LASSO, which is seen to be robust against outliers in the response variable but not against leverage points (outliers in covariates) and also has a low efficiency under pure data. A robust version (RLARS) of the LARS algorithm, related to the least angle regression for calculating the LASSO solution path, has also been developed by Khan et al. (2007). Alfons et al. (2013) have proposed a sparse version of the least trimmed squares method (sLTS) to achieve strong robustness against outliers in both response and covariate spaces with high dimensionality; this method considers a trimmed version of the squared-error loss in (2.2) removing a few largest residuals from its computation and, thus, allows a trade-off between efficiency and robustness. An extension of sLTS using the elastic-net penalty (enetLTS), instead of the LASSO penalty, has also been studied for both linear and logistic regression models by Kurnaz et al. (2018). All these methods, however, can be seen as special cases of the class of penalized M-estimators.

The M-estimation is a popular tool for robust inference in classical statistical models, where an estimate of the underlying model parameter is obtained by solving an appropriate estimating equation, generalizing from the ML score equation (Huber, 1981; Hampel et al., 1986). For the GLMs (2.3), and denoting $z_i = (y_i, x_i)$ for each i, the M-estimator of β is defined as the solution of an estimating equation of the form $\sum_{i=1}^{n} \psi(z_i, \beta) = 0_p$, where ψ is an appropriately chosen function taking values in $\mathbb{R}^p$. Alternatively, assuming that an antiderivative of ψ exists and is given by $\rho(\cdot)$ a.s, the M-estimator can also be equivalently defined as

$$\hat{\beta}_M = \arg\min_{\beta} \sum_{i=1}^{n} \rho(z_i, \beta). \tag{2.9}$$

The ρ-function in (2.9) is chosen appropriately so that it behaves like a loss function. The M-estimator and its properties are thus characterized by either the ψ-function or the ρ-function with the second one being a more general formulation. In particular, taking the negative log-likelihood and the score function as the choices of ρ and ψ, respectively, we see that the ML estimate is a special M-estimator. An M-estimator becomes robust if the corresponding ψ-function downweights the contributions from outlying observations. In the particular case of linear regression, these ψ and ρ functions are chosen as $\psi(z_i, \beta) = \psi(y_i - x_i^t\beta)$ and $\rho(z_i, \beta) = \rho(y_i - x_i^t\beta)$, thus a robust M-estimator in LRM downweights the effect of only those outlying observations that lead to large residual values (in absolute). In order to make a regression M-estimator robust also against leverage points, one needs to add an appropriate weight function for each covariate value, say $w(x_i)$, multiplied with the above ψ or ρ functions; see Rousseeuw and Leroy (2005) for details. A popular choice for ψ-function, proposed by Huber, is given by $\psi(z) = zI(|z| \leq k) + k\,sign(z)I(|z| > k)$ with k controlling the trade-offs between robustness and efficiency. Different other choices of ψ-functions are available in literature; see, for example, Hampel et al. (1986).

Using definition (2.9), the M-estimators can easily be extended for our high-dimensional contexts by just adding an appropriate sparsity-induced penalty in the objective function. Thus, given a ρ-function, the penalized M-estimator of $\boldsymbol{\beta}$ under the high-dimensional GLM (2.3) is defined as

$$\hat{\beta}_M(\lambda) = \arg\min_{\beta} \left(\sum_{i=1}^{n} \rho(z_i, \beta) + \sum_{j=1}^{p} p_\lambda(|\beta_j|) \right). \tag{2.10}$$

Comparing it with (2.6), we can see that a general regularized estimator will be an M-estimator, provided the loss function L_n can be written as the sum of terms depending only on the individual observations (additive in data). Subsequently, the properties of the penalized M-estimators are derived under different assumptions. In particular, we would like to mention the works of Negahban et al. (2012) who derived the asymptotic properties of these penalized M-estimators, and Loh (2013) and Loh and Wainwright (2015) who studied both their statistical and algorithm properties focusing on the local optima obtained in any practical implementation. The influence function of the penalized M-estimators, in its general form, has been investigated by Avella-Medina (2017) to examine their (local) robustness properties.

As mentioned previously, several existing robust high-dimensional methods, namely LAD-LASSO, RLARS, sLTS, and enetLTS, are all indeed the penalized M-estimators with different choices of the penalty and the ρ-functions. More recently, Avella-Medina and Ronchetti (2018) presented another important penalized M-estimators under high-dimensional GLMs, using the *quasi-likelihood loss* function (Cantoni and Ronchetti, 2001) and the general nonconcave penalties satisfying Assumption (P). They have also established that the resulting estimator indeed satisfies the oracle properties and is stable in a neighborhood of the model.

2.5 Robust minimum divergence methods for high-dimensional regressions

The second approach to parametric robust inference is the minimum divergence procedures where the parameter (of interest) is estimated by minimizing a suitable measure of discrepancy (divergence) between the observed sample and the assume model. Particularly, density-based divergences and the corresponding minimum divergence estimators have recently become very popular, under the classical low-dimensional set-ups, due to their strong robustness along with high (sometime full) asymptotic efficiency at the model; see Pardo (2006) and Basu et al. (2011) for details. Motivated by their success, these minimum divergence estimators are extended for robust and sparse estimation under the high-dimensional regime, where one needs to add an appropriate penalty function $p_\lambda(s)$ to achieve the required sparsity as in (2.6). But, the loss function $L_n(\beta)$ in (2.6) is now obtained from the empirical divergence measure between the data and the model distributions to gain robustness in the line of classical minimum divergence methods. Note that, in this approach, it is always necessary to assume a parametric model that is believed to fit the majority of the data well.

The first penalized minimum distance method has been studied by Lozano et al. (2016) for the LRM (2.1) with normally distributed errors. They considered the LASSO penalty in (2.6) along with the robust integrated squared-error loss between the true conditional density of Y given $X = x$ and the corresponding parametric model density, $f(y; x^t\beta)$. In particular, assuming $f(\cdot\,; x^t\beta)$ to be the $\mathcal{N}\left(x^t\beta, \sigma^2\right)$ density, and introducing a tuning parameter C for the unknown σ^2, Lozano et al. (2016) defined the resulting MD-LASSO estimator of β as

$$\hat{\beta}_{MD}(\lambda) = \arg\min_{\beta}\left(-c\log\left[\sum_{i=1}^{n}\exp\left(-\frac{1}{2c}\left(y_i - x_i^t\beta\right)^2\right)\right] + \lambda\|\beta\|_1\right).$$

It has been established that the MD-LASSO can tolerate at least 50% arbitrarily corrupted observations indicating strong robustness and still produces consistent estimators having bounded l_2 norm error. But the efficiency of the MD-LASSO is seen to be significantly low under pure data.

A penalized minimum divergence procedure that allows controlling the trade-offs between the robustness and efficiency has been proposed by Kawashima and Fujisawa (2017) under the high-dimensional LRM (2.1). They have also used the LASSO penalty, but the loss function is obtained from the γ-divergence or the logarithmic density power divergence (LDPD) of Jones et al. (2001) and is given by

$$L_n(\beta) = -\frac{1}{\gamma}\log\frac{1}{n}\sum_{i=1}^{n} f(y_i; x_i^t\beta)^{\gamma} + \frac{1}{1+\gamma}\log\frac{1}{n}\sum_{i=1}^{n}\int f(y; x_i^t\beta)^{\gamma+1}dy, \tag{2.11}$$

where $\gamma \geq 0$ is a tuning parameter controlling the robustness-efficiency trade-off. At $\gamma = 0$ it leads to (in a limiting sense) the nonrobust likelihood-based loss, and the robustness increases as γ (>0) increases with decreasing efficiency. The LDPD is also closely related the Renyi divergence measure, as explored by Castilla et al. (2020a). They have derived the asymptotic and robustness theory for the resulting estimators under the high-dimensional LRM (2.3) with general location-scale type error distributions and a general class of nonconcave penalties satisfying (P). Note that, the loss function in (2.11) is equally valid for the high-dimensional GLMs (2.3), although their detailed performance is yet to be studied.

Another popular divergence measure, used widely in the classical robust inference, is the density power divergence (DPD) (Basu et al., 1998, 2011; Ghosh and Basu, 2013, 2016). For the high-dimensional LRM (2.1), Zang et al. (2017) have used the sparsified DPD loss function along with the grouped LASSO penalty to obtain robust inference for genomic data. Recently, Ghosh and Majumdar (2020) have combined the strengths of the nonconcave penalties and the DPD loss function for simultaneous variable selection and robust estimation of β under the high-dimensional LRM (2.1) with general location-scale errors and illustrated its superior performance compared to other existing robust inference procedures. We describe this DPD-based procedure, in detail, in the following subsections.

2.5.1 The minimum penalized density power divergence estimator

The DPD between two densities g and f, with respect to some common dominating measure, is defined initially by Basu et al. (1998) as given by

$$d_\alpha(g,f) = \int \left\{ f^{1+\alpha}(x) - \left(1 + \frac{1}{\alpha}\right) f^\alpha(x) g(x) + \frac{1}{\alpha} g^{1+\alpha}(x) \right\} dx \qquad \alpha > 0, \tag{2.12}$$

$$d_0(g,f) = \int g(x) \log \frac{g(x)}{f(x)} dx, \tag{2.13}$$

where $\alpha \geq 0$ is a tuning parameter controlling the trade-off between robustness and efficiency of the resulting estimator. Basu et al. (1998) initially proposed it for the classical low-dimensional parametric set-ups with independent and identically distributed data. The resulting minimum DPD estimator (MDPDE) of the unknown model parameter was shown to have a very intuitive justification as the solution of a weighted extension of the ML score equation with the weights being the α-power of the model density at each data point. So, as α increases the outlying observations get downweighted more, leading to increased robustness with a slight loss in efficiency; all weights are one at $\alpha = 0$ so that the corresponding MDPDE coincides with the ML estimate.

While considering the LRM (2.1) with normal error, but with $p < n$, Durio and Isaia (2011) defined the MDPDE of (β, σ) as the minimizer of the extended DPD-based loss function

$$L_n^{(\alpha)}(\beta, \sigma) = \frac{1}{(2\pi)^{\alpha/2}\sigma^{\alpha}\sqrt{1+\alpha}} - \frac{1+\alpha}{\alpha}\frac{1}{(2\pi)^{\alpha/2}\sigma^{\alpha}}\frac{1}{n}\sum_{i=1}^{n}\exp\left(-\alpha\frac{(y_i - x_i^t\beta)^2}{2\sigma^2}\right). \tag{2.14}$$

That the above loss function can be used for fixed-design cases, besides the situations with random covariates, is later justified by Ghosh and Basu (2013) while extending the MDPDE for general independent but nonhomogeneous (INH) set-ups. This idea has been further extended to develop robust and efficient estimators under different important parametric regression models; see, for example, Ghosh and Basu (2016, 2019); Castilla et al. (2018, 2021) among others.

Ghosh and Majumdar (2020) have extended the above idea to the high-dimensional LRM (2.1) to propose the MPDPDE assuming the error component ε_i to have a density $(1/\sigma)f(\epsilon/\sigma)$, where f is any univariate density with zero mean and unit variance such that $M_f^{(\alpha)} = \int f(\epsilon)^{1+\alpha}d\epsilon < \infty$. Then, the MPDPDE of $\theta = (\beta, \sigma)$ is defined as the minimizer of the general penalized objective function, as in (2.6), with general penalty function p_λ satisfying (P) and the DPD-based loss function given by

$$L_n^{(\alpha)}(\theta) = L_n^{(\alpha)}(\beta, \sigma) = \frac{1}{\sigma^{\alpha}}M_f^{(\alpha)} - \frac{1+\alpha}{\alpha}\frac{1}{n\sigma^{\alpha}}\sum_{i=1}^{n}f^{\alpha}\left(\frac{y_i - x_i^t\beta}{\sigma}\right) + \frac{1}{\alpha}. \tag{2.15}$$

As $\alpha \downarrow 0$, $L_n^{\alpha}(\theta)$ coincides (in a limiting sense) with the negative log-likelihood, the corresponding MPDPDE becomes the (nonrobust) likelihood-based estimator studied in Fan and Li (2001), Fan and Lv (2011), and Kim et al. (2008). Under appropriate conditions on the fixed-design matrix and the penalty function, Ghosh and Majumdar (2020) proved that this MPDPDE possesses large-sample oracle properties in an ultrahigh-dimensional regime where p increases exponentially with n. Its robustness is also justifies through the IF analyses, and an efficient computational algorithm has been proposed using the Concave–Convex procedure (Yuille and Rangarajan, 2003).

However, in practice, with $p \gg n$, the computation of the MPDPDE becomes extremely challenging and time-inefficient while using the nonconcave penalties such as SCAD or MCP. To expedite the practical computation, Ghosh et al., 2020a have developed a close approximation of these extremely robust and efficient MPDPDEs, given by the AW-DPD-LASSO estimator, which is defined as the minimizer of the DPD loss function with an adaptively weighted LASSO penalty. Besides easy computability, these AW-DPD-LASSO estimators are also shown to satisfy desired robustness, consistency, and oracle property under appropriate conditions. The AW-PDPD-LASSO estimator is also studied for the high-dimensional logistic regression models by Basu et al. (2021) and (Ghosh et al., 2022).

Finally, we note that the MPDPDE can also be extended to the high-dimensional GLMs (2.3) provided a DPD-based loss function, analogues to (2.15), can be defined for them. For the classical low-dimensional GLMs, the MDPDE has already been studied by Ghosh and Basu (2016) who defined the corresponding DPD-based loss function as

$$L_n^{(\alpha)}(\beta) = \frac{1}{n(1+\alpha)}\sum_{i=1}^{n}\left[\int f(y, x_i^t\beta)^{\alpha+1}dy - \frac{1+\alpha}{\alpha}f(y_i, x_i^t\beta)^{\alpha} + \frac{1}{\alpha}\right]. \tag{2.16}$$

The same loss function can also be used to define the MPDPDE under high-dimensional GLMs (3), along with general nonconcave penalties as in (2.6). In the next subsection, extending the arguments from Ghosh and Majumdar (2020), we derive the asymptotic properties for the general MPDPDE under the ultrahigh-dimensional GLMs.

2.5.2 Asymptotic properties of the MDPDE under high-dimensional GLMs

Consider the set-up and notation of the high-dimensional GLMs (2.3), as described in Section 2.2.2, and assume that $\log p = O(n^l)$ for some $l \in (0,1)$. Suppose that the true value $\boldsymbol{\beta}_0 = (\beta_{01}, \ldots, \beta_{0p})$ of $\boldsymbol{\beta}$ is sparse having only s $(\ll n)$ nonzero elements and define the true active set $S = \{j : \beta_{0j} \neq 0\} as \lhd$. For any p-vector $\mathbf{w} = (w_1, \ldots, w_p)$, we consider the partition $\mathbf{w} = \left(\mathbf{w}_S^t, \mathbf{w}_N^t\right)^t k$ with $\mathbf{w}_S = \left(w_j : j \in S\right)^t$ and $N = S^c$ and also denote the corresponding partition of the design matrix as $\boldsymbol{X} = \begin{bmatrix} \boldsymbol{X}_S & \boldsymbol{X}_N \end{bmatrix}$ with X_S containing s columns of $\boldsymbol{X}$ having indices in S. The goal is to simultaneously identify s, S, and $\boldsymbol{\beta}_{0S}$. Without loss of generality, we may assume $S = 1, 2, \ldots, s\}$ and so $\boldsymbol{\beta}_0 = (\boldsymbol{\beta}_{0S}^t, 0_{p-s}^t)^t$.

Now recall that, for a given $\alpha > 0$, the MPDPDE of $\boldsymbol{\beta}$ is defined as in (2.6) with the α-dependent loss function $L_n^{(\alpha)}(\boldsymbol{\beta})$ given in (2.15) and a general nonconcave penalty p_λ; denote the corresponding (penalized) objective function as $Q_{n,\lambda}^{(\alpha)}(\boldsymbol{\beta})$. Let us assume that p_λ satisfies Assumption (P) and define, for any $\boldsymbol{v} = \left(v_1, \ldots, v_q\right)^t \in \mathbb{R}^q$, the vector $\boldsymbol{P}_\lambda^*(\boldsymbol{v}) = \left(p_\lambda'(|v_1|), \cdots, p_\lambda(|v_q|)\right)^t$ and the matrix $\boldsymbol{P}_\lambda^{**}(\boldsymbol{v}) = \text{diag}\left\{p_\lambda'(|v_1|), \cdots, p_\lambda(|v_q|)\right\}$. Then, one can easily check that the estimating equation of this MPDPDE of $\boldsymbol{\beta}$, with tuning parameter $\alpha \geq 0$, is given by

$$\sum_{i=1}^{n}\psi_\alpha(y_i, \boldsymbol{x}_i^t\boldsymbol{\beta})\boldsymbol{x}_i + \boldsymbol{P}_\lambda^*(\boldsymbol{\beta}) = 0_p, \tag{2.17}$$

where $\psi_\alpha(y, \eta) = (y - g^{-1}(\eta))f(y;\eta)^\alpha - \xi_\alpha(\eta)$ with $\xi_\alpha(\eta) = \int (y - g^{-1}(\eta))f(y;\eta)^{1+\alpha}dy$. This also shows that the MPDPDE is indeed an M-estimator with a model-dependent ψ-function (ψ_α), and it coincides with the penalized likelihood-based estimator at $\alpha = 0$.

In order to derive the asymptotic properties of the MPDPDE under GLMs (2.3), for a fix $\alpha \geq 0$, let us define the $n \times n$ diagonal matrices $\boldsymbol{\Sigma}_{n,\alpha}(\boldsymbol{\beta}) = \text{diag}\left\{-\frac{\partial}{\partial\eta}\psi_\alpha(y_i, \eta)|_{\eta = x_i^t\beta} : i = 1, \ldots, n\right\}$ and $\boldsymbol{\Sigma}_{n,\alpha}^*(\boldsymbol{\beta}) = \text{diag}\{\psi_\alpha(y_i, \boldsymbol{x}_i^t\boldsymbol{\beta})^2 :$

$i = 1,\ldots,n\}$. Then, put $\boldsymbol{\Psi}_{n,\alpha}(\boldsymbol{\beta}) = \left(\boldsymbol{X}_S^t\boldsymbol{\Sigma}_{n,\alpha}(\boldsymbol{\beta}_0)\boldsymbol{X}_S\right)$, $\tilde{\boldsymbol{\Psi}}_{n,\alpha}(\boldsymbol{\beta}) = \left(\boldsymbol{X}_N^t\boldsymbol{\Sigma}_{n,\alpha}(\boldsymbol{\beta}_0)\boldsymbol{X}_S\right)$ and $\boldsymbol{\Omega}_{n,\alpha}(\boldsymbol{\beta}) = (\boldsymbol{X}_S^t\boldsymbol{\Sigma}_{n,\alpha}^*(\boldsymbol{\beta}_0)\boldsymbol{X}_S)$. Note that, these quantities previously appeared in the asymptotic variance of the MDPDE under low-dimensional GLMs (Ghosh and Basu, 2016). In our present context, we assume the following conditions, where p, s, and λ all (implicitly) depend on n, $d_n = \min_{j\in S}|\beta_{0j}|/2$, $b_s \in \mathbb{R}^+$ is a divergent sequence depending on s, and $\Lambda_{\max}$ and $\Lambda_{\min}$ denote the maximum and minimum eigenvalues of their argument, respectively.

(A1) The l_2-norm of each column of $\boldsymbol{X}$ is $O(\sqrt{n})$.

(A2) $\left\|\boldsymbol{\Psi}_{n,\alpha}(\boldsymbol{\beta}_0)^{-1}\right\|_\infty = O\left(\frac{b_s}{n}\right)$, $\left\|\tilde{\boldsymbol{\Psi}}_{n,\alpha}(\boldsymbol{\beta}_0)\boldsymbol{\Psi}_{n,\alpha}(\boldsymbol{\beta}_0)^{-1}\right\|_\infty < \min\left\{\frac{Cp'_\lambda(0+)}{p'_\lambda(d_n)}, O(n^{\tau_1})\right\}$ and $\max_{\boldsymbol{\delta}\in\mathcal{N}_0}\max_{1\le j\le p+1}\left\{\Lambda_{\max}\left(\boldsymbol{X}_S\left[\nabla_{\boldsymbol{\delta}}^2\Gamma_{j,\alpha}(\boldsymbol{\delta})\right]\boldsymbol{X}_S^t\right)\right\} = O(n)$, for some constants $C\in(0,1)$, $\tau_1\in[0,0.5]$, where $\mathcal{N}_0 = \left\{\boldsymbol{\delta}\in\mathbb{R}^s:\|\boldsymbol{\delta}-\boldsymbol{\beta}_{S0}\|_\infty \le d_n\right\}$, $\boldsymbol{\Gamma}_\alpha(\boldsymbol{\delta}) = \left(\Gamma_{1,\alpha}(\boldsymbol{\delta}),\ldots,\Gamma_{p,\alpha}(\boldsymbol{\delta})\right)^t = \sum_{i=1}^n\psi_\alpha\left(y_i;x_{iS}^t\boldsymbol{\delta}\right)$. and $\nabla_{\boldsymbol{\delta}}^2$ denotes the second-order derivative with respect to $\boldsymbol{\delta}$.

(A3) For some $\tau\in(0,0.5]$, we have $d_n \ge \log n/n^\tau$ and $b_s = o(\min\{n^{1/2-\tau}\sqrt{\log n}, n^\tau/s\log n\})$. Further, with $s = O(n^{\tau_0})$, we define $\tau^* = \min 0.5, 2\tau-\tau_0\} - \tau_1$. Then, we have $p_\lambda'(d_n) = o\left(\frac{\log n}{b_s n^\tau}\right)$, $\lambda \ge (\log n)^2 n^{-\tau^*}$, and $\max_{\boldsymbol{\delta}\in\mathcal{N}_0}\zeta(p_\lambda;\boldsymbol{\delta}) = o(\max_{\boldsymbol{\delta}\in\mathcal{N}_0}\Lambda_{\min}[n^{-1}\boldsymbol{\Psi}_{n,\alpha}((\boldsymbol{\delta}^t,\boldsymbol{0}_{p-s}^t))])$, where ζ denote the local concavity of the penalty p_λ as defined by Ghosh and Majumdar (2020), Definition 1. Also, the maximum (in absolute) element of the design matrix $\boldsymbol{X}$ is of order $o\left(n^{\tau^*}/\sqrt{\log n}\right)$.

(A4) For any $\boldsymbol{a}\in\mathbb{R}^n$ and $0<\varepsilon<\|\mathbf{a}\|/\|\mathbf{a}\|_\infty$, there exists a $c_1>0$ such that

$$P\left(\left|\sum_{i=1}^n a_i\psi_\alpha(y_i,\boldsymbol{x}_i^t\boldsymbol{\beta}_0)\right| > \|\boldsymbol{a}\|\epsilon\right) \le 2e^{-c_1\epsilon^2}.$$

(A5) $p'_\lambda(d_n) = O\left((sn)^{-1/2}\right)$, $\max_{1\le i\le n}E\left|\psi_\alpha(y_i,\boldsymbol{x}_i^t\boldsymbol{\beta}_0)\right|^3 = O(1)$, $\sum_{i=1}^n\left[\boldsymbol{x}_{iS}^t\boldsymbol{\Omega}_{n,\alpha}(\boldsymbol{\beta}_0)^{-1}\boldsymbol{x}_{iS}\right]^{3/2} = o(1)$, and $\min_{(\boldsymbol{\delta},\sigma)\in\mathcal{N}_0}\Lambda_{\min}\left[\boldsymbol{\Omega}_{n,\alpha}((\boldsymbol{\delta}^t,\boldsymbol{0}_{p-s}^t))\right] \ge cn$, where $\mathcal{N}_0$ as defined in (A2).

Assumptions (A1)–(A5) are indeed extensions of those used in Ghosh and Majumdar (2020) and Fan and Lv (2011), having similar justifications. Under these assumptions, we have derived the weak oracle property and asymptotic normality of the MNPDPDE of $\boldsymbol{\beta}$ as presented below.

Theorem 2.1: Under the set-up of ultrahigh-dimensional GLMs (2.3) with $s = o(n)$ and $\log p = O(n^{1-2\tau^*})$, let us assume that Assumptions (A1)–(A5) and (P) hold for some fixed $\alpha\ge 0$. Then, there exist MNPDPDEs $\hat{\boldsymbol{\beta}} = \left(\hat{\boldsymbol{\beta}}_S^t, \hat{\boldsymbol{\beta}}_N^t\right)$ of $\boldsymbol{\beta}$, with $\hat{\boldsymbol{\beta}}_S\in\mathbb{R}^s$, such that $\hat{\boldsymbol{\beta}}$ is a (strict) local minimizer of $Q_{n,\lambda}^{(\alpha)}(\boldsymbol{\beta})$ and satisfies the following.

1. $\hat{\boldsymbol{\beta}}_N = \boldsymbol{0}_{p-s}$, and $\left\|\hat{\boldsymbol{\beta}}_S - \boldsymbol{\beta}_{0S}\right\|_\infty = O(n^{-\tau}\log n)$, with probability greater than or equal to $P_n = 1-(2/n)(1+s+(p-s)\exp[-n^{1-2\tau^*}])$.

2. Let $\mathbf{A}_n \in R^{q\times(s+1)}$ such that $\mathbf{A}_n\mathbf{A}_n^t \to \mathbf{G}$ as $n \to \infty$, where $\mathbf{G}$ is symmetric and positive definite. Then, with probability tending to one, $\boldsymbol{A}_n\boldsymbol{\Omega}_{n,\alpha}(\beta_0)^{-1/2}\boldsymbol{\Psi}_{n,\alpha}(\beta_0)\left(\hat{\beta}_S - \beta_{0S}\right) \longrightarrow^{\mathcal{D}} N_q(\mathbf{0}_q, \boldsymbol{G})$.

The rate of consistency in the above theorem (Part 1) can be improved further, in the line of Ghosh and Majumdar (2020), by replacing (A2)–(A3) by their stronger versions as given below.

(A2*) $\max_{\delta\in\mathcal{N}_0}\Lambda_{\min}\left[\boldsymbol{\Psi}_{n,\alpha}\left(\left(\boldsymbol{\delta}^t, \mathbf{0}_{p-s}^t\right)\right)\right] \geq cn$, $\max_{\delta\in\mathcal{N}_0}\max_{1\leq j\leq p+1}\left\{\Lambda_{\max}\left(\boldsymbol{X}_S\left[\nabla_{\delta}^2\Gamma_{j,\alpha}(\boldsymbol{\delta})\right]\boldsymbol{X}_S^t\right)\right\} = O(n)$, and $\left\|\tilde{\boldsymbol{\Psi}}_{n,\alpha}(\beta_0)\right\|_{2,\infty} = O(n)$, for some $c>0$, where $\mathcal{N}_0, \mathbf{G}_\alpha(\mathbf{d})$ are as defined in (A2) and $||\boldsymbol{A}||_{2,\infty} = \max_{||\boldsymbol{v}||_2=1}||\boldsymbol{A}\boldsymbol{v}||_\infty$ for a matrix $\boldsymbol{A}$. Further, $E\left\|\sum_{i=1}^n \psi_\alpha(y_i;\boldsymbol{x}_i^t\beta_0)\boldsymbol{x}_{iS}\right\|_2^2 = O(sn)$ where the expectation is taken with respect to the model distribution (with parameter β_0) of $\boldsymbol{y}$ given $\boldsymbol{X}$.

(A3*) $p_\lambda'(d_n) = O(n^{-1/2})$, $d_n \gg \lambda \gg \min\left\{s^{1/2}n^{-1/2}, n^{\frac{\tau-1}{2}}\sqrt{\log n}\right\}$, and $\max_{\delta\in\mathcal{N}_0}\zeta(p_\lambda;\boldsymbol{\delta}) = o(1)$, for some $\tau\in(0, 0.5]$. Also, the maximum (in absolute) element of $\boldsymbol{X}$ is of $o(n^{(1-\tau)/2}/\sqrt{\log n})$.

The following theorem now presents the strong oracle consistency of the MNPDPDE of β using the above (A2*) and (A3*) in place of (A2) and (A3), respectively.

Theorem 2.2: Under the set-up of ultrahigh-dimensional GLMs (3) with $s \ll n$ and $\log p = O(n^{\tau^*})$ for some $\tau^* \in (0, 0.5)$, let us assume that Assumptions (A1), (A2*), (A3*), (A4), (A5) and (P) hold for some fixed $\alpha \geq 0$. Then, there exist MNPDPDEs $\hat{\beta} = \left(\hat{\beta}_S^t, \hat{\beta}_N^t\right)^t$ of β, with $\hat{\beta}_S \in \mathbb{R}^s$, such that $\hat{\beta}$ is a (strict) local minimizer of $Q_{n,\lambda}^{(\alpha)}(\beta)$ and satisfies the following results with probability tending to 1 as $n \to \infty$.

1. $\hat{\beta}_N = \mathbf{0}_{p-s}$, and $\left\|\hat{\beta} - \beta_0\right\|_2 = O(\sqrt{s/n})$.
2. Let $\boldsymbol{A}_n \in R^{q\times(s+1)}$ such that $\boldsymbol{A}_n\boldsymbol{A}_n^t \to \boldsymbol{G}$ as $n \to \infty$, where $\boldsymbol{G}$ is symmetric and positive definite. Then, $\boldsymbol{A}_n\boldsymbol{\Omega}_{n,\alpha}(\beta_0)^{-1/2}\boldsymbol{\Psi}_{n,\alpha}(\beta_0)\left(\hat{\beta}_S - \beta_{0S}\right) \longrightarrow^{\mathcal{D}} N_q(\mathbf{0}_q, \boldsymbol{G})$.

Our Theorems 2.1–5.2, stating the properties of the MPDPDE under the ultrahigh-dimensional GLMs, clearly complement the corresponding results for the quasi-likelihood based penalized M-estimators derived by Avella-Medina and Ronchetti (2018).

2.6 A real-life application: identifying important descriptors of amines for explaining their mutagenic activity

We consider a dataset (Debnath et al., 1992) containing the values of 275 structural and chemical descriptors for each amines in a congeneric set of 95 such compounds and their chemical mutagenic activities on a pathogenic Gram-negative bacteria, the *Salmonella typhimurium* strain TA98. These data report the number of revertants per nmol when each amine compound is applied to a test culture. Our objective is to identify the important descriptors of the amine compounds in order to explain these numbers of revertants per nmol as a measure of their mutagenic activities. There are four types of descriptors—158 Topochemical (TC), 108 Topostructural (TS), and 6 Quantum Chemical (QC) descriptors along with another 3 descriptors on its three-dimensional (3D) structure. These data have been investigated in many QSAR studies; see, for example, Majumdar et al. (2013), Basak and Majumdar (2015), Majumdar and Basak (2018), and the references therein.

Here we consider all 275 descriptors together to fit a sparse LRM with the response being the logarithm of the revertants counts. Since we have data on 95 amine compounds only, we are in high-dimensional set-up ($p = 275$ and $n = 95$). After appropriate robust standardization of all covariates and the response variable, using their respective median and median absolute deviation, different classical (nonrobust) and robust inference methodologies are applied for the selection of important covariates (descriptors) and simultaneous estimation of the regression coefficients; note that, the estimated regression coefficients help us to get an idea about the extent of their relationship with the response (relative importance) in measuring the mutagenic activities. In particular, we consider the usual LASSO procedure with squared-error loss (LS-LASSO), as the most common nonrobust procedures, and different robust penalized M-estimators, mentioned in Section 2.4; we also consider the minimum penalized divergence methods based on the DPD and the LDPD measures along with the l_1 penalty, which we refer to as the DPD-LASSO(α) and LDPD-LASSO(γ), respectively, for the tuning parameters α and γ.

However, in the high-dimensional contexts, the results obtained by any such sparse learning procedure based on only one given dataset should not be used directly, since they may be affected by the issues of overfitting, sampling fluctuations, and false discoveries (positives or negatives). To overcome these problems, some suitable cross-validation techniques are often used to validate the list of selected important variables. We will suggest the use of a recently developed method, namely the *stability selection* proposed by Meinshausen and Bühlmann (2010), which is theoretically guaranteed to appropriately control the number of false discoveries in a finite-sample settings and reduce overfitting. For this purpose, we randomly divide the sample data into two halves (of size $n/2$ each) and apply a sparse variable selection procedure to both these parts of the data. Repeating this process several (say $B = 50$) times, we compute the percentage of times a covariate (feature variable) is selected by the given procedure among the total $2B$ applications, which we will refer to as the *stability percentage* (SP). If a covariate is truly important for the response, it should be selected in most of the times and should generate a high SP value. A *good* robust variable selection procedure should also select an important covariate most of the times, whatever be the contamination structure there, leading to a high SP value for that particular covariate. We apply this procedure for our dataset, with each of the variable selection methods, to select the "most stable" model structure for identifying the truly important covariates (descriptors), which would have stable nonzero effects on the response for any similar datasets.

It has been observed that, among all robust procedures, the DPD-LASSO method provides most stable results for any $\alpha > 0$ with the SP values being as high as 95%–100%.

Table 2.1 Results from the final (second-stage) LRM of the mutagenic activities on 19 descriptors, selected by the robust DPD-LASSO procedure with stability percentage (SP) $\geq 60\%$. The estimated regression coefficients (Coeff.) and their standard errors (SE) are computed using the S-estimation approach, and the relative importance (RI) of each descriptors is computed as the re-percentage of standardized coefficient estimates (adjusted $R^2 = 0.96$).

Descriptors	SP	Coeff.	SE	RI	Descriptors	SP	Coeff.	SE	RI
Intercept	100%	1.113	0.663	–	LUMO1	81%	−0.151	0.089	−3.8%
SC3	82%	0.563	0.094	13.3%	SIC6	82%	0.368	0.078	10.4%
K2	96%	−0.173	0.361	−1.1%	CIC3	82%	0.427	0.092	10.3%
J	95%	−0.746	0.136	−12.2%	TETS2	83%	−0.006	0.158	−0.1%
ASN2	75%	−0.564	0.692	−1.8%	GMAX	81%	−0.023	0.012	−4.1%
DSN2	85%	0.578	0.461	2.8%	HMIN	86%	−0.037	0.014	−6.0%
ANS4	73%	0.159	0.784	0.4%	SAASC	66%	−0.037	0.012	−7.0%
NRINGS	98%	−0.219	0.183	−2.7%	SSNH2	63%	0.033	0.012	5.9%
KP1	73%	−0.054	0.343	−0.3%	NUMHBA	71%	−0.054	0.115	−1.1%
KP3	94%	−0.125	0.142	−1.9%	SHHBD	90%	−0.049	0.007	−14.9%

So, these DPD-LASSO results are used to select the final set of important descriptors, using the cutoff of SP value $\geq 60\%$; this leads to the same list of 19 descriptors for all $\alpha > 0.1$. We then fit a robust (low-dimensional) LRM with these 19 selected covariates using the efficient and robust S-estimation approach (Rousseeuw and Yohai, 1984) and report the results in Table 2.1. To illustrate the power of such robust modeling, we have also repeated the above procedure for the usual LS-LASSO estimator, which is able to select only 9 descriptors (namely, K10, DN212, DN2Z2, FW, SHSNH2, GMAX, HMIN, GMIN, and SHHBD) having SP $\geq 40\%$ and the resulting second-stage LRM modeling has adjusted R^2 value of 0.83 only. On the other hand, the final model based on the robust DPD-LASSO leads to an adjusted R^2 value of 0.96, indicating the usefulness and significantly superior behavior of the robust approach in any real-life application involving potential data contamination.

In our specific example, from Table 2.1, we can see that the 3D descriptors are not extremely important in the presence of TC and TS descriptors and only one QC descriptor (LUMO1) is chosen to be important with relatively low importance (3.8%). These are in line with the conclusions from previous QSAR studies performed with these data. Further, 9 TC and 9 TS descriptors are found to be important for explaining the mutagenic activities with the TS descriptors being more important than TC ones (total relative importances are 60% and 36%, respectively). Individually, two TC descriptors (SC3, J) have high relative importance of 13.3% and 12.2%, respectively, whereas the most important TS descriptors are (in decreasing order of RI) SHHBD (highest RI of 14.9%), SIC6, CIC3, and SAASC. However, among these important descriptors only SHHBD was selected by usual nonrobust LASSO estimate and others remain hidden due to potential noises in the data. The robust DPD-LASSO can successfully reveal the truly important descriptors leading to accurate and stable insights about the underlying research problem.

2.7 Concluding remarks

In this chapter, we have discussed important statistical procedures for variable selection and parameter estimation under high-dimensional parametric regression models and their robustness concerns in the presence of data contamination. Several robust extensions are reviewed, along with a detailed treatment of the minimum penalized DPD estimator and its application in a real-life data example.

All the robust variable selection procedures involve different complex loss and/or penalty functions; hence, the algorithmic challenges increase in the respective optimization problems for large high-dimensional datasets. Several appropriate algorithms have been developed for these robust and nonrobust high-dimensional procedures. But, they often work well only for moderately high-dimensional datasets (similar to our real data example) and fail in many large and complex real-life datasets. The problem becomes even more pernicious for more complex learning problems beyond regressions. For such cases, one can alternatively use some data-reduction techniques to reduce the number of covariates and then apply appropriate variable selection procedures. Although the PCA or the ICA (and their appropriate robust extensions) can be used for the purpose of data reduction (Filzmoser and Nordhausen, 2020), they are also computationally challenging for large high-dimensional datasets and they do not pick the variable itself; Basak et al. (1988), however, presented a possible explanation of the resulting PCs (linear combination of variables) particularly in the context of chemometrics (e.g., chemical descriptors as in our data example). A popular method for initial screening of the variable itself is the sure independence screening (SIS) developed by Fan and Lv (2008) and Fan and Song (2010) for the LRMs and the GLMs, respectively; also see Saldana and Feng (2018). However, the SIS is indeed based on the correlation measure or the ML estimate of marginal regression models and, hence, is extremely nonrobust against data contamination. A robust and efficient parametric extension of SIS has recently been proposed by Ghosh and Thoresen (2021) and Ghosh et al., 2020b. Under appropriate conditions, both the SIS and its robust extension satisfy the desired sure screening property, that is, they both select the true active set asymptotically with probability tending to one.

However, there are still several open challenges remaining in the context of high-dimensional data analysis, where more research is needed to develop appropriate robust extensions. These include the problem of network selection, classification, clustering, and other complex procedures with high-dimensional data. Efficient computational algorithms are also needed to be discovered in order to apply these robust procedures to large-scale practical datasets. More importantly, further research is also needed to develop robust hypothesis testing and interval estimation procedures under high dimensionality. All these potential future research works would eventually help us to generate more correct and stable research insights by robustly analyzing the huge amount of noisy data available in modern era.

Appendix: A list of useful R-packages for high-dimensional data analysis

The following R-packages, available from the CRAN repository, provide the ready-to-use implementation of different high-dimensional statistical procedures that can be applied directly, even by any nonstatisticians, for real-life data analyses.

Nonrobust procedures:

Package	Description
`lars`	Efficient computation of LASSO, via least angle regression, for the LRM
`glmnet`	LASSO and elastic-net estimator for GLMs
`HDCI`	LASSO estimate and their bootstrap confidence intervals
`penalized`	LASSO and Fused LASSO for GLMs
`ncvreg`	Penalized likelihood-based estimation with SCAD/MCP penalties for GLMs

Robust procedures:

Package	Description
robustHD	RLARS and sLTS for LRM
enetLTS	sLTS and enetLTS for LRM and logistic regression
pense	Penalized S-estimator for LRM with l_1 and elastic-net penalties
flare	LASSO, LAD-LASSO, Dantzig Selector, and LASSO with l_q loss for LRM
gamreg	Minimum penalized LDPD estimator with l_1 penalty

R codes for the minimum penalized DPD estimators under the LRM and the logistic regression are available upon request from the authors of Ghosh and Majumdar (2020) and Ghosh et al., 2020a, respectively. The R codes for robust DPD based adaptive lasso estimators are available in the GitHub repository \texttt {awDPDlasso}[2] and those for the robust extensions of SIS are available from the repository \texttt{dpdSIS}[3].

The R-package \texttt{stabs} can be used for stability selection with any robust or nonrobust procedure.

Acknowledgments

The author wish to express his sincere gratitude and thanks to Prof. Subhash Basak (Natural Resources Research Institute, University of Minnesota Duluth, USA) and Dr. Subhabrata Majumda (AT&T Lab, USA) for sharing the dataset used in this chapter, and also to Mr. Gregory Gunnwald (Natural Resources Research Institute, University of Minnesota Duluth, USA) and Claudiu Lungu (Faculty of Chemistry and Chemical Engineering, Babes-Bolyai University, Romania) for their help in generating the molecular descriptors within the dataset.

This research is partially supported by an INSPIRE Faculty research grant and a grant (No. SRG/2020/000072) from Science and Engineering Research Board, both under the Department of Science and Technology, Government of India, India.

References

Alfons, A., Croux, C., Gelper, S., 2013. Sparse least trimmed squares regression for analyzing high-dimensional large data sets. Ann. Appl. Stat. 7, 226–248.

Avella-Medina, M., 2017. Influence functions for penalized M-estimators. Bernoulli 23, 3178–3196.

Avella-Medina, M., Ronchetti, E., 2018. Robust and consistent variable selection in high-dimensional generalized linear models. Biometrika 105 (1), 31–44.

[2] https://github.com/MariaJaenada/awDPDlasso.

[3] https://github.com/abhianik/dpdSIS.

Basak, S.C., Majumdar, S., 2015. Prediction of mutagenicity of chemicals from their calculated molecular descriptors: a case study with structurally homogeneous versus diverse datasets. Curr. Comput. Aided Drug Des. 11 (2), 117–123.

Basak, S.C., Magnuson, V.R., Niemi, G.J., Regal, R.R., 1988. Determining structural similarity of chemicals using graph-theoretic indices. Discret. Appl. Math. 19 (1–3), 17–44.

Basu, A., Harris, I.R., Hjort, N.L., Jones, M.C., 1998. Robust and efficient estimation by minimising a density power divergence. Biometrika 85, 549–559.

Basu, A., Shioya, H., Park, C., 2011. Statistical Inference: The Minimum Distance Approach. Chapman & Hall/CRC, Boca de Raton.

Basu, A., Ghosh, A., Jaenada, M., Pardo, L., 2021. Robust adaptive Lasso in high-dimensional logistic regression with an application to genomic classification of cancer patients. arXiv preprint, arXiv:2109.03028.

Bickel, P.J., Ritov, Y., Tsybakov, A.B., 2009. Simultaneous analysis of lasso and Dantzig selector. Ann. Stat. 37, 1705–1732.

Bühlmann, P., Meier, L., 2008. Discussion of One-step sparse estimates in nonconcave penalized likelihood models" (auths H. Zou and R. Li). Ann. Stat. 36, 1534–1541.

Bühlmann, P., Van de Geer, S., 2011. Statistics for High-Dimensional Data: Methods, Theory and Applications. Springer Science & Business Media.

Bunea, F., Tsybakov, A., Wegkamp, M., 2007. Sparsity oracle inequalities for the Lasso. Electron. J. Stat. 1, 169–194.

Candes, E., Tao, T., 2007. The Dantzig selector: statistical estimation when p is much larger than n. Ann. Stat. 35, 2313–2351.

Cantoni, E., Ronchetti, E., 2001. Robust inference for generalized linear models. J. Amer. Statist. Assoc. 96 (455), 1022–1030.

Castilla, E., Ghosh, A., Martin, N., Pardo, L., 2018. New robust statistical procedures for the polytomous logistic regression models. Biometrics 74 (4), 1282–1291.

Castilla, E., Ghosh, A., Jaenada, M., Pardo, L., 2020a. On regularization methods based on Rényi's pseudodistances for sparse high-dimensional linear regression models. arXiv preprint, arXiv:2007.15929.

Castilla, E., Ghosh, A., Martin, N., Pardo, L., 2021. Robust semiparametric inference for polytomous logistic regression with complex survey design. Adv. Data Anal. Classification 15, 701–734.

Debnath, A.K., Debnath, G., Shusterman, A.J., Hansch, C., 1992. A QSAR investigation of the role of hydrophobicity in regulating mutagenicity in the Ames test: 1. Mutagenicity of aromatic and heteroaromatic amines in *Salmonella typhimurium* TA98 and TA100. Environ. Mol. Mutagenesis 19 (1), 37–52.

Durio, A., Isaia, E.D., 2011. The minimum density power divergence approach in building robust regression models. Informatica 22 (1), 43–56.

El Karoui, N., Bean, D., Bickel, P.J., Lim, C., Yu, B., 2013. On robust regression with high-dimensional predictors. Proc. Nat. Acad. Sci. USA 110 (36), 14557–14562.

Fan, J., 1997. Comments on "Wavelets in statistics: A review" by A. Antoniadis. J. Italian Stat. Soc. 6, 131–138.

Fan, J., Li, R., 2001. Variable Selection via Nonconcave Penalized Likelihood and its Oracle Properties. J. Amer. Statist. Assoc. 96, 1348–1360.

Fan, J., Li, R., 2006. Statistical challenges with high dimensionality: feature selection in knowledge discovery. In: Sanz-Sole, M., Soria, J., Varona, J.L., Verdera, J. (Eds.), Proceedings of the International Congress of Mathematicians. European Mathematical Society, Zurich, pp. 595–622.

Fan, J., Lv, J., 2008. Sure independence screening for ultrahigh dimensional feature space. J. Royal Stat. Soc. B 70 (5), 849–911.

Fan, J., Song, R., 2010. Sure independence screening in generalized linear models with NP-dimensionality. Ann. Stat. 38 (6), 3567–3604.

Fan, J., Lv, J., 2011. Nonconcave penalized likelihood with NP-dimensionality. IEEE Trans. Info. Theory 57 (8), 5467–5484.

Filzmoser, P., Nordhausen, K., 2020. Robust linear regression for high dimensional data: an overview. Wiley Interdisciplinary Reviews: Computational Statistics e1524.

Ghosh, A., Basu, A., 2013. Robust estimation for independent non-homogeneous observations using density power divergence with applications to linear regression. Electron. J. Stat. 7, 2420–2456.

Ghosh, A., Basu, A., 2016. Robust estimation in generalized linear models: the density power divergence approach. Test 25 (2), 269–290.

Ghosh, A., Thoresen, M., 2018. Non-concave penalization in linear mixed-effect models and regularized selection of fixed effects. AStA Adv. Stat. Anal. 102 (2), 179–210.

Ghosh, A., Basu, A., 2019. Robust and efficient estimation in the parametric proportional hazards model under random censoring. Stat. Med. 38 (27), 5283–5299.

Ghosh, A., Jaenada, M., Pardo, L., 2022. Classification of COVID19 patients using robust logistic regression. Journal of Statistical Theory and Practice 16, 67. Available from: https://doi.org/10.1007/s42519-022-00295-3. In this issue.

Ghosh, A., Majumdar, S., 2020. Ultrahigh-dimensional robust and efficient sparse regression using non-concave penalized density power divergence. IEEE Trans. Info. Theory 66 (12), 7812–7827.

Ghosh, A., Thoresen, M., 2021. A robust variable screening procedure for ultra-high dimensional data. Stat. Methods Med. Res. 30 (8), 1816–1832.

Ghosh, A., Jaenada, M., Pardo, L., 2020a. Robust adaptive variable selection in ultra-high dimensional regression models. arXiv preprint. Available from: https://doi.org/10.48550/arXiv.2004.05470.

Ghosh, A., Ponzi, E., Sandanger, T., Thoresen, M., 2022. Robust sure independence screening for non-polynomial dimensional generalized linear models. To appear in Scandinavian Journal of Statistics. Available from: https://doi.org/10.48550/arXiv.2005.12068.

Hampel, F.R., Ronchetti, E., Rousseeuw, P.J., Stahel, W., 1986. Robust Statistics: *The Approach Based on Influence Functions*. John Wiley & Sons, New York, USA.

Hastie, T., Tibshirani, R., Wainwright, M., 2015. Statistical Learning with Sparsity: the Lasso and Generalizations. CRC press.

Huang, J., Ma, S., Zhang, C.H., 2008. Adaptive Lasso for sparse high-dimensional regression models. Statistica Sinica 18, 1603–1618.

Huber, P.J., 1981. Robust Statistics. Wiley, New York.

Jones, M.C., Hjort, N.L., Harris, I.R., Basu, A., 2001. A comparison of related density-based minimum divergence estimators. Biometrika 88 (3), 865–873.

Kawashima, T., Fujisawa, H., 2017. Robust and sparse regression via γ-divergence. Entropy 19 (11), 608.

Khan, J.A., van Aelst, S., Zamar, R.H., 2007. Robust linear model selection based on least angle regression. J. Amer. Statist. Assoc. 102, 1289–1299.

Kim, Y., Choi, H., Oh, H.S., 2008. Smoothly clipped absolute deviation on high dimensions. J. Amer. Statist. Assoc. 103, 1665–1673.

Kurnaz, F.S., Hoffmann, I., Filzmoser, P., 2018. Robust and sparse estimation methods for high-dimensional linear and logistic regression. Chemo. Int. Lab. Sys. 172, 211–222.

Loh, P.L., 2013. Local Optima of Nonconvex Regularized M-estimators (Doctoral dissertation). University of California, Berkeley Spring.
Loh, P.L., Wainwright, M.J., 2015. Regularized M-estimators with nonconvexity: statistical and algorithmic theory for local optima. J. Machine Learn. Res. 16, 559–616.
Loubes, J.M., van de Geer, 2002. Adaptive estimation in regression, using soft thresholding type penalties. Statistica Neerlandica 56, 453–478.
Lozano, A.C., Meinshausen, N.-, Yang, E., 2016. Minimum Distance Lasso for robust high-dimensional regression. Electron. J. Stat. 10, 1296–1340.
Majumdar, S., Basak, S.C., 2018. Beware of external validation!-a comparative study of several validation techniques used in QSAR modelling. Curr. Comput. Aided Drug Des. 14 (4), 284–291.
Majumdar, S., C Basak, S., D Grunwald, G., 2013. Adapting interrelated two-way clustering method for quantitative structure-activity relationship (QSAR) modeling of mutagenicity/non-mutagenicity of a diverse set of chemicals. Curr. Comput. Aided Drug Des. 9 (4), 463–471.
Meinshausen, N., 2007. Relaxed Lasso. Comput. Stat. Data Anal. 52, 374–393.
Meinshausen, N., Bühlmann, P., 2010. Stability selection. J. Royal Stat. Soc. B 72, 417–473.
Negahban, S.N., Ravikumar, P., Wainwright, M.J., Yu, B., 2012. A unified framework for high-dimensional analysis of M-estimators with decomposable regularizers. Stat. Sci. 27 (4), 538–557.
Pardo, L., 2006. Statistical Inference Based on Divergence Measures. CRC press.
Rousseeuw, P., Yohai, V., 1984. Robust regression by means of S-estimators. Robust and Nonlinear Time Series Analysis. Springer, New York, NY, pp. 256–272.
Rousseeuw, P.J., Leroy, A.M., 2005. Robust Regression and Outlier Detection, Vol. 589. John Wiley & sons.
Saldana, D.F., Feng, Y., 2018. SIS: an R package for sure independence screening in ultrahigh-dimensional statistical models. J. Stat. Software 83 (2), 1–25.
Schelldorfer, J., Bühlmann, P., Van de Geer, S., 2011. Estimation for high-dimensional linear mixed-effects models using l_1- penalization. Scand. J. Stat. 32 (2), 197–214.
Schelldorfer, J., Meier, L., Bühlmann, P., 2014. GLMM Lasso: an algorithm for high-dimensional generalized linear mixed models using l_1- penalization. J. Comput. Graphical Stat. 23, 460–477.
Städler, N., Bühlmann, P., van de Geer, S., 2010. l_1- penalization for mixture regression models (with discussion). Test 19, 209–285.
Tibshirani, R., 1996. Regression shrinkage and selection via the Lasso. J. Royal Stat. Soc. B 58 (1), 267–288.
Tibshirani, R., Saunders, M., Rosset, S., Zhu, J., Knight, K., 2005. Sparsity and smoothness via the fused Lasso. J. Royal Stat. Soc. B 67, 91–108.
van de Geer, S.A., 2008. High-dimensional generalized linear models and the lasso. Ann. Stat. 36, 614–645.
van de Geer, S.A., Bühlmann, P., 2009. On the conditions used to prove oracle results for the Lasso. Electron. J. Stat. 3, 1360–1392.
Wainwright, M.J., 2009. Sharp thresholds for high-dimensional and noisy sparsity recovery using l_1- Constrained Quadratic Programming. IEEE Trans. Info. Theory 55 (5), 2183–2202.
Wainwright, M.J., 2019. High-Dimensional Statistics: A Non-asymptotic Viewpoint, Vol. 48. Cambridge University Press.
Wang, H., Li, G., Jiang, G., 2007. Robust regression shrinkage and consistent variable selection through the LAD-Lasso. J. Bus. Econ. Stat. 25, 347–355.

Yuan, M., Lin, Y., 2006. Model selection and estimation in regression with grouped variables. J. Royal Stat. Soc. B 68 (1), 49–67.

Yuille, A.L., Rangarajan, A., 2003. The concave-convex procedure. Neural Comput 15 (4), 915–936.

Zhang, C.H., 2010. Nearly unbiased variable selection under minimax concave penalty. Ann. Stat. 38, 894–942.

Zhao, P., Yu, B., 2006. On model selection consistency of Lasso. J. Machine Learn. Res. 7, 2541–2563.

Zhang, C.H., Huang, J., 2008. The sparsity and bias of the LASSO selection in high-dimensional linear regression. Ann. Stat. 36, 1567–1594.

Zang, Y., Zhao, Q., Zhang, Q., et al., 2017. Inferring gene regulatory relationships with a high-dimensional robust approach. Genet. Epidemiol. 41 (5), 437–454.

Zou, H., 2006. The adaptive lasso and its oracle properties. J. Amer. Stat. Assoc. 101, 1418–1429.

Fairness, explainability, privacy, and robustness for trustworthy algorithmic decision-making

3

Subhabrata Majumdar[1,2]
[1]AI Vulnerability Database, Seattle, WA, USA, [2]Bias Buccaneers, Seattle, WA, USA

3.1 Introduction

In the current digital world, statistics and machine learning (ML) techniques that use large quantities of data are being deployed by companies across a broad range of industries. Especially in the last decade or so, this has been enabled by the increasing sophistication of high-performance computing and the democratization of data and data-analytical tools. While this automation of decision-making has resulted in significant increase of efficiency and revenues in business processes, some unintentional negative impacts of such people-facing implementations have recently come to light. For example, Amazon had to abandon an ML system aimed at streamlining its hiring process by shortlisting resumes since it was discriminating against female applicants due to gender imbalance in historical data (Cook, 2018). A number of targeting options in Facebook's advertising platform were correlated with sensitive features like gender and race (Perez, 2018). As a result certain categories of targeted ads disproportionately left out minority groups. Several other examples exist that have put forth questions on lapses in transparency, privacy or security, reliability, and robustness (Cheng et al., 2021).

Incorporating such value-driven qualities into ML systems has been the objective of a growing field of research in recent past. This field is often referred to by terms such as trustworthy ML or responsible ML (Toreini et al., 2020; Xiong et al., 2021). While the initial push started as a spurt of action among computer science researchers, due to the practical nature of problems tackled, a number of interdisciplinary dimensions emerged soon to make the developed solutions relevant to actual stakeholders. Keeping such practical motivations in mind, in this chapter, we survey the landscape of trustworthy ML research and fundamental concepts. We devote each section of the chapter to one major aspect of trustworthy ML—fairness, explainability, privacy, and robustness. To focus on the implementation aspects of each method, at the end of a section, we provide pointers to open-source computational resources for the interested reader.

3.2 Fairness in machine learning

In the context of big data and ML, the concepts of fairness and bias are heavily related and may carry a number of implications based on the specific application.

Big Data Analytics in Chemoinformatics and Bioinformatics. DOI: https://doi.org/10.1016/B978-0-323-85713-0.00017-7

While there are many different kinds of bias (e.g., estimation bias, confirmation bias, cognitive bias) with not necessarily negative connotations, we focus on demographic bias due to inherent issues in the data and/or ML model outcomes that perpetuate historical and systemic inequities. Broadly speaking, fairness can be construed as the equal treatment of similar individuals—within or irrespective of whether they belong to specific demographic groups. In spite of being intuitive, some theoretical assumptions are required for a strict mathematical formulation of the above, and multiple definitions of fairness exist based on the requirements of the specific application (Mitchell et al., 2021). However, a common conceptual thread running through such definitions is the shared objective that the deviation (or statistical bias) of one or more *parity metrics* should be minimized across individuals or individual groups of interest.

In this section, we present an overview of formal notions of such parity metrics and fairness definitions (Section 3.2.1), bias mitigation methodology (Section 3.2.2), and tools and frameworks to implement such methods in practice (Section 3.2.3). We summarize the overarching concepts at a fairly high level, owing to the fact that ML bias and fairness has been a hotly researched area in the recent past. Interested readers can check fairness-specific literature surveys for more granular discussions and references (Mitchell et al., 2021; Mehrabi et al., 2019a; Shrestha and Yang, 2019).

3.2.1 Fairness metrics and definitions

Parity measurement metrics quantify the extent a fairness notion is adhered to for an attribute or prediction outcome under consideration. While a number of such metrics exist in the literature (Mehrabi et al., 2019a; Verma and Rubin, 2018), for brevity we define below some of the most widely used metrics. For all definitions, we denote by $Y, X, S, \hat{Y}$ the random variables denoting, respectively, the binary output feature, input feature(s), sensitive feature, and predicted output from an ML model. Also denote the probability of an event $A \in \mathcal{A}$, a set of events, by $P(A)$.

Definition 1.: The prediction $\hat{Y}$ satisfies *equalized odds* (Hardt et al., 2016) with respect to sensitive attribute S and output Y if

$$P(\hat{Y} = 1|S = 0, Y = y) = P(\hat{Y} = 1|S = 1, Y = y); y = 0, 1.$$

In other words, $\hat{Y}$ and S are independent conditional on Y.

Definition 2.: The prediction $\hat{Y}$ satisfies *demographic parity* (Bellamy et al., 2018) with respect to sensitive attribute S if

$$P(\hat{Y} = 1|S = 0) = P(\hat{Y} = 1|S = 1).$$

A similar definition with Y in place of $\hat{Y}$ denotes the demographic parity of the actual output feature values in the data (rather than output values predicted from the model).

Definition 3.: The prediction $\hat{Y}$ is said to achieve *counterfactual fairness* (Kusner et al., 2017) if for any value of the input(s), say $X = x$, the probability that $\hat{Y} = y$ for any y is the same across the values of S. This ensures that all individuals are treated similarly irrespective of their demographic group membership.

It is possible that a fairness metric conforms to only certain definitions of fairness. For example, equalized odds and demographic parity ensure *group fairness*, that is, similar treatment/outcomes getting assigned to groups of people defined by one of more sensitive demographic features (such as race, gender). On the other hand, counterfactual fairness implies *individual fairness*, which refers to similar treatment of similar individuals irrespective of their sensitive feature values (Bellamy et al., 2018). A third notion of fairness also exists at the intersection of these two, called *subgroup fairness*, and can involve simultaneously optimizing for multiple metrics (Kearns et al., 2018, 2019).

3.2.2 Bias mitigation in machine learning models

While fairness concerns tend to stem from systemic problems and data quality issues, rectifying such problems is often nontrivial, cost-prohibitive, and even impossible in practice (Holstein et al., 2019). Apropos of a typical ML model building pipeline (See Fig. 3.1), research on bias-aware ML methods can be divided into three stages: preprocessing, in-processing, and postprocessing. While fair versions of many ML or statistical techniques such as principal component analysis (PCA) (Samadi et al., 2018), clustering (Backurs et al., 2019), community detection (Mehrabi et al., 2019b), and causal models (Zhang et al., 2017a) exist in the literature, a disproportionate amount of research on fairness methodology is on supervised ML models (Mehrabi et al., 2019a)—owing possibly to their widespread use in real applications. For this reason, we summarize below the three stages assuming a supervised ML model with continuous or discrete outputs; see Table 3.1 for information on exemplar methods applicable to each of these stages.

3.2.2.1 Preprocessing

These model-agnostic methods aim to address fairness issues in data before it is fed into an ML model by transforming one or more features. Such transformations can either be predetermined functions (DIR in Table 3.1), or learned from the data (LFR, OP), and can operate on the output Y and/or input X.

3.2.2.2 In-processing

Such algorithms incorporate one or more metrics directly into the model training process, utilizing techniques such as adversarial training (Zhang et al., 2018) and regularization (Celis et al., 2018). While conceptually these methods are fairly general, model-specific implementations may be very different. The metrics can be either fairness-specific (AD, PR) or related to general model performance (MA).

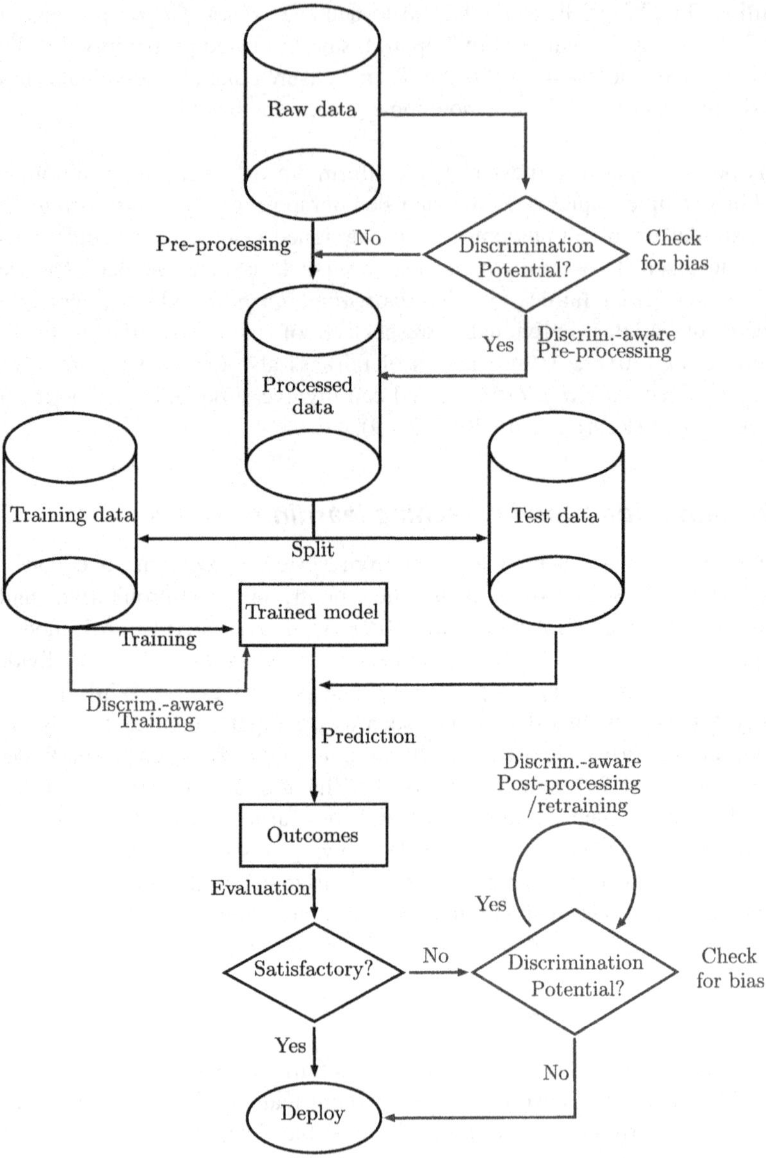

Figure 3.1 A general ML pipeline that shows opportunities for bias detection and mitigation in pre-, in-, and postprocessing stages (marked in blue). *ML*, Machine learning.

3.2.2.3 Postprocessing

Similar to preprocessing techniques, these methods do not require access to the trained model. However, instead of the training data, they operate on the predicted outputs of the ML model and attempt to mitigate bias in the predictions—ensured

Table 3.1 Mitigation algorithms for every stage of machine learning model building.

Stage	Technique	Problem type	Data required	Fairness level	Applicable fairness metrics
Preprocessing	DIR (Feldman et al., 2015)	Classification	S, X	Group	Disparate impact
	LFR (Zemel et al., 2013)	Classification	Y, S, X	Group, individual	Statistical parity
	OP (Calmon et al., 2017)	Classification	Y, S, X	Group, individual	General strategy
	Reweighing (Kamiran and Calders, 2012)	Classification	Y, S, X	Group	Statistical parity difference
In-processing	AD (Zhang et al., 2018)	Optimization	Y, S, X	Individual	Equality of odds, demographic parity, equality of opportunity
	PR (Kamishima et al., 2012)	Optimization	Y, S, X	Individual	PI, Normalized PI
	MA (Celis et al., 2018)	Classification	Y, S, X	Group	Accuracy, precision, recall, True/false positive
Postprocessing	RO (Kamiran et al., 2012)	Classification	$\hat{P}, S, X$	Group, individual	Equality of odds, demographic parity, Equality of opportunity
	EO (Hardt et al., 2016)	Classification, binary output	$\hat{Y}, S$	Group	Accuracy, precision, recall, true/false positive
	CEO (Pleiss et al., 2017)	Classification, probability output	$\hat{P}, S$	Group	AUC , lift, capture rate

For classification models with probability output, $\hat{P} \equiv P(\hat{Y} = 1)$.

AD, Adversarial debiasing; *AUC*, area under curve; *CEO*, calibrated equalized odds; *DIR*, disparate impact remover; *EO*, equalized odds; *LFR*, learning fair representations; *MA*, metaalgorithm for fair classification; *OP*, optimized preprocessing; *PI*, prejudice index; *PR*, prejudice remover; *RO*, reject option classification.

using equity in performance (EO, CEO) or fairness (RO) metrics. Such methods are particularly useful in third-party situations when the modeler does not have access to the training data, model, or both.

3.2.3 Implementation

There are a number of toolkits, built on top of numerical programming languages such as Python and R, which package existing fairness-related methods aimed toward calculation of metrics and bias mitigation. AI Fairness 360 (AIF360) (Bellamy et al., 2018) is perhaps the most well-known of them. Among other such packages, Aequitas (Stevens et al., 2018), Fairness Measures (Zehlike et al., 2017), FairML (Adebayo, 2016), FairTest (Tramer et al., 2017), and Themis (Galhotra et al., 2017) are capable of bias detection, while Fairlearn (Dudik et al., 2020) and Themis-mL (Bantilan, 2018) can perform both bias detection and mitigation. These packages are largely open source—thus technically expandable to incorporate new metrics and mitigation techniques in an on-demand basis.

The abovementioned methods and packages provide parts of the technical apparatus to integrate fairness monitoring into different stages of the ML pipeline. However, implementing them into real-world projects is challenging. Besides the obvious limitation of not being able to verify or fulfill theoretical conditions for individual detection and mitigation techniques in practice, several other challenges exist, with often domain-specific nuances (Holstein et al., 2019; Veale et al., 2018). In a survey of industry ML practitioners, Holstein et al. (2019) identified a number of such challenges as the main impediment to developing and deploying of fairness-aware ML processes:

1. lack of guidance in data collection and identification of sensitive features,
2. blind spots in detecting bias concerns due to lack of team diversity and domain knowledge,
3. use case diversity and lack of adequate tools for the specific project domain, and
4. the need for human oversight for bias risk assessment in the different stages.

To address technical concerns in data collection, blind spots, and lack of specific guidance, a number of recently proposed methods enable documentation and lineage tracking of ML life cycles to aid in future reuse. These include Datasheets (Gebru et al., 2018), data nutrition labels (Holland et al., 2018), FactSheets (Arnold et al., 2019), and Model Cards (Mitchell et al., 2019). To address scalability challenges of bias detection and mitigation in large-scale ML workflows, the LinkedIn Fairness Toolkit (Vasudevan and Kenthapadi, 2020) provides an open-source Scala/Spark library implementing a number of fairness metrics.

Effectively integrating human oversight for fairness-aware ML is a more challenging proposition. Depending on the fairness risk of a project and the potential adverse impact or such risks, this can necessarily be a deliberative and slow process. A number of human-in-the-loop strategies help in such situations. For example, structured algorithmic audits (Raji et al., 2019) and codesigned fairness checklists (Madaio et al., 2020) can ensure that deployed ML models conform to company values and principles *and* meet performance metrics. Performing such risk assessments at multiple stages of the ML workflow, guided by documented information from past similar projects and the oversight of in-house subject matter experts (Dodwell et al., 2020) makes the final deployed system increasingly more likely to function in a responsible manner while satisfying business goals.

3.3 Explainable artificial intelligence

In high-stakes automated decision-making, such as disease diagnosis or recidivism prediction, the need for explaining and elucidating decisions of the ML model involved is a crucial factor in eliciting the trust of stakeholders and regulatory authorities. However, owing to their complexity and scale, production-grade ML systems—mostly "black-box" models that are easily amenable to experimentation but not explanation—suffer from lack of transparency to their inner workings that produce user-facing decisions (Du et al., 2020; Carvalho et al., 2019; Adadi and Berrada, 2018). Motivated by such needs, the field of explainable artificial intelligence (XAI) attempts to deal with the broad problem of comprehending an ML model and its (potential) predictions.

XAI consists of the study and research on several related but distinct concepts, such as interpretability, explainability, intelligibility, all pertaining to making ML models more comprehensible to stakeholders and end users. To clarify this ambiguity, we begin with an overview of formalisms of the concepts (Section 3.3.1). Following this, we review the major technical concepts and their applications (Section 3.3.2) and finish with an impact assessment of XAI methods (Section 3.3.3). Similar to Section 3.3.2, we keep the discourse at a high level and refer the interested reader to a number of high-quality resources for literature surveys (Carvalho et al., 2019; Adadi and Berrada, 2018; Gilpin et al., 2018; Holzinger et al., 2019; Covert et al., 2020) and conceptual details (Lipton, 2016; Doshi-Velez and Kim, 2017).

3.3.1 *Formal objectives of explainable artificial intelligence*

Following philosophical notions of what constitutes an explanation (Bromburger, 1992), and their interpretations in the context of ML, explainability of an ML model refers to the answers to "why" questions based on its existing or potential predicted outcomes (Carvalho et al., 2019; Gilpin et al., 2018). Such answers need to achieve both *interpretability* and *completeness*. In other words, an answer needs to be (1) *comprehensible*, that is, good enough to explain the mechanisms of a potentially complex ML model to a potentially nontechnical audience, and (2) *correct*, that is, an accurate enough description of how the model actually works. This is not an easy task. While one simple sweeping answer to diverse "why" questions is easily comprehensible, it may not be an accurate representation of a model and can even be overly *persuasive* to elicit undue trust of a human evaluator (Herman, 2017). On the other hand, accurately explaining the complexities and edge cases of a predictive model runs the risk of information overload. At the heart of effective explanation methods is a trade-off between these two objectives (Carvalho et al., 2019; Gilpin et al., 2018; Covert et al., 2020; Doshi-Velez and Kim, 2017)—customized to the problem (or problem domain) at hand and the target audience of the explanation.

3.3.1.1 Why explain?

There are a number of (not necessarily disjoint) motivations to develop explainability techniques. The first is three are due to the survey on XAI methods by Adadi and Berrada (2018), while the fourth one is motivated by the need to move from a deductive to inductive reasoning of explanations (Holzinger et al., 2019; Moraffah et al., 2020).

1. **Justification**: explanations can help justify dubious or negative decisions, defend algorithmic decision-making, or comply with rules and regulations—such as the "right to explanation" under the European Union General Data Protection Regulation (GDPR),[1] or credit reporting reason codes.[2]
2. **Control and improvement**: insights into how an ML model is making predictions help pinpoint the reasons behind anomalous or erroneous behavior and enable efficient troubleshooting in future iterations.
3. **Discovery**: explanations can help discover the limitations or errors in our decision-making and enrich human knowledge by articulating insights on predictions where the model performs better than human benchmarks.
4. **Causation**: designing and conducting experiments based on explanations of model outcomes, then analyzing that observational data have the potential to form and validate hypotheses on cause–effect relationships—thus moving forward from typical association-based data analyses (Holzinger et al., 2019; Pearl, 2018), (Section 3.3.2.3).

3.3.1.2 Terminologies

We conclude the XAI formalisms with reconciliation of a few terms. In the XAI literature, multiple words are used interchangeably, such as "interpretable" and "explainable." While the specific words do have slightly different connotations, in XAI parlance, they often denote similar notions of model comprehension, along with terms such as understandability, comprehensibility, and intelligibility. One factor of the different terms is the target audience and context. For example, user-group-specific Google search trends suggest that technical ML community of researchers and practitioners prefer using the word "interpretable," while "explainable" is more preferred in public discourse (Adadi and Berrada, 2018). For ease of narrative, we use "explainability" throughout the next section and return to this topic in Section 3.3.3.

[1] https://gdpr-info.eu.

[2] https://www.reasoncode.org/reasoncode101.

3.3.2 Taxonomy of methods

While explainability methods can be divided according to their applicability in the three stages of an ML pipeline (Fig. 3.1), pre-model explainability closely maps to data explainability and transformations—consisting of unsupervised methods such as PCA and K-means (Carvalho et al., 2019). In this review, we focus on recent developments of in-model and post-model XAI methodology.

3.3.2.1 In-model versus post-model explanations

The above two categories are closely tied to models that are being explained. In-model explainability pertains to the explanation method being tied to the model *by definition*. White-box/glass-box ML models that have an open architecture thus possess in-model or *intrinsic* explainability. On the other hand, black-box/opaque model structures that are difficult or impossible to represent explicitly—because of issues like scale and propriety—are more amenable to being explained by post-model or *post hoc* explainability techniques.

There is another dimension to this dichotomy: model-specific versus model-agnostic. By definition, intrinsic methods are model-specific—explanations generated by an explainable model (e.g., linear regression, decision trees, least absolute shrinkage and selection operator (LASSO) are specific to that model. On the other hand, post hoc explanation techniques *tend to be* model-agnostic (Carvalho et al., 2019; Adadi and Berrada, 2018). Post hoc techniques (e.g., local interpretable model-agnostic explanations (LIME, Ribeiro et al., 2016), Shapley values (SHAP, Lundberg and Lee, 2017), model understanding through subspace explanations (MUSE, Lakkaraju et al., 2019)) can thus be applied to models that are explainable themselves. Independence of the base model and the resulting modularity is a clear advantage of post hoc methods. However, the use of a second model introduces another scope of error in the explanation process apart from errors pertaining to the main model. In feature-rich large datasets, multiple models often exist that have similar prediction performances—the so-called "Rashomon effect" (Fisher et al., 2019; Rudin, 2019). In such situations, depending on driving factors such as the consequences of a false prediction and access to domain knowledge, the development of explainable models can be pursued that achieve similar performance as a black-box model *and* satisfy a specific definition of interpretability relevant to the problem (Rudin, 2019).

3.3.2.2 Global and local explanations

Intrinsic and post hoc explainable methods can produce explanations at either global or local level. Global explanation methods aim to produce an overall *comprehensible* overview of an ML model. Such overviews often (Carvalho et al., 2019) take the form of feature summaries (Ribeiro et al., 2016; Lundberg and Lee, 2017), model internals (linear models, LASSO, decision trees), representative datapoints (tabular LIME), and surrogate models to explain explanations (Lakkaraju et al., 2019). Note that the emphasis here is on a human being able to comprehend the overview. Thus global explanations need not necessarily be holistic. They can be *modular* instead, in the sense that a model interpretation can be decomposed into chunks (subsets of samples and/or features) that hold specific meanings to the aimed user (Lipton, 2016; Lou et al., 2016).

Local explanations aim to explain one single sample or small groups of samples. The general idea here is that on small neighborhoods in the sample space, variation in the behavior of a trained model is low. As a result, simple, interpretable supervised models can be trained on tightly clustered data-points, taking model predictions as labels. Following LIME (Ribeiro et al., 2016) and SHAP (Lundberg and Lee, 2017), a large number of local methods have been proposed for producing post hoc explanations of black-box models. As summarized by a survey of post hoc methods (Covert et al., 2020), a common principle local explanation methods are based on is measuring the effect of removing one or more features from a model on its predictions.

3.3.2.3 Causal explainability

Conventional ML models are based solely on observational data and are thus only able to infer *associations* instead of true cause-effect relationships between features. Causal explainability methods based on evaluating counterfactual situations—proposing and evaluating alternate model outcomes under alternate situations such as different input features, training setups—provide tools that can address such shortcomings (Moraffah et al., 2020; Guo et al., 2020; Pearl, 2000).

A major conceptual framework in causal inference is that of Structural Causal Models (SCMs).

Definition 4.: A *Structural Causal Model* (Pearl, 2000) is defined by the 4-tuple (X,U,f,P_u), where

- X is a finite set of endogenous variables that are usually observable,
- U is a finite set of exogenous variables that are usually unobserved or noise, and
- $f = \{f_1,\ldots,f_n\}$ is a set of functions representing causal mechanisms.

$$x_i = f_i(Pa(x_i), u_i),$$

for $x_i \in X; Pa(x_i) \subseteq (X\ x_i), \cup\ U,$

- P_u is a probability distribution over U.

Methods in causal explainability (and causal inference in general) are mainly based on modeling complex ML models such as DNNs as SCM, incorporating causal reasoning and human intelligibility (Moraffah et al., 2020; Harradon et al., 2018). Moraffah et al. (2020) divide such methods into four categories; we summarize them in Table 3.2.

Table 3.2 Categorization of causal explainability methods.

Category	Purpose	References
Causal explainability of models	Explain the effects of model component(s) on its predictions	Narendra et al. (2018), Chattopadhyay et al. (2019), Zhao and Hastie (2021), Parafita and Vitriá (2019), Madumal et al. (2020)
Counterfactual explanations	Generate explanations for outcomes in alternate input or training scenarios	Wachter et al. (2017), Goyal et al. (2019), Moore et al. (2019), Mothilal et al. (2020), Rathi (2019), Hendricks et al. (2018)
Causal fairness	Use of explainable causal models to ensure fairness	Kusner et al. (2017), Kilbertus et al. (2017), Madras et al. (2019), Zhang and Bareinboim (2018)
Verifying causal relationships in data	Verify causal assumptions between features, ensure interpretability using causal inference	Kim and Bastani (2019), Caruana et al. (2015)

3.3.3 Do explanations serve their purpose?

Model explanations need to make sense to whoever the explanations are aimed at. Based on the expertise level of these stakeholders/end users (such as data scientist, project manager, business leader) and the use case being analyzed, different explanation methods may be prioritized. Motivated by this reasoning, Doshi-Velez and Kim (2017) asked the question that whether all XAI methods are "equally interpretable" and proposed three avenues of evaluating them—grounded on (1) application, (2) human user, and (3) the function of the model. The first two involve a human in the loop in evaluating either the implementation of the explanation by a domain expert (application grounded) or the explanations directly (human user grounded). The third one evaluates explanation methods that have already received some form of human vetting, through specific functional metrics.

In practice, however, there is admittedly a disconnect between research on proposing new explainability methods versus assessing existing methods on the above criteria; just 5% of XAI methods proposed until now deal with impact assessment of XAI (Carvalho et al., 2019; Adadi and Berrada, 2018). This is an important shortcoming. The inherent subjective nature of the area (that should be clear to the reader by now) means that *explainability lies in the eyes of the stakeholder*—it is imperative to ensure that an XAI method is actually serving its purpose of being useful to the users it is producing explanations for.

3.3.3.1 From explanation to understanding

To develop effective explanation methods that translates to understanding of an ML model and its actions for the human in the loop, insights from other fields like psychology and philosophy can be borrowed (Miller, 2019; Miller et al., 2017). This means making an effort to produce explanations that are human-like or at the least human-friendly. To mimic the sparse and prototype-based nature of human reasoning, human-like explanations (Kim et al., 2016; Gurumoorthy et al., 2019) need to be (1) *contrastive*: explain why one event happened instead of another, (2) *selective*: avoid information overload by focusing on a small number of causes, and (3) *relatable*: appeal to the mental model of explainee and let them draw inference. Human-friendly methods, on the other hand, tend to focus on producing intelligible interpretations and visualizations of complex ML models or their outcomes. Apart from ML expertise, conceptualizing and implementing such techniques often and should involve concepts from the field of human–computer interaction (HCI) (Wortman Vaughan and Wallach, 2020). Relevant works in the intersection of XAI and HCI that build explanation interfaces include (Bauer and Baldes, 2005), eXplainable AI for Designers (XAID) (Zhu et al., 2018), Rivelo (Tamagnini et al., 2017), and Gamut (Hohman et al., 2019). The Weight of Evidence framework and metaalgorithm of Alvarez-Melis et al. (2019) is one of the first attempts to produce explanations that are themselves human-oriented. Moving beyond model intelligibility, a number recent tools provide error analysis interfaces to help users explore the deficiencies of an ML model in detail (Barraza et al., 2019; Amershi et al., 2015; Ren et al., 2016).

The final step in ensuring that the product of XAI research and efforts of translating that to understanding of a model serve their purpose is user evaluation. While research in this work is extremely sparse, some very recent studies reveal interesting insights on the impact of XAI methods. A study (Kaur et al., 2020) on data science practitioners comprising a survey (sample size $N = 197$) and contextual inquiry ($N = 11$) revealed that users tend to put too much trust on automated explanations and are inclined to trust a model based on its positive explanation without doing a detailed check. As is expected in this context, Lakkaraju and Bastani (2020) showed that it is in fact possible to *mislead* users with explanations. Rogue black-box explanations generated using their proposed mechanism that *did not* include any sensitive features were able to successfully mislead domain experts into trusting a black-box model that actually used those sensitive features to make predictions. In perhaps the largest user study of its kind, Poursabzi-Sangdeh et al. (2021) rigorously evaluated aspects of model explainability using a randomized experiment on 3800 participants. As part of a number of surprising outcomes, model interpretations seemed to make a user unduly believe in the efficacy of a badly performing model and correctness of its mispredictions. Their observations also indicate that highly detailed explanations impede users' ability to detect unusual input feature values.

3.3.3.2 Implementations and tools

We conclude this section with an overview of available tools for the interested reader to implement XAI methods. Implementations of a number of well-known methods are available in standard computer languages, such as in R[3], Python[4], and Julia[5]. Going a step further, two open-source projects also offer expandable platforms for practitioners to implement their own XAI methods, in addition to a number of built-in options for existing methods—AI Explainability 360 (Arya et al., 2019) by IBM and InterpretML (Nori et al., 2019) by Microsoft. A recent review article summarizes and compares a number of R packages for XAI (Maksymiuk et al., 2020). Given the subjectivity and importance of the role of explainability in ML systems, these resources are valuable in driving the adoption of XAI methods in practice.

3.4 Notions of algorithmic privacy

As ML algorithms get developed and deployed in the real world with increasing frequency, the longstanding problem of how to preserve the privacy of individuals in any data-analytic exercise has become more and more important. While traditional approaches based on anonymization are somewhat effective in concealing direct identifiers such as name or address, reidentification of subjects using auxiliary data available elsewhere is a very real threat (Vadhan, 2017; Narayanan et al., 2016)—more so in the 21st-century world where access to information is democratized.

At a high level, algorithmic privacy aims to provide meaningful answers to specific questions (or "queries") about the population of interest (or the representative sample) *without* disclosing any individual's information. A popular way of achieving this is to pose such queries to a trusted curator with full access to the data, which computes and releases an answer that is "safe enough" according to a predefined privacy guarantee. Given the diverse domains (such as medical or social sciences, advertising, communication) or modalities (graph, streaming data, manifolds), large real-world datasets can be associated with, providing such secure answers is not easy in general.

While there are other ways to ensure inferential privacy and maintaining accuracy of the answer at the same time (such as du Pin Calmon and Fawaz, 2012), in this section, we focus on the area of *differential privacy* (DP) that has seen notable developments in the last decade. We begin with an overview of basic definitions and concepts (Section 3.4.1), then review major streams of DP research (Section 3.4.2). We finish with an overview of privacy frameworks that extend or build up on the concept of DP and some real-world examples (Section 3.4.3). For mathematical and methodological details, we refer the interested reader to a number of technical resources (Vadhan, 2017; Dwork and Roth, 2014; Kamath and Ullman, 2020; Wood et al., 2018; Gong et al., 2020).

[3] https://uc-r.github.io/iml-pkg.

[4] https://www.analyticsvidhya.com/blog/2020/03/6-python-libraries-interpret-machine-learning-models.

[5] https://github.com/interpretable-mL/IML.jl.

3.4.1 Preliminaries of differential privacy

The fundamental idea of DP is randomization—the curator introduces enough *data-independent* noise in the query output such that the noisy output is "similar" to the original output but does not give away information on the inclusion of individual data-points in computing the output. A fixed *privacy budget* $\varepsilon > 0$ quantifies this similarity, with values closer to 0 denoting higher degrees of similarity. Noise distributions are specific to the class of queries being answered, and there is an inherent information loss that needs to be traded off against an individual's privacy risk. We formalize these shortly (see also Fig. 3.2).

Consider datasets $X \in \mathcal{X}^n$ comprising of n samples, each drawn from domain X, a function M_q: $\mathcal{Q} \times \mathcal{X}^n \rightarrow \mathcal{R}$ that takes in a query $q \in \mathcal{Q}$ on the dataset and gives a *randomized* answer in the range $\mathcal{R}$. Also, suppose $Y \in \mathcal{X}^n$ is another dataset that differs from X in one element—call it an adjacent dataset of X, and denote by $X \sim Y$.

Definition 5: (Dwork et al., 2006). The mechanism M_q is called ε-differentially private, or ε-DP in short, if for any pair of adjacent datasets X, Y the following holds for any measurable $R \subseteq \mathcal{R}$:

$$P\,(M_q(X) \in R) \leq e^{\varepsilon} P(M_q(Y) \in R). \tag{3.1}$$

The mechanism M_q is called (ε, δ)-differentially private, or (ε, δ)-DP, if:

$$P\,(M_q(X) \in R) \leq e^{\varepsilon} P(M_q(Y) \in R) + \delta. \tag{3.2}$$

Notice that (ε, δ)-DP is a weaker condition than ε-DP, and ε-DP is the same as (ε, 0)-DP. Typically, ε, δ are small but nonnegligible, so that the difference between the two probabilities in (3.1) and (3.2) is minimal (Fig. 3.2). Consequently, whether a particular sample belongs to X or not, the answer to q remains almost the same. A

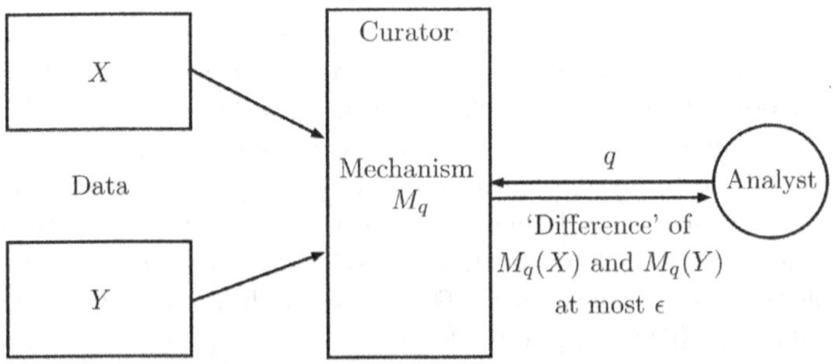

Figure 3.2 A schematic of differential privacy. The answers to a query q, obtained using the randomized privacy algorithm M_q, are very similar for two datasets X, Y which differ in only one sample.

malicious analyst would thus not be able to get any information about this sample by observing the dissimilarities between $M_q(X)$ and $M_q(Y)$.

There are two broad classes of queries q: numeric and nonnumeric queries, pertaining to numeric (such as mean, median, quantiles), or nonnumeric (such as maximum, minimum, top 10%) answers to q, respectively. Formally, a query $q: \mathcal{X}^n \to \mathcal{R}$ maps a dataset to an output, and a mechanism M_q introduces noise in the result-generating procedure of the query. For numerical queries, two well-known procedures introduce noise as a postprocessing step.

Definition 6.: For a numeric query $q: \mathcal{X}^n \to \mathbb{R}^d$, $d > 0$ giving real-valued results, *global sensitivity* is defined as

$$\Delta_p(q) \equiv \Delta(q, \|\cdot\|_p) = \max_{X \sim Y} \|q(X) - q(Y)\|_p, \tag{3.3}$$

for the ℓ_p-norm $\|\cdot\|_p$ and the maximum taken over all possible adjacent X, Y.

Definition 7: (Dwork et al., 2006). For a numeric query $q: \mathcal{X}^n \to \mathbb{R}^d$, the *Laplace mechanism* is defined as $M_{q,\varepsilon} \equiv M_q$ such that

$$M_q(X) = q(X) + (\eta_1, \ldots, \eta_d),$$

where η_i, $i = 1, \ldots, d$ are independently and identically distributed (hereafter i.i.d.) as Laplace $(\Delta_1(q)/\varepsilon)$ random variables, with probability density function

$$p(\eta_i) \propto \exp\left(-\frac{\varepsilon|\eta_i|}{\Delta_1(q)}\right).$$

Definition 8: (Blum et al., 2005). For a numeric query $q: \mathcal{X}^n \to \mathbb{R}^d$, the *Gaussian mechanism* is defined as $M_{q,\varepsilon'} \equiv M_q$ such that

$$M_q(X) = q(X) + (\gamma_1, \ldots, \gamma_d),$$

where γ_1, $i = 1, \ldots, d$ are i.i.d. Gaussian random variables

$$N\left(0, \frac{2\log(2/\delta)\Delta_2^2(q)}{\varepsilon^2}\right).$$

Given ε, $\delta > 0$, the Laplace mechanism satisfies ε-DP, while the Gaussian mechanism satisfies (ε, δ)-DP.

The situation is somewhat complex for nonnumeric queries due to their generic nature. Each (nonnumeric) element $r \in \mathcal{R}$ is first assigned a utility score $u(X, r)$ specific to the dataset X, quantifying the preference of r as the answer to $q(X)$. While a direct (nonprivate) answer means choosing the r with the highest utility, privacy mechanisms randomize this choice. The *exponential mechanism* (McSherry and Talwar, 2007) is one such well-known mechanism that preserves ε-DP.

Definition 9: (McSherry and Talwar, 2007). For a nonnumeric query q with utility u: $\mathcal{X}^n \times \mathcal{R} \rightarrow \mathbb{R}$, the *exponential mechanism* is defined as the randomized choice of an answer r:

$$P\left(M_q(X) = r\right) \propto \exp\left(\frac{\varepsilon u(X, r)}{2\Delta(u)}\right).$$

with $\Delta(u) = \max_{r \in \mathcal{R}} \max_{X \sim Y} |u(X, r) - u(Y, r)|$ being the global sensitivity of the utility function.

Consequently, high-utility answers are more likely to be chosen compared to those with lower utility. However, the inherent randomness results in a privacy guarantee on the eventual answer.

Finally, a combination of DP mechanisms is of practical interest, such as transformations on the mechanism M_q and answering multiple queries (Vadhan, 2017; Dwork and Roth, 2014). To this end, the following basic results hold:

Lemma 1: (Vadhan, 2017). If Mq: $\mathcal{Q} \times \mathcal{X}^n \rightarrow \mathcal{R}$ is (ε, δ)-DP, and F: $\mathcal{R} \rightarrow \mathcal{R}'$ is any random function, then $F o\, M_q$: $\mathcal{Q} \times \mathcal{X}^n \rightarrow \mathcal{R}'$ is (ε, δ)-DP.

Lemma 2: (McSherry, 2009). If $M_{1q}, \ldots, M_{kq}$, $k > 0$ are (ε, δ)-DP mechanisms, then their serial combination $M_q = (M_{1q}, \ldots, M_{kq})$ is $(k\varepsilon, k\delta)$-DP.

Lemma 3: (McSherry, 2009). Given a dataset $X \in \mathcal{X}^n$, consider any partition $\{X_1, \ldots, X_k\}$, with $X_k \in X^{nk}$. If M_{kq}: $\mathcal{Q} \times \mathcal{X}^{nk} \rightarrow \mathcal{R}$ are ε-DP mechanisms on the corresponding data partitions, then for any (fixed or random) function, their parallel combination is ε-DP on the full dataset:

$$M_q(X) = G(M_{1q}(X_1), \ldots M_{kq}(X_k)).$$

3.4.2 Privacy-preserving methodology

While the above principles provide privacy guarantees for broad classes of queries, implementing them for specific ML algorithms and model outputs is nontrivial. Laplace and exponential mechanisms rely on the global sensitivity being a common upper bound to all possible numeric or nonnumeric queries. For the specific dataset being analyzed, or the class of datasets that are of interest, this bound can be tightened—potentially reducing the injected random noise while still providing the same DP guarantees.

ML models often learn patterns from the data in an iterative process, with numerous calls to the training data. These calls can be seen as queries that may be combined or nested. For example, in any gradient descent-based optimization algorithm, each gradient computation can be seen as a query that can potentially be perturbed. These computations combine to produce a trained ML model, which can also be seen as a larger query that produces outputs or predictions. Depending on the specifics of the model or dataset, privacy noise injected at different stages of the training process can result in different magnitudes of perturbation in the final output.

3.4.2.1 Local sensitivity and other mechanisms

The concept of *local sensitivity* (Nissim et al., 2007) is based on the idea that tuning noise levels *locally* to the dataset being analyzed, instead of fixing a common upper bound (with global sensitivity) can potentially result in more accurate outputs with the same DP guarantee.

Definition 10.: Consider a query $q \in \mathcal{Q} \rightarrow \mathcal{R}$, operating on a dataset X. Then we define the following.

- If q is numeric, the local sensitivity (with ℓ_p norm) is

$$\Delta_p(q, X) = \max_{Y:X \sim Y} \left\| q(X)\text{-}q(Y) \right\|_p.$$

- If q is nonnumeric, then for utility function u, the local sensitivity is

$$\Delta(u, X) = \max_{r \in \mathcal{R}} \max_{Y:X \sim Y} |u(X, r)\text{-}u(Y, r)|.$$

Note that in the above definition, the maximums are taken over all datasets Y adjacent to a fixed X, calibrating the noise magnitude to the data at hand. This is challenging, since the privacy noise level needs to be independent of the data, simply substituting the maximum value calculated from the samples in X does not satisfy DP guarantees (Nissim et al., 2007; Farias et al., 2020). Navigating this problem is comparatively easier for numeric queries—Nissim et al. (2007) proposed the smooth sensitivity framework and implemented it to design algorithms that solve:—means and Gaussian mixture learning problems while preserving (ε, δ)-DP. Building upon this fundamental work, local sensitivity–based methodology with DP guarantees have been proposed—such as in PCA (Gonem and Gilad-Bachrach, 2018), answering subgraph counting queries (Karwa et al., 2014) and deep learning algorithms (Sun et al., 2020).

For nonnumeric queries, the very recently proposed *local dampening* mechanism incorporates local sensitivity to design private algorithms (Farias et al., 2020).

Their generic method uses attenuated versions of utility functions in combination with the exponential mechanism, and is ε-DP. Compared to exponential mechanism, a local dampening-based approach results in significant reduction of privacy budget in high-influence node detection problems on graphs, and higher accuracy in decision-tree based ML models (Farias et al., 2020).

A number of alternative mechanisms can be used in place of the traditional choices of Laplace/Gaussian/exponential mechanisms. Bun and Steinke (2019) proposed a local sensitivity framework that extends to three more noise distributions, as compared to only Laplace noise in smooth sensitivity (Nissim et al., 2007). Ladder functions (Zhang et al., 2015) and the staircase mechanism (Geng et al., 2015) are alternatives to the exponential mechanism for certain nonnumeric queries. Very recently, McKenna and Sheldon (2020) proposed a simple modification of the exponential mechanism, called permute-and-flip that significantly improves accuracy in private median estimation.

3.4.2.2 Algorithms with differential privacy guarantees

References to DP answers for a number of simpler queries such as median, mean, or distribution calculations, and broad classes of methods such as hypothesis testing and graph analysis can be found in Kamath and Ullman (2020). For more complex ML models, there are two broad categories of perturbations, which add noise (1) directly into the training steps, or (2) take the outputs or objective functions of a nonprivate model and perturb them (Gong et al., 2020).

DP versions of traditional "shallow" ML and statistical models have been proposed and refined over the past few years (Gong et al., 2020). This includes decision trees (Blum et al., 2005; Liu et al., 2018), Naïve Bayes (Vaidya et al., 2013; Li et al., 2018a), bagging (Dwork et al., 2010), random forest (Jagannathan et al., 2012; Rana et al., 2015; Fletcher and Zahidul Islam, 2015), clustering (Nissim et al., 2007; Su et al., 2016; Schellekens et al., 2019), PCA (Blum et al., 2005; Gonem and Gilad-Bachrach, 2018; Chaudhuri et al., 2013), and online learning (Jain et al., 2012; Li et al., 2018b). At a more fundamental level, a number of papers focus on incorporating DP into generic computational algorithms that can be used in ML model training, for example, Markov Chain Monte Carlo (Yildirim and Ermis, 2019; Mikko, 2019), Hamiltonian Monte Carlo (Lode, 2019), expectation maximization (EM) (Park et al., 2017), and Stochastic gradient descent (SGD) (Song et al., 2013; Rajkumar and Agarwal, 2012; Abadi et al., 2016). All of these incorporate privacy perturbations into model training. Among methods that perturb model outputs/objective function, notable methods include DP generalized linear models (Zhang et al., 2012) and their variants such as M-estimators (Lei, 2011) and LASSO (Talwar et al., 2015), support vector machines (SVM) (Zhang et al., 2019a; Jain and Thakurta, 2013), and empirical risk minimization (ERM) in general (Chaudhuri et al., 2011; Kifer et al., 2012; Wang et al., 2019).

Incorporating privacy guarantees into computation-intensive deep learning methods produces some unique challenges related to the large-scale nature of models, high noise during model training due to a large number of model parameters, and the black-box nature of such models. For training-level perturbations, instead of using existing DP versions of SGD (Song et al., 2013; Rajkumar and Agarwal, 2012; Abadi et al., 2016), it is possible to use distributed computing to speed up model training process *while* conforming to DP guarantees (Shokri and Shmatikov, 2015; Zhang et al., 2017b; Papernot et al., 2018). Moving beyond classification tasks, McMahan et al., (2018) proposed using Gaussian mechanism to ensure user-level privacy in Long Short-Term Memory Networks, while Beaulieu-Jones et al. (2019) and Xie et al. (2018) use gradient perturbation to achieve DP in Generative Adversarial Networks (GAN). Finally, DP work on output perturbation of deep learning models is extremely scarce, owing to the instability of objective functions or final solution in this regime (Gong et al., 2020). The three existing methods in this domain pertain to deep autoencoders (Phan et al., 2016), convolutional deep belief networks (Phan et al., 2017), and deep network embeddings (Xu et al., 2018).

3.4.3 Generalizations, variants, and applications

3.4.3.1 Pufferfish

The concept of Pufferfish (Kifer and Machanavajjhala, 2014) expands on the notions of DP to propose a larger class of privacy mechanisms that are able to counter many different types of malicious attacks. Based on domain knowledge, Pufferfish lets experts decide what secrets they want to protect, what secrets they wish to be indistinguishable, and what types of attacks they want to protect against. These three entities are formalized by the specifications of a Pufferfish framework:

- the set of secrets, denoted by $\mathscr{S}$;
- the set of secret pairs, denoted by $\mathscr{S}_P$; and
- data evolution scenarios: denoted by D, a set of probability distributions over the data domain $\mathscr{X}^n$—representing the set of attackers to protect the secret pairs from.

Definition 11.: Given a framework ($\mathscr{S}$, $\mathscr{S}_P$, D) a mechanism M_q is ε-Pufferfish for the query q if, for all distributions $\delta \in D$ and a dataset X drawn from δ, all secret pairs $(s_i, s_j) \in \mathscr{S}_P$ such that $P(s_i \mid \delta) \neq 0$, $P(s_j \mid \delta) \neq 0$, and all $r \in \mathscr{R}$, we have

$$e^{-\varepsilon} \leq \frac{P(M_q(X) = r)|s_i, \delta)}{P(M_q(X) = r)|s_j, \delta)} \leq e^{\varepsilon}. \tag{3.4}$$

Note the obvious parallels with the definition of ε-DP (Definition 5). Indeed, DP is a special case of Pufferfish, where statements of the form "sample i has value x," $i \in \{1,\ldots,n\}$, $x \in X$ are the secrets, secret pairs are pairs of such statements for the same sample but different values, and D is composed of size-n data distributions with independent samples (Kifer and Machanavajjhala, 2014).

As the first practical instantiation of Pufferfish that is different from DP, Song et al. (2017) proposed the Wasserstein mechanism to ensure randomization-based privacy guarantees for correlated data (such as time series). Subsequent works on a similar theme include FGS-Pufferfish privacy for temporally correlated trajectories

(Ou et al., 2018), private monitoring of web browsing activity (Liang et al., 2020), and Pufferfish for correlated categorical data (Xi et al., 2020). In later research, the authors of Pufferfish proposed a further generalization called Blowfish (He et al., 2014) that allows to incorporate formal policies and data constraints to fine-tune a privacy mechanism even more.

3.4.3.2 Other variations

Among other variations of the basic DP framework, perhaps the most intuitive and straightforward is that of Group Differential Privacy (GDP) (Dwork and Roth, 2014). Simply stated, instead of adjacent datasets, the definition of ε-GDP or (ε, δ)-GDP requires the same respective bounds in (3.1) or (3.2) hold over all datasets that differ in a number of prespecified group memberships of their samples. A number of applications of GDP propose privacy mechanisms for correlated data: such as network data (Chen et al., 2014), temporal correlations (Cao et al., 2017), and multilevel graphs (Palanisamy et al., 2017).

Pufferfish formalizes the role of data distributions in the study of privacy. Alternatively, *divergences*—the concept of "distances" between probability distributions—can be used to embed knowledge or assumptions on data distributions into privacy definitions. Rényi Differential Privacy (Mironov, 2017) is one such method that uses the Rényi divergence to propose a relaxation of conventional DP:

Definition 12: (Mironov, 2017). For two probability distributions P, Q taking values in the range of query results $\mathcal{R}$, and $\alpha \geq 1$, define the *Rényi divergence* as

$$D_\alpha(P||Q) = \begin{cases} \left\{ \mathbb{E}_{x\sim P} \log \frac{P(x)}{Q(x)}, \right. & \text{if } \alpha = 1 \\ \frac{1}{\alpha - 1} \log \mathbb{E}_{x\sim Q} \left(\frac{P(x)}{Q(x)} \right)^{\alpha} & \text{if } > 1 \\ \sup_{x \in \mathcal{R}} \log \frac{P(x)}{Q(x)}, & \text{if } \alpha = \infty \end{cases}$$

A mechanism M_q is said to satisfy ε-Rényi DP of order α, (α, ε)-Rényi DP in short, if for any pair of adjacent datasets X, Y, the following holds:

$$D_\alpha(M_q(X)||M_q(Y)) \leq \varepsilon.$$

Compared to traditional DP, this relaxed definition provides better privacy guarantees under composition of multiple heterogeneous queries (Mironov, 2017).

Other proposals that extend DP using notions related to data distributions include concentrated DP (Dwork and Rothblum, 2016; Bun and Steinke, 2016), capacity-bounded DP (Chaudhuri et al., 2019), Poisson subsampled Rényi DP (Zhu and Wang, 2019), Bootstrap DP (O'Keefe and Charest, 2019) and *f*-DP (Dong et al., 2021).

Table 3.3 Open-source tools on differential privacy.

Library	Platform	Link
Diffprivlib, by IBM	Python	https://github.com/IBM/differential-privacy-library
Diffpriv	R	https://cran.r-project.org/web/packages/diffpriv/index.html
Google's differential privacy library	C++, Go, Java	https://github.com/google/differential-privacy
Opacus, by Facebook	Python (PyTorch)	https://github.com/pytorch/opacus
SmartNoise Core	Python	https://github.com/opendifferentialprivacy/smartnoise-core
WhiteNoise	Microsoft Azure ML	https://medium.com/microsoftazure/whitenoise-an-open-source-library-to-protect-data-with-differential-privacy-fed740e29a49
OpenDP	Python	https://github.com/opendifferentialprivacy

Going in a different route than the divergence-based notion, the f -DP framework (Dong et al., 2021) focuses on a hypothesis testing interpretation of DP guarantees and implements privacy amplification methods using subsampling.

3.4.3.3 Implementations

A number of public-facing implementations of DP algorithms concern the generation of private, synthetic data on population commuting patterns using source data collected by the US Census Bureau (Machanavajjhala et al., 2008), the RAPPOR algorithm implemented in the Google Chrome browser (Erlingsson et al., 2014), and the scalable local DP algorithms by Apple (Differential Privacy Team, 2017). The geo-indistinguishability framework provides a principled approach for location privacy (Andrés et al., 2013; Oya et al., 2017). Finally, a number of open-source tools are available for users to apply well-known DP algorithms in the literature in their own data analytic tasks; we list them in Table 3.3 as references for the interested reader.

3.5 Robustness

The concept of robustness in statistics and data-analytic exercises in general dates back decades ago to (Huber, 1981). Broadly speaking, robust methods refer to techniques that are unaffected by the presence of outliers or other departures from model assumptions in the data used to implement them. Robust versions of many statistical methods have been proposed, and it is an active field of research—which will be discussed in another chapter of this book.

As more and more large-scale ML systems get deployed and updated in an automated manner, there is need for a different kind of robustness. This is robustness specifically against samples that are not "similar" to the typical data a model was trained on. Oftentimes, such examples are tweaked by malicious actors (called adversaries) in a targeted manner to make an ML model perform badly. Thus this notion of robustness—hereafter called *adversarial robustness*—has a notion of security and reliability attached to it. Note that adversarial robustness is different from the traditional notion above. It concerns robustness with respect to perturbations in both the *test data* and training data, as compared to training data alone, and data contaminationations are generally tailored to the problem in hand.

In this section, we discuss the high-level concepts and research directions in adversarial robustness, with appropriate references as necessary. As in other sections, we refer interested readers to a number of survey articles for more details (Serban et al., 2020; Silva and Najafirad, 2020; Chakraborty et al., 2018).

3.5.1 Adversarial attacks

Adversarial attacks on ML model have three categorizations—*who* they attack, *how* the attack happens, and *why* the attacks occur (Silva and Najafirad, 2020; Chakraborty et al., 2018). We consider only black-box models because of their ubiquitous nature in current ML, and since due to the transparency of white-box model crafting targeted attacks are much easier. In the first category, *evasion attacks* occur by maliciously adjusting testing samples, *data poisoning* attacks perturb the training data to corrupt the model training process, and *exploratory attacks* query a black-box model to reverse engineer the training algorithm. As the second categorization, attacks can be orchestrated in a number of ways, depending on what information the attacker has access to. If an adversary has access to the training data or model, they can either modify the training data directly (data modification), add bad training data (data injection), or corrupt the trained model itself (logic corruption). On the other hand, in testing time, adversaries can use the following kind of attacks on a black-box model:

1. **Adaptive attack:** Adversary labels a carefully constructed set of input feature through querying the target model, fits another ML model to predict these outputs, and tailors adversarial examples by focusing on areas where the second model has high error rates. Examples of adaptive attacks include (Fredrikson et al., 2015; Rosenberg et al., 2018).
2. **Nonadaptive attack:** Adversary has some prior knowledge about the training data distribution, which they use to generate the input data and predictions to fit a second-level model as above. Examples include Tramèr et al. (2016) and Papernot et al. (2016).
3. **Strict attack:** Such attacks occur when the adversary uses actual input–output pairs from the original model to craft their attack. An example of strict attack is the method by Hitaj et al. (2017).

Finally, as the third characterization, there are multiple possible goals of adversarial attacks. In decreasing order of specificity, they may aim for (1) general confidence reduction by deteriorating performance of the ML model as a whole, (2) misclassification of all input examples, (3) targeted misclassification of all input examples into specific classes, and (4) targeted misclassification of specific input examples into specific classes.

Not all attacks are equally difficult to perform, or equally effective. As a rule thumb, increasingly complex attacks are more difficult to perform and tend to be more targeted to specific examples or tasks. We refer the reader to Figure 5 in Chakraborty et al. (2018) for a full comparison.

3.5.2 Defense mechanisms

We now discuss three broad classes of defense mechanisms against adversarial attacks: adversarial training (or retraining), regularization, and certified defenses. As somewhat expected, each of these strategies is effective against certain types of adversarial attacks and comes with their own performance guarantees.

3.5.2.1 Adversarial (re)training

Adversarial training is a popular method of adversarial defense, where the modeler wants to ensure robustness against certain types of adversaries during the model training phase. This can be done in a number of ways. Adversarial perturbations can be directly added to the training data, without any change to the training algorithm (Goodfellow et al., 2015; Wong et al., 2019). Methods that include defense measures incorporated into the learning model itself include minimizing the loss function over a grid of small perturbations around input points (Madry et al., 2018), ensemble training (Tramer et al., 2018), training using an adversary critic (Matyasko and Chau, 2018), ME-Net (Yang et al., 2019), and misclassification-aware training (Wang et al., 2020).

3.5.2.2 Use of regularization

A number of adversarially robust optimization methods aim to limit the effect of small perturbations on input or outputs through controlling gradient updates during iterative model training. To this end, they use different kinds of norm bounds (such as ℓ_2, ℓ_∞) as additional constraints added to the overall loss function, or changing the gradient itself. Such algorithms include Parseval networks (Cisse et al., 2017), DeepDefense (Yan et al., 2018), and TRADES (Zhang et al., 2019b). We also refer the reader to Table 3 in Silva and Najafirad (2020) for a number of (re)training and regularization methods that use norm constraints.

3.5.2.3 Certified defenses

Compared to the above two, the strength of this class of methods is that they provide probabilistic guarantees for the robustness of the resulting model. One of the first works in this domain is Reluplex (Katz et al., 2017), which provides a baseline framework for small neural networks with ReLU activation function. Subsequent research expanded this idea for more general networks (Gehr et al., 2018) and other activation functions (Singh et al., 2018). Recent research has also established theoretical guarantees for other quantities, such as a lower bound on the minimum necessary perturbation required to affect predictions of a model (Hein and Andriushchenko, 2017) and upper bounds on adversarial loss (Raghunathan et al., 2018; Wong and Kolter, 2018).

3.5.3 Implementations

We conclude with two implementations of adversarially robust algorithms that are available as open-source packages. The Adversarial Robustness Toolbox (ART)[6] was released by IBM in 2019 and enables researchers and practitioners to evaluate and defend ML models evasion, poisoning, extraction, and exploratory attacks (Buesser and Goldsteen, 2020). The second package, AdvBox[7] was released by Baidu in January 2020 and provides similar functionalities. Both toolboxes are in Python. As the field of adversarial ML matures with the increasing need for safe and reliable ML-based applications in the real world, efforts like these are essential toward faster adoption of the latest research.

3.6 Discussion

In this chapter, we have provided an overview of the technical methods for codifying human-centered values into large-scale ML systems. While there has been much research on this area in the last few years, a clear divide exists between theory and practice. Implementation frameworks of trustworthy ML methods are few and far between, and there are several challenges to develop and deploy such solutions in the wild (Holstein et al., 2019; Dodwell et al., 2020; Brennen, 2020; Andrus et al., 2021). To this end, it is critical to think about the cascading effects of big data collection processes and algorithmic systems built on such data that affect society at large—and vice versa. We sincerely hope that this chapter motivates practitioners and domain experts working on diverse application areas in leveraging their expertise to build participatory data-driven solutions that contribute toward the benefit of humankind.

References

Abadi, M., Chu, A., Goodfellow, I., et al., 2016. Deep learning with differential privacy. In: Proc. the 2016 ACM SIGSAC Conference on Computer and Communications Security, pp. 308–318.

Adadi, A., Berrada, M., 2018. Peeking inside the Black-Box: a survey on explainable artificial intelligence (XAI). IEEE Access. 6, 52138–52160.

[6] https://github.com/Trusted-AI/adversarial-robustness-toolbox.

[7] https://github.com/advboxes/AdvBox.

Adebayo, J.A., 2016. FairML: Toolbox for Diagnosing Bias in Predictive Modeling (Master's thesis). MIT, United States. https://github.com/adebayoj/fairml.

Alvarez-Melis, D., Daumé, H., Vaughan, J.W., Wallach, H., 2019. Weight of evidence as a basis for human-oriented explanations. In: Human-Centric Machine Learning (HCML) Workshop@ NeurIPS 2019. arXiv:1910.13503.

Amershi, S., Chickering, M., Drucker, S.M., et al., 2015. ModelTracker: redesigning performance analysis tools for machine learning. In: Proc. the 2015 CHI conference on human factors in computing systems (CHI), pp. 337–346.

Andrés, M.E., Bordenabe, N.E., Chatzikokolakis, K., Palamidessi, C., 2013. Geo-indistinguishability: differential privacy for location-based systems. In: Proc. the 2013 ACM SIGSAC Conference on Computer & Communications Security, pp. 901–914.

Andrus, M., Spitzer, E., Brown, J., Xiang, A., 2021. "What we can't measure, we can't understand": Challenges to demographic data procurement in the pursuit of fairness. In: Proc. the 2021 Conference on Fairness, Accountability, and Transparency.

Arnold, M., Bellamy, R., Hind, M., et al., 2019. FactSheets: increasing trust in AI services through supplier's declarations of conformity. IBM J. Res. Dev. 63 (4/5), 6:1–6:13.

Arya, V., Bellamy, R.K.E., Chen, P.-Y., et al., 2019. One explanation does not fit all: A toolkit and taxonomy of AI explainability techniques. arXiv:1909.03012.

Backurs, A., et al., 2019. Scalable fair clustering. In: International Conference on Machine Learning, vol. 97, pp. 405–413.

Bantilan, N., 2018. Themis-ML: a fairness-aware ML interface for end-to-end discrimination discovery and mitigation. J. Technol. Hum. Serv. 36 (1), 15–30.

Barraza, R., Eames, R., Balducci, Y.E., et al., 2019. Error terrain analysis for machine learning: Tool and visualizations. In: ICLR Workshop on Debugging Machine Learning Models.

Bauer, M., Baldes, S., 2005. An ontology-based interface for machine learning. In: Proc. the 10th International Conference on Intelligent User Interfaces, pp. 314–316.

Beaulieu-Jones, B.K., Wu, Z.S., Williams, C., et al., 2019. Privacy-preserving generative deep neural networks support clinical data sharing. Circ. Cardiovasc. Qual. Outcomes 12 (7), e005122.

Bellamy, R.K.E., Dey, K., Hind, M., et al., 2018. AI Fairness 360: An Extensible Toolkit for Detecting, Understanding, and Mitigating Unwanted Algorithmic Bias. arXiv:1810.01943.

Blum, A., Dwork, C., McSherry, F., Nissim, K., 2005. Practical privacy: the SuLQ framework. In: Proc. the Twenty-Fourth ACM SIGMOD-SIGACT-SIGART Symposium on Principles of Database Systems, pp. 128–138.

Brennen, A., 2020. What do people really want when they say they want "explainable AI?" We asked 60 stakeholders. In: Extended Abstracts of the 2020 CHI Conference on Human Factors in Computing Systems.

Bromburger, S., 1992. On What We Know We Don't Know: Explanation, Theory, Linguistics, and How Questions Shape Them. University of Chicago Press.

Buesser, B., Goldsteen, A., 2020. Adversarial Robustness Toolbox: One Year Later with v1.4. https://www.ibm.com/blogs/research/2020/10/adversarial-robustness-toolbox-one-year-later-with-v1-4.

Bun, M., Steinke, T., 2016. Concentrated differential privacy: simplifications, extensions, and lower bounds. In: Theory of Cryptography, Springer Berlin Heidelberg, Berlin, Heidelberg, pp. 635–658.

Bun, M., Steinke, T., 2019. Average-case averages: private algorithms for smooth sensitivity and mean estimation. Adv. Neural Inf. Process. Syst. 32, 181–191.

Calmon, F., et al., 2017. Optimized pre-processing for discrimination prevention. Adv. Neural Inf. Process. Syst. 30, 3992–4001.

Cao, Y., Yoshikawa, M., Xiao, Y., Xiong, L., 2017. Quantifying differential privacy under temporal correlations. In: 2017 IEEE 33rd International Conference on Data Engineering (ICDE), pp. 821–832.

Caruana, R., Lou, Y., Gehrke, J., et al., 2015. Intelligible models for healthcare: predicting pneumonia risk and hospital 30-day readmission. In: Proc. the 21th ACM SIGKDD International Conference on Knowledge Discovery and Data Mining, pp. 1721–1730.

Carvalho, D.V., Pereira, E.M., Cardoso, J.S., 2019. Machine learning interpretability: a survey on methods and metrics. Electronics 8 (832), 1–34.

Celis, L., et al., 2018. Classification with fairness constraints: a meta-algorithm with provable guarantees. In: Proc. the Conference on Fairness, Accountability, and Transparency, pp. 319–328.

Chakraborty, A., Alam, M., Dey, V., Chattopadhyay, A., Mukhopadhyay, D., 2018. Adversarial Attacks and Defences: A Survey. arXiv:1810.00069.

Chattopadhyay A., Manupriya, P., Sarkar, A., Balasubramanian, V.N., 2019. Neural network attributions: a causal perspective. In: Proc. the 36th International Conference on Machine Learning, vol. 97, pp. 981–990.

Chaudhuri, K., Sarwate, A.D., Sinha, K., 2011. Differentially private empirical risk minimization. J. Mach. Learn. Res. 12, 1069–1109.

Chaudhuri, K., Sarwate, A.D., Sinha, K., Near-Optimal, A., 2013. Algorithm for differentially-private principal components. J. Mach. Learn. Res. 14, 2905–2943.

Chaudhuri, K., Imola, J., Machanavajjhala, A., 2019. Capacity bounded differential privacy. Adv. Neural Inf. Process. Syst. 32, 3474–3483.

Chen, R., Fung, B.C., Yu, P.S., Desai, B.C., 2014. Correlated network data publication via differential privacy. VLDB J. 23 (4), 653–676.

Cheng L., Varshney, K.R., Liu, H., 2021. Socially Responsible AI Algorithms: Issues, Purposes, and Challenges. arXiv:2101.02032.

Cisse, M., Bojanowski, P., Grave, E., et al., 2017. Parseval networks: improving robustness to adversarial examples. In: Proc. the 34th International Conference on Machine Learning, vol. 70, pp. 854–863.

Cook, J., 2018. Amazon scraps 'sexist AI' recruiting tool that showed bias against women (oct). Available from: <https://www.telegraph.co.uk/technology/2018/10/10/amazon-scraps-sexist-ai-recruiting-tool-showed-bias-against/>.

Covert, I.C., Lundberg, S., Lee, S.-I., 2020. Feature removal is a unifying principle for model explanation methods. In: NeurIPS 2020 ML-Retrospectives, Surveys & Meta-Analyses Workshop. arXiv:2011.03623.

Differential Privacy Team, 2017. Apple, Learning with Privacy at Scale. Available from: <https://docs-assets.developer.apple.com/mL-research/papers/learning-with-privacy-at-scale.pdf>.

Dodwell, E., Flynn, C., Krishnamurthy, B., et al., 2020. Towards Integrating Fairness Transparently in Industrial Applications. arXiv:2006.06082.

Dong, J., Roth, A., Su, W.J., 2021. Gaussian differential privacy. J. R. Stat. Society: Ser. B arXiv: 1905.02383.

Doshi-Velez, F., Kim, B., 2017. Towards a Rigorous Science of Interpretable Machine Learning. arXiv:1702.08608.

du Pin Calmon, F., Fawaz, N., 2012. Privacy against statistical inference. In: 2012 50th Annual Allerton Conference on Communication, Control, and Computing (Allerton), pp. 1401–1408.

Du, M., Liu, N., Hu, X., 2020. Techniques for interpretable machine learning. Commun. ACM 63 (1), 68–77.

Dudik, M., et al., 2020. Fairlearn. Available from: <https://github.com/fairlearn/fairlearn>.

Dwork, C., McSherry, F., Nissim, K., Smith, A., 2006. Calibrating noise to sensitivity in private data analysis. Theory of Cryptography. Springer Berlin Heidelberg, Berlin, Heidelberg, pp. 265–284.

Dwork, C., Rothblum, G.N., Vadhan, S., 2010. Boosting and differential privacy. In: 2010 IEEE 51st Annual Symposium on Foundations of Computer Science, pp. 51–60.

Dwork, C., Roth, A., 2014. The algorithmic foundations of differential privacy. Found. Trends Theor. Computer Sci. 9 (3–4), 211–407.

Dwork, C., Rothblum G.N., 2016. Concentrated Differential Privacy. arXiv:1603.01887.

Erlingsson, Ú., Pihur, V., Korolova, A., 2014. RAPPOR: randomized aggregatable privacy-preserving ordinal response. In: Proc. the 2014 ACM SIGSAC Conference on Computer and Communications Security, Scottsdale, AZ, USA, November 3–7, 2014, pp. 1054–1067.

Farias, V.A.E., Brito, F.T., Flynn, C., et al., 2020. Local dampening: differential privacy for non-numeric queries via local sensitivity. Proc. VLDB Endow. 14 (4), 521–533.

Feldman, M., Friedler, S.A., Moeller, J., et al., 2015. Certifying and removing disparate impact. In: Proc. the 21th ACM SIGKDD International Conference on Knowledge Discovery and Data Mining, pp. 259–268.

Fisher, A., Rudin, C., Dominici, F., 2019. All models are wrong, but many are useful: learning a variable's importance by studying an entire class of prediction models simultaneously. J. Mach. Learn. Res. 20 (177), 1–81.

Fletcher, S., Zahidul Islam, M., 2015. A differentially private decision forest. In: Proc. the 13th Australasian Data Mining Conference (AusDM 2015), pp. 99–108.

Fredrikson, M., Jha, S., Ristenpart, T., 2015. Model inversion attacks that exploit confidence information and basic countermeasures. In: Proc. the 22nd ACM SIGSAC Conference on Computer and Communications Security, pp. 1322–1333.

Galhotra, S., Brun, Y., Meliou, A., 2017. Fairness testing: testing software for discrimination. In: ESEC/FSE-2017.

Gebru, T., Morgenstern, J., Vecchione, B., et al., 2018. Datasheets for Datasets. arXiv:1908.09635.

Gehr T., M. Mirman, Drachsler-Cohen, D., et al., 2018. AI2: safety and robustness certification of neural networks with abstract interpretation. In: 2018 IEEE Symposium on Security and Privacy (SP), pp. 3–18.

Geng, Q., Kairouz, P., Oh, S., Viswanath, P., 2015. The staircase mechanism in differential privacy. IEEE J. Sel. Top. Signal Process. 9 (7), 1176–1184.

Gilpin, L.H., Bau, D., Yuan, B.Z., et al., 2018. Explaining explanations: an overview of interpretability of machine learning. In: 2018 IEEE 5th International Conference on Data Science and Advanced Analytics (DSAA), Turin, Italy, pp. 80–89.

Gonem, A., Gilad-Bachrach, R., 2018. Smooth sensitivity based approach for differentially private PCA. In: Proc. Algorithmic Learning Theory, pp. 438–450.

Gong, M., Xie, Y., Pan, K., et al., 2020. A survey on differentially private machine learning [review article]. IEEE Comput. Intell. Mag. 15 (2), 49–64.

Goodfellow, I., Shlens, J., Szegedy, C., 2015. Explaining and harnessing adversarial examples. In: International Conference on Learning Representations.

Goyal, Y., Shalit, U., Kim, B., 2019. Explaining classifiers with causal concept effect (CaCE). arXiv:1907.07165.

Guo, R., Cheng, L., Li, J., et al., 2020. A survey of learning causality with data: problems and methods. ACM Comput. Surv. 53 (4).

Gurumoorthy, K.S., Dhurandhar, A., Cecchi, G.A., Aggarwal, C.C., 2019. Efficient data representation by selecting prototypes with importance weights. In: 2019 IEEE International Conference on Data Mining, ICDM 2019, Beijing, China, November 8–11, 2019, pp. 260–269.

Hardt, M., Price, E., Srebro, N., 2016. Equality of opportunity in supervised learning. Adv. Neural Inf. Process. Systems 29, 3315–3323.

Harradon, M., Druce, J., Ruttenberg, B., 2018. Causal learning and explanation of deep neural networks via autoencoded activations. arXiv:1802.00541.

He, X., Machanavajjhala, A., Ding, B., 2014. Blowfish privacy: tuning privacy-utility trade-offs using policies. In: Proc. the 2014 ACM SIGMOD International Conference on Management of Data, pp. 1447–1458.

Hein, M., Andriushchenko, M., 2017. Formal guarantees on the robustness of a classifier against adversarial manipulation. Adv. Neural Inf. Process. Syst. 30, 2266–2276.

Hendricks, L.A., Hu, R., Darrell, T., Akata, Z., 2018. Generating Counterfactual Explanations with Natural Language. arXiv:1806.09809.

Herman, B., 2017. The promise and peril of human evaluation for model interpretability. In: Proc. NIPS 2017 Symposium on Interpretable Machine Learning. arXiv:1711.07414.

Hitaj, B., Ateniese, G., Perez-Cruz, F., 2017. Deep models under the GAN: information leakage from collaborative deep learning. In: Proc. the 2017 ACM SIGSAC Conference on Computer and Communications Security, pp. 603–618.

Hohman, F., Head, A., Caruana, R., et al., 2019. Gamut: a design probe to understand how data scientists understand machine learning models. In: Proc. the 2019 CHI Conference on Human Factors in Computing Systems, pp. 1–13.

Holland, S., Hosny, A., Newman, S., et al., 2018. The dataset nutrition label: a framework to drive higher data quality standards. arXiv:arxiv.org/abs/1805.03677.

Holstein, K., Vaughan, J., Daume, H., et al., 2019. Improving fairness in machine learning systems: what do industry practitioners need? In: Proc. the 2019 CHI Conference on Human Factors in Computing Systems.

Holzinger, A., Langs, G., Denk, H., et al., 2019. Causability and explainability of artificial intelligence in medicine. WIREs Data Min. Knowl. Discov. 9 (e1312), 1–13.

Huber, P.J., 1981. Robust Statistics. John Wiley & Sons, Inc, New York.

Jagannathan, G., Pillaipakkamnatt, K., Wright, R.N., 2012. A practical differentially private random decision tree classifier. Trans. Data Priv. 5 (1), 273–295.

Jain, P., Kothari, P., Thakurta, A., 2012. Differentially private online learning. In: Proc. the 25th Annual Conference on Learning Theory, pp. 24.1–24.34.

Jain, P., Thakurta, A., 2013. Differentially private learning with kernels. In: Proc. the 30th International Conference on Machine Learning, vol. 28, pp. 118–126.

Kamath, G., Ullman, J., 2020. A Primer on Private Statistics. arXiv:2005.00010.

Kamiran, F., Calders, T., 2012. Data preprocessing techniques for classification without discrimina-tion. Knowl. Inf. Syst. 33 (1), 1–33.

Kamiran, F., Karim, A., Zhang, X., 2012. Decision theory for discrimination-aware classification. In: IEEE 12th International Conference on Data Mining, pp. 924–929.

Kamishima T., et al., 2012. Fairness-aware classifier with prejudice remover regularizer. In: Machine Learning and Knowledge Discovery in Databases, pp. 35–50.

Karwa, V., Raskhodnikova, S., Smith, A., Yaroslavtsev, G., 2014. Private analysis of graph structure. ACM Trans. Database Syst. 39 (3).

Katz, G., Barrett, C., Dill, D.L., et al., 2017. Reluplex: an efficient SMT solver for verifying deep neural networks. Computer Aided Verification. Springer International Publishing, pp. 97–117.

Kaur, H., Nori, H., Jenkins, S., et al., 2020. Interpreting interpretability: understanding data scientists' use of interpretability tools for machine learning. In: Proc. the 2020 CHI Conference on Human Factors in Computing Systems, no. 92, pp. 1–14.

Kearns, M., Neel, S., Roth, A., Wu, Z., 2018. Preventing fairness gerrymandering: auditing and learning for subgroup fairness. In: International Conference on Machine Learning, pp. 2569–2577.

Kearns, M., Neel, S., Roth, A., Wu, Z., 2019. An empirical study of rich subgroup fairness for machine learning. In: Proc. the Conference on Fairness, Accountability, and Transparency, pp. 100–109.

Kifer, D., Smith, A., Thakurta, A., 2012. Private convex empirical risk minimization and high- dimensional regression. In: Proc. the 25th Annual Conference on Learning Theory, vol. 23, pp. 25.1–25.40.

Kifer, D., Machanavajjhala, A., 2014. Pufferfish: a framework for mathematical privacy definitions. ACM Trans. Database Syst. 39 (1).

Kilbertus, N., Carulla, M.R., Parascandolo, G., et al., 2017. Avoiding discrimination through causal reasoning. Adv. Neural Inf. Process. Syst. 30, 656–666.

Kim, B., Khanna, R., Koyejo, O.O., 2016. Examples are not enough, learn to criticize! Criticism for Interpretability. Adv. Neural Inf. Process. Syst. 29, 2280–2288.

Kim, C., Bastani, O., 2019. Learning Interpretable Models with Causal Guarantees. arXiv:1901.08576.

Kusner, M., Loftus, J., Russell, C., Silva, R., 2017. Counterfactual fairness. Adv. Neural Inf. Process. Syst. 30, 4066–4076.

Lakkaraju, H., Bastani, O., 2020. "How do i fool you?": manipulating user trust via misleading black box explanations. In: Proc. the AAAI/ACM Conference on AI, Ethics, and Society, New York, NY, USA, pp. 79–85.

Lakkaraju, H., Kamar, E., Caruana, R., Leskovec, J., 2019. Faithful and customizable explanations of black box models. In: Proc. the 2019 AAAI/ACM Conference on AI, Ethics, and Society, pp. 131–138.

Lei, J., 2011. Differentially private M-estimators. Adv. Neural Inf. Process. Syst. 24, 361–369.

Li, T., Li, J., Liu, Z., et al., 2018a. Differentially private Naive Bayes learning over multiple data sources. Inf. Sci. 444, 89–104.

Li, C., Zhou, P., Xiong, L., Wang, Q., Wang, T., 2018b. Differentially private distributed online learning. IEEE Trans. Knowl. Data Eng. 30 (8), 1440–1453.

Liang, W., Chen, H., Liu, R., et al., 2020. A Pufferfish privacy mechanism for monitoring web browsing behavior under temporal correlations. Comput. Secur. 92, 101754.

Lipton, Z.C., 2016. The mythos of model interpretability. in: 2016 ICML Workshop on Human Interpretability in Machine Learning. arXiv:1606.03490.

Liu, X., Li, Q., Li, T., Chen, D., 2018. Differentially private classification with decision tree ensemble. Appl. Soft Comput. 62, 807–816.

Lode, L., 2019. Sub-sampled and Differentially Private Hamiltonian Monte Carlo (Master's thesis). University of Helsinki, Finland.

Lou, Y., Caruana, R., Gehrke, J., 2016. Intelligible models for classification and regression. In: Proc. the 18th ACM SIGKDD international conference on Knowledge discovery and data mining, pp. 150–158.

Lundberg, S., Lee, S.-I., 2017. A unified approach to interpreting model predictions. In: Proc. the 31st International Conference on Neural Information Processing Systems, pp. 4768–4777.

Machanavajjhala, A., Kifer, D., Abowd, J., et al., 2008. Privacy: theory meets practice on the map. In: 2008 IEEE 24th International Conference on Data Engineering, pp. 277–286.

Madaio, M., Stark, L., Vaughan, J., Wallach, H., 2020. Co-Designing checklists to understand organizational challenges and opportunities around fairness in AI. In: Proc. the 2020 CHI Conference on Human Factors in Computing Systems.

Madras, D., Creager, E., Pitassi, T., Zemel, R., 2019. Fairness through causal awareness: learning causal latent-variable models for biased data. In: Proc. the Conference on Fairness, Accountability, and Transparency, pp. 349–358.

Madry, A., Makelov, A., Schmidt, L., et al., 2018. Towards deep learning models resistant to adversarial attacks. In: International Conference on Learning Representations.

Madumal, P., Miller, T., Sonenberg, L., Vetere, F., 2020. Explainable reinforcement learning through a causal lens. In: Proc. the 34th AAAI Conference on Artificial Intelligence, pp. 2493–2500.

Maksymiuk, S., Gosiewska, A., Biecek, P., 2020. Landscape of R packages for eXplainable artificial intelligence. arXiv:2009.13248.

Matyasko, A., Chau, L.-P., 2018. Improved network robustness with adversary critic. In: Proc. the 32nd International Conference on Neural Information Processing Systems, pp. 10601–10610.

McKenna, R., Sheldon, D.R., 2020. Permute-and-flip: a new mechanism for differentially private selection. In: Advances in Neural Information Processing Systems.

McMahan, H.B., Ramage, D., Talwar, K., Zhang, L., 2018. Learning differentially private recurrent language models. In 6th International Conference on Learning Representations, ICLR 2018, Vancouver, BC, Canada, April 30 - May 3, 2018, Conference Track Proceedings. OpenReview.net, 2018.

McSherry, F.D., 2009. Privacy integrated queries: an extensible platform for privacy-preserving data analysis. In: Proc. the 2009 ACM SIGMOD International Conference on Management of Data, pp. 19–30.

McSherry, F., Talwar, K., 2007. Mechanism design via differential privacy. In: 48th Annual IEEE Symposium on Foundations of Computer Science (FOCS'07), pp. 94–103.

Mehrabi, N., Morstatter, F., Saxena, N., et al., 2019a. A Survey on Bias and Fairness in Machine Learning. arXiv:1908.09635.

Mehrabi, N., Morstatter, F., Peng, N., Galstyan, A., 2019b. Debiasing community detection: the importance of lowly connected nodes. In: Proc. the 2019 IEEE/ACM International Conference on Advances in Social Networks Analysis and Mining, pp. 509–512.

Mikko, A.H., Joonas, J., Onur, D., Antti, H., 2019. Differentially private Markov chain Monte Carlo. In: NeurIPS-2019.

Miller, T., 2019. Explanation in artificial intelligence: insights from the social sciences. Artif. Intell. 267, 1–38.

Miller, T., Howe, P., Sonenberg, L., 2017. Explainable AI: beware of inmates running the asylum. In: Proc. IJCAI Workshop Explainable AI (XAI), pp. 36–42.

Mironov, I., 2017. Rényi differential privacy. In: 2017 IEEE 30th Computer Security Foundations Symposium (CSF), pp. 263–275.

Mitchell, M., Wu, S., Zaldivar, A., et al., 2019. Model cards for model reporting. In: Proc. the Conference on Fairness, Accountability, and Transparency, pp. 220–229.

Mitchell, S., Potash, E., Barocas, S., et al., 2021. Algorithmic fairness: choices, assumptions, and definitions. Annu. Rev. Stat. Appl. 8 (1).

Moore, J., Hammerla, N., Watkins, C., 2019. Explaining deep learning models with constrained adversarial examples. In: PRICAI 2019: Trends in Artificial Intelligence, pp. 43–56.

Moraffah, R., Karami, M., Guo, R., et al., 2020. Causal Interpretability for machine learning – problems, methods and evaluation. ACM SIGKDD Explor. Newsl. 22 (1), 18–33.

Mothilal, R.K., Sharma, A., Tan, C., 2020. Explaining machine learning classifiers through diverse counterfactual explanations. In: Proc. the 2020 Conference on Fairness, Accountability, and Transparency, pp. 607–617.

Narayanan, A., Huey, J., Felten, E.W., 2016. A precautionary approach to big data privacy. Data Protection on the Move: Current Developments in ICT and Privacy/Data Protection. Springer, Netherlands, pp. 357–385.

Narendra, T., Sankaran, A., Vijaykeerthy, D., Mani, S., 2018. Explaining deep learning models using causal inference. arXiv:1811.04376.

Nissim, K., Raskhodnikova, S., Smith, A., 2007. Smooth sensitivity and sampling in private data analysis. In: Proc. the Thirty-Ninth Annual ACM Symposium on Theory of Computing, pp. 75–84.

Nori, H., Jenkins, S., Koch, P., Caruana, R., 2019. InterpretML: A Unified Framework for Machine Learning Interpretability. arXiv:1909.09223.

O'Keefe, C.M., Charest, A.-S., 2019. Bootstrap differential privacy. Trans. Data Priv. 12, 1–28.

Ou, L., Qin, Z., Liao, S., et al., 2018. An optimal pufferfish privacy mechanism for temporally correlated trajectories. IEEE Access. 6, 37150–37165.

Oya, S., Troncoso, C., Pérez-González, F., 2017. Is geo-indistinguishability what you are looking for? In: Proc. the 2017 on Workshop on Privacy in the Electronic Society, pp. 137–140.

Palanisamy, B., Li, C., Krishnamurthy, P., 2017. Group differential privacy-preserving disclosure of multi-level association graphs. in: 2017 IEEE 37th International Conference on Distributed Computing Systems (ICDCS), pp. 2587–2588.

Papernot, N., McDaniel, P., Goodfellow, I., 2016. Transferability in Machine Learning: From Phenomena to Black-Box Attacks using Adversarial Samples. arXiv:1605.07277.

Papernot, N., Song, S., Mironov, I., et al., 2018. Scalable Private Learning with PATE. ICLR, 2018.

Parafita, A., Vitriá, J., 2019. Explaining visual models by causal attribution. In: 2019 IEEE/CVF International Conference on Computer Vision Workshop (ICCVW), pp. 4167–4175.

Park, M., Foulds, J., Choudhary, K., Welling, M., 2017. DP-EM: differentially private expectation maximization. In: Proc. the 20th International Conference on Artificial Intelligence and Statistics, vol. 54, pp. 896–904.

Pearl, J., 2000. Causality: Models, Reasoning and Inference. Cambridge University Press.

Pearl, J., 2018. Theoretical impediments to machine learning with seven sparks from the causal revolution. In: Proc. the Eleventh ACM International Conference on Web Search and Data Mining.

Perez, S., 2018. Facebook is removing over 5, 000 ad targeting options to prevent discriminatory ads (aug). Available from: <https://techcrunch.com/2018/08/21/facebook-is-removing-over-5000-ad-targeting-options-to-prevent-discriminatory-ads>.

Phan, N., Wang, Y., Wu, X., Dou, D., 2016. Differential privacy preservation for deep autoencoders: an application of human behavior prediction. In: Proc. the Thirtieth AAAI Conference on Artificial Intelligence, pp. 1309–1316.

Phan, N., Wu, X., Dou, D., 2017. Preserving differential privacy in convolutional deep belief networks. Mach. Learn. 106 (9–10), 1681–1704.

Pleiss, G., et al., 2017. On fairness and calibration. In: Advances in Neural Information Processing Systems, pp. 5680–5689.

Poursabzi-Sangdeh, F., Goldstein, D.G., Hofman, J.M., et al., 2021. Manipulating and measuring model interpretability. In: Proc. the 2021 CHI Conference on Human Factors in Computing Systems. arXiv:1802.07810.

Raghunathan, A., Steinhardt, J., Liang, P., 2018. Certified defenses against adversarial examples. In: International Conference on Learning Representations.

Raji, I., Smart, A., White, R., et al., 2019. Closing the AI accountability gap: defining an end-to-end framework for internal algorithmic auditing. In: Proc. the Conference on Fairness, Accountability, and Transparency.

Rajkumar, A., Agarwal, S., 2012. A differentially private stochastic gradient descent algorithm for multiparty classification. In: Proc. the Fifteenth International Conference on Artificial Intelligence and Statistics, pp. 933–941.

Rana, S., Gupta, S.K., Venkatesh, S., 2015. Differentially private random forest with high utility. In: 2015 IEEE International Conference on Data Mining, pp. 955–960.

Rathi, S., 2019. Generating Counterfactual and Contrastive Explanations using SHAP. arXiv:1906.09293.

Ren, D., Amershi, S., Lee, B., et al., 2016. Squares: supporting interactive performance analysis for multiclass classifiers. IEEE Trans. Vis. Comput. Graph. 23 (1), 61–70.

Ribeiro, M.T., Singh, S., Guestrin, C., 2016. Why should i trust you?: explaining the predictions of any classifier. In: Proc. the 22nd ACM SIGKDD International Conference on Knowledge Discovery and Data Mining, pp. 1135–1144.

Rosenberg, I., Shabtai, A., Rokach, L., Elovici, Y., 2018. Generic black-box end-to-end attack against state of the art api call based malware classifiers. Research in Attacks, Intrusions, and Defenses. Springer International Publishing, pp. 490–510.

Rudin, C., 2019. Stop explaining black box machine learning models for high stakes decisions and use interpretable models instead. Nat. Mach. Intell. 1, 206–215.

Samadi, S., Tantipongpipat, U., Morgenstern, J.H., et al., 2018. The price of fair PCA: one extra dimension. Adv. Neural Inf. Process. Systems 31, 10976–10987.

Schellekens, V., Chatalic, A., Houssiau, F., et al., 2019. Differentially private compressive k-means. In: ICASSP 2019 – 2019 IEEE International Conference on Acoustics, Speech and Signal Processing (ICASSP), pp. 7933–7937.

Serban, A., Poll, E., Visser, J., 2020. Adversarial examples on object recognition: a comprehensive survey. ACM Comput. Surv. 53 (3).

Shokri, R., Shmatikov, V., 2015. Privacy-preserving deep learning. In: 2015 53rd Annual Allerton Conference on Communication, Control, and Computing (Allerton), pp. 909–910.

Shrestha, Y., Yang, Y., 2019. Fairness in algorithmic decision-making: applications in multi-winner voting, machine learning, and recommender systems. Algorithms 12 (199), 1–28.

Silva, S.H., Najafirad, P., 2020. Opportunities and Challenges in Deep Learning Adversarial Robust- ness: A Survey. arXiv:2007.00753.

Singh, G., Gehr, T., Mirman, M., et al., 2018. Fast and effective robustness certification. In: Proc. the 32nd International Conference on Neural Information Processing Systems, pp. 10825–10836.

Song, S., Chaudhuri, K., Sarwate, A.D., 2013. Stochastic gradient descent with differentially private updates. In: 2013 IEEE Global Conference on Signal and Information Processing, pp. 245–248.

Song, S., Wang, Y., Chaudhuri, K., 2017. Pufferfish privacy mechanisms for correlated data. In: Proc. the 2017 ACM International Conference on Management of Data, pp. 1291–1306.

Stevens, A., et al., 2018. Aequitas: bias and fairness audit. In: Tech Report, Center for Data Science and Public Policy, the University of Chicago. https://github.com/dssg/aequitas.

Su, D., Cao, J., Li, N., et al., 2016. Differentially private k-means clustering. In: Proc. the Sixth ACM Conference on Data and Application Security and Privacy, pp. 26–37.

Sun, L., Zhou, Y., Yu, P.S., Xiong, C., 2020. Differentially Private Deep Learning With Smooth Sensitivity. arXiv:2003.00505.

Talwar, K., Thakurta, A., Zhang, L., 2015. Nearly-optimal private LASSO. In: Proc. the 28th International Conference on Neural Information Processing Systems, vol. 2, pp. 3025–3033.

Tamagnini, P., Krause, J., Dasgupta, A., Bertini, E., 2017. Interpreting black-box classifiers using instance-level visual explanations. In: Proc. the 2nd Workshop on Human-In-the-Loop Data Analytics.

Toreini, E., Aitken, M., Coopamootoo, K., et al., 2020. The relationship between trust in AI and trustworthy machine learning technologies. In: Proc. the 2020 Conference on Fairness, Accountability, and Transparency, pp. 272–283.

Tramèr, F., Zhang, F., Juels, A., et al., 2016. Stealing machine learning models via prediction APIs. In: Proc. the 25th USENIX Conference on Security Symposium, pp. 601–618.

Tramer, A., et al., 2017. FairTest: discovering unwarranted associations in data-driven applications. In: Euro S&P-2017.

Tramer, F., Kurakin, A., Papernot, N., et al., 2018. Ensemble adversarial training: attacks and defenses. In: International Conference on Learning Representations.

Vadhan, S., 2017. The complexity of differential privacy. Tutorials on the Foundations of Cryptography: Dedicated to Oded Goldreich. Springer International Publishing, pp. 347–450.

Vaidya, J., Shafiq, B., Basu, A., Hong, Y., 2013. Differentially private naive Bayes classification. In: 2013 IEEE/WIC/ACM International Joint Conferences on Web Intelligence (WI) and Intelligent Agent Technologies (IAT), vol. 1, pp. 571–576.

Vasudevan, S., Kenthapadi, K., 2020. LiFT: a scalable framework for measuring fairness in ML applications. In: Proc. the 29th ACM International Conference on Information and Knowledge Management.

Veale, M., Cleek, M.V., Binns, R., 2018. Fairness and accountability design needs for algorithmic support in highstakes public sector decision-making. In: CHI '18: Proceedings of the 2018 CHI Conference on Human Factors in Computing Systems.

Verma, S., Rubin, J., 2018. Fairness definitions explained. In: FairWare '18: Proceedings of the International Workshop on Software Fairness.

Wachter, S., Mittelstadt, B.D., Russell, C., 2017. Counterfactual Explanations Without Opening the Black Box: Automated Decisions and the GDPR. arXiv:1711.00399.

Wang, D., Chen, C., Xu, J., 2019. Differentially private empirical risk minimization with non-convex loss functions. In: Proc. the 36th International Conference on Machine Learning, vol. 97, pp. 6526–6535.

Wang, Y., Zou, D., Yi, J., et al., 2020. Improving adversarial robustness requires revisiting misclassified examples. In: International Conference on Learning Representations.

Wong, E., Kolter, J.Z., 2018. Provable defenses against adversarial examples via the convex outer adversarial polytope. In: Proc. the 35th International Conference on Machine Learning, ICML 2018, Stockholmsmässan, Stockholm, Sweden, July 10–15, 2018, pp. 5283–5292.

Wong, E., Rice, L., Kolter, J.Z., 2019. Fast is better than free: revisiting adversarial training. In: International Conference on Learning Representations.

Wood, A., Altman, M., Bembenek, A., et al., 2018. Differential privacy: a primer for a non-technical audience. Vanderbilt J. Entertain. Technol. Law 21 (1), 209–275.

Wortman Vaughan, J., Wallach, H., 2020. A Human-Centered Agenda for Intelligible Machine Learning. Available from: <http://www.jennwv.com/papers/intel-chapter.pdf>.

Xi, Z., Sang, Y., Zhong, H., et al., 2020. Pufferfish privacy mechanism based on multi-dimensional markov chain model for correlated categorical data sequences. Parallel Architectures, Algorithms and Programming. Springer Singapore, Singapore, pp. 430–439.

Xie, L., Lin, K., Wang, S., Wang, F., Zhou, J., 2018. Differentially Private Generative Adversarial Network. arXiv:1802.06739.

Xiong, P., Buffett, S., Iqbal, S., et al., 2021. Towards a Robust and Trustworthy Machine Learning System Development. arXiv:2101.03042.

Xu, D., Yuan, S., Wu, X., et al., 2018. DPNE: differentially private network embedding. Adv. Knowl. Discov. Data Min. 235–246.

Yan, Z., Guo, Y., Zhang, C., 2018. Deep defense: training DNNs with improved adversarial robustness. In: Proc. the 32nd International Conference on Neural Information Processing Systems, pp. 417–426.

Yang, Y., Zhang, G., Katabi, D., Xu, Z., 2019. ME-Net: towards effective adversarial robustness with matrix estimation. In: Proc. the 36th International Conference on Machine Learning, vol. 97, pp. 7025–7034.

Yildirim, S., Ermis, B., 2019. Exact MCMC with differentially private moves. Stat. Comput. 29, 947–963.

Zehlike, M., et al., 2017. Fairness Measures: Datasets and Software for Detecting Algorithmic Discrimination. Available from: <http://fairness-measures.org>.

Zemel, R., et al., 2013. Learning fair representations. In: Proc. the 30th International Conference on Machine Learning, vol. 28, pp. 325–333.

Zhang, J., Zhang, Z., Xiao, X., et al., 2012. Functional mechanism: regression analysis under differential privacy. Proc. VLDB Endow. 5 (11), 1364–1375.

Zhang, J., Cormode, G., Procopiuc, C.M., et al., 2015. Private release of graph statistics using ladder functions. In: Proc. the 2015 ACM SIGMOD International Conference on Management of Data, pp. 731–745.

Zhang, L., Wu, Y., Wu, X., 2017a. A Causal Framework for Discovering and Removing Direct and Indirect Discrimination, In: IJCAI-2017.

Zhang, X., Ji, S., Wang, H., Wang, T., 2017b. Private, yet practical, multiparty deep learning. In: 2017 IEEE 37th International Conference on Distributed Computing Systems (ICDCS), pp. 1442–1452.

Zhang, J., Bareinboim, E., 2018. Fairness in decision-making—the causal explanation formula. In: AAAI-2018.

Zhang, B., Lemoine, B., Mitchell, M., 2018. Mitigating unwanted biases with adversarial learning. In: AAAI Conference on AI, Ethics and Society.

Zhang, Y., Hao, Z., Wang, S., 2019a. A differential privacy support vector machine classifier based on dual variable perturbation. IEEE Access. 7, 98238–98251.

Zhang, H., Yu, Y., Jiao, J., et al., 2019b. Theoretically principled trade-off between robustness and accuracy. In: Proc. the 36th International Conference on Machine Learning, pp. 7472–7482.

Zhao, Q., Hastie, T., 2021. Causal interpretations of black-box models. J. Bus. Econ. Stat. 39 (1), 272–281.

Zhu, J., Liapis, A., Risi, S., et al., 2018. Explainable AI for designers: a human-centered perspective on mixed-initiative co-creation. In: 2018 IEEE Conference on Computational Intelligence and Games (CIG), pp. 1–8.

Zhu, Y., Wang, Y.-X., 2019. Poission subsampled rényi differential privacy. In: Proc. the 36th International Conference on Machine Learning, pp. 7634–7642.

Section 2

Chemistry & chemoinformatics section

How to integrate the "small and big" data into a complex adverse outcome pathway?

4

Marjan Vračko
Theory Department, Kemijski inštitut/National Institute of Chemistry, Ljubljana, Slovenia

4.1 Introduction

Adverse outcome pathway (AOP) is a toxicology paradigm that was introduced a decade ago (Ankley et al., 2010). According to this viewpoint, a toxic effect is the result of causally related events that occur at various biological levels: molecular, cell organelles, cellular, tissue, organ, individual organism, and population. When a xenobiotic enters the body, it can cause a number of metabolic reactions. This also implies that all possible metabolic products should be considered. The second critical piece of information is based on the distribution and kinetics of molecules in organisms. When discussing AOP, two additional terms must be defined: the Molecular Initiating Event (MIE), which describes the first chemical interaction of a molecule with bio-targets, and the key events (KEs). They are decisive events in the spread of a negative effect on the higher biological level. To emphasize that an AOP is a series of causally linked events, knowledge of KE is required. Completing the entire AOP scheme for a molecule is a difficult task that could take decades or generations. The Organization for Economic Cooperation and Development (OECD) has launched a very promising and optimistic initiative to keep track of AOP that have been developed or are in the process of development on the web platform wikiaop (https://aopwiki.org/). Some examples are presented in Section 4.2.

A toxic effect of a chemical is reported in traditional toxicology as an effect of a chemical on specific organisms. This concept predates the concept of AOP, and an evolution of experimental protocols and acceptance criteria for obtained results can be seen. During the traditional toxicology era, a large amount of data was collected and stored in various data bases (also known as "small data"). Many quantitative structure–activity relationship (QSAR) models have been developed on this basis with the goal of expanding toxicity knowledge to a larger chemical space. A QSAR model, in essence, expresses the mathematical relationship between a chemical structure represented by numerical descriptors and a toxic property. The OECD adopted the principles for validation of (Q)SAR models to provide guidelines and assistance to model developers and end users (OECD, 2004). The five guiding principles are as follows: (1) a defined endpoint, (2) an

Big Data Analytics in Chemoinformatics and Bioinformatics. DOI: https://doi.org/10.1016/B978-0-323-85713-0.00027-X

unambiguous algorithm, (3) a defined domain of applicability, (4) appropriate goodness-of-fit, robustness, and predictability measures, and, if possible, (5) a mechanistic interpretation. The relationship was primarily expressed as a linear equation, and it was discovered using the multiple linear regression (MLR) method. In recent decades, linear methods have frequently been replaced by mathematically more complex methods such as artificial neural networks (Drgan et al., 2016). All of these models have one thing in common: the relationship is determined using statistical methods. Various methods and strategies have been proposed to justify the selection of final models, which are commonly referred to as the model validation procedure. The fundamental flaw in these approaches is a lack of mechanistic interpretation of predictions, which leads to a lack of understanding of toxic mechanisms. This issue has been a source of contention since the inception of "the QSAR area" (Hansch et al., 1962; Hansch and Leo, 1979). According to Piir et al.'s study, approximately 42% of all models are documented in a reproducible manner, yielding the second best result of 52% for eco-toxicity endpoints and the best result of 56% for environmental fate endpoints. According to the study, more than 82% of QSAR eco-toxicity endpoint models mention fish toxicity (40%), algae toxicity (22%), and Daphnia toxicity (20%) 211 multiple linear models were the most frequently used models in the eco-toxicity group, which was represented by 261 QSAR articles (JRC QSAR Model Database), Artificial neural networks came in second with 23 reported models. This trend is similar to that seen in a decade-old review by Netzeva et al. (2007), which looked at aquatic toxicity data sources, QSARs, and testing strategies. This review includes over 60 MLR models targeting fish toxicity endpoints and over 90 MLR equations for various aquatic toxicity endpoints. A small number of models employing advanced modeling techniques are mentioned. The most frequently used models were Hansch-type models that predicted toxicity as a function of the octanol/water partition coefficient (Hansch et al., 1962). Other parameters, such as electronic and steric properties, hydrogen bonding, and so on, are required for specific toxicants. Many theoretical descriptors have been developed to describe structures in a hierarchical manner, beginning with the structure of molecules and progressing to quantum chemical electronic structure (Basak et al., 1998; Basak et al., 2003). Attempts have been made to incorporate biological data into QSAR models (Basak, 2013; Basak and Vracko, 2020). To assess the environmental hazard, various toxicity data are available: Tetrahymena pyriformis, Daphnia magna, zebra fish, worm, mice, and so on. To assess the risk for human organ specific toxicities/activities, such as liver toxicity, heart toxicity, skin sensitizing potential, cancer potential, and so on, are reported. On the other hand, there is a large repository of "big data," which refers to high-throughput screening and -omics results. It is difficult to combine "big and small" data from different levels of biological organization into a single toxic endpoint. At the end of the Introduction, it is important to note that standard toxicological data were not obtained in order to complete the AOP scheme. The next task will be to illuminate and integrate existing data into AOP schemes.

4.2 State and review

Wikiaop has more than 306 AOP, MIE, and 2200 KE relationships. The AOP Knowledge Base has been launched as a result of a collaboration between the OECD, the US Environmental Protection Agency, and the European Commission Joint Research Centre. This is a web-based platform that gathers all knowledge on how chemicals can cause adverse effects and serves as a focal point for AOP development and dissemination. The first module is the AOP Wiki, which is an interactive and virtual encyclopedia for AOP development. It is organized in accordance with the original OECD "Guidance document and a template for developing and assessing adverse outcome pathways" (OECD, 2017) and the more recent Handbook for AOP developers. In OECD (2012), published first AOP: The AOP for Skin Sensitisation Initiated by Covalent Binding to Proteins.

The AOP is a broad concept that can be used to analyze chemicals, nanoparticles, mixtures, and other stressors such as radiation (Song et al., 2020). Some recently published reports are discussed further below. For nanomaterials, the AOP concept for toxicity assessment has been widely used. Because the properties of nanoparticles, including their toxicity, depend on their size, shape, and surface charge, they may differ from those of bulk material. As a result, the traditional concept of molecular structure–property relationship is no longer valid. However, it is to determine which MIEs or KEs share both material forms. Halappanavar et al. (2020) discussed the AOPs associated with lung toxicity caused by nanomaterials. The authors looked at several AOP that stem from the same MIE, namely the interaction between nanomaterials and lung cells. One example is the interaction of AOP 173 substances with lung resident cell membrane components, which results in fibrosis. It reports the interaction of stressors with the lung resident cell membrane as MIE, which is followed by several KEs that lead to fibrosis and a decrease in lung volume. The AOP has been fully developed and is applicable to a variety of materials. Another example is AOP 303, which inhibited phagocytosis and caused lung cancer. It is mostly applicable to materials with a high length/diameter ratio, such as nanowires, nanorods, nanotubes, and asbestos. The KEs that lead to cancer are: increased secretion of proinflammatory mediators (cytokines) (KE1), increased influx of leukocytes into the lungs (KE2), metabolism modification and production of reactive oxygen species (ROS) (KE3), mutation, and cancer. Other AOPs reported include: AOP NO ID substance interaction with lung epithelial and macrophage cell membrane leading to lung fibrosis (similar to AOP 173); AOP 1.25 increased substance interaction with alveolar cell membrane leading to emphysema; AOP 237 cellular sensing of stressor leading to plaque progression; and AOP 302 disruption of lung surfactant function leading to acute inhalation toxicity. The authors then examined the entire network of the AOPs mentioned in order to identify common KEs. Clewell et al. (2020) describe the use of AOP in conjunction with the aggregate exposure pathway (AEP) for phthalates. The AEP framework represents a structured and organized method of collecting exposure information from various sources and directing it to the target site. Phtalates have been extensively studied due to their endocrine system activity, which can result in adverse effects such as liver tumors or disruption in the development of sexual organs. The AEP for di(2-ethylhexyl) phthalate (DEHP), di-*n*-butyl phthalate (DnBP), and some of their metabolites is presented by the authors. The testosterone-mediated male sexual development of rats is the proposed AOP. The MIE is cPLA2 enzyme inhibition, followed by three KEs: AA release inhibition, steroidogenic gene transcription reduction, and testosterone production reduction. The exposure was calculated using real-world conditions and various sources such as outdoor air, dust, personal care products, and so on. The authors propose estimating cumulative risk for phthalates by combining the AOP and AEP. Both schemes enable risk estimation for additional chemicals found in the environment, but they share the AOP or parts of it. Gundacker and Ellinger (2020) provide an overview of how to study the AOP43 disruption of VEGFR signaling, which leads to developmental defects, using human placenta model cells. The MIE occurs when a vascular disrupting chemical interacts with the vascular endothelial growth factor receptor, followed by three KEs: KE 28 reduction, angiogenesis, KE 110 impairment, endothelial network, and KE 298 insufficiency, vascular. Jin et al. (2021) presented a detailed analysis of the AOP network to study Benzo(a)pyrene, 2,3,7,8-tetrachlorodibenzoparadioxin, valproic acid, and quercetin in liver and lung diseases. The authors compiled data from various sources and established the AOP network, which was extensively analyzed. They describe a link between lung and liver AOPs. The activation of the aryl hydrocarbon receptor was demonstrated using MIE. Spinu et al. (2020) present a strategy for using AOPs to predict chemical hazards. In this strategy, also known as quantitative adverse outcome pathway (qAOP), the chemicals are mapped to AOP space, where the common events and final adverse outcomes are identified in the following stage. Models cover a wide range of adverse outcomes, such as organ failures in human toxicology or eco-toxicological endpoints. The entire AOP system that is available on the internet is already a highly branched system that includes various data, hypotheses, and so on. The qAOP strategy is based on computational analysis and the selection of networks that are relevant for the group of chemicals under consideration. From a mathematical standpoint, the methods include statistical and/or mechanistic approaches. A software list with 20 different tools is provided. Finally, the authors propose twelve principles that an ideal qAOP should follow. They must ensure, similarly to the principles for developing QSAR models, that models are transparent, reproducible, flexible, and so on. Knapen et al. (2020) present the AOP concept in their study of thyroid hormone disruption in the regulation of swim bladder inflation reduction in fish. As previously stated, the AOP system accessible via the internet is a well-branched system. Wang (2020) conducted a semantic analysis of the system to investigate the coherence of all issues. Maki et al. (2020) disrupted the genes encoding the enzyme responsible for melanin production (tyrosinaze) in fish fathead minnow. The study backs up AOP 292—*Inhibition of throsinase leads to a decrease in fish population.* Depigmentation in the skin and retina is to blame for the population shift. The corresponding genes were mutated in the experiment using the CRISPR/Cas9 gene editing technique. They demonstrated that a precisely targeted mutation of specific sequences results in specific phenotypic changes. On the other hand, there is a wealth of information on the effects of chemicals on fish development. There is still work to be done to connect the chemicals and their negative effects. Wang et al. (2018) describe the use of the AOP concept in the assessment of ecological risk for bisphenol A and 4-nonylphenol. Both have been shown to be endocrine disruptors. A wide range of aquatic organisms have been considered, including plants (algae), invertebrates, and vertebrates (fish). Most of them have a wealth of data, particularly for traditional toxicological endpoints such as individual survival, development, and reproduction ability, as well as offspring development. Both chemicals function as estrogen receptor agonists at the molecular level. The authors divided the traditional data into those that were linked to endocrine-related AOP and those that were not. Furthermore, the authors analyzed available toxicological data and incorporated it into a single parameter predicted-no-effect concentration, which is a widely accepted measure of water quality. It exemplifies the incorporation of traditional toxicological data and a novel AOP concept into a risk characterization scheme. The omics (genomics, proteomics, etc.) techniques have become increasingly important in toxicology over the last few decades. The strategy of this research is to identify genes or proteins that are linked to chemical exposure and specific toxic endpoints. Proteomics is of particular interest because proteins are typically the first functional biomolecules affected by chemicals and are located in the first stage of the AOP hierarchy. The pioneer work is reported by Brockmeier et al. (2017). Wang et al. (2020) provide an overview of eco-toxico-proteomic evaluations of microplastics and plastic additives in aquatic organisms. They examined tissues from various species (vertebrates and invertebrates) and identified proteins that are most affected after exposure to hazardous chemicals. Similar strategy is described in report by Vračko et al. (2018). The authors present a chemometrical analysis of proteomics data from three cell lines: Caco-2 and HT29-MTX co-culture, primary small airway epithelial cells (SAEC), and THP-1 macrophage-like cells. The cells were exposed to two kinds of nanomaterials: TiO_2 nanobelts and multiwalled carbon nanotubes under different exposure scenarios (two concentrations, 10 and 100 μg/mL, and two-time regimes, 3 and 24 hours). When considering the clustering structure of all samples, they discovered that the time of exposure is the most important factor. They discovered that the proteins of the histone family may be important in the long-term exposure response of cells using chemometric methods. Basak et al. (2016) examined proteomics data from Caco-2/HT29-MTX cells in order to identify highly perturbed proteins specific for either type of nanoparticle. Eight and 25 proteins were identified for multiwalled carbon nanotubes and TiO_2 nanobelts, respectively. An earlier study examined the proteomics data of rat hepatocytes exposed to 14 halocarbons. Proteomics were submitted for chemometrical analysis as two-dimensional maps generated by the two-dimensional electrophoresis technique. 1401 spots were assigned to each map (Vračko et al., 2006). Aside from proteomics, six in vitro cell endpoints were measured: cell viability, membrane integrity (LDH assay), total cellular thiols, lipid peroxidation, reactive oxygen species production, and catalase activity. The authors used the genetic algorithm to identify the ten most important spots for six endpoints. This study must be viewed as an attempt to identify proteins that are significantly affected when cells are exposed to xenobiotics. More targeted studies, however, are required to incorporate the findings into an AOP scheme. Ma et al. studied zebrafish exposed to silver nanoparticles in vivo (AgNP).

They discovered an increase in the level of reactive oxygen species as a result of AgNP accumulation. The study also found that the activation of the genes bax, caspase-9, and caspase-3 is responsible for the mitochondrial-mediated apoptosis pathway. Khan et al. (2019) studied the effects of graphene oxide nanoparticles on the marine bivale. They assigned three biomarkers that are proposed in the development of AOPs: the level of malondialdehyde, which results from oxidation of cell membranes and measures oxidative stress in cells; the activity of glutathione-s-transferase, which is a cell response to chemical stress; and the total protein assay, which indicates changes in cell protein status after exposure. According to the AOP system, the toxicological endpoint of graphene oxide exposure is cell death. Furthermore, Khan et al. (2020) propose a systematic approach to develop AOPs on the basis of biomarkers. As an ecological indicator, bivalves are used as model organisms. The report describes the development of AOP 97, in which inhibition of the 5-hydroxytryptamine transporter (5-HTT; SERT) leads to population decline. A MIE is the inhibition of the mentioned transporter. Corsi et al. describe a strategy for integrating monitoring data, high-throughput (HTP) data from in vitro tests (ToxCast), and existing knowledge about potential AOPs into a complex hazard assessment system. They assessed and prioritized 49 compounds from the ToxCast assay and matched them to 23 potential AOPs from AOP-Wiki.

4.3 Binding affinity to androgen nuclear receptor evaluated with respect to carcinogenic potency data

Hormones are endogenic molecules that regulate organism development and physiology. According to the US EPA, an endocrine disruptor is an exogenous agent that interferes with the production, release, transport, metabolism, binding action, or elimination of natural blood-borne hormones in the body that are responsible for homeostasis, reproduction, and developmental process regulation (Diamanti-Kandarakis et al., 2009). Altering the function of the endocrine system may have negative consequences in organisms, including cancer. The first molecular interaction between a chemical and an organism is usually binding to nuclear endocrine receptors such as androgen-AR, estrogen-ER, glucocorticoid-GR, liver X_LXR, mineralocorticoid-MR, peroxisome proliferator-PPAR, progesterone-PR, retinoid X-RXR, and thyroid-TR. The EPA's ToxCast and Tox21 projects provide HTP in vitro estrogen screening data (CERAP) (Mansouri et al., 2016; Judson et al., 2015) and androgen receptor (AR) (CoMPARA) (Mansouri et al., 2020). The estrogen data have been extensively analyzed with different in silico methods (Ruiz et al., 2017). The reported exercise included two sets of compounds: 1654 agonists and 1522 antagonists, the activities of which to AR were measured in 11 high-throughput screening assays. On the basis of the measured activity, which was expressed as area under the curve in any of the in vitro assays, the compounds are binary classified as active (1) or nonactive (0) (Kleinstreuer et al., 2017).

Cancer, as a negative outcome, is linked to organs and organisms, and thus serves as an end-point in the AOP scheme. It is a complicated toxicological endpoint. Tumor initiation and progression are regulated by various KE and mechanisms, some of which are related to endocrine disruption (Severson and Zwart, 2017). There is a great deal of evidence that xenobiotics cause cancer, but data suitable for in silico modeling are scarce. Many QSAR models and expert systems for specific chemical classes, such as amines, nitro compounds, and poly-aromatic hydrocarbons, have been developed (Benfenati et al., 2009; Richard and Benigni, 2002; Passerini, 2003; Woo et al., 2005). The model for carcinogenicity Computer Assisted Evaluation of industrial chemical Substances According to Regulations (CAESAR), which is applied in the presented case study, is available in Virtual models for property Evaluation of chemicals within a Global Architecture (VEGA) HUB; https://www.vegahub.eu/. The model was built on the training set of 805 compounds from Carcinogenic Potency Database (CPDB). More details of the model are reported by Fjodorova et al. (2010). The model's output for the compound includes the following information: carcinogenicity prediction expressed as a binary classification (nonpositive/positive), information on applicability domain expressed with eight indices, and the similarity set. This set consists of the model training set's six most similar compounds. It is important to note that the experimental and predicted values for the similarity set are provided, allowing for additional justification of predicted classification. The definition of similarity is the critical question. The index calculated as a weighted sum of factors describing the constitutional and structural properties of molecules is used in the applied model to express similarity (Floris et al., 2014).

The proposed strategy is to analyze QSAR predictions for large data sets using small similarity sets (small sets of similar structures) and chemometrical clustering methods (Vračko and Bobst, 2014; Vračko and Bobst, 2015). An example of representation for the compound indole is shown in Fig. 4.1. The six compounds construct the similarity set and serve as descriptors in the clustering procedure. The entire HTP screened data set is thus represented by molecules from the model's training set (CPDB). With the right chemometrical tools, this entire representation space can be analyzed to reveal its clustering structure or to reduce its dimension. The principal component analysis (Vračko and Bobst, 2014) and self-organizing maps seem to be suitable for these tasks (Vračko and Drgan, 2017). Furthermore, the methods enable to identify the cluster's indicators, i.e., the compounds from model's training set, which played the essential role in mathematical clustering algorithm. The procedure is schematically shown in Fig. 4.2. Table 4.1 shows three clusters from ToxCast EPA project database represented by cluster indicators from model's training set (Vračko and Drgan, 2017).Please, print the Figure 4.1 in color.

As schematically shown in Fig. 4.3 one can estimate the relationship between both data sets: the CoMPARA and the CMTPI. The model reported 768 compounds as having similarity sets for all compounds in the CoMPARA data set. This represents approximately 95% of the entire training set of the VEGA (CAESAR) model. However, the distribution is not uniform. Table 4.2 shows the similar compounds with the highest frequency, i.e., the count of its appearing in the similarity sets. From 25 mostly frequented compounds 25 are noncarcinogenic (72 %).

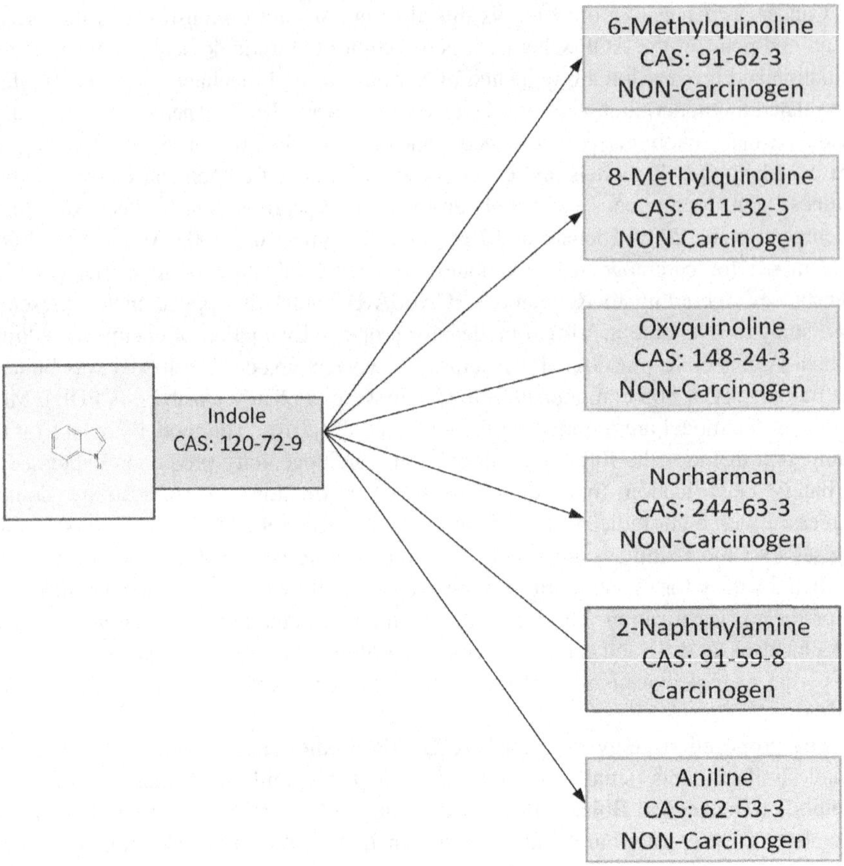

Figure 4.1 Carcinogenicity prediction for indole and six compounds from VEGA (CAESAR) training set, which represent the indole in the clustering procedure.

A comparison of two data sets, for example, a large data set obtained from HTP screening assays and a small data set containing traditional toxicity data, is shown in the example. It should be noted that the predictions are based on training set data. The example provided does not propose a specific AOP, but rather enables data clustering and focusing on specific points in the big-data environment. It could serve as a starting point for AOP development.

4.4 Conclusion and future directions

The AOP concept has been introduced a decade ago (Ankley et al., 2010). It represents a series of biological events that result in negative consequences. It is depicted as a series of causally linked events that occur at the molecular, cell, tissue, organ,

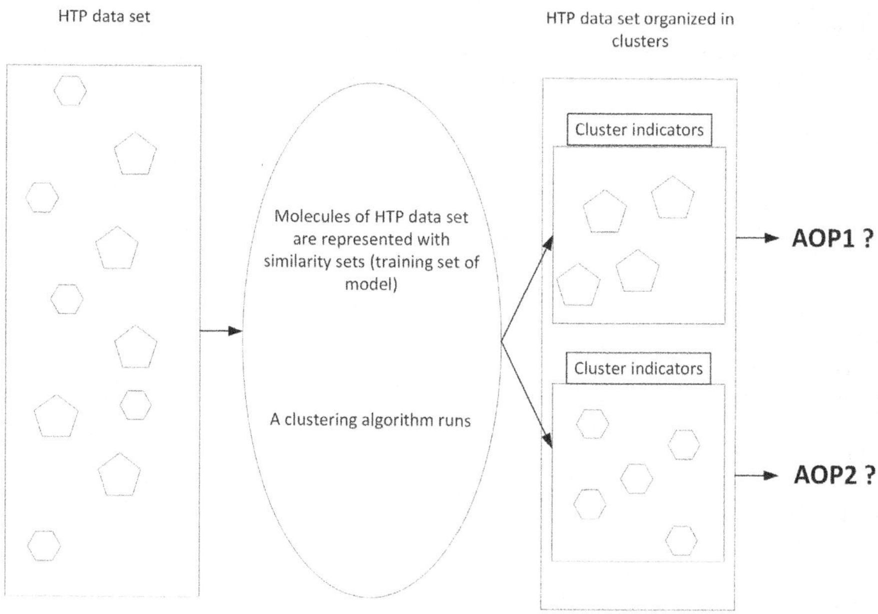

Figure 4.2 Scheme of the proposed clustering scheme for high-throughput (HTP) data.

Table 4.1 Clusters of compounds, which show in vitro activity to AR.

Compound	Predicted/Exp.[b]	AR activity
Agonists		
Cluster [a]		
DB00621	Carc	1
DB00687	Non-carc	1
Prednisone	Carc	1
Triamcinolone	Non-carc	1
Dexamethasone sodium phosphate	Non-carc	1
Cyproterone acetate	Non-carc	1
Melengestrol acetate	Non-carc	1
DB01185	Non-carc	1
DB01406	Carc	1
DB01541	Carc	1
Corticosterone	Non-carc	1
DB00624	Carc	1
DB00858	Carc	1
DB01481	Carc	1
DB01564	Carc	1
DB02998	Carc	1
DB06412	Non-carc	1
5-alpha-Dihydrotestosterone	Carc	1
Progesterone	Carc	1
4-Androstene-3,17-dione	Carc	1
17-Methyltestosterone	Carc	1
17 beta Trenbolone	Carc	1
17-alpha-Hydroxyprogesterone	Non-carc	1
Norethindrone	Carc	1
Norgestrel	Carc	1

(*Continued*)

Table 4.1 (Continued)

Testosterone propionate	Non-carc	1
Diisobutyl adipate	Non-carc	0
Gibberellic acid	Non-carc	0
Niclosamide	Carc	0
4-Butylphenol	Non-carc	0
(2-Dodecenyl)succinic anhydride	Non-carc	0
Lovastatin	Non-carc	0
Sethoxydim	Carc	0
Chlorfenapyr	Non-carc	0
Nilutamide	Carc	0
Benoxacor	Carc	0
Kresoxim-methyl	Carc	0
Dapsone[a]	Carc/carc	0
Finasteride	Non-carc	0
DB00984	Non-carc	0
DB08804	Non-carc	0
Mifepristone	Carc	0
Trimethoxypropylsilane	Carc	0
Antagonists		
Cluster 1		
o,p'-DDD	Non-carc	1
p,p'-DDD [c]	Non-carc/Non-carc	1
Dicofol [c]	Carc/Non-carc	1
p,p'-DDE[c]	Non-carc/Non-carc	1
Dichlorophen	Carc	1
2,2-Bis(4-hydroxyphenyl)-1,1,1-trichloro	Non-carc	1
o,p'-DDT	Carc	1
3,3',5,5'-Tetrabromobisphenol A	Non-carc	1
Bisphenol AF	Non-carc	1
Cluster 2		
Chlorobenzilate [c]	Non-carc/Non-carc	1
Imazalil	Non-carc	1
Econazole nitrate	Non-carc	1
Fenbuconazole	Non-carc	1
Propiconazole	Carc	1
Tebuconazole	Non-carc	1
Bromuconazole	Carc	1
Triticonazole	Carc	1
Metconazole	Carc	1
Hexaconazole	Carc	1
Ipconazole	Carc	1
Diniconazole	Non-carc	1
Norgestrel	Carc	0
Cyclopentanol	Carc	0
MK-578	Non-carc	0
Sodium fluoroacetate|Fluoroacetic acid	Carc	0
Cyclanilide	Non-carc	0
1,5,9-Cyclododecatriene	Carc	0

[a]Dapsone is a cluster indicator.
[b]Predicted with the model VEGA (CAESAR). Model provides the experimental value when applicable.
[c]Cluster indicators for cluster 1 and 2, respectively.

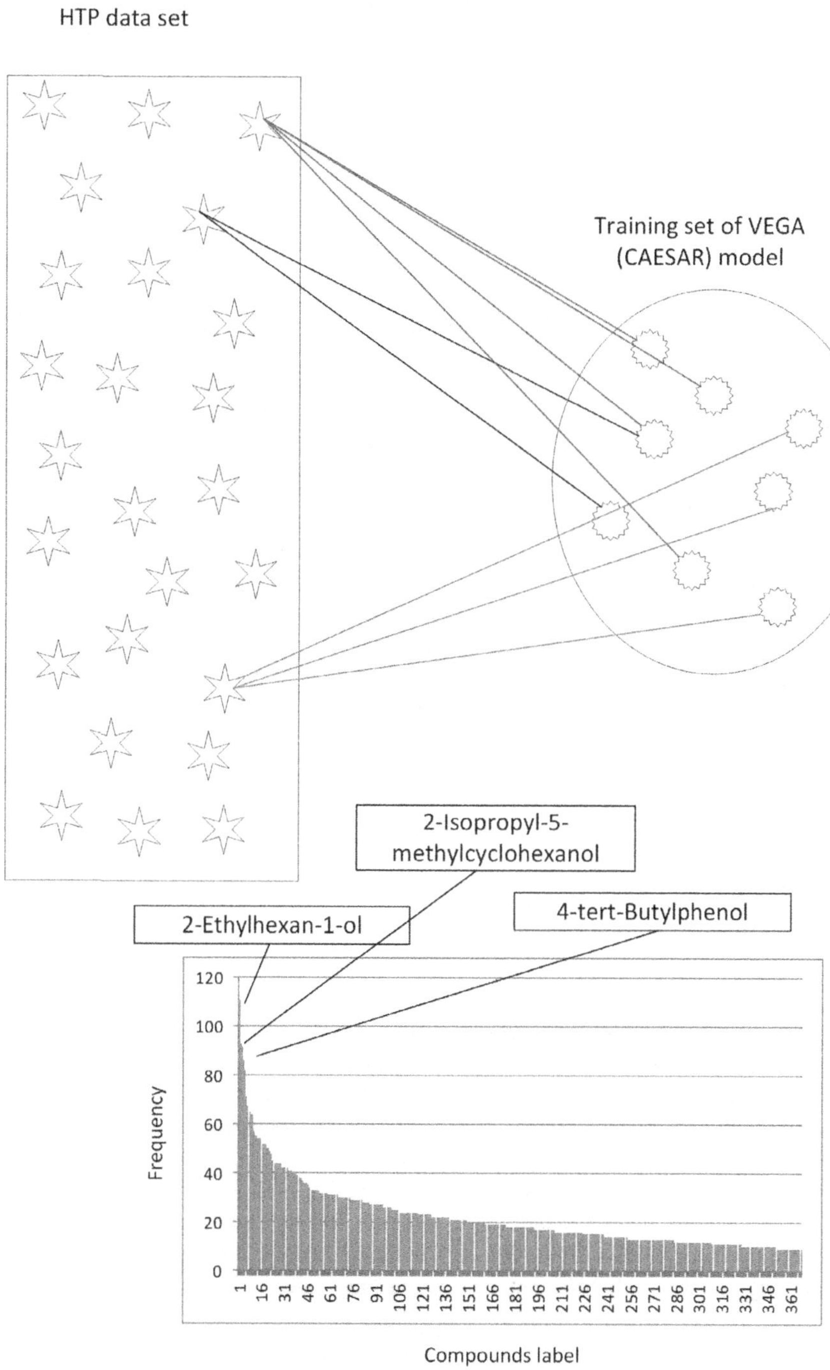

Figure 4.3 The high throughput (HTP) data set is projected onto training set of "standards" model. The lower picture shows the frequency of the compounds from quantitative structure–activity relationship (QSAR) model. Three compounds with the highest frequency are indicated (see Table 4.2).

Table 4.2 The compounds of representation space with the frequency larger than 35.

Chemical	Experimental[a]	Predicted[a]	Frequency
2-Ethylhexan-1-ol	Non-carc	Carc	111
2-Isopropyl-5-methylcyclohexanol	Non-carc	Carc	92
4-*tert*-Butylphenol	Non-carc	Non-carc	84
2-*tert*-Butyl-4-methylphenol	Non-carc	Non-carc	68
Di(2-ethylhexyl) adipate	Non-carc	Non-carc	64
Dimethyl p-phthalate	Non-carc	Non-carc	55
Benzyl butyl phthalate	Carc	Carc	51
Isoacetophorone	Carc	Carc	50
Bisphenol A	Non-carc	Non-carc	48
Cyclohexanone	Non-carc	Carc	43
Bis(2-ethylhexyl) phthalate	Carc	Carc	43
Dimethoxane	Carc	Non-carc	42
Dehydroepiandrosterone	Carc	Carc	42
2,6-Di-*tert*-butyl-4-methylphenol	Non-carc	Non-carc	41
Diethylene glycol	Carc	Carc	41
Benzoic acid	Non-carc	Carc	40
Trifluralin	Non-carc	Non-carc	39
1,2-propanediol	Non-carc	Non-carc	38
Ethyl 4,4'-dichlorobenzilate	Non-carc	Non-carc	36
Acetic acid benzyl ester	Non-carc	Non-carc	36
4-Hexylresorcinol	Non-carc	Non-carc	35
Methyl linoleate	Non-carc	Non-carc	52
Fenvalerate	Non-carc	Non-carc	49
Diallyl phthalate	Non-carc	Non-carc	41
Indomethacin	Carc	Carc	40

[a]Predicted with the model VEGA (CAESAR). The experimental values are reported by the same model.

organism, and population levels. The MIE and the KE are two other terms related to AOP. The AOP concept is mechanical. Its goal is to provide a detailed biological and mechanistic picture of a stressor's adverse effect. Visually, one assembles a large puzzle in the hopes of obtaining a comprehensive view of a toxic effect. Nowadays, there is an abundance of big data that is constantly being generated. Big-data encompasses -omics data, HTP data from in vitro measurements of a large set of chemicals, and pathways demonstrating the biological effect on a biochemical level. Various computational techniques for data mining, result clustering, graphical interfaces, and so on have been developed concurrently. Because big data is frequently treated with different algorithms, the calculated results can be interpreted in a variety of ways. Furthermore, there is an abundance of "small data." The "small data" are traditional toxicological data in which a toxic effect is determined directly after organisms are exposed to stressors. It is important to note that these data have been organized into smaller data sets, collected under different laboratory conditions and protocols, and are thus not completely comparable. Typically, data are only available for a limited set of chemicals, so QSAR and similar techniques have been developed and applied to broaden our knowledge to a larger chemical space. A major challenge for current and future generations is to integrate and leverage all knowledge for the development of AOP schemes. An effort must be made to examine all of the data. To condense the entire landscape of information into useful knowledge, specialized knowledge is required (Watanabe-Sailor et al., 2020).

References

Ankley, G.T., Bennett, R.S., Erickson, R.J., Hoff, D.J., Hornung, M.W., Johnson, R.D., et al., 2010. Adverse outcome pathways: a conceptual framework to support ecotoxicology research and risk assessment. Environ. Toxicol. Chem. 29, 730–741.

Basak, S.C., 2013. Mathematical descriptors for the prediction of property, bioactivity, and toxicity of chemicals from their structure: a chemical-cum-biochemical approach. Curr. Comput.-Aided Drug. Des. 9, 449–462.

Basak, S.C., Grunwald, G.D., Host, G.E., Niemi, G.J., Bradbury, S.P., 1998. A comparative study of molecular similarity, statistical, and neural methods for predicting toxic modes of action. Environ. Toxicol. Chem. 17, 1056–1064.

Basak, S.C., Balasubramanian, K., Gute, B.G., Mills, D., Gorczynska, A., Roszak, S., 2003. Prediction of cellular toxicity of halocarbons from computed chemodescriptors: ahierarchical QSAR approach. J. Chem. Inf. Comput. Sci. 43, 1103–1109.

Basak, S.C., Vracko, M., Witzmann, F.A., 2016. Mathematical nanotoxicoproteomics: quantitative characterization of effects of multi-walled carbon nanotubes (MWCNT) and TiO_2 nanobelts (TiO_2 – NB) on protein expression patterns in human intestinal cells. Curr. Comput. Aided Drug. Des 12, 259–264.

Basak, M., Vracko, M.G., 2020. Editorial: parsimony principle and its proper use/ application in computer-assisted drug design and QSAR. Curr. Comput.-Aided Drug. Des 16, 1–5.

Benfenati, E., Benigni, R., DeMarini, D.M., 2009. Predictive models for carcinogenicity and mutagenicity: Frameworks, state-of-the-art, and perspectives. J. Environ. Sci. Health C. 27, 57–90.

Brockmeier, E.K., Hodges, G., Hutchinson, T.H., Butler, E., Hecker, M., Tollefsen, K.E., et al., 2017. The role of omics in the application of adverse outcome pathways for chemical risk assessment. Toxicol. Sci. 158, 252–262.

Clewell, R.A., Leonard, J.A., Nicolas, C.I., Campbell, J.L., Yoon, M., Efremenko, A.Y., et al., 2020. Application of a combined aggregate exposure pathway and adverse outcome pathway (AEP-AOP) approach to inform a cumulative risk assessment: a case study with phthalates. Toxicol. Vitro 66, 104855.

Diamanti-Kandarakis, E., Bourguignon, J., Giudice, L.C., Hauser, R., Prins, G.S., Soto, A. M., Zoeller, et al., 2009. Endocrine-disrupting chemicals: an endocrine society scientific statement. Endocr. Rev. 30 (4), 293–342.

Drgan, V., Župerl, Š., Vračko, M., Como, F., Novič, M., 2016. Robust modelling of acute toxicity towards fathead minnow (*Pimephales promelas*) using counter-propagation neural networks and genetic algorithm. SAR & QSAR Environ. Res. 27, 501–519.

Fjodorova, N., Vračko, M., Novič, M., Roncaglioni, A., Benfenati, E., 2010. New public QSAR model for carcinogenicity. Chem. Cent. J. 4 (Suppl 1), S3.

Floris, M., Manganaro, A., Nicolotti, O., Medda, R., Mangiatordi, G.F., Benfenati, E., 2014. A generalizable definition of chemical similarity for read-across. J. Cheminform 6, 39.

Gundacker, C., Ellinger, I., 2020. The unique applicability of the human placenta to the adverse outcome pathway (AOP) concept: the placenta provides fundamental insights into human organ functions at multiple levels of biological organization. Repro. Toxicol. 96, 273–281.

Halappanavar, S., van den Brule, S., Nymark, P., Gaté, L., Seidel, C., Valentino, S., Zhernovkov, V., et al., 2020. Adverse outcome pathways as a tool for the design of testing strategies to support the safety assessment of emerging advanced materials at the nanoscale. Part. Fibre Toxicol. 17, 16.

Hansch, C., Leo, A., 1979. Substituent Constants for Correlation Analysis in Chemistry and Biology. John Wiley & Sons, New York.

Hansch, C., Maloney, P.P., Fujita, T., Muir, R.M., 1962. Correlation of biological activity of phenoxyacetic acids with Hammett substituent constants and partition coefficients. Nature 194, 178–180.

Jin, Y., Feng, M., Ma, W., Wei, Y., Qi, G., Jiao Luo, J., et al., 2021. A toxicity pathway-oriented approach to develop adverse outcome pathway: AHR activation as a case study. Environ. Pollut. 268, 115733.

JRC QSAR Model database. User Manual, Version 2, European Commission, DG Joint Research Centre, Institute for Health and Consumer Protection, Systems Toxicology Unit. <https://qsardb.jrc.ec.europa.eu/qmrf> (accessed 27.11.20).

Judson, R.S., Magpantay, F.M., Chickarmane, V., Haskell, C., Tania, N., Taylor, J., et al., 2015. Integrated model of chemical perturbations of a biological pathway using 18 in vitro high-throughput screening assays for the estrogen receptor. Toxicol. Sci. 148, 137–154.

Khan, B., Adeleye, A.S., Burgess, R.M., Russo, S.M., Ho, K.T., 2019. Effects of graphene oxide nanomaterial exposures on the marine bivalve, *Crassostrea virginica*. Aquat. Toxicol. 216, 105297.

Khan, B., Ho, K.T., Burgess, R.M., 2020. Application of biomarker tools using bivalve models toward the development of adverse outcome pathways for contaminants of emerging concern. Environ. Toxicol. Chem. 39 (8), 1472–1484.

Kleinstreuer, N.C., Ceger, P., Watt, E.D., Martin, M., Houck, K., Browne, P., et al., 2017. Development and validation of a computational model for androgen receptor activity. Chem. Res. Toxicol. 30, 946–964.

Knapen, D., Stinckens, E., Cavallin, J.E., Ankley, G.T., Holbech, H., Villeneuve, D.L., et al., 2020. Toward an AOP network-based tiered testing strategy for the assessment of thyroid hormone disruption. Environ. Sci. Technol. 54, 8491–8499.

Maki, J.A., Cavallinc, J.E., Lott, K.G., Saari, T.W., Ankley, G.T., Villeneuve, D.L., 2020. A method for CRISPR/Cas9 mutation of genes in fathead minnow (*Pimephales promelas*). Aquat. Toxicol 222, 105464.

Mansouri, K., Abdelaziz, A., Rybacka, A., Roncaglioni, A., Tropsha, A., Varnek, A., et al., 2016. CERAPP: collaborative estrogen receptor activity prediction project. Environ. Health Perspect. 124, 1023–1033.

Mansouri, K., Kleinstreuer, N., Abdelaziz, A.M., Alberga, D., Alves, V.M., Andersson, P., et al., 2020. CoMPARA: collaborative modeling project for androgen receptor activity. Environ. Health Persp 128, 027002-1-17.

Netzeva, T., Pavan, M., Worth, A., 2007. Review of data sources, QSARs and integrated testing strategies for aquatic toxicity. EUR 22943 EN-2007 (accessed 27.11.20).

OECD, 2004. Validation of (Q)SAR Models. <https://www.oecd.org/chemicalsafety/risk-assessment/validationofqsarmodels.htm> (accessed 27.11.20).

OECD, 2012. The adverse outcome pathway for skin sensitisation initiated by covalent binding to proteins, part 1: scientific evidence. Series on testing and assessment. No.168. ENV/JM/MONO(2012)10/PART1.

OECD, 2017. Revised guidance document on developing and assessing adverse outcome pathways series on testing & assessment. No. 184. <https://www.oecd-ilibrary.org/environment/oecd-series-on-adverse-outcome-pathways_2415170x> (accessed 16.12.20).

Passerini, L., 2003. QSARs for individual classes of chemical mutagens and carcinogens. In: Benigni, R. (Ed.), The Quantitative Structure-Activity Relationship (QSARs). CRC Press, Boca Raton, FL, pp. 81–123.

Richard, A.M., Benigni, R., 2002. AI and SAR approaches for predicting chemical carcinogenicity: survey and status report. SAR. QSAR Environ. Res 13, 1–19.

Ruiz, P., Sack, A., Wampole, M., Bobst, S., Vracko, M., 2017. Integration of in silico methods and computational systems biology to explore endocrine-disrupting chemical binding with nuclear hormone receptors. Chemosphere 178, 99–109.

Severson, T.M., Zwart, W., 2017. A review of estrogen receptor/androgen receptor genomics in male breast cancer. Endocrine-Related Cancer 24, R27–R34.

Song, Y., Xie, L., Lee, Y.K., Brede, D.A., Fern Lyne, F., Kassaye, Y., et al., 2020. Integrative assessment of low-dose gamma radiation effects on *Daphnia magna* reproduction: toxicity pathway assembly and AOP development. Sci. Tot. Environ 705, 135912.

Spinu, N., Cronin, M.T.D., Enoch, S.J., Madden, J.C., Worth, A.P., 2020. Quantitative adverse outcome pathway (qAOP) models for toxicity prediction. Arch. Toxicol. 94, 1497–1510.

Vračko, M., Bobst, S., 2015. Prediction of mutagenicity and carcinogenicity using in silico modelling: a case study of polychlorinated biphenyls. SAR. QSAR Environ. Res. 26, 667–682.

Vračko, M., Drgan, V., 2017. Grouping of CoMPARA data with respect to compounds from carcinogenic potency database. SAR. QSAR Environ. Res 28, 801–813.

Vračko, M., Basak, S.C., Geiss, K., Witzmann, F., 2006. Proteomic maps-toxicity relationship of halocarbons studied with similarity index and genetic algorithm. J. Chem. Inf. Model 46, 130–136.

Vračko, M., Basak, S.C., Witzmann, F., 2018. Chemometrical analysis of proteomics data obtained from three cell types treated with multi-walled carbon nanotubes and TiO_2 nanobelts. SAR. QSAR Environ. Res 29, 567–577.

Vračko, M., Bobst, S., 2014. Performance evaluation of CAESAR-QSAR output using PAHs as a case study. J. Chemom. 28, 100–107.

Wang, R.-L., 2020. Semantic characterization of adverse outcome pathways. Aquat. Toxicol. 222, 105478.

Wang, Y., Na, G., Zong, H., Ma, X., Yang, X., Mu, J., et al., 2018. Applying adverse outcome pathways and species sensitivity–weighted distribution to predicted-no-effect concentration derivation and quantitative ecological risk assessment for bisphenol a and 4-nonylphenol in aquatic environments: a case study on Tianjin City, China. Environ. Toxicol. Chem. 37 (2), 551–562.

Wang, L., Zhao, Y., Shi, Z., Li, Z., Liang, X., 2020. Ecotoxicoproteomic assessment of microplastics and plastic additives in aquatic organisms: a review. Comp. Biochem. Physiol.—Part. D. 36, 100713.

Watanabe-Sailor, K.H., Aladjov, H., Bell, S.M., Burgoon, L., Cheng, W.-Y., Conolly, R., et al., 2020. Big data integration and interface. In: Neagu, D., Richarz, A.-N. (Eds.), Big Data in Predictive Toxicology. Royal Society of Chemistry, Chambridge, pp. 264–306.
Woo, Y.-T., Ali, D.Y., Y., D., 2005. OncoLogic: a mechanism-based expert system for predicting the carcinogenic potential of chemicals. In: Helma, C. (Ed.), Predictive Toxicology. CRC Press, Boca Raton, FL, pp. 385–413.

Big data and deep learning: extracting and revising chemical knowledge from data

5

Giuseppina Gini[1], *Chiakang Hung*[1] *and Emilio Benfenati*[2]
[1]Politecnico di Milano, DEIB, Piazza Leonardo da Vinci, Milano, Italy, [2]Laboratory of Environmental Chemistry and Toxicology, Istituto di Ricerche Farmacologiche Mario Negri IRCCS, Milano, Italy

5.1 Introduction

Computational toxicology is a large spectrum of methods aimed at assessing the toxicity/safety of chemical substances using computers and models instead of wet experiments. Computational toxicology is the result of two research lines: mathematical chemistry and the 3Rs (Replace, Reduce, Refine) principle, defined in the fifties of the last century, and devoted to finding ways to reduce animal use in assessing chemicals.

Mathematical chemistry (Basak, 2013) and chemoinformatics, started with the definition of molecular structures and mainly worked on defining numerical chemical descriptors, that is, human-engineered features representing various characteristics such as geometry, topology, and energy distribution of molecules. In a few decades of research, thousands of molecular descriptors have been defined, usually in relation to the rapid development of predictive methods for toxicology and new molecule design (Todeschini and Consonni, 2009).

Chemical properties can be studied using physical models and simulation. Those methods are common for the design of new molecules, but are not affordable when the properties of interest are biological effects; in fact, knowledge about the living systems (for instance, the receptor and the mechanism used to cause the effect) is seldom available. For modeling biological effects, (quantitative) structure–activity relationship ((Q)SAR) models are commonly used. QSARs are based on the postulate that similar molecules exhibit similar physical and biological activities (Johnson and Maggiora, 1990), and are the main nonphysical predictive models in use (Gini, 2016).

QSAR methods built on molecular descriptors appeared around the middle of the last century. Those models were initially simple regressions, using very few, and possibly simple, chemical descriptors. Choosing informative descriptors for the task at hand is a key aspect and requires deep insights into chemical and biological properties, as interactions between molecules, reactions and enzymes involved, and metabolic degradation of the molecules should be somehow represented.

Big Data Analytics in Chemoinformatics and Bioinformatics. DOI: https://doi.org/10.1016/B978-0-323-85713-0.00030-X

Usually building QSARs starts with computing a large number of chemical descriptors, then applying a method to reduce them to a few, and adopting a computational technique to build the model that predicts new assays.

SAR is instead based on the idea that specific functional groups, often defined by experts and called structural alerts (SAs), are responsible for molecular behavior, so their presence is checked to predict the activity. SAR and QSAR are usually mixed, as many QSAR methods use fingerprints, and often the explanation of QSAR models considers the functional subgroups too.

For regulatory purposes, toxicologists use expert-intensive methods like read-across, which are accepted, and often integrated into final decisions by panels of experts (Benfenati et al., 2019). In read-across, the target molecule is compared with a couple (or a few) of similar molecules for which the test has been performed, and the property value is predicted. The problem with this method is human bias; different experts select different similar molecules, or look for different substructures as responsible for the toxicity, or give a different weight to properties (Benfenati et al., 2016).

Human bias is reduced when using QSAR models that learn from a larger population of molecules. However, two factors make QSARs not widely accepted: the lack of confidence in statistical methods, and the shortage of data available that makes it problematic to cover the entire chemical space. Toxicity data are available for just a very small fraction of the entire chemical space of the molecules of biological interest. It has been estimated that this chemical space exceeds 10^{60} molecules (Kirkpatrick and Ellis, 2004), while the toxicological characterization of chemicals is available for a few thousand molecules.

Laboratory tests are expensive, time-consuming, and in some cases forbidden by regulations. This situation pushes for the adoption of computational methods that make use of available data and knowledge and can integrate results from living organisms and cellular lines. Adopting (Q)SAR becomes a necessity, and making QSARs effective is more and more pursued.

Machine learning (ML) algorithms are being adopted to replace the statistical methods, as ML can be fully data-driven and does not need to make a hypothesis about the mathematical function that at best explains the data. Connectionist models, based on distributed representation and parallel computation of the process, as neural networks (NNs), received attention and criticism in QSAR.

The situation is changing as large quantities of in vitro test values are made available, and they challenge traditional modeling methods. In the last decade very big NNs, called deep neural networks (DNNs) provided the top models in the Merck Kaggle challenge in 2012 and in the Tox21 challenge in 2014 for activity prediction. Using DNNs to solve some of the open issues in computational toxicology is a very active field.

There are still major challenges to obtain toxicology models that can replace animal experiments as follows:

- One challenge is to improve the process of constructing QSAR models. Big data are more often available, and the present way to build QSAR models takes too much time and human expertise. Instead of computing thousands of numerical descriptors and selecting a few ones as features, a simplified process would be to use directly as input the molecular structure and let the algorithm to automatically extract the features.

- Another challenge is to avoid, or to reduce, the bias of the expert that characterizes the SAR approach; usually the functional subgroups responsible for the activity are proposed by experts on the basis of the study of some mechanism of action. However, they are not a complete list and are not universally applicable. For instance, Toxtree (Benigni and Bossa, 2008) uses a few tens of functional subgroups as mutagenicity alerts, and indicates the percentage of their presence in toxic molecules; in some cases, the presence is less than 50%. The role of SAs in fact is to give the expert a reason for toxicity in case this appears, but other mechanisms at the organ or cellular level can make such alerts transformed into other chemicals before they act.
- The last challenge is to provide an explanation of the result obtained by the model (Gini, 2018). Considering that toxicology models are used in many processes and regulations, some kind of explanation to ground the result obtained into general expert knowledge is important. The common way of interpreting QSAR problems by a priori choosing simple and interpretable descriptors can work for explanatory QSARs, but not for predictive QSARS.

Other open problems to take into account are data availability and data quality. Open and public databases are often available. However, a lot of data are proprietary, often for the same endpoint results using different standards can be found, and different databases may contain different values for the same experimental test. This leads to extra efforts to create a reliable data set. In the following, the data set for predicting Ames test results is taken from the literature; it includes a large set of proprietary data provided by the Japan Ministry of Industry for the first Ames International challenge (Honma et al., 2019).

Section 5.2 presents the basic definitions of the DL architectures, which are special organizations of multilayers neural networks, such as recurrent neural network (RNN), convolutional neural network (CNN), and graph convolutional neural network (GCN) (Wu et al., 2019) used to generalize CNN to handle graph data. How to use DNNs on chemical input is shortly reviewed in Section 5.3. Section 5.4 discusses a case study on mutagenicity models developed in the three before mentioned architectures. Section 5.5 shows how DL methods can be interpreted. The discussion continues in Section 5.6, where the results obtained are put into a perspective.

5.2 Basic methods in neural networks and deep learning

5.2.1 Neural networks

NNs are biologically inspired programming paradigms equivalent to a mathematical function that maps a given input to the desired output. A NN is made up of processing nodes, called neurons, arranged in layers; each neuron receives inputs and converts them to a single output, which is sent to the neurons of the next layer. The number of hidden layers determines the complexity of the function that the network can approximate; nets with one hidden layer can approximate any nonlinear function (Chen and Chen, 1995). Weights on the connections are learned to adjust the

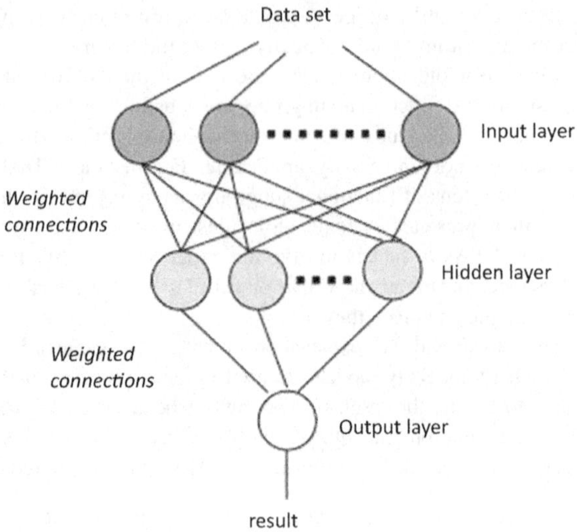

Figure 5.1 A feedforward neural network. Neurons in each layer are fully connected through weighted connections to neurons in the next layer.

output of the network iteratively during training. In the basic feedforward architecture (Fig. 5.1) each neuron is connected to all the neurons of the next layer; the connections go in one direction, from input toward output neurons.

- The top layer, *input layer*, contains neurons that just pass the information to the next layer.
- The bottom layer, *output layer*, contains the output neurons, which give the result of the network computation.
- The middle layer, *hidden layer*, contains neurons that make intermediate processing, and then transfer the weights to the following layer. Usually, a bias neuron is added to each hidden layer.

The output of a node is the result of applying the *activation function* $h(x)$ to its input. Activation functions should be continuous and continuously differentiable. Only nonlinear activation functions allow NNs to compute nontrivial functions.

Currently, the most used activation function is rectified linear unit (ReLU), as in Eq. (5.1)

$$\text{Re}LU(x) = \max(0, x) \tag{5.1}$$

ReLU is not continuously differentiable. Its derivative is 0 for $x < 0$, undefined for $x = 0$, and 1 for $x > 0$.

Leaky ReLU, Eq. (5.2), adds a small slope to the negative values to have a nonzero gradient.

$$\text{Leaky Re}LU\ (x) = \text{if } x > 0 \text{ then } x, \text{ else } 0.001\ (x) \tag{5.2}$$

5.2.2 Neural network learning

NNs are mainly supervised methods and are trained using labeled data to approximate the wanted function. During training, weights and thresholds are changed so to reduce the error on the training data.

NNs are usually trained using backpropagation of the gradient descent of the loss function with respect to the weights (Werbos, 1994). Gradient descent is an iterative optimization method that tries to minimize the objective function (error function) by changing the rate of inclination of a slope. It is applied many times using the training data, with the aim of finding the function that fits at best the data. Gradient descent with backpropagation is not guaranteed to find the global minimum of the error function, but only a local minimum.

The weights updates are done by Eq. (5.3)

$$W_{ij}(t+1) = W_{ij}(t) + \eta\left(\frac{dC}{dw_{ij}} + x(t)\right) \tag{5.3}$$

where wij is the weight of the connection i to j, η is the learning rate, C is the loss function, $x\ (t)$ is a stochastic term.

After a random initialization of the weights, training data are passed through the net and the output is computed. The error gradient is then used in a backward direction to change the weights.

Different loss functions can be used, such as mean square error (MSE) for classifiers and cross-entropy for models whose output is a probability value [Eqs. (5.4) and (5.5)].

$$\text{MSE} = \frac{1}{2 * n}\sum_{x} ||y(x) - a||^2 \tag{5.4}$$

where w is the collection of weights of the network, n is the total number of training inputs, and a is the vector of outputs from the network when x is input.

$$\text{Cross} - \text{entropy} = -\ 1n\sum_{x}(y \ln a + (1-y)\ln(1-a)) \tag{5.5}$$

where: a is the output of the neuron, y is the true label, and ln is the natural log.

For training different settings of the hyperparameters are usually tried in order to find the best net.

One epoch is when the entire data set is passed once forward and backward through the net. A too-small number of epochs makes the net to underfit the data, a too-big value makes it to overfit. Big data sets are divided into batches as follows:

- Batch size: it is the number of data present in a batch.
- Iterations: it is the number of batches needed to complete one epoch.
- Learning rate: determines the step size at every iteration. It is denoted by the symbol η, and has a value in (0, 1). η controls how much to change the model in response to the

estimated error each time the model weights are updated as in Eq. (5.3). Smaller η require more training epochs.

Different regularization methods are used to avoid overfitting. For big networks and large data sets, the most used is dropout during training. It means that nodes of the networks, with a predefined probability, are randomly deleted (in practice set to zero) so to avoid that the net learns too much of the data and eventually becomes unable to generalize. Dropout has also a theoretical interpretation in terms of uncertainty estimation. Gal and Ghahramani (2016) presented some theoretical justification, showing that a NN with arbitrary depth and nonlinearity, with dropout applied before every weight layer, is mathematically equivalent to an approximation of the probabilistic deep Gaussian process. They showed that model uncertainty can be obtained from NN models using dropout at inference time and not just during training and that the dropout probability characterizes the posterior uncertainty. Of course, this is just an approximation of Bayesian reasoning. Optimization algorithms, as Adam (Kingma and Ba, 2017), help to minimize the loss function.

5.2.3 Deep learning and multilayer neural networks

Big networks with many inner layers can approximate more complex functions. However using feedforward fully connected neural nets with many layers is not efficient, for many reasons. First of all, determining the number of neurons in each layer is usually done using trial and errors, and the process is very long. The effectiveness of DNNs depends on taking advantage of the structure of the data to find the net architecture. Deep learning (DL) is a field in machine learning based on NN that work as a representation learning (Bengio et al., 2013). The most known of such networks are shortly illustrated.

5.2.3.1 Convolutional neural network

A CNN is a network that breaks down an input, typically an image, into smaller pieces and extracts the feature to be used to make a classification decision (LeCun and Bengio, 1995).

CNNs combine convolutional layers, pooling layers, fully connected layers, and activation layers (Fig. 5.2). CNNs are effective in using the geometric information contained in the input.

- Convolutional layers use a set of fixed-sized weight matrices, called filters, which perform element-wise multiplication on the image pixels. Weights are learned. Convolutional layers use strides (positive numbers) to accelerate the dimension reduction; the default stride 1 moves one pixel at a time, and higher strides move more pixels.
- Pooling layers usually come after or in between convolutional layers to reduce the dimension of the original input. Average pooling smoothens the features taking the average, while max pooling picks the largest value to extract distinctive features.
- Fully connected layers flatten the previous layer and connect to all nodes of the previous layer to each of its nodes. Then the output is fed into a nonlinear layer.

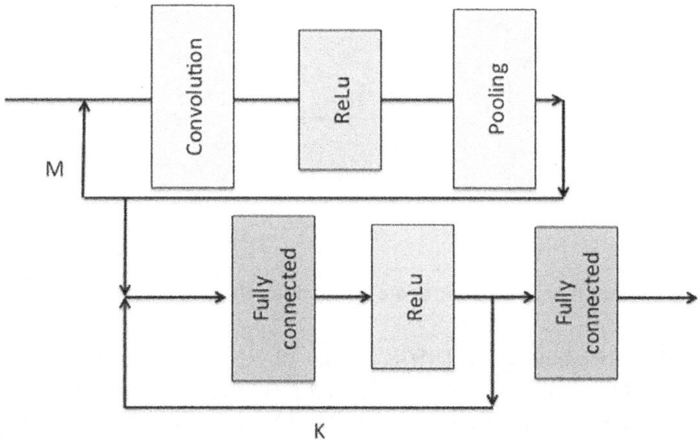

Figure 5.2 The convolutional neural network. *M* and *K* are the number of times each block is performed.

CNNs are better than humans in image classification (LeCun et al., 2015); a possible explanation is they make use of the two main bio-inspired principles: population coding and statistical coding. Population coding is the mechanism used in the brain to represent sensory information through a group of spatially organized neurons, in which neighboring neurons have similar activity. Statistical coding, a special kind of population coding, is observed in the areas of the brain devoted to the preprocessing of the sensory data, by reducing the dimensionality of the input space.

5.2.3.2 Recurrent neural network

Feedforward networks can receive a fixed-sized number of data to the hidden layers. The inability to handle variable length input, as in the case of texts, and the necessity to consider long-term dependencies give rise to using recurrent units with feed-forward NN.

RNNs (Williams et al., 1986), in Fig. 5.3, use the same function and the same set of parameters for every time step: at each time step, the previous hidden state and the current input are fed through the function to update the hidden state. The loss is defined as the sum of the loss from each time step. RNNs are effective in using the temporal interconnections present in the data.

Backpropagation through time is used for training. During backpropagation, gradients in RNN tend to vanish or explode as time length increases. Exploding gradient means a big increase in the norm of the gradient because of long-term components growing exponentially larger than the short-term ones. The vanishing gradient is the opposite. Different techniques are available (Bengio et al., 2013) to solve them. In particular, using ReLU as the activation function prevents the derivative from shrinking the gradients when x is larger than zero.

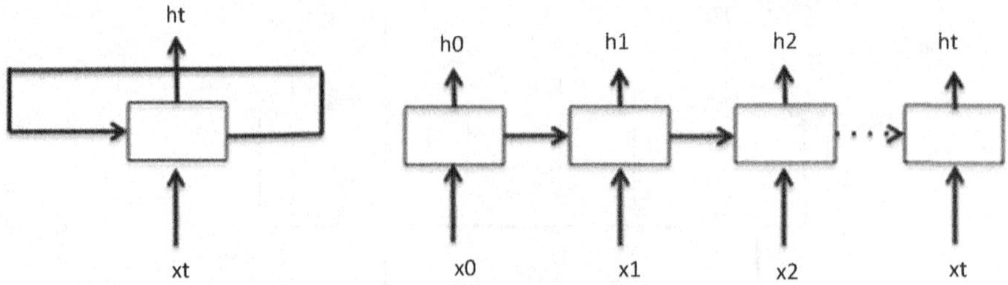

Figure 5.3 Recurrent neural network as a cycle and unfolded at each time step.

Long short-term memory (LSTM) is a kind of RNN developed by Hochreiter and Schmidhuber (1997). It has a more complicated recurrent unit, containing an input gate, a forget gate, and an output gate to control information that gets passed on and stored.

In bidirectional RNN two directions of the next time of the hidden states are used; one goes from old to new and the other goes from new to old. In this way, the learning process can both maintain the past and look to future information.

5.2.3.3 Graph convolutional neural networks

A graph is a couple $G = <V, E>$, where V is a finite set of nodes and E, the set of edges, is a subset of V^2. A fully connected feedforward NN can represent graphs as sets of nodes. However, this representation does not consider the information of the edges, which connect a few of the nodes in a nonregular way.

Extracting features from graphs using CNNs is not easy, as the graph nodes do not have the regularity of pixels. However, the advantages of CNNs, that is, local connection, shared weights and multilayer refinement can be useful for graphs too. A GCN (Zhou et al., 2019) encodes the nodes of the graph into an embedding space that approximates similarity in the original network. Mathematically a GCN applies convolutions; while CNNs apply two-dimensional convolutions to the image pixels, GCNs apply convolutions to node features. It means a GCN passes a filter over the graph, looking for essential vertices and edges. Graph filtering refines the node feature, graph and pooling generates a smaller graph.

Images are represented as matrices, the products of pixels and weights. Graphs too are represented as matrices, the product of adjacency matrix and weights. A GCN in general considers an adjacency matrix concatenated with a matrix of node features, and instead of a window to select the neighboring pixels it uses an aggregation function. The matrix of the node features contains node information for features such as atom weights, etc. Neighborhood aggregation function aggregates the node features using mean, concatenate, or sum functions.

A sequence of those filter layers produces an effect similar to CNN (Kipf and Welling, 2017). The pooling layer converts a graph into subgraphs to represent higher-level representations. It is computationally expensive to use all hidden states from all the nodes for prediction, hence the need of downsampling. The pooling operation, as max, mean, or sum, can be combined with attention mechanisms.

Readout generates graph-level representation by summing up all the hidden states in the graphs, and a fully connected layer is used to generate the output. Fig. 5.4 shows the similarity between GCN and CNN.

GCNs generalize convolutions in the graph domain by operating on groups of spatially close node neighbors. One of the challenges of this approach is to define an operator which works with different-sized neighborhoods and maintains the weight-sharing property of CNNs. Hamilton et al. (2017) introduced GraphSAGE, a method for computing node representations by sampling a fixed-size neighborhood of each node, and then by aggregating them, using for instance the mean over all the sampled neighbors' feature vectors.

5.2.4 Attention mechanism

The attention mechanism, observed in animals' visual cortex, means focusing on a specific part of the sensorial data to help interpret the scene. The same process is used while reading a sentence: first we focus on the subject and the verb, then we move to articles and adjectives. In the same way, a NN can define the important parts of the input it receives in order to make a correct prediction. Cho et al. (2015) first proposed an RNN with the addition of an attention mechanism in order to reduce the computation resources needed to achieve good accuracy. The idea is to use a layer connected to the RNN that receives the context vector and calculates the weight this has on the final prediction. Then softmax transforms it in a probability.

The attention mechanisms can deal with variable-sized inputs, focusing on the most relevant parts of the input to make decisions. The attention mechanism used to compute a representation of a single sequence, named self-attention, has been useful in sentence representation and translation.

The attention mechanisms can deal with variable-sized inputs, focusing on the most relevant parts of the input to make decisions. The attention mechanism used to compute a representation of a single sequence (named self-attention) has been useful in sentence representation and translation.

Velickovic et al. (2017) introduced the graph attention network (GAT), which is an attention-based architecture to perform node classification of graph-structured data. The idea is to compute the hidden representations of each node in the graph, by attending over its neighbors, following a self-attention strategy.

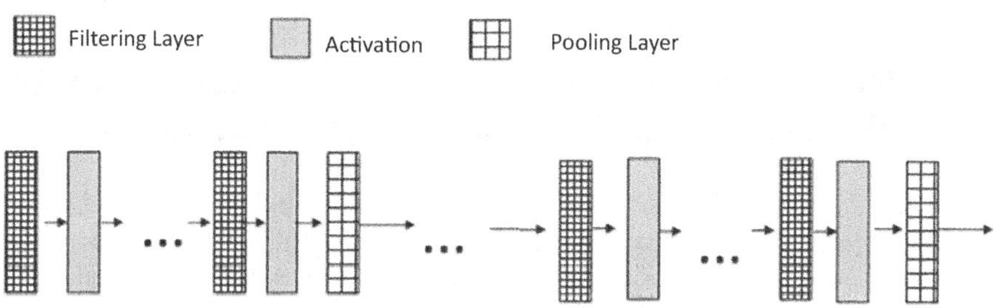

Figure 5.4 The structure of a graph convolutional neural network.

A single graph attentional layer is used in the GAT architecture. The input to the attentional layer is a set of node features; the output is a new set of node features (of potentially different cardinality). The attention mechanism is a single-layer feedforward NN. In order to obtain sufficient expressive power to transform the input features into higher-level features, a learnable linear transformation is required. To that end, Leaky ReLU, parameterized by a weight matrix, is applied to every node at the beginning. Then the e_{ij} coefficients that indicate the importance of node j's features to node i, are computed for a limited number of neighboring j nodes. To make coefficients comparable across the various nodes, the e_{ij} are normalized using the softmax function, so producing the new coefficients $a_{ij} = \text{softmax}(e_{ij})$

Softmax takes as input a vector **z** of K real numbers and normalizes it into a probability distribution consisting of K probabilities proportional to the exponentials of the input numbers. It turns the vector of K real values zi into a vector of K real values that sum up to 1, thus constituting a probability distribution [Eq. (5.6)].

$$s(z)i = \frac{e^{zi}}{\sum_{j=1}^{K} e^{zj}} \tag{5.6}$$

The normalized attention weights are used as the coefficients of the hidden states in the update function.

To stabilize this learning process, multihead attention is used. It means that the eij are calculated K (for $K \geq 1$) times and then the features are concatenated or averaged.

The benefits of GAT architecture are many; it is computationally efficient in storage and number of operations [as it is linear in (vertices + edges)], is not dependent on graph size, it is localized, it is inductive because it does not depend on global graph structure.

Many DNNs end in a penultimate layer that outputs real numbers. Softmax converts those numbers to a normalized probability distribution, which can be more easily interpreted. For this reason, it is usual to append a softmax function as the final layer of the NN, even in case the attention mechanism is not used.

5.3 Neural networks for quantitative structure–activity relationship: input, output, and parameters

In QSAR modeling NNs have been often used as algorithms to create models (Devillers, 1996; Gini and Katrizky, 1999). In classical QSAR, this means assigning to the input neurons the descriptors values. As chemical descriptors are produced in large numbers, usually a selection method is used to select the relevant ones to construct the model (Fig. 5.5).

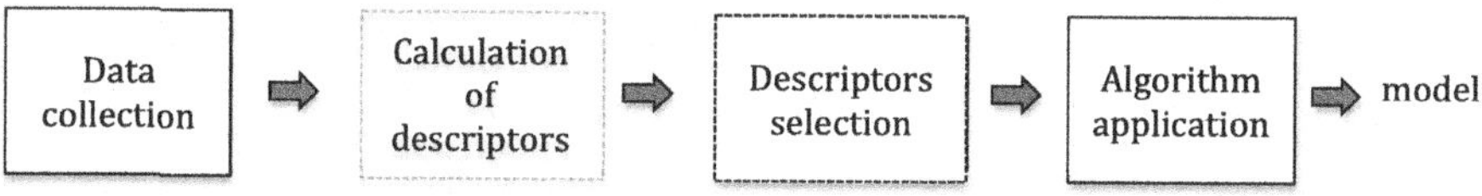

Figure 5.5 The steps in creating a quantitative structure–activity relationship model using machine learning methods.

5.3.1 Input

NNs are used with numerical values as input. While chemical descriptors are numbers, whole molecules are structured data. Developing NNs for graph input has been for a long time a research target. Early examples of graph input to NN for chemical problems are in (Micheli et al., 2001). Those early methods did not have a large impact for two reasons: the network computation was heavy, and the input was ad hoc format.

Today DNNs allow avoiding chemical descriptors as well as crafting ad hoc representations. Three easily available formats for molecules can be used: SMILES (Weininger et al., 1989), images, and chemical graphs. DNNs that learn directly useful features from data have been successfully developed for chemical applications, especially in drug discovery (Zhang et al., 2017).

The ambitious goal of DNN-based QSAR is designing a new architecture, which depends only on raw data and no other a priori or expert knowledge. The knowledge self-generated during training can be extracted in order to compare it with the existing one or to analyze it as new knowledge.

5.3.2 Chemical graphs and their representation

In mathematical chemistry a chemical graph, as in Fig. 5.6, represents the structural formula of a chemical compound. A chemical graph is an undirected labeled graph, whose vertices correspond to the atoms and edges correspond to the chemical bonds. Vertices are labeled with the kinds of corresponding atoms and edges are labeled with the types of bonds. Visiting the chemical graph in top-down order produces a string, called SMILES, that is equivalent to the chemical graph (Weininger et al., 1989). The reverse is not true, as more SMILES correctly represent the same chemical graph.

Another problem is that a small change in the structure can produce a large change in the SMILES, as in Fig. 5.7. In this case the similarity of the chemical graphs seems lost (Gini, 2020).

5.3.2.1 SMILES as input

NNs able to receive directly the information from public repositories are an important tool, and SMILES are commonly available in chemical data sets. RNN are apt to work on strings. In order to avoid ambiguity, a bidirectional cell in the RNN allows having both the previous and the future input, in order to easily distinguish

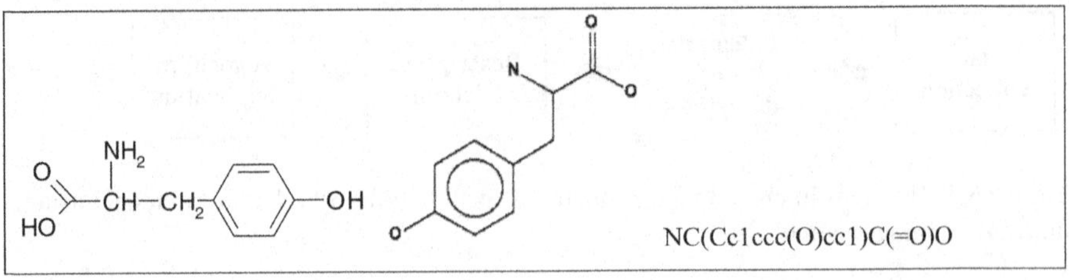

Figure 5.6 From *left* to *right*: a chemical structure, its H-depleted chemical graph, where unmarked vertices represent Carbon atoms, and its SMILES.

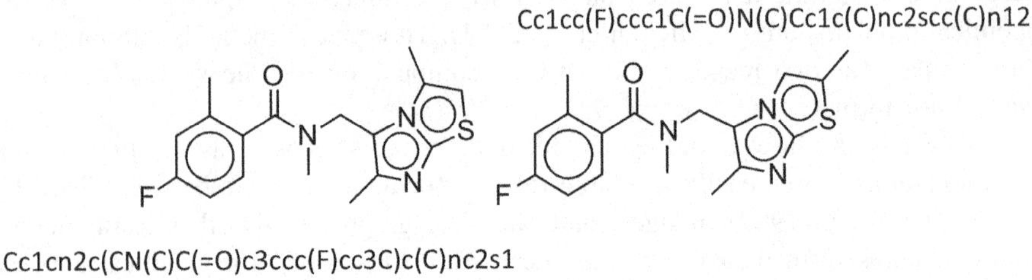

Figure 5.7 Two similar molecules and their different SMILES.

the differences. Strings in the input are transformed into numbers (word embedding), usually by a NN, using a technique called Word2Vec. Examples of DNNs with SMILES as input are in Goh et al. (2018) and Gini et al. (2019).

5.3.2.2 Images of two-dimensional structures as input

From a data set composed of SMILES strings, it is easy to generate a data set of images using the public library RDKit. Two examples are in Fig. 5.8. It is obvious that a low resolution (for instance, 80×80 pixels) is enough for the first molecule, critical for the second one. As the image generated by RDKit uses black for the bonds, blue and red for atoms different from Carbon (not represented), the color images in RGB at low resolution require $80 \times 80 \times 3$ pixels, as reported in (Gini et al., 2019; Gini and Zanoli, 2020; Goh et al. (2017)).

5.3.2.3 Chemical graphs as input

Graphs and SMILES contain the same information. However, graphs are not affected by the lack of unicity of SMILES. This is not a problem when training, as it is easy to use the same algorithm to generate canonical SMILES, but it can be a problem when using the trained net. Moreover, using chemical graphs presents the advantage of reasoning directly on topological similarity. GCNs (Wu et al., 2019) can accept chemical graphs represented as (largely empty) adjacency matrices,

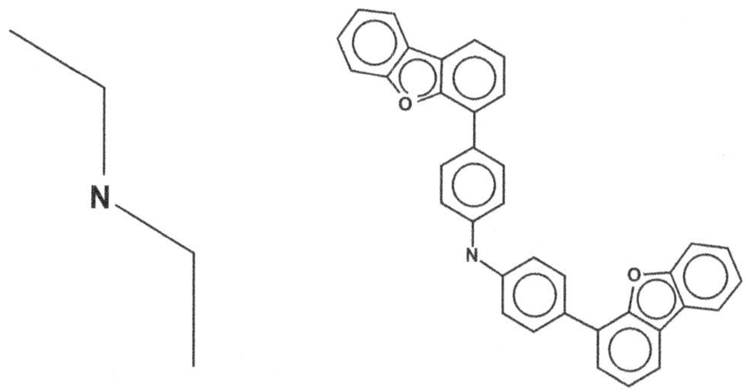

Figure 5.8 Two examples of drawings generated by RDKit.

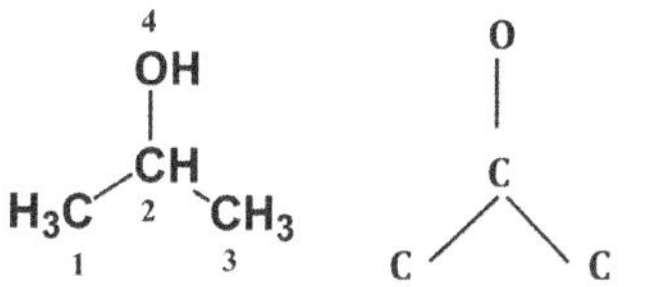

1	1	0	0
1	1	1	1
0	1	1	0
0	1	0	1

6	3	4	0
6	1	4	0
6	3	4	0
8	1	2	0

Figure 5.9 A chemical graph, adjacency matrix, and feature matrix (the columns indicate atom type, #H, valence, aromaticity).

where the rows and the columns are the node labels and the edge are represented by entry 1 if two nodes share an edge. Each node (atom) can have a set of features, as a number of hydrogen bonds, valence, aromaticity, and so on. The feature values are inserted in a feature matrix. In GCN the adjacency matrix of the molecule and the feature matrix (Fig. 5.9) are concatenated to create the input.

GCN performs on the matrices the same operations that CNN performs on the images; instead of extracting edges and shapes, it extracts subgraphs.

5.3.3 Output

The output of the networks can be real numbers or class labels. Even in case a class label is desired, this label can be obtained from a regression net that predicts numbers in (0, 1). The advantage is that it is possible to set different thresholds to get the classes, and the output of the model can be interpreted as the probability for the compound to belong to a specific class.

Two loss functions during the training process are needed: MSE if the output is a class, and cross-entropy if the output is a real number.

5.3.4 Performance parameters

As for any model the subdivision of training and test sets, and the use of *n*-fold cross-validation are common methods to assess the prediction ability. The performance of the obtained classifier is measured on the test and training sets, or in cross-validation, using a bunch of the following parameters. Accuracy [Eq. (5.7)] alone is not a reliable metric because it will yield misleading results if the data set is unbalanced.

$$\text{accuracy} = \frac{\text{TruePositives} + \text{TrueNegatives}}{\text{TruePositives} + \text{TrueNegatives} + \text{FalseNegatives} + \text{FalsePositives}} = \frac{\text{TP} + \text{TN}}{\text{TP} + \text{TN} + \text{FN} + \text{FP}}$$

Other parameters, in Eq. (5.8), are sensitivity (which is the proportion of real positive cases that are correctly predicted), specificity, precision (or positive predictive value), and *F1*, which is the harmonic mean of precision and sensitivity.

$$\text{sensitivity} = \text{recall} = \frac{\text{TP}}{\text{TP} + \text{FN}}$$

$$\text{specificity} = \frac{\text{TN}}{\text{TN} + \text{FP}}$$

$$\text{precision} = \frac{\text{TP}}{\text{TP} + \text{FP}}$$

$$F1 = \frac{2 * \text{TP}}{2 * \text{TP} + \text{FP} + \text{FN}} \tag{5.8}$$

The Matthews correlation coefficient (MCC), in Eq. (5.9), is a well-balanced measure; it is a correlation coefficient between the observed and predicted binary classifications, and it returns a value between -1 and $+1$. A coefficient of $+1$ represents a perfect prediction, 0 means no better than random prediction and -1 indicates total disagreement between prediction and observation.

$$\text{MCC} = \frac{\text{TP} * \text{TN} - \text{FP} * \text{FN}}{\sqrt{(\text{TN} + \text{FP})(\text{TN} + \text{FN})(\text{TP} + \text{FP})(\text{TP} + \text{FN})}} \tag{5.9}$$

5.4 Deep learning models for mutagenicity prediction

Mutagenicity refers to a chemical or physical agent's capacity to cause mutations, that is, genetic alterations. The most adopted in vitro test for mutagenicity is the Ames test (Ames, 1984) on different engineered strains of the bacteria *Salmonella typhimurium*; it is largely used for the registration of chemicals, and to screen impurities in pharmaceuticals and metabolites in pesticides.

5.4.1 Structure–activity relationship and quantitative structure–activity relationship models for Ames test

Many SAs have been defined for mutagenicity both by experts and by data-mining tools.

Toxtree is the open implementation of the Benigni and Bossa (2008) rules for mutagenicity and carcinogenicity. Toxtree processes a query chemical and answers whether SAs for carcinogenicity are found or not. SAs for carcinogenicity are used, as the action mechanism of genotoxic carcinogenicity apply also to the mutagenic activity in bacteria. SAs can only point out chemicals potentially toxic, whereas no conclusions about nontoxic chemicals are possible. Thus the SAs are not a discriminant model, differently from QSAR models. The majority of SAs are direct-acting carcinogens, while others relate to genotoxic carcinogens that become toxic after metabolic transformations (Plošnik et al., 2016). The situation is more complicated, as other factors may influence the activity of a chemical, as molecular weight, reactivity, solubility, physical state, and the molecule's geometry. Moreover, the SAs-based models do not work equally efficiently in discriminating between active and inactive chemicals in the same chemical class; the reason is that the SA models lack subrules detailed enough to describe these modulations. Indeed o reason on difficult chemical classes, in a read-across approach, a few hundred SAs have been created both by expert judgments and statistical tools (Gini et al., 2014). SARpy is such a statistical tool that automatically extracts substructures from data sets (Gini et al., 2013). The substructures are accepted as positive SAs after statistical analysis, and substructures never found in toxic chemicals are named negative SAs.

More than 60 QSAR models, either free or commercial, are available to predict the output of the Ames test. Most of them are built using a data set of about 6000 molecules of chemicals of industrial interest, made freely available by Kazius et al. (2005) and Hansen et al. (2009).

For assessing some of those models, a comparison has been done using a new large set provided by Masamitsu Honma, from the Division of Genetics and Mutagenesis in the National Institutes of Health Sciences (NIHS) of Japan, who distributed to participating institutes about 12,000 newly tested substances, most of them negatives, for the first AMES/QSAR International challenge (Honma et al., 2019). This comparison (Benfenati et al., 2018) considered a few Ames prediction models, constructed on the Kazius and Hansen data sets, and used as a test set the new NIHS data. Table 5.1 indicates some of the considered models, the algorithm used, and the number of predicted classes.

The results of those seven models on a test set of 2427 molecules of the NIHS data, which were not part of the training set of any of the considered models, are in Table 5.2.

The coverage for two of the models is not full, as they do not take decisions for dubious molecules. Those performances can characterize the state of the art of predictive models for the Ames test. The MCC value is not higher than 0.31.

Table 5.1 Seven mutagenicity models tested on an a large test set in Benfenati et al. (2018).

Model name and provider	Method	Predicted classes
CAESAR (IRFMN-in VEGA)	Chemical descriptors for SVM + SA search	Pos, Neg, Suspect Pos
SARpy (IRFMN-in VEGA)	SA search	Pos, Neg, Possible Neg
ISS (IRFMN-in VEGA)	SA search	Pos, Neg, Unknown
KNN (IRFMN-in VEGA)	k-NN	Pos, Neg
SAm (Nestlè)	SA search	Pos, Neg
AIm (Nestlè)	k-NN	Pos, Neg
AZAMES (Swetox)	Chemical descriptors - conformal prediction	Pos, Neg, Both, Unknown

Table 5.2 Results of seven of the tested models on a test set of 2427 molecules.

Model/ performance	CAESAR	ISS	SARpy	KNN	SAm	Aim	AZAMES
Total predictions	2427	2427	2427	2427	2427	*2399*	*2415*
Accuracy	0.67	0.70	0.66	0.67	0.79	0.76	**0.85**
Sensitivity	**0.65**	0.59	0.57	0.51	0.56	0.40	0.37
Specificity	0.68	0.72	0.68	0.70	0.82	0.81	**0.92**
MCC	0.23	0.22	0.18	0.15	**0.31**	0.17	0.29

Note: In bold the best value for each parameter.

5.4.2 Deep learning models for Ames test

Three kinds of DL models for the Ames test are presented, each using as input SMILES, images of the two-dimensional structures, or chemical graphs.

The data sets used for all of them is a large data set, which includes public and proprietary data (NIHS data), plus data collected through PubMed, and whose statistical details are in Zanoli (2018). It is quite unbalanced, as reported in Table 5.3.

5.4.2.1 Learning from SMILES

The basic RNN structure cannot remember input values for a long gap of time/words; LSTM units make RNN capable of remembering values over arbitrary time. SmilesNet (Gini, 2018; Zanoli, 2018) is a bidirectional RNN-LSTM model that analyzes SMILES. It was trained with the data set of 24,003 molecules, randomly divided into training (80%) and test (20%) sets.

Table 5.3 Characteristics of Ames-Zanoli dataset.

	Number	Percentage
Positive	8127	33.85%
Negative	15,876	66.14%
Total	24,003	100%

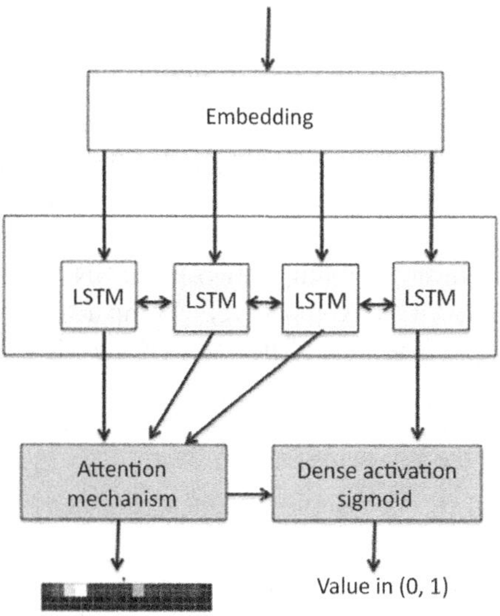

Figure 5.10 SmilesNet architecture.

In SmilesNet (Fig. 5.10) the input SMILES go through word embedding, then into a bidirectional RNN that gives inputs to the attention mechanism and to the final dense layer for prediction. The attention mechanism highlights the parts of the string that are important for reaching the result.

The dense layer uses the softmax function to produce a value in (0, 1). Setting a threshold at 0.5 (or another value) transforms the result into a binary classification.

The best net, optimized using Adam, has the following parameters: 100 epochs, 400 neurons, batch size 32, learning rate 0.001, dropout probability 0.2. On the test set of 20% chemicals, the performances of SmilesNet are in Table 5.4, which shows also the results of the two models presented next.

5.4.2.2 Learning from images

Toxception (Gini and Zanoli, 2020) is a CNN using the images of the molecules generated from the canonical SMILES written by VEGA (Benfenati et al., 2013). The net uses the inception architecture, which contains blocks of filters of different dimensions, including 1 × 1, and the concept of the residual network (He et al., 2016).

The data set has been divided into 80% for training and 20% for testing. To reduce computation time the size of the input image is reduced from the original 299 × 299 × 3 pixels to 80 × 80 × 3. The best network, optimized using Adam, has the following parameters: neurons in the first layer 16, 200 epochs, batch size 32, learning rate 0.001, dropout probability 0.2. The results of Toxception on the test set of 20% randomly selected molecules are in Table 5.4.

5.4.2.3 Integrating features from SMILES and images

SmilesNet and Toxception behave in the same way: they extract the features from data and use them in a final layer to produce the classification. They can be used as standalone models or be combined. Gini et al. (2019) found that the results improve just by taking the features from the trained nets and using them to learn the final classification through a simple two-layer NN, as illustrated in Fig. 5.11. This final model, called C-Tox, positively compares to both Toxception and SmilesNet, as reported in Table 5.4. The value of the MCC for those three models is quite good.

The features extracted by the networks produce also an explanation of the results, as discussed in the following section.

Table 5.4 Performances of the three deep learning models on the testing set.

Model	MCC	Specificity	Sensitivity
Toxception	0.53	0.62	0.87
SmilesNet	0.63	0.74	0.82
C-Tox	0.70	0.76	0.83

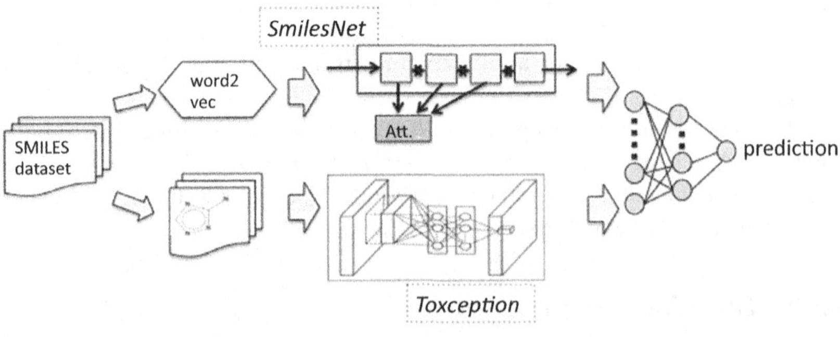

Figure 5.11 C-Tox combines the features extracted by Toxception and SmilesNet (removing the final fully connected layer) and produces a classification. The extracted features are visualized in Toxception and SmilesNet.

5.4.2.4 Learning from chemical graphs

The last developed DL model (Hung, 2020) uses directly chemical graphs in the GCN architecture as illustrated in Fig. 5.12. It has convolutional layers, attention layers, readout layers to sum up the hidden states, and a fully connected NN to produce the prediction.

In the used data set the longest SMILES contains 503 characters, so the largest graph has less than 503 nodes, and this number characterizes the convolutional layer.

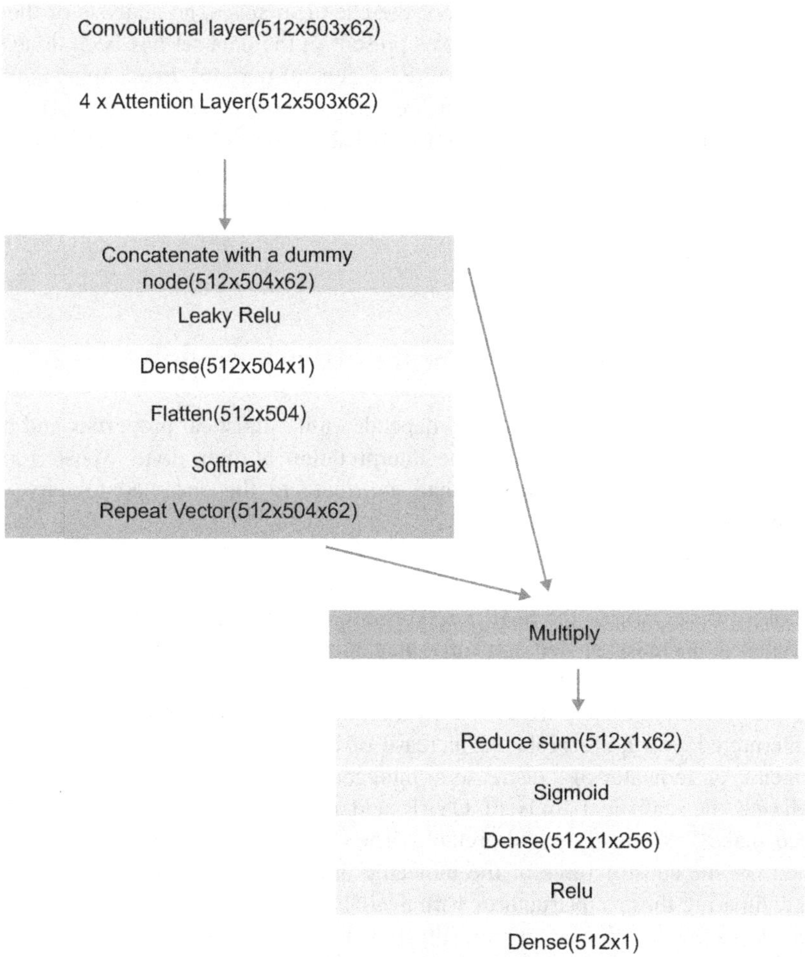

Figure 5.12 G-Ames architecture. It has a convolutional layer (batch_size × 503 × 62), four attention layers, and dense fully connected layers to give the output. A concatenation with a dummy node is used to extract from the attention layer the relevant subgraphs.

The training set of G-Ames is composed of 80% of the data set of 24,003 SMILES; the test set is 20%, which corresponds to 4800 SMILES. Data are randomized with seed 0 using the library NumPy. The values of the G-Ames hyperparameters have been experimentally determined using Adam. They are number of layers 2, batch size 512, and learning rate 0.01. In the attention layer, which computes the hidden state of each node using weighted neighborhood aggregation, the value of the attention head is 2. The pooling strategy at readout is the sum, which performs the weighted sum of the hidden state of each node, given the attention score function. The attention score function uses both the hidden state of the nodes and the target state, using a dummy node to link all the nodes, and feeds the fully connected layer. The statistics of G-Ames are in Table 5.5.

To better understand the coverage of the chemical space, an analysis of the prediction accuracy of the chemical classes present in the data set has been done. The classes are determined using ClassyFire, a general-purpose free application that considers a hierarchy of chemical classes (Djoumbou Fenang et al., 2016). The chemical characterization of the test set is in Table 5.6.

In G-Ames the best-predicted class is organic thiophosphoric acids, the worst predicted class is lipids and lipid-like. The performances for those two chemical classes are in Fig. 5.13.

5.5 Interpreting deep neural network models

The acceptance of a predictive model depends on its statistical properties and on its interpretability. In modern QSAR the interpretation is often done a posteriori, by extracting the functional elements that contribute to the observed toxicity value (Polishchuk, 2017).

Simple structural elements, as coded in chemical descriptors, are routinely used to interpret QSAR models. Parts of the SMILES can be used instead. Using simple symbols to code parts of the SMILES, Toropov et al. (2012) observed how they are correlated to increase or decrease of mutagenicity of the molecules. In particular, the presence of the attribute "(", which means the presence of any branching in an aromatic system, and the attribute "2", which means presence of any two rings, can be interpreted as a quite probable increase of mutagenicity. Also, the presence of nitrogen is an indicator of a decrease in mutagenicity.

Mixing the statistical tools of QSAR and the expert knowledge of SAR can indeed make DNN models interpretable. The method requires extracting from the trained net the substructures of the molecule that mainly contribute to its toxicity, then comparing those substructures with available knowledge.

In the following, those steps are illustrated for the three kinds of DNN before being presented. In fact, any CNN, RNN, or GCN extracts the features that are then used to classify the input. Those features are in nonlinear relationships with the property. The combination of the features, more than the presence of a specific one, characterizes the output result. Therefore the knowledge extracted could describe

Table 5.5 Performance of G-Ames on the test set.

TP	FP	FN	TN	Acc	MCC	Specificity	Precision	Recall	F1
1172	500	416	2712	0.81	0.58	0.84	0.70	0.74	0.72

Table 5.6 Percentages of chemical classes in the test set.

Class	Percentage in the test set
Organoheterocyclic	23.19
Benzene and substituted derivatives	17.27
Other benzenoids	10,6
Others	9.79
Carboxylic acids and derivatives	8.1
Ethers	4.9
Lipids and lipid-like molecules	3.85
Organonitrogen compounds	3.85
Benzoic acids	3.27
Other organic oxygen compounds	3.2
Phenylpropanoids and polyketides	3.2
Other organic acids and derivatives	2.43
Organohalogen compounds	1.98
Biphenyls and derivatives	1.73
Phenols	1.06
Alcohols and polyols	0.79
Hydrocarbons	0.58
Organic thiophosphoric acids	0.08

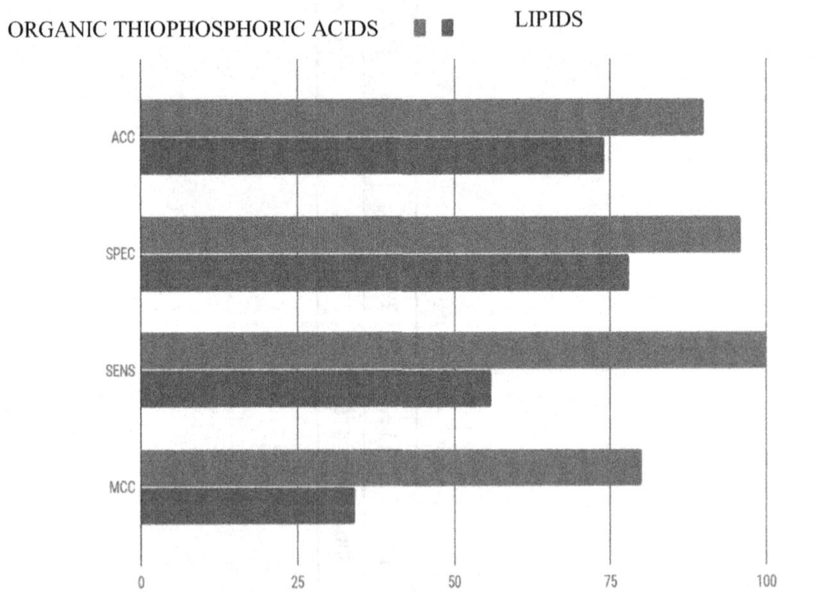

Figure 5.13 The performances of the best and worst predicted chemical classes in G-Ames.

the complex phenomenon of toxicity but it is not apt to become a set of SAs as in SAR methods.

Anyhow, comparing the obtained features with other known SAs can help understanding the role of substructures in toxicity. In particular, the SAs considered in the following are the ones obtained by Toxtree (Benigni and Bossa, 2008) and SARpy (Gini et al., 2013). SAs of Toxtree are explaining both in vivo and in vitro carcinogenicity and mutagenicity data. In databases including chemicals from diverse chemical classes, the Toxtree models agree around 65% with rodent carcinogenicity data, and 75% with Salmonella mutagenicity data. SAs of SARpy are derived from Salmonella mutagenicity data only.

5.5.1 Extracting substructures

Using images as input means that initially the neurons represent pixels, and after convolutions, pooling and ReLU the neurons represent features distorted in a nonlinear way and in lower number (LeCun et al., 2015). Going deeper into the network, the neurons get information from larger parts of the image and from other neurons so learning more complicated features like groups of atoms.

The weight vectors passed to the output neurons contain a value in (0, 1), which gives the predicted output. The final layer in Toxception contains a correlation matrix, whose values can be visualized by highlighting which parts of the molecule have higher values (Gini and Zanoli, 2020).

From the attention layer of SmilesNet, it is possible to get the correlation values of each SMILES character in the classification layer. As the inputs are SMILES, it is possible to apply directly to them the correlation matrix of the attention mechanism, so obtaining the weight that each character has on the final prediction. Substrings, containing at least three characters, have been considered. Each substring appears in the data set many times, often both in toxic and nontoxic molecules; this finding is expected as the presence of SAs alone can give ambiguous results (Gini et al., 2014).

For G-Ames the extraction of subgraphs is even easier. The effect of using convolutions and pooling on the graph is that of transforming groups of near nodes to one node, as in Fig. 5.14.

The extraction of subgraphs in GCN has been automatically done by the following procedure:

- For each SMILES, a weight matrix is extracted from the attention part of the readout layer. The attention part consists of an FC layer, which performs weighted sum, and a softmax layer that transforms the weights to make them sum to 1.
- For each SMILES the fragments with maximum weights are extracted. The maximum length of the fragment can be assigned by the user.
- For each fragment, the atoms that have a threshold greater than 0.5 are extracted.

G-Ames outputs the image of the test molecule with the identified substructure in red (Fig. 5.15).

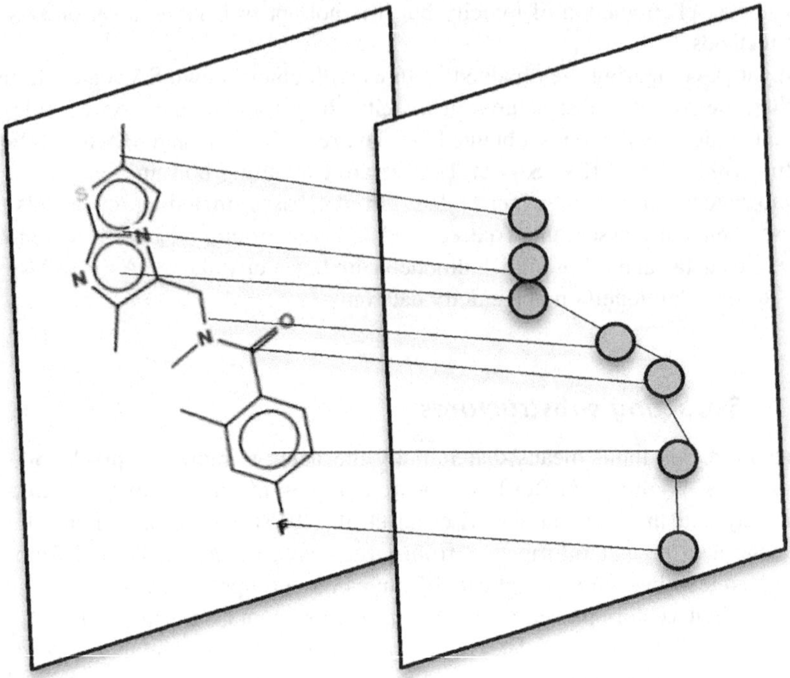

Figure 5.14 Graph reduction in the convolution layers of graph convolutional neural network.

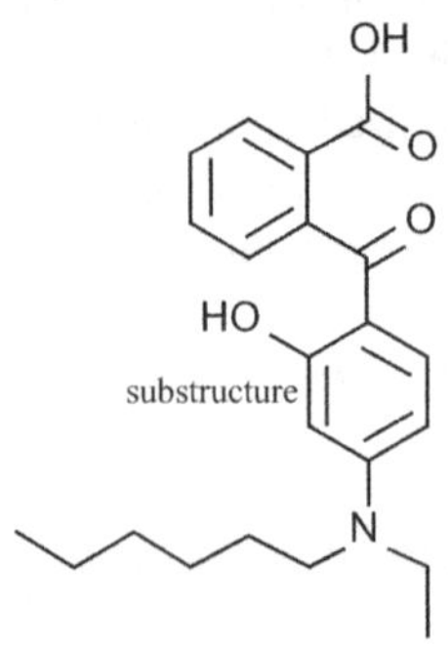

Figure 5.15 Visualization (dark color here, in red in the program interface) of a SA identified by G-Ames in the SMILES: O = C(O)c2ccccc2(C(=O)c1ccc(cc1(O))N(CC)CCCCCC).

5.5.2 Comparison of substrings with SARpy SAs

SARpy mines data sets to extract SAs candidates (in number of thousands) and selects the ones (in number of hundreds) that are statistically stronger. SARpy prefers long substructures, so to cope with the activity landscape, while SmilesNet returns all substrings found with a high weight. G-Ames returns all subgraphs up to a maximum number of nodes.

The extracted substructures cannot be automatically taken as SAs. However, they can be compared against the set of known mutagenicity SAs. In case of coincidence, the confidence in the model can improve. In the reverse case, the absence of coincidence does not imply that the model is wrong; however, other considerations will be necessary to explain the result.

Table 5.7 illustrates four molecules. Each row considers a positive molecule (predicted with toxicity probability >0.5), a substring found by SmilesNet, and the most similar substructure alert (using Tanimoto similarity) found by SARpy. The four molecules are predicted as mutagenic by Toxtree which recognizes them, respectively, as primary aromatic amine, hydrazine, epoxide, and arizidine, aromatic nitroso group.

SmilesNet and SARpy can extract toxic fragments and fragments associated with the absence of toxicity; the same substring may appear in molecules with different toxicity probability, as the prediction depends on the whole structure, not only on the presence of the SA.

Table 5.8 illustrates some examples of the coherence of the alerts found by SmilesNet and SARpy in correctly predicting the mutagenicity property.

Note: More cases are in Gini (2020).

5.5.3 Comparison of substructures with Toxtree

Another way to look at the results of the DL models is to see whether they find some expert-defined SAs, as the ones available in Toxtree. Twenty-five Toxtree SAs can explain the 1290 substrings extracted from SmilesNet. An example is in Fig. 5.16, while Table 5.9 shows those SAs associated with the number of similar substrings, and their probability of being mutagen, with three different thresholds. Although SAs code a rule of toxicity, they have been found mostly in experimentally nonmutagenic molecules (Honma et al., 2019). Indeed, this low accuracy is reflected in the numerical distribution of the substring's probability of toxicity.

Using G-Ames, an analysis has been done considering the chemical classes, where the extracted subgroups are compared with Toxtree SAs. TP (positive according to Toxtree as they contain at least a SA), FP, TN, FN are counted, and the statistics of the local models are reported in Table 5.10.

Fig. 5.17 shows the SA (QSA18_Ames.Polycyclic Aromatic Hydrocarbons) identified in Toxtree in the SMILES O=C5OC2(c4ccc(cc4(Oc1cc(c(cc12)Nc3ccccc3)C))N(CCCC)CCCC)c6ccccc56. G-Ames identifies nearly the same substructure for fragment-size of 20. For fragment-size of 10, the substructure extracted is smaller but the same SA can be identified.

Table 5.11 shows some examples of the substructures extracted by G-Ames, their corresponding Toxtree SAs, and the predicted mutagenicity class.

The substructures that are predicted as mutagenic correspond to known SAs in Toxtree. The number of occurrences in the positive substructures for each Toxtree SA is in Table 5.12. 22 SAs of Toxtree are not found by G-Ames, as indicated in

Table 5.7 Examples of substrings extracted by SmilesNet compared with SARpy SAs.

MOLECULE	Structure	p tox	Substructure in SmilesNet	SARpy SA
Nc1sccn1	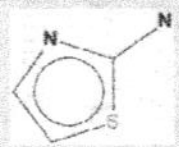	0.88	Nc1nccs1 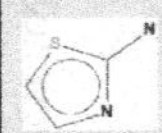	Nc1nccs1
O = C(NNc1ccc(cc1)C)C(C)(CO)CO	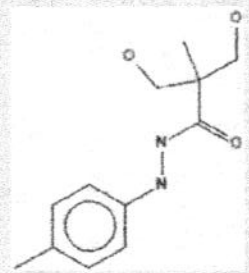	0.92	cccccNNC = O 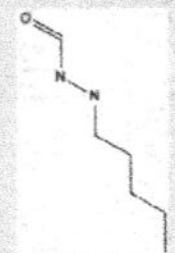	O = CNNc1ccccc1
C1CN1c2nc(nc(n2)N3CC3)N4CC4	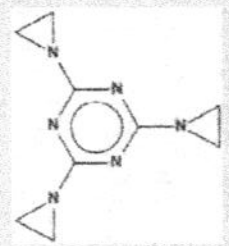	0.88	cN1*C1 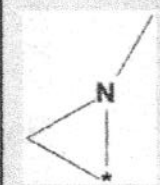	C1CN1
O = Nc1c2ccccc2c3ccc4cccc5ccc1c3c45	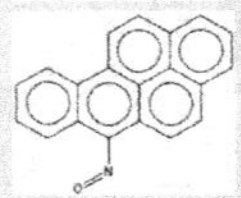	0.86	cccN = O 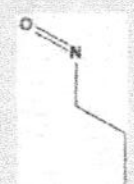	O = Nc1ccccc1

Table 5.8 Coherence between the substring found by SmilesNet and the SARpy equivalent fragment for correctly predicted positive and negative molecules.

Molecule in SMILES	Predicted SmilesNet probability	substring = SARpy fragment	SARpy_prediction
NC(N) = S	0.163	NC = S	Negative
O = S(=O)(OCC(F)(F)F)C	0.243	CC(F)F	Negative
NC1 = NNC = N1	0.327	c1nc[nH]n1	Negative
O = C(OCC)C (=CNC1CC1)C(=O)c2cc (F)c(F)cc2(F)	0.347	NC1CC1	Negative
OC(=O)CS	0.406	O = CCS	Negative
N#CCCSCc1nc(N = C(N)N) sc1	0.470	CC#N	Negative
Nc1sccn1	0.884	Nc1nccs1	Positive
CCCI	0.980	CCI	Positive

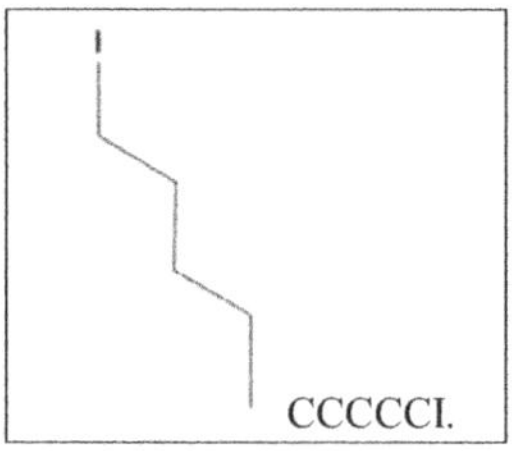

Figure 5.16 An example of substring extracted by SmilesNet. The fragment CCCCCI is associated with a probability 0.98 of being mutagen, and 0.02 of being not mutagenic. In Toxtree it is considered as SA8: Aliphatic halogens.

Table 5.12. It is possible that the test set does not include SMILES with these alerts. Another possible reason is that many SMILES experimentally nonmutagenic are identified as mutagenic in Toxtree, so they are not counted in the occurrences.

Table 5.13 considers the statistics on the test set using only the Toxtree SAS. Note that there are many SMILES in the Ames test set that have experimental values different from the ones predicted by Toxtree.

Finally, a comparison of the substrings found by SmilesNet and the substructures found by G-Ames is illustrated in Table 5.14 for some molecules of Tables 5.7 and 5.8. For the considered molecules the predictions agree but the substructures can be quite different. The reasons are many. Only the most important substructure is considered, but a few are indeed extracted for each molecule. Moreover, they may change according to the parameters used in the extraction, and on the attention weights that are computed in different ways.

Table 5.9 Toxtree SAs matched with SmilesNet substrings.

Toxtree SAs	Matched SmilesNet fragments	Prob < 0.5	Prob > = 0.5	Prob > = 0.8
SA1 Acyl halides	11	2	9	1
SA10 alfa, beta unsaturated carbonyls	77	37	40	4
SA11 Simple aldehyde	151	54	97	35
SA13 Hydrazine	150	84	66	16
SA14 Aliphatic azo and azoxy	17	15	2	0
SA15 Isocyanate and isothiocyanate groups	15	8	7	3
SA16 Alkyl carbamate and thiocarbamate	11	4	7	2
SA2 Alkyl(C < 5) or benzyl ester of sulfonic or phosphonic acid	30	21	9	0
SA21 Alkyl and aryl N-nitroso groups	48	30	18	0
SA22 Azide and triazene groups	31	25	6	0
SA23 Aliphatic N-nitro	5	5	0	0
SA25 Aromatic nitroso group	10	4	6	4
SA27 Nitro aromatic	3	3	0	0
SA28 Primary aromatic amine, hydroxyl amine, and its derived esters (with restrictions)	167	74	93	26
SA28bis Aromatic mono- and dialkylamine	16	6	10	6
SA28ter Aromatic N-acyl amine	37	0	37	15
SA29 Aromatic diazo	1	0	1	0
SA3 N-methylol derivatives	6	2	4	0
SA4 Monohaloalkene	20	11	9	0
SA5S or N mustard	2	0	2	0
SA61 Alkyl hydroperoxides	8	4	4	0
SA64 Hydroxamic acid derivatives	9	5	4	0
SA7 Epoxides and aziridines	64	33	31	7
SA8 Aliphatic halogens	354	195	159	44
SA9 Alkyl nitrite	8	7	1	0

Note: The columns "Prob" show the number of matched substrings that have that probability of being mutagens.

A comparison of the substructures obtained from the different models shows that some molecules are differently classified and/or are classified giving more importance to different subparts. This is expected, as the result is obtained from different nonlinear transformations of the input. Big SAs can better account for the property, in case of SAs-based models, but short substructures and their combination are of interest in many QSARs that use fingerprints. The fragments extracted from DNNs are somehow similar to fingerprints more than to full SAs. The previously reported paper (Toropov et al., 2012) illustrates a similar situation, where small structural features can help in explaining the behavior.

Table 5.10 The statistics of Toxtree SAs grouped by the chemical classes of the molecules.

Statistics\ classes	Alcohols and polyols	Benzene and substituted derivatives	Benzoic acids and derivatives	Carboxylic acids and derivatives	Ethers
Accuracy	0.92	0.69	0.66	0.86	0.79
MCC	0.54	0.27	0.17	0.52	0.37
Specificity	0.97	0.75	0.75	0.93	0.92
Precision	0.67	0.45	0.39	0.65	0.61
Recall	0.5	0.43	0.54	0.56	0.41
F1	0.67	0.39	0.45	0.65	0.61

Statistics \classes	Lipids and lipid-like molecules	Organo halogen	Organo heterocyclic	Organo nitrogen	Phenols	Phenyl propanoids and polyketides
Accuracy	0.77	0.78	0.63	0.82	0.68	0.71
MCC	0.18	0.57	0.16	0.58	0.38	0.20
Specificity	0.94	0.90	0.68	0.88	0.64	0.83
Precision	0.47	0.84	0.37	0.73	0.5	0.42
Recall	0.19	0.64	0.49	0.69	0.77	0.37
F1	0.47	0.84	0.37	0.73	0.5	0.42

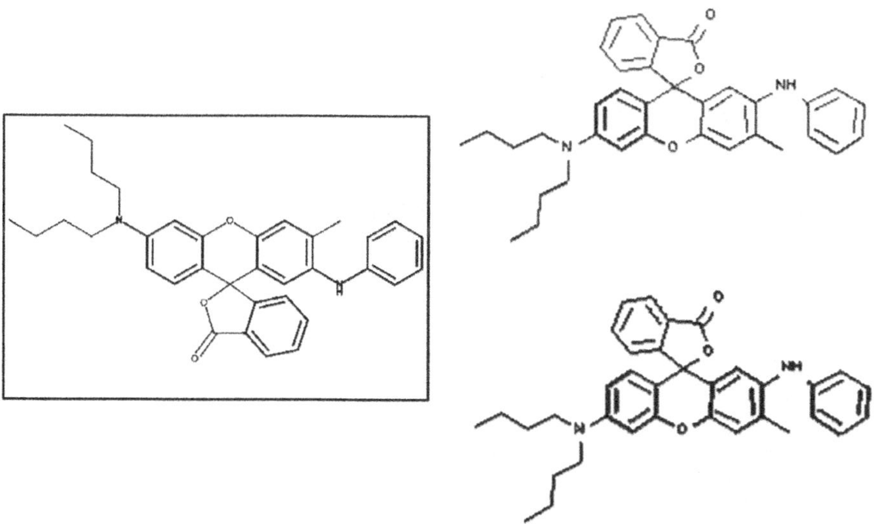

Figure 5.17 On the *left* the image generated by Toxtree for the SMILES: O = C5OC2(c4ccc (cc4(Oc1cc(c(cc12)Nc3ccccc3)C))N(CCCC)CCCC)c6ccccc56. On the *right*, in dark, the two substructures identified by G-Ames with fragment-size of 10 and 20, respectively.

Table 5.11 Substructure extracted by the model validated by Toxtree.

SMILES extracted by G-Ames	Matched SA in Toxtree	Model predicted value
Cc1ccc([N +](=O)[O-])cc1	SA27_Ames	Mutagenic
S/C(Cl) = C(\Cl)CCl	SA8_Ames	Mutagenic
ccccOCCO	SA24_Ames	Mutagenic
c1ccc2ncccc2c1	SA69_Ames	Mutagenic
O = CCO	SA11_Ames	Mutagenic
C\C = N/N	SA13_Ames	Mutagenic
Nc1ccccc1	SA28_Ames	Mutagenic
O = C1OCC(C(Br)Br) = C1Cl	SA65_Ames	Mutagenic
CC = CC(=O)Cl	SA1_Ames	Mutagenic
COC(=O)[C@H](C)CN = [N +] = [N-]	SA22_Ames	Mutagenic
O = C1OCC(C(Br)Br) = C1Cl	SA8_Ames, SA65_Ames	Mutagenic
CC = CC(=O)Cl	SA1_Ames, SA10_Ames	Mutagenic
Oc1ccccc1	None	Nonmutagenic
FC(F)(F)c1ccccc1	None	Nonmutagenic
NCc1ccccn1	None	Nonmutagenic
FC(F)c1ccccc1	None	Nonmutagenic
C[N +](C)(C)CCCN	None	Nonmutagenic
CCOCCC(=O)C(F)F	None	Nonmutagenic
CN = Nc1ccccc1	None	Nonmutagenic
N#Cc1ccc([nH])c(n)c1	None	Nonmutagenic
FCOC(F)(F)C(F)(F)Br	None	Nonmutagenic
NCC(F)(F)F	None	Nonmutagenic

5.6 Discussion and conclusions

This section recalls the main points presented and unifies them in the prospective of understanding the role of DL in the QSAR domain.

The mutagenicity models presented in the previous sections show some novelties in input, output, and methods with respect to the state of the art of QSAR modeling. In particular:

1. The input takes directly the chemical two-dimensional structure. SMILES (used in SmilesNet) and their equivalent chemical graphs (used in Toxception) are commonly used and easily understandable. Available chemical libraries automatically compute adjacency and feature matrices, used by G-Ames. This last format allows avoiding any problem with SMILES input, that is, the nonunique SMILES representation for the same molecule, and the possible lack of SMILES similarity for similar molecules. Moreover, SMILES are not affected by low image resolution as in the case of using images.
2. There is no need of computing chemical descriptors; the feature matrix can contain simple information and eventually be empty.
3. Neither fingerprints nor functional subgroups are a priori selected.
4. Results are possibly expressed with uncertainty estimation.

Table 5.12 Toxtree SAs identified by G-Ames in the Ames test set.

Toxtree SAs	**Occurrences identified by G-Ames**	**Toxtree SAs**	**Occurrences identified by G-Ames**
SA10_Ames	26	*SA37_Ames*	*0*
SA11_Ames	29	SA38_Ames	8
SA12_Ames	*0*	*SA39_Ames*	*0*
SA13_Ames	10	*SA3_Ames*	*0*
SA14_Ames	6	SA4_Ames	6
SA15_Ames	*0*	*SA57_Ames*	*0*
SA16_Ames	2	*SA58_Ames*	*0*
SA18_Ames	*0*	*SA59_Ames*	*0*
SA19_Ames	*0*	SA5_Ames	2
SA1_Ames	12	*SA60_Ames*	*0*
SA21_Ames	34	SA61_Ames	1
SA22_Ames	11	*SA62_Ames*	*0*
SA23_Ames	5	*SA63_Ames*	*0*
SA24_Ames	21	SA64_Ames	2
SA25_Ames	11	SA65_Ames	4
SA26_Ames	*0*	*SA66_Ames*	*0*
SA27_Ames	115	*SA67_Ames*	*0*
SA28_Ames	77	*SA68_Ames*	*0*
SA28bis_Ames	11	SA69_Ames	4
SA28ter_Ames	*0*	*SA6_Ames*	*0*
SA29_Ames	*0*	SA7_Ames	34
SA2_Ames	8	SA8_Ames	91
SA30_Ames	*0*	SA9_Ames	1

Note: In *italics* SAs not found.

Table 5.13 Statistics on the test set of applying only the SAs identified by Toxtree.

Acc	**MCC**	**Specificity**	**Precision**	**Recall**	**F1**
0.68	0.23	0.74	0.38	0.51	0.38

Note: The values are worse than the results of G-Ames.

5. Results are analyzed in "universal" chemical classes, that is, in classes that are not constructed around given functional subgroups relevant to the property under study.
6. Functional subgroups found as important are automatically extracted for helping in the model interpretation. Their concordance with available knowledge can improve confidence in the results.
7. The models work as QSARs, so they produce an output for every chemical.

The DL models before presented are trained without adding any external knowledge besides the SMILES and the experimental Ames test results. This choice has advantages and drawbacks.

Table 5.14 Comparison of substructures extracted from SmilesNet (SN) and G-Ames (G) for molecules predicted in the same class by both the models.

SMILES	SN pred	G pred	SN substring	G substructure	comment
O = Nc1c2ccccc2c3ccc4cccc5ccc1c3c4	1	1	cccN = O	cccccccccc	
O = C(Nc1c(nc(cc1SC)C)SC)CBr	1	1	= C(Nc1c(nc (cc1	CcccSC	
Nc1ccc2nc3c(N)cccc3nc2c1	1	1	Nc1ccc2nc3	Nc1ccccc1	
O = Nc1c2ccccc2c3ccc4cccc5ccc1c3c45	1	1	O = Nc1c2c	cccccccccc	
Nc2ccc(c1ccc(N)cc1C)c(c2)C	1	1	Nc2cc	cc(c1ccc(N)cc1C)	
IC(I)I	1	1	C(I)I	IC(I)I	SmilesNet <G-Ames
NC(N) = S	0	0	NC = S	NC(N) = S	SmilesNet <G-Ames
O = S(=O)(OCC(F)(F)F)C	0	0	CC(F)F	CS(=O)(=O)OCC (F)(F)F	
NC1 = NNC = N1	0	0	CC(F)F	Nc1nc[nH]n1	
O = C(OCC)C(=CNC1CC1)C(=O)c2cc (F)c(F)cc2(F)	0	0	NC1CC1	C = CN	
OC(=O)CS	0	0	O = CCS	C	No meaningful substructure in G-Ames
N#CCCSCc1nc(N = C(N)N)sc1	0	0	CC#N	CCCS	

Advantages are obvious, since the steps of feature computation and reduction are not needed, and no fingerprints with a priori choice of substructures are computed. The knowledge extracted from the network is auto-generated and not biased by a priori choices.

Drawbacks are also obvious since DL models should be "opened" to extract the features and so to explain somehow what the network has learned. Another possible drawback is that the training of the network requires a long computation time and is better on hardware-enhanced computers. The main equations of a NN are matrix/vector operations, which are apt for execution on massively parallel hardware architectures, such as GPUs, to speed up the training process. Anyhow, GPUs and the needed open-source software are easily available. Another drawback is that DL models are good if they can use large training sets. Finally, NN architecture design is still an art, not a science; playing with the DL libraries requires some training.

The characteristics of C-Tox, SmilesNet, and Toxception models, presented in Section 5.4, match the points 2, 3, and 6 listed above. The data interpretation of those models uses the attention layer, takes the segments with high weights, then transformed into SMILES for human understanding. The G-Ames model has all the mentioned characteristics.

Another DL model for mutagenicity has been developed at MultiCase company (Chakravarti and Radha Mani Alla, 2019) using the NIHS mutagenicity data set. It also divides the external test set into chemical classes, using 53 chemical classes (not indicating how the classes have been defined), and calculates the prediction sensitivity and specificity within the classes. It is not possible to make a full comparison with the results of Section 5.4, as the classes are differently defined. It also uses the attention layer to extract substructures, but no statistical analysis of the found alerts is presented.

All the DL models have statistics not worse and generally better than the traditional models.

5.6.1 A future for deep learning models

QSAR is based on the hypothesis that the chemical structure is responsible for the activity; it follows that similar molecules are expected to have similar properties. This simple assertion is contradicted in many cases, where similar molecules behave in a different way. The similarity is such a subtle property that it cannot be defined in a universal way and independently on the use (Gini, 2020). DL makes the similarity hypothesis integrated into the process of feature extraction. It is also interesting to observe that even using SMILES as input the statistical results are very similar to the ones obtained by graphs. This is a further confirmation that the feature extraction of DL makes the similarity hypothesis less crucial.

The hidden neurons of DNNs may represent previously known SAs, as illustrated for SARpy and Toxtree, which usually are humanly engineered by experts. We may expect that hidden neurons contain also novel groups, yet undiscovered.

Cichy and Kaiser (2019) say that the general idea that DNNs results cannot be explained should be corrected:

> *"The answer to a question such as "why does a DNN unit behave such and such" is not "because it represents this or that feature of the world," but "because the unit needs to respond such in order to fulfill its function in enabling a particular objective, such as object recognition." That is, the nature of the explanation is teleological."*

They finally underline that DNNs have an important aspect that lacks in many other models: they can be used for exploration. The exploratory power of DNNs is a strong point in favor of using them in cases when a theory is missing, as in most cases of toxicity.

The concluding sentence is taken from Buckner and Garson (2019):

> *"Over the centuries, philosophers have struggled to understand how our concepts are defined. It is now widely acknowledged that trying to characterize ordinary notions with necessary and sufficient conditions is doomed to failure. Connectionist models seem especially well suited to accommodating graded notions of category membership of this kind. Nets can learn to appreciate subtle statistical patterns that would be very hard to express as hard and fast rules."*

References

Ames, B.N., 1984. The detection of environmental mutagens and potential. Cancer 53, 2030–2040.

Basak, S.C., 2013. Philosophy of mathematical chemistry: a personal perspective. HYLE–Int. J. Philos. Chem. 19 (1), 3–17.

Benfenati, E., Manganaro, A., Gini, G., 2013. VEGA-QSAR: AI inside a platform for predictive toxicology, Wokshop Popularize Artif. Intell. (PAI) 2013, Torino Dec. 5, 2013, pp. 21–28, http://ceur-ws.org/Vol-1107/.

Benfenati, E., Belli, M., Borges, T., Casimiro, E., Cester, J., Fernandez, A., et al., 2016. Results of a round-robin exercise on read-across. SAR. QSAR Env. Res. 27 (5), 371–384.

Benfenati, E., Golbamaki, A., Raitano, G., Roncaglioni, A., Manganelli, S., Lemke, F., et al., 2018. A large comparison of integrated SAR/QSAR models of the Ames test for mutagenicity. SAR. QSAR Env. Res. 29 (8), 591–611.

Benfenati, E., Chaudhry, Q., Gini, G., Dorne, J.L., 2019. Integrating in silico models and read-across methods for predicting toxicity of chemicals: a step-wise strategy. Environ. Int. 131, 105060.

Bengio, Y., Courville, A., Vincent, P., 2013. Representation learning: a review and new perspectives. IEEE Trans. PAMI 35 (8), 1798–1828.

Benigni, R., Bossa, C., 2008. Structure alerts for carcinogenicity, and the Salmonella assay system: a novel insight through the chemical relational databases technology. Mutat. Res. 659 (3), 248–261.

Buckner, C., Garson, J., 2019. Connectionism. The Stanford Encyclopedia of Philosophy. <https://plato.stanford.edu/archives/fall2019/entries/connectionism/>.

Chakravarti, S.K., Radha Mani Alla, S., 2019. Descriptor free QSAR modeling using deep learning with long short-term memory neural networks. Front. Artif. Intell. 2 (17).

Chen, T., Chen, H., 1995. Universal approximation to nonlinear operators by neural networks with arbitrary activation functions and its application to dynamical systems. IEEE Trans. Neural Netw. 6 (4), 911–917.

Cho, K., Courville, A.C., Bengio, Y., 2015. Describing multimedia content using attention-based encoder-decoder networks. IEEE Trans. Multimed. 17 (11), 1875–1886.

Cichy, R.M., Kaiser, D., 2019. Deep neural networks as scientific models. Trends Cog Sci. 23 (4), 305–317.

Devillers, J. (Ed.), 1996. Neural Networks in QSAR and Drug Design. Academic Press, San Diego, CA.

Djoumbou Fenang, Y., Esner, R., Knox, C., Chepeley, L., Hastings, J., Owen, G., et al., 2016. ClassyFire: automated chemical classification with a comprehensive, computable taxonomy. J. Chemoinf 8, 61.

Gal, Y., Ghahramani, Z., 2016. Dropout as a bayesian approximation: representing model uncertainty in deep learning. In: Proceedings of The 33rd International Conference on Machine Learning, PMLR 48:1050-1059.

Gini, G., Katrizky, A. (Eds.), 1999. Predictive toxicology of chemicals: experiences and impact of AI tools. In: Papers from the AAAI Spring Symposium on Predictive toxicology SS-99-01. AAAI Press, Menlo Park, CA.

Gini, G., Ferrari, T., Cattaneo, D., Bakhtyari, N.G., Manganaro, A., Benfenati, E., 2013. Automatic knowledge extraction from chemical structures: the case of mutagenicity prediction. SAR. QSAR Env. Res. 24 (5), 365–383.

Gini, G., Franchi, A.M., Manganaro, A., Golbamaki, A., Benfenati, E., 2014. ToxRead: a tool to assist in read across and its use to assess mutagenicity of chemicals. SAR. QSAR Env. Res, 25 12, 999–1011.

Gini, G., 2016. QSAR methods. In: Benfenati, E. (Ed.), In Silico Methods for Predicting Drug Toxicity. Springer, Clifton, N.J., pp. 1–20.

Gini, G., 2018. QSAR: what else? In: Nicolotti, O. (Ed.), Computational Toxicology. Methods in Molecular Biology. Humana Press, New York, NY, pp. 79–105.

Gini, G., 2020. The QSAR similarity principle in the deep learning era: confirmation or revision? Found. Chem. 22, 383–402.

Gini, G., Zanoli, F., Gamba, A., Raitano, G., Benfenati, E., 2019. Could deep learning in neural networks improve the QSAR models? SAR. QSAR Env. Res. 30 (9), 617–642.

Gini, G., Zanoli, F., 2020. Machine learning and deep learning methods in ecotoxicological QSAR modeling. In: Roy, K. (Ed.), Ecotoxicological QSARs. Springer Nature, Berlin-Heidelberg.

Goh, G., Hodas, N., Siegel, C., Vishnu, A., 2018. SMILES2vec: an interpretable general-purpose deep neural network for predicting chemical properties, arXiv:1712.02034v2 [stat.ML].

Goh, G., Siegel, C., Vishnu, A., Hodas, N.O., Baker, N., 2017. Chemception: a deep neural network with minimal chemistry knowledge matches the performance of expert developed QSAR/QSPR models. Arvix.org/abs/1706.06689.

Hamilton, W.L., Ying, R., Leskovec, J., 2017. Inductive representation learning on large graphs. In: Proceedings Neural Information Processing Systems (NIPS).

Hansen, K., Mika, S., Schroeter, T., Sutter, A., ter Laak, A., Steger-Hartmann, T., et al., 2009. Benchmark data set for in silico prediction of Ames mutagenicity. J. Chem. Inf. Model. 49 (9), 2077–2081.

He, K., Zhang. X., Ren, S., Sun, J., 2016. Deep residual learning for image recognition. In: Proceedings IEEE Conference on Computer Vision and Pattern Recognition (CVPR), pp. 770–778.

Honma, M., Kitazawa, A., Cayley, A., Williams, R.V., Barber, C., Hanser, T., et al., 2019. Improvement of quantitative structure-activity relationship (QSAR) tools for predicting Ames mutagenicity: outcomes of the Ames/QSAR International Challenge Project. Mutagenesis 34 (1), 3–16.

Hung, C. 2020. Bayesian Graph Neural Network with uncertainty estimation to predict mutagenicity of chemicals. Master Thesis in Computer Science and Engineering. Politecnico di Milano, Italy.

Johnson, A.M., Maggiora, G.M., 1990. Concepts and Applications of Molecular Similarity. John Willey & Sons, New York.

Kazius, J., McGuire, R., Bursi, R., 2005. Derivation and validation of toxicophores for mutagenicity prediction. J. Med. Chem. 48, 312–320.

Kingma, D.P., Ba, J., 2017. Adam: a method for stochastic optimization, arXiv:1412.6980[cs.LG].

Kipf, T.N., Welling, M., 2017. Semi-supervised classification with graph convolutional networks. In: Proceedings International Conference on Learning Representations (ICLR 2017).

LeCun, Y., Bengio, Y., 1995. Convolutional networks for images, speech, and time series. In: Arbib, M.A. (Ed.), The Handbook of Brain Theory and Neural Networks, vol. 3361, no. 10.

Kirkpatrick, P., Ellis, C., 2004. Chemical space. Nature 32 (16), 823.

LeCun, Y., Bengio, Y., Hinton, G., 2015. Deep learning. Nature 521, 436–444.

Hochreiter, S., Schmidhuber, J., 1997. Long short-term memory. Neural Comput. 9 (8), 1735–1780.

Micheli, A., Sperduti, A., Starita, A., 2001. Analysis of the internal representations developed by neural networks for structures applied to quantitative structure-activity relationship studies of benzodiazepines. J. Chem. Inf. Comput. Sci. 41, 202–218.

Plošnik, A., Vračko, M., Sollner Dolenc., M., 2016. Mutagenic and carcinogenic structural alerts and their mechanisms of action. Arh. Hig. Rada Toksikol. 2016 (67), 169–182.

Polishchuk, P.G., 2017. Interpretation of QSAR models: past, present and future. J. Chem. Inf. Model.

RDKit: Open-Source Cheminformatics Software. <https://www.rdkit.org>.

Todeschini, R., Consonni, V., 2009. Molecular Descriptors for Chemoinformatics. Wiley-VCH.

Toropov, A.A., Toropova, A.P., Benfenati, E., Gini, G., Leszczynska, D., Leszczynski, J., 2012. Calculation of molecular features with apparent impact on both activity of mutagens and activity of anticancer agents. Anti-Cancer Agents Med. Chem. 12.

Velickovic, P., Cucurull, G., Casanova, A., Romero, A., Lio, P., Bengio, Y., 2017. Graph attention networks. In: Proceedings ICLR.

Weininger, M., Weininger, A., Weininger, J.L., 1989. SMILES. 2. Algorithm for generation of unique SMILES notation. J. Chem. Inf. Model. 29, 97–101.

Werbos, P.J., 1994. The Roots of Backpropagation: From Ordered Derivatives to Neural Networks And Political Forecasting. Wiley, New York.

Williams, R.J., Hinton, G.E., Rumelhart, D.E., 1986. Learning representations by back-propagating errors. Nature. 323, 533–536.

Wu, Z., Pan, S., Chen, F., Long, G., Zhang, C., Yu, P.S., 2019. A comprehensive survey on graph neural networks. Preprint arXiv:1901.00596v3 [cs.LG].

Zanoli, F., 2018. T-Tox: a new deep learning model to predict mutagenicity of chemicals. Master thesis in Computer Science and Engineering. Politecnico di Milano, Italy.

Zhang, L., Tan, J., Han, D., Zhu, H., 2017. From machine learning to deep learning: progress in machine intelligence for rational drug discovery. Drug. Discovery Today 22 (1), 1680–1685.

Zhou, J., Cui, G., Zhang, Z., Yang, C., Liu, Z., Wang, L., et al., 2019. Graph neural networks: a review of methods and applications. arXiv:1812.08434v4 [cs.LG].

Retrosynthetic space modeled by big data descriptors

6

Claudiu N. Lungu[1,2]
[1]Department of Chemistry, Faculty of Chemistry and Chemical Engineering, Babes-Bolyai University, Cluj, Romania, [2]Department of Surgery, Faculty of Medicine and Pharmacy, University of Galati, Galati, Romania

6.1 Introduction

Retrosynthesis is a methodology for the rational planning of organic syntheses. Its impacts on everyday life are IK: from food to cutting-edge materials. Today retrosynthesis techniques are involved in the development of advanced materials. Currently, computational chemistry and specifically computational molecular design suggest new molecules. The actual existence of these molecules can only be proven by the success of their synthesis (Rodrigo et al., 2017).

Historically breakthrough organic synthesis was achieved by R.B. Woodward, the father of organic synthesis in the remarkable era of 1950–60. Woodward was distinguished with the Nobel Prize for the synthesis of strychnine. The total synthesis of vitamin B_{12} in 1973 is another notable point (Jeffrey and Woodward, 2017).

E.J. Corey was inspired by the strategy of Woodward et al., who later developed the theory of organic synthesis. Corey postulated that the synthesis (synthetic planning) must start with the final product and work backward towards a simple starting material. The imaginary reactions resulting are called antithetical reactions. Corey was the first organic chemist to use computers in synthesis (Joel et al., 2018).

The technique is used mainly in complex organic molecules where the target compound is downsized progressively to a more straightforward structure. The distinct simplified structures are called retrons. The pathway that simplifies the complex structure is chosen in such a way to ultimately leads to a simple commercially available material (Shuai Yuan et al., 2018). The process described above, called retrosynthetic analysis, actually generates the synthetic plan with it as the road map of the synthesis (Zhongliang et al., 2020).

According to the synthetic plan, the synthesis is classified as linear synthesis and convergent synthesis. In linear synthesis, the target compound is synthesized using a series of linear cascade transformations. However, linear synthesis is subject to failure due to a lack of flexibility, while each component is dependent on the precedent one. The linear synthesis is extended in length compared to a convergent synthesis (John, 2017). Errors occurring on the pathway are usually irreversible, leading to low synthesis yield or complete failure (Nair et al., 2019).

Big Data Analytics in Chemoinformatics and Bioinformatics. DOI: https://doi.org/10.1016/B978-0-323-85713-0.00018-9

In convergent synthesis, key fragments are synthesized separately, independently from the other steps. Then, the fragments are put all together at a later stage to converge into the target molecule.

Convergent synthesis is shorter and more efficient compared to linear ones. Moreover, the yield of convergent synthesis is notably higher (John et al., 2019).

As mentioned above, the process of designing a retrosynthetic road map is called retrosynthetic analysis. The retrosynthetic analysis of a target molecule is based on the viable yield, of known chemical reactions to confer a realistic dimension to the synthesis process (Christos et al., 2020).

An assessment of all retrosynthetic pathway hypotheses must be performed to determine the best feasible pathway that can be translated into a genuine synthesis.

Synthetic planning involves the conversion of simple commercially available molecules into target molecules using known chemical reactions and reagents.

Both retrosynthetic analysis and synthetic planning require knowledge and experience. A sound synthetic plan should apply convergent synthesis whenever possible as compared to a linear synthesis (Laura et al., 2020).

In order to ensure the success of the retrosynthesis, some logic operations have been defined.

Disconnection is the process of systematically breaking the target molecule using a combination of functional group interconversion and disconnection. Disconnection refers to breaking a carbon–carbon bond of a compound in order to generate simpler fragments. The process is repeated until the target is reduced to a simple starting molecule (Shelby et al., 2020).

Functional group interconversion or converting one functional group to another does not significantly impact the synthesis, but it favors the disconnection of the intermediates (Anne-Gaëlle et al., 2012).

Lastly, a retrosynthetic tree, a complex arrangement of all possible directions of compound synthesis, leads to the identification of several starting materials. Each structure resulting from a disconnection becomes a target molecule for further analysis. Analysis can be repeated for each starting point generating the second level of precursors (Jesús Naveja et al., 2019). Each precursor is checked for availability. Repeated disconnections result in many routes that are all called retrosynthetic trees. So, the retrosynthetic tree is a graph of several synthetic routes and distinct synthetic precursors (Daria et al., 2019).

6.2 Computer-assisted organic synthesis

As stated before, computational synthesis was proposed by Corey; subsequently, ~~since~~; numerous computer programs have been used. Those programs are roughly divided into passive programs and active programs. Passive software uses compound libraries fitted with structural filters to screen through the library for a similar compound. Active programs are fitted with algorithms able to search in a database of molecules and reactions.

Computer software used for computer-assisted organic synthesis (CAOS) facilitates the design and prediction of chemical reactions CAOS issues can be reduced by identifying a series of chemical reactions that produce the desired starting molecule from starting materials. CAOS-solving algorithms typically use two databases. The first database of a known chemical reaction and the second one of known starting material. One example of software package is ICSYNTH−software that generates synthetic pathways (Bøgevig et al., 2015), SYLVIA can prioritize thousands of structures and evaluate the ease of synthesis (Gasteiger et al., 1992), CHIRON—an interactive computer program for stereochemical analysis and heuristic synthesis planning (Ramachandran et al., 2011), ChemPlanner—helps chemists design viable synthetic routes for their target molecules (Linda, 2017), Chematica—a software/database that uses algorithms and a collective database to predict and provide synthesis pathways for molecules (Tomasz et al., 2018).

All retrosynthetic computational strategies are based on mining big data: large molecular databases (large collections of individual compounds) and extensive reaction libraries (large collections of molecules interconnected by the chemical reactivity rationals) (Feng et al., 2018). Both types of extensive data libraries are explored using algorithms and classifiers to reduce big data size and dimensionality progressively. Namely, big data is sampled using these methodologies. Sampling big data is crucial. The question that has to be asked about big data sets is whether it is necessary to look at the actual data to formulate conclusions about the properties of the data or whether a sample is good enough to characterize the accurate data. Big data contains a term related to size, an essential distinction of big data. However, sampling facilitates the selection of the data points from within the more critical data set to estimate the characteristics of the whole population. For example, microfluidics and microarray data in experimental drug design and biology are available at short time intervals. It may not be necessary to look at all the data for an accurate prediction, but a sample may be sufficient (Yuan et al., 2018).

One retrosynthesis method to apply to big data is retrosynthesis based on molecular similarity. Similarity scores based on the molecular graph are computed for an extensive database of compounds to find similar reaction substituents. Coley et al. demonstrated that molecular similarity is surprisingly effective for ranking and identifying the retrosynthetic steps based on the analogy of the preceding reactions.

In 1969, Corey stated that a successful program must be interactive and capable of a series of computational tasks. Firstly, the program must generate trees that are limited in size but incorporate as many valuable pathways as necessary. Also, the software must allow the operator to intervene at any step. Finally, the operator must also decide the depth of the analysis and solution search. So the logic-centered part of the analysis is performed by the computer and the complex information center part by the operator (chemist).

The analysis begins by defining structural features within the target which are of synthetic interest then a reduction in molecular complexity is performed.

Structural features defined are chains, rings, appendages, functional groups, asymmetric centers, attached groups, chemical reactivity, sensitivity, and instability (Corey et al., 1974).

Reduction in molecular complexity is performed by combining the following: scission of rings, disconnection of chains, removal of functionality, modification or removal of sites with high chemical reactivity or instability, simplification of stereochemistry, and removal of asymmetric centers.

However, some subgoals aid in simplification without themselves being simplifiers: functional group interconversion, introduction, the introduction of groups for stereochemical of regiochemical control, internal rearrangement to modify rings, chains, and functional groups.

Corey represented molecules as graphs, with atoms being the nodes and bonds being the branches. Algorithms were implemented to recognize computationally functional groups, symmetry, stereochemistry. Also, besides topological descriptors (as the number of rings and functional groups), electronic descriptors algorithms were introduced to recognize electronic group properties.

For example, the ring perception algorithm designed to find all cycles in the chemical graph performs as follows: (1) algorithm arbitrarily chooses an atom as the origin, and a path grows out along the molecular network until the path doubles back on itself; (2) If the ring does not duplicate an already recorded one, it is placed in the ring list; (3) When all paths from the origin have been traversed, all atoms in the structure have not been converted; then the structure consists of more than one fragment; (4) A new origin is chosen in the following fragment, and the process is repeated. The number of the chemical rings is given by the relation nRing = nb-na-nf, where nb = number of bonds, na = number of atoms, and nf = number of fragments in the structure.

Corey stated that the chemist easily recognizes a real ring, and all the other pseudo rings are simple combinations of one or more real rings.

Corey developed a set of fundamental Heuristics by choosing the most potent and general principles/reactions available in organic synthesis at that time (Corey et al., 1985).

Deep neural networks have been applied to assume that computers do not need to follow human-defined reaction rules, so the chemical reaction can be reinterpreted using millions of reaction examples. This method is called end-to-end, while the scientists only provide computers with the reactant (starting point) and the other is the product (final point). These methods are entirely data-driven. Many approaches have been proposed in recent years. However, these approaches are dependent on big data that is being analyzed and used as a knowledge base (Feng et al., 2018).

Topology-oriented retrosynthesis, proposed by Corey, begins with network analysis and identifying the maximally bridged ring (heteroatoms-topological center). Once the topological center has been identified, the bonds within that ring are analyzed to determine which are considered strategic. When broken in the retrosynthetic sense, the bonds directly lead back to a less bridged, simplified precursor, such as significant ($>$7-membered) rings or acyclic fragments containing chiral centers. Suppose multiple bonds in a ring are identified as strategic. In that case, one is selected for disconnection based on complementary strategies for retrosynthesis, such as functional groups or known transforms. The precursor identified in this way can then be computed until a precursor lacking any bridged ring systems is found (Nathanyal et al., 2019).

This approach to retrosynthesis focuses primarily on the topology of the target, rendering consideration of functional group manipulation secondary, avoiding any secondary or unnecessary steps. However, the rules Corey laid out for network analysis, while well-suited to their original goal of automating retrosynthesis, should be applied more flexibly when analyzing complex compounds, especially natural products (Akhil et al., 2018).

Not only should single-bond disconnections be considered, but also two-bond disconnections as well. The bond network analysis aids in the development of a retrosynthetic strategy for the caged, architecturally complex scaffold. In addition to containing a caged, highly bridged framework, the functional groups in the natural products are primarily peripheral, allowing for an analysis that almost completely disregards functional groups, making it an ideal candidate for bond-network analysis (Wang et al., 2018).

6.2.1 Retrosynthetic space explored by molecular descriptors using big data sets

Molecular descriptors are suitable for planning a retrosynthesis of a particular compound. A retrosynthetic space can be defined using chemoinformatics with chemical space and molecular descriptors.

Molecular descriptors are used to explore and characterize large data sets. For example, a specific molecular descriptor can obtain multidimensional representations of chemical space. Principal component analysis represents and characterize the chemical space.

Around 90 topological descriptors were computed for 2692 chemicals. The first 10 principal components explained over 92.6%% of the variance in data. Moreover, it was stated that the intrinsic dimensionality of the structure space was much less than 90% (Basak et al., 1987). As stated above, chemical space is a concept in cheminformatics referring to the property space spanned by all possible molecules and chemical compounds, which are defined by a set of rules and constitution factors (type of bonds, number of bonds). It consists of millions of molecules that are chemically possible.

Several theoretical spaces can be derived from the available chemical space. In pharmacology, chemical space is often referred to as pharmacologically active molecules. This chemical, pharmacological space is 10^{60} molecules (Ruddigkeit et al., 2012). No method can accurately define the exact size of the space. The size estimation is based on molecular descriptors like the Lipinski rules, the number of carbon, hydrogen, oxygen, nitrogen, and sulfur atoms.

Intrinsic exploration of chemical space is possible by generating computational databases of molecules, which can be represented by projecting multidimensional property space of molecules in lower dimensions (Kirkpatrick and Ellis, 2009). Furthermore, the generation of chemical spaces may involve creating stoichiometric combinations of electrons and atomic nuclei to cover all possible topological isomers for the given set of construction principles.

In the real world, chemical reactions allow moving in chemical space. The mapping between chemical space and molecular properties is often not unique, meaning that there can be very different molecules exhibiting very similar properties. Materials design and drug discovery both involve the exploration of chemical space with the ultimate goal to create a molecule with a desirable property and toxicity profile, for example, an antidepressant that does not lead to liver failure or a useful organophosphate that does not cause delayed neurotoxicity (Bade et al., 2010).

Like bioactivity, which molecular descriptors can describe, the retrosynthetic possibility of a molecule can also be described by molecular descriptors. Such descriptor built on the retrosynthesis rule is incorporated in various software packages. For example, one molecular modeling software that covers such descriptors is Molecular Operating Environment (MOE) software (MOE, 2015).

6.2.2 The exploration of chemical retrosynthetic space using retrosynthetic feasibility functions

These functions are designed to assess the synthetic feasibility of a molecule. The results of applying these functions onto a molecular graph are expressed as a binary score between 0 and 1, with 1one meaning full feasibility and 0 referring to a compound with very low synthesis feasibility. Another way to express the results is by using a score involving heavy atoms of the molecule and atoms that coincide with reaction centers. These reaction centers indicate problematic parts of the molecule. The heavy atom centers indicate where the chemical space of retrosynthesis can be explored.

These functions used classical organic synthesis operations, The algorithm performs repeated bond disconnections according to the retrosynthetic rules until no more disconnections are made. Each disconnection operation breaks down the molecule into one or more fragments that are more disconnected. The fragments are first transformed into molecules by satisfying valences, adding hydrogen, or for hydrolysis reaction adding a hydroxyl group to cap the fragments. Then, each molecular fragment resulting from the disconnection operation is converted from scientific data format (SDF) or molecular format (MOL) to simplified molecular-input line-entry system (SMILES) format (Weininger, 1988).

Further, the SMILES template screening is conducted in the Starting materials database (SMD). The SMD is to apply a disconnection procedure to a molecular database until no further disconnection is possible. In completing the procedure, a score is used for each heavy atom. The score is calculated from the original molecule (the trunk of the synthesis tree). If an original atom ends up in a final impossible to disconnect fragment that is found in the database, then the atom is considered covered. The score of the whole molecule is expressed as the percentage of covered heavy atoms in the original molecule.

Some atoms and bonds are vital in exploring the retrosynthetic space as follows: imine, (thio) amide, (thio)esters, enamine, acetal, hemiacetal, glycine C-N, glycerol, disulfide bonds, biphenyl bonds, aldol, mono aryl diketone groups, amino benzyl, aryl ether, aminopyridine bonds, alkyl imide bonds. Also, electrophilic, nucleophilic aromatic substitution, lithiation, Claisen reaction, and expoide opening are essential.

In illustrating the characteristics of chemical retrosynthetic space, a computational study was carried on an extensive data set of chemical compounds (big data set). The set was retrieved from the CHEMBL database. A set of 1941411 distinct compounds were considered (https://www.ebi.ac.uk/chembl/ accession date 19.03.21). All compounds were energetically minimized using AMBER10 EHT force field, charges corrected, and protonated at constant conditions.

A series of 31 chemical descriptors (Table. 6.1) were computed for all compounds to characterize the retrosynthetic space. In addition, the rsynth descriptor discussed above was also computed and chosen as the active property.

The layout of retrosynthetic space in terms of atom composition illustrated by the atom type is as follows: correlation of rsynth with the number of heavy atoms retrieves an $y = -9.51625(\text{synth}) + 34.59$ with $r = -0.1318$ and $r^2 = 0.0174$; the correlation of rsynth with the number of atoms (two heavy atoms in close contact) retrieves $y = -9.7271(\text{rsynth}) + 37.1573$ with $r = -0.1280$ and $r^2 = 0.0164$, respectively; correlation with the number of H-bond acceptor atoms retrieves $y = -1.69172(\text{rsynth}) + 5.31534$ with $r = -0.0971$ and $r^2 = 0.00094$; correlation with the number of aromatic atoms retrieves $y = 4.34871(\text{rsynth}) + 11.0211$ with $r = 0.1604$ and $r^2 = 0.0257$; the number of basic atoms gives a correlation of $y = 0.03071\ (\text{rsynth}) + 0.0246093$ with $r = 0.0249$ and $r^2 = 0.0006$; the total number of atoms has $y = -22.5163(\text{rsynth}) + 65.6136$ with $r = -0.1564$ and $r^2 = 0.0244$; number of H bond donor has $y = -2.44752(\text{rsynth}) + 3.2684$ with $r = -0.1560$ and $r^2 = 0.0244$; number of hydrophobic atoms have $y = -5.8721\ (\text{rsynth}) + 23.2406$ with $r = -0.1461$ and $r^2 = 0.0213$.

Halogens present also poor correlations: the number of boron at $y = 0.00320348$ with $r = -0.0166$, $r^2 = 0.0003$; number of bromine atoms $y = 0.0151289(\text{rsynth}) + 0.0450875$ with $r = 0.0142$, $r^2 = 0.0002$; number of chlorine atoms retrieve an $y = -0.00672121(\text{rsynth}) + 0.254502$ with $r = -0.0028$ and $r^2 = 0.0$; number of floride atoms present $y = -0.195077(\text{rsynth}) + 0.502434$ with $r = -0.0417$ and $r^2 = 0.0017$; number of phosphor atoms retrive $y = 0.0375178(\text{rsynth}) = 0.00840954$ and $r = 0.0275$ with $r^2 = 0.008$; However, the number of carbon atoms retrieve also a poor correlation of $y = -6.92039(\text{rsynth}) + 25.3021$ with $r = -0.1450$ and $r^2 = 0.0210$ also, the number of hydrogen atoms has $y = -13\ (\text{rsynth}) + 31.0237$ ewith $r = -0.1738$ and $r^2 = 0.0032$; number of O atoms has $y = -2.52119(\text{rsynth}) + 4.68637$ with $r = -0.1517$ and $r^2 = 0.0230$; number of sulfur atoms has $y = 0.126431(\text{rsynth}) + 0.33648$ with $r = 0.0450$ and $r^2 = 0.0020$.

Electrostatics alone also has a poor direct correlation with retrosynthetic space. The difference of bonded atom polarizabilities has $y = -14.3444(\text{rsynth}) + 43.2898$ with $r = -0.1390$ and $r^2 = 0.0193$.

Table 6.1 Chemical descriptors computed for the compounds set.

Nr.	Descriptors
1	Number of heavy atoms
2	Number of atoms
3	Number of H-bond acceptor atoms
4	Number of aromatic atoms
5	Number of basic atoms
6	Total number of atoms
7	Number of H bond donor
8	Number of hydrophobic atoms
9	Number of boron
10	Number of bromine atoms
11	Number of chlorine atoms
12	Number of fluorine atoms
13	Number of phosphor atoms
14	Number of carbon atoms
15	Number of hydrogen atoms
16	Number of O atoms
17	Number of sulfur atoms
18	Difference of bonded atom polarizabilities
19	Number of rotatable single bounds
20	Number of aromatic bounds
21	Number of bounds
22	Number of double bounds
23	Number of rotable bounds
24	Maximum chain length
25	Number of clusters
26	Principal moment of inertia for *x*, *y*, *z* axes
27	Number of chiral centers
28	Water-accessible surface area
29	SlogP
30	Log S
31	Density of compounds
32	Diameter of molecules
33	Reactivity of the molecules
34	Information content index
35	Mean atom information content

Structurally the number of rotatable single bounds retrieve an $y = -2.91095$ (rsynth) + .98506 and $r = -0.0918$ with $r^2 = 0.0084$. The number of aromatic bounds has $y = 4.41328$ (rsynth) + 11.2712 with $r = 0.1570$ and $r^2 = 0.0246$. The number of bounds has $y = -22.7271(\text{rsynth}) + 68.181$ with $r = -0.1544$ and $r^2 = 0.0238$. The number of double bounds has $y = -2.067(\text{rsynth}) + 3.04442$ with $r = -0.1786$ and $r^2 = 0.0319$. The number of rotable bounds has $y = -3.888365$ (rsynth) + 9.40342 with $r = -0.1027$ and $r^2 = 0.0105$.

The maximum chain length present also a low correlation with retrosynthetic capability with $y = -0.393731(\text{rsynth}) + 3.24867$ with $r = -0.0409$ and $r^2 = 0.0017$. Clusters are also implied in retrosynthetic space with $y = 462322$ (rsynth) + 763859 with $r = 0.2021$ and $r^2 = 0.0412$. Principal moment of inertia for x, y, and z have the same correlation with synth for x and y axes. $y = -11981.3$ (rsynth) = 11928.9 with $r = -0.0386$ and $r^2 = 0.0015$, and 0 for z axes. The number of chiral centers has $y = 2.92387(\text{rsynth}) + 2.67906$ with $r = -0.2161$ and $r^2 = 0.0467$.

From the solvent point of view, the water-accessible area has $y = -58.2987$ (rsynth) + 453.082 with $r = -0.0772$ and $r^2 = 0.0060$. Lipophilicity like SlogP have $y = 0.196199(\text{rsynth}) + 3.41594$ with $r = 0.0202$ and $r^2 = 0.0004$; log S has $y = 1.11765(\text{rsynth}) - 5.70687$ with $r = 0.1065$ and $r^2 = 0.0113$. Density of compounds has $y = -0.133913$ (rsynth) + 1.98073 with $r = -0.0671$ and $r^2 = 0.0045$. Diameter of molecules has $y = -2.364(\text{rsynth}) + 16.9013$ with $r = -0.0693$ and $r^2 = 0.0048$. The reactivity of the molecules has $y = -0.0421124(\text{rsynh}) + 0.26602$ with $r = -0.0240$ and $r^2 = 0.0006$.

Regarding the information content index computed after Shannon's formula $y = -32.1578(\text{rsynth}) + 103.276$. with $r = -0.1349$ and $r^2 = 0.0182$. Mean atom information content has $y = 0.0640222(\text{rsynth}) + 1.58535$ with $r = 0.0834$ with $r^2 = 0.0070$.

As shown above, the retrosynthetic characteristic of a compound has a poor correlation with an individual-specific characteristic like composition, topology, structure, or density.

Like any other property retrosynthetic feasibility of a compound has a natural distribution in a big data set; in Fig. 6.1, the value of rsynth is represented across the whole data. There are, as observed, few compounds with zero synthesis feasibility. Instead, most of the compounds have intermediate values.

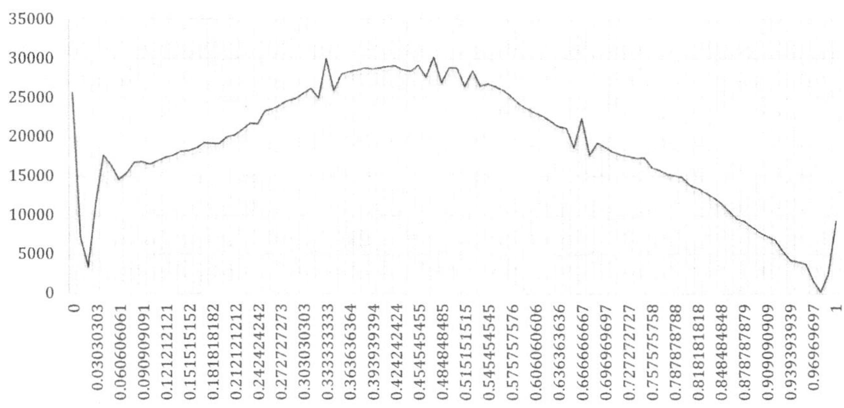

Figure 6.1 Values of rsynth descriptor across the whole molecules population. It is observed that most of the compounds have medium values. Only a few compounds have an excellent rsynth value between 0.9 and 1 from the retrosynthetic point of view.

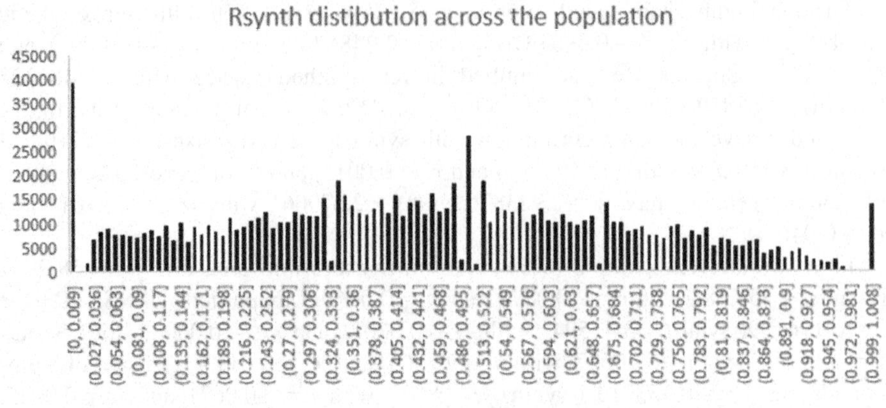

Figure 6.2 Histogram of rsynth distribution across the big data set of chemical compounds with minimal range of 0 maximum range of 1 median of 0.440, mean of 0.444, a standard deviation of 0.264, skewness of 0.116, and excess kurtosis of −0.757.

The same graph is represented as a histogram in Fig. 6.2 shows the interval distribution of rsynth.

In order to show the two-dimensional dimensionality of retrosynthetic space, a series of rhombelanes were studied (Diudea et al., 2018).

A series of complex and straightforward rhombelanes were considered in order to perform a computational retrosynthesis due to their drug-like properties. Computational retrosynthetic analysis techniques were used. The class of compounds was evaluated using retrosynthesis descriptors.

The functions used classical organic synthesis operations. The function algorithm performs repeated bond disconnections according to the retrosynthetic rules until no more disconnections are possible. Each disconnection operation breaks down the molecule into one or more fragments that are further disconnected. The fragments are first transformed into molecules by satisfying valences, adding hydrogen, or for hydrolysis reaction adding a hydroxyl group to cap the fragments. Then, each resulting fragment molecule from the disconnection operation is converted from SDF or mol format to SMILES format.

Further, a screening using a template of the SMILE is conducted in the SMD. The SMD is to apply a disconnection procedure to a molecular database until no further disconnection is possible. In completing the procedure, a score is used for each heavy atom. The score is calculated from the original molecule (the trunk of the synthesis tree). If an original atom ends up in a final impossible to disconnect fragment that is found in the database, then the atom is considered covered. The score of the whole molecule is expressed as the percentage of covered heavy atoms in the original molecule.

In order to represent a two-dimensional retrosynthetic space some molecular descriptors were also computed with respect to both ionizable and neutral compounds, respectively: molecular refractivity (Mr), log solubility in water (logS), log octanol−water partition coefficient (log$P_{o/w}$), log solubility in water (hlogS),

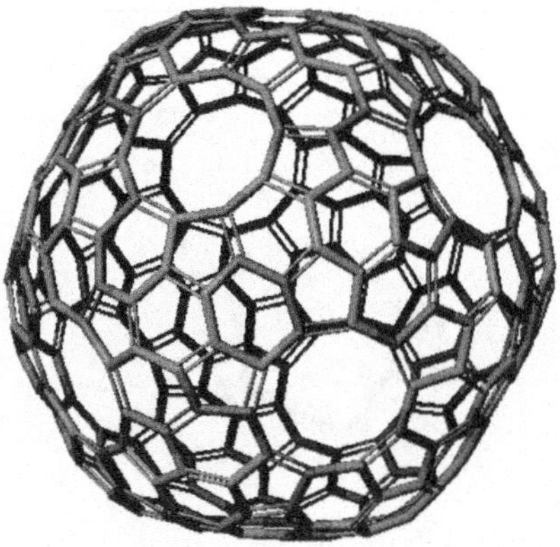

Figure 6.3 Rhombelane $C_{180}O_{120}$.

octanol water partition coefficient (hlogP), octanol–water distribution coefficient (hlogD), sum of atomic polarizabilities (apol).

The molecular formulas of discussed compounds are $C_{36}O_{24}$, $C_{72}O_{48}$, $C_{108}O_{72}$, $C_{57}H_{18}O_{21}S_6$, $C_{39}H_{16}O_{15}S_6$ $C_{73}H_{34}O_{21}S_6$, $C_{260}H_{120}$, $C_{180}O_{120}$. The last compound is represented in Fig. 6.3.

For the represented structures, molecular descriptors were computed to explore their chemical space. In computing molecular descriptors, MOE software was used. Chemical space was represented as a radar plot in Fig. 6.4.

It is observed that the shape of the surface remains relatively constant in a congeneric series of compounds but increases with the number of atoms and consecutively with the mass of compounds. As observed in Fig. 6.4, the surface of the congeneric rhombelanes has some shared graph coordinates:rsynth, Mr, logP (o/w), h_logP,h_logD. This property is not observed in a diverse data series (like the big data set). The dynamic of molecular descriptors that characterize de surface on a radar plot is erratic.

In order to explore the dimensionality of retrosynthetic space, a quantitative structure–activity relationship (QSAR) model was computed using molecular descriptors and rsynth descriptor value as the active property.

6.3 Quantitative structure–activity relationship model

When computing the QSAR model, various methodologies were used to explore the considerable amount of data.

For the partial least square methodology rsynth was selected as the activity field. The observed population was set to be the entire 1.941.411 compounds. The root

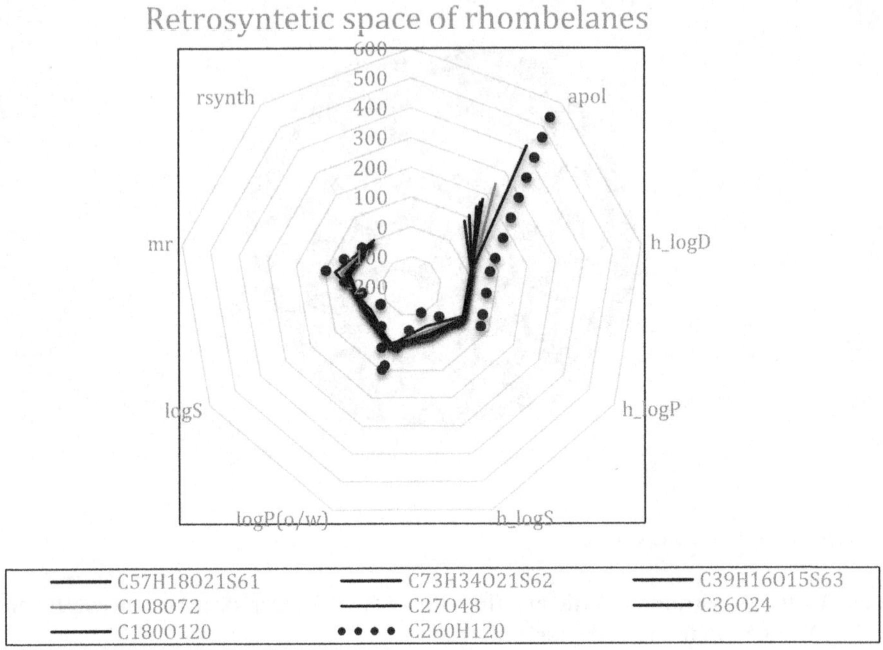

Figure 6.4 Retrosynthetic space of rhombelanes. The retrosynthetic space of the rhombelane with the most significant molecular mass ($C_{260}H_{120}$) is represented in a dotted line.

mean square error (RMSE) obtained was 0.23864 with a correlation coefficient r^2 of 0.06070. As observed, the results are less than modest. When principal component regression methodology was used on the same activity and population, an RMSE of 0.23866 and r^2 of 0.06059 were obtained

Judging by the results, a QSAR based on binary methodology was used. The binary methodology assumes that the data can take only two values, either 1 or 0, that is, the compound can either be obtained by retrosynthesis or not. This corresponds to a high-throughput screening in which the result for each molecule is active (1) or inactive (0). The binary transformation of the continuous data is performed by using a threshold criterion. If Y is a random variable with a value of either 0 or 1 and $X = (X_1, \ldots, X_n)$ is a random variable over n-vectors, that is, a random molecular descriptor, then by using the conditional distribution $\Pr(Y|X)$ in order to determine the probability of a molecule x is active within $\Pr(Y = 1|X = x)$. By using the Bayes theorem, the following can be written:

$p(x) = \Pr(Y = 1|X = x) = \frac{f(x,1)a}{f(x,1)a + f(x,0)(1-a)}$, while rearranging the equation can be rewritten as $p(x) = \left[1 + \frac{f(x,0)}{f(x,1)} \quad \frac{1-a}{a}\right]$-1. Furthermore, it can be easily assumed that each descriptor has a mean 0 and variance 1. Assumed that the individual molecular descriptors, X_i, are mutually independent, the following results:

$p(x) = \left[1 + \frac{1-a}{a}\prod_{j=1}^{n}\frac{fj(xj,0)}{fj(xj,1)}\right]^{-1}$, where $fj(x,y) = \Pr(Xj = x|Y = y)$. Distribution of the form $\Pr(Xj = x|Y = y)$ and the prior a must be estimated. The random variable Y takes the values 0 or 1. Furthermore, a can take the maximum values $S = y_1 + \ldots + y_m$, for $a = \frac{S}{m}$. It results in an unbiased estimate with the smallest possible variance overall unbiased estimators. For the small sample size, it is possible to estimate $a = 0$. The biased Bayes estimate is used under a uniform prior, which is $a = (S+1)\backslash(m+2)$, with remain in the same interval (0,1).

If m_1 is the number of active molecules in the data set, and the number of inactive molecules is denoted as $m_o = m - m_1$, the following can be written: $p(x) \approx \left[1 + \frac{m0+1}{m1+1}\prod_{j=1}^{n}\frac{fj(xj,0)}{fj(xj,1)}\right]^{-1}$. Suppose $z_1, \ldots, z_m$, are m samples of continuous random variable Z. In that case, f can be estimated by accumulated histogram observed samples on a set of B bins $(b_0, b_1], \ldots, (b_B - 1, b_B]$ defined by $B+1$ numbers $bj < bj + 1$, with b_0 equal to minus infinity and b_B equal to plus infinity. The usual procedure for counting the number of observations among m samples in bin $j > 0$ is: $Bj = \sum_{j=1}^{m}\delta\left(zi \in \left(bj - 1, bj\right)\right) = \sum_{i=1}^{m}\int_{bj-1}^{bj}\delta(x - zi)dx$. To reduce the sensitivity of bin boundaries, the delta function observed density observation is replaced with a standard random variable with mean zi with the variance S^2; lastly, the following equation results:

$Bj = \sum_{i=1}^{m}\int_{bj-1}^{bj}\frac{1}{s\sqrt{2\pi}}\exp\left[-\frac{1}{2}\frac{(x-zi)^2}{S^2}\right]dx = \frac{1}{2}\sum_{i=1}^{m}\left[\mathrm{erf}\left(\frac{bj-zi}{s\sqrt{2}}\right) - \mathrm{erf}\left(\frac{bj-1-zi}{S\sqrt{2}}\right)\right]$. Thus, each descriptor distribution can be modeled $fj(x,y)$ for y equal 0 and 1 and the n descriptors. Finally, two distributions are estimated for each descriptor: one for the active molecules in the training set and one for the inactive molecules.

When applied this methodology to the data set with the active variable rsynth, a model with the accuracy of 0.966559 where the following formula calculates accuracy: $\frac{TP+TN}{TP+TN+FP+FN}$, where TP = Tru Positives, TN = True Negatives, FP = False Positives, FN = False Negative or Number of correct predictions divided by the total number of prediction, and a significance p-value of 0.998 is observed, the 37 descriptors used to build the binary model are listed together with their importance score: 0.112234 number of clusters; 0.027999 mean atom information content; 0.025114 density; 0.011359 number of chlorine atoms; 0.010975 number of bromine atoms; 0.007323 number of aromatic atoms; 0.007137 number of aromatic bonds; 0.006495 number of fluorine atoms; 0.005901 maximum single bound chain length; 0.005659 number of sulfur atoms; 0.005508 fraction of rotable single bounds; 0.002732 water accesable surface area; 0.001503 number of chiral centers; 0.001424 diameter; 0.001304 number of hydrogen bound donor atoms; 0.001057 number of hydrogen atoms; 0.001009 number of double bounds; 0.000983 number of oxygen atoms; 0.000952 number of rotable single bounds; 0.000857 number of heavy atoms; 0.000853logS; 0.000851SlogP; 0.000840 atom information content; 0.000801 number of hydrogen bound acceptor atoms; 0.000783 principal moment of inertia on X axis; 0.000778 number of phosphor atoms; 0.000718 number of

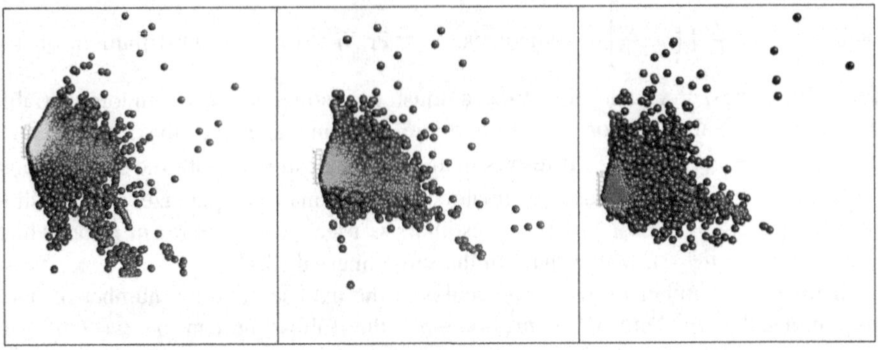

Figure 6.5 (A) Three-dimensional representation of retrosynthetic space using the number of heavy atoms and number of carbon atoms as a second and third coordinate with rsynth descriptor as the primary coordinate. (B) Three-dimensional representation of retrosynthetic space using the number of carbon atoms and total atoms as a second and third coordinate with rsynth descriptor as the primary coordinate. (C). Three-dimensional representation of retrosynthetic space using logS and ASA as a second and third coordinate with rsynth descriptor as the primary coordinate. Balls space-filling was used to represent the distinct molecules to get a clear view of the generalized shape. Additionally, retrosynthetic rules were applied to this space (big data set). As a result, hypothetically starting compounds will be generated from which the whole data set can be reconstructed. Namely, applying retrosynthetic rules to a data set, a reduction in the data dimensionality is to be observed.

atoms; 0.000691 principal moment of inertia on *Y* axis; 0.000665 number of bounds; 0.000664 difference of bonded atoms polarizabilities; 0.000659 number of carbon atoms; number of hydrophobic atoms.

In Fig. 6.5, retrosynthetic space is represented using rsynth descriptor as the primary coordinate. As stated before, the total number of molecules used in this representation is over 1.900.000.

6.4 Dimensionality reduction using retrosynthetic analysis

Retrosynthetic analysis fragments a molecule by breaking certain bonds estimated to be implied in common organic chemical reactions. Each resulting fragment is represented by assigning a SMILE code that retains the chemical context of the broken bond. By applying this algorithm to a data set, a population of individual fragments is obtained. Statistics of these fragments are gathered and used to generate new chemical structures.

As an example, consider the dimethyl-ether molecule. When *the ether bond is broken in silica*, both atoms implied in the bound will be tagged with the basic -ether label. In the SMILES coding, the oxygen atom will be coded as {OH; ether], and the carbon atom will be named as [CH3; ether]. So in retrosynthetic computational analysis, fragments of molecules are recombined to produce novel compounds that satisfy a specific algorithm. For example, MOE software uses an algorithm that first chooses a random sample from a database. Secondly, a random atom from a molecule is selected, and a substitution point is identified. Next, a random data set is used to identify a matching atom. If a substitution point is identified, then a new molecule is generated. If the generated molecule passes a particular filter (reactive/Lead like), then the screening of protonate bases and deprotonate acids against the inputted target compound. is performed. Finally, if the resulting screen is reasonable, the output molecule is generated. The retrosynthesis algorithm repeats the above procedure until a predetermined output limit is reached or a predetermined number of attempts have been made.

In Fig. 6.6 retrosynthetic analyses of all data sets are represented. The entire set of compounds was reduced to 3553 compounds. Retrosynthetic spaces representation was performed using the same methodology as in Fig. 6.5

Lastly, the retrosynthetic analysis was carried on to a third stage, where 15 final compounds resulted. Retrosynthetic space after a drastic dimensional reduction is represented in Fig. 6.7.

In Fig. 6.8 rsynth descriptor values across the molecules population after retrosynthetic analysis is presented. It is observed that the synth behavior is similar with a relatively small population.

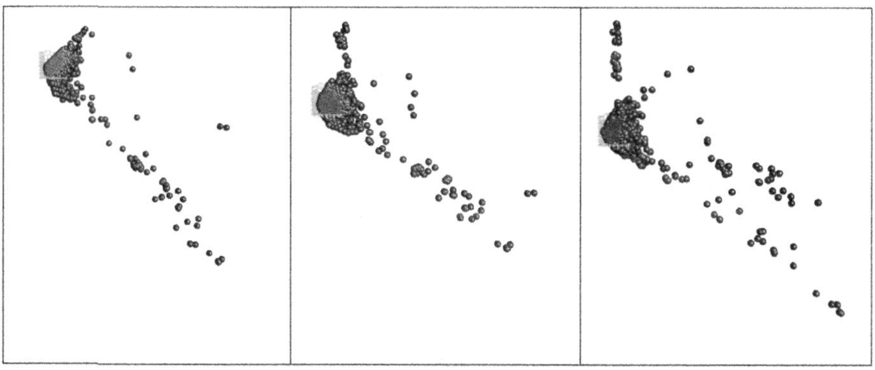

Figure 6.6 Retrosynthetic space reduction resulted after applying the retrosynthetic rules to the original big data set. (A) Three-dimensional representation of retrosynthetic space using the number of heavy atoms and number of carbon atoms as a second and third coordinate with rsynth descriptor as the primary coordinate. (B) Three-dimensional representation of retrosynthetic space using the number of carbon atoms and total atoms as a second and third coordinate with rsynth descriptor as the primary coordinate. (C) Three-dimensional representation of retrosynthetic space using logS and ASA as a second and third coordinate with rsynth descriptor as the primary coordinate. Balls space-filling was used to represent the distinct molecules to get a clear view of the generalized shape.

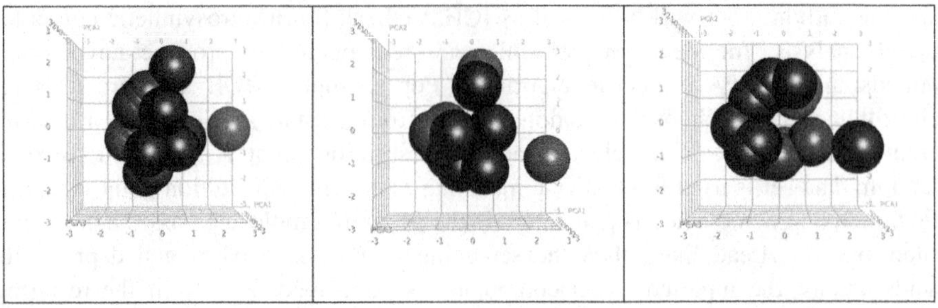

Figure 6.7 Second-generation retrosynthesis. Retrosynthetic space reduction resulted after applying the retrosynthetic rules to the original big data set. (A) Three-dimensional representation of retrosynthetic space using the number of heavy atoms and number of carbon atoms as a second and third coordinate with rsynth descriptor as the primary coordinate. (B) Three-dimensional representation of retrosynthetic space using the number of carbon atoms and total atoms as a second and third coordinate with rsynth descriptor as the primary coordinate. (C) Three-dimensional representation of retrosynthetic space using logS and ASA as a second and third coordinate with rsynth descriptor as the primary coordinate. Balls space-filling was used to represent the distinct molecules to get a clear view of the generalized shape.

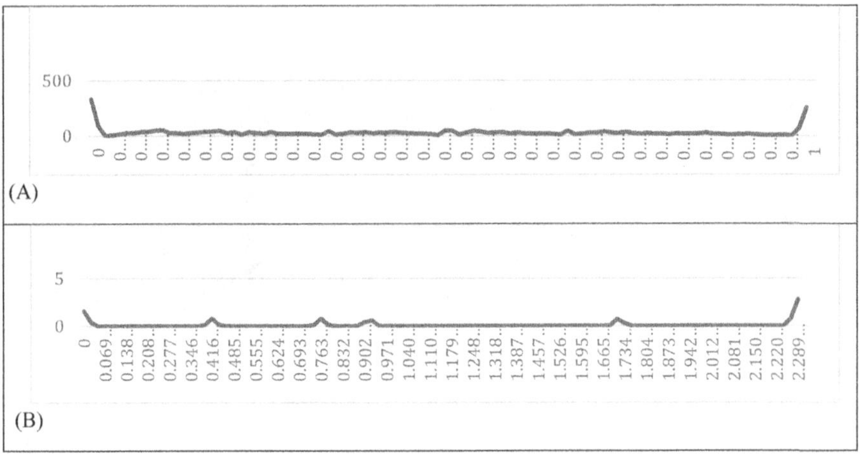

Figure 6.8 Rsynth descriptors values across the molecular population resulted after retrosynthetic analysis. (A) Rsynth value after the retrosynthetic analysis of all data series. (B) Rsynth value after retrosynthetic analysis of the previous results.

6.5 Discussion

The chemical space is composed of a vast number of molecules. Using lead-like and drug-like filters, namely Lipinski, rule the chemical space is reduced to the pharmacological space (the number of compounds in the druggable pharmacological space) which is estimated to be $10^{60}-10^{63}$ molecules (Mahendra et al., 2017). This space characterized the number of active molecules that can potentially produce a pharmacological response. Furthermore, this space was obtained by limiting the type of atoms to carbon, hydrogen, oxygen, nitrogen, and sulfur. In addition to this pharmacological space, the concept of known drug space (KDS) was introduced. The KDS is defined by the molecular descriptors computed for the Food and Drug Administration (FDA)-approved molecules (Mirza et al., 2009). The retrosynthetic space is bigger and thus includes the KDS. As KDS, the retrosynthetic space was explored using molecular descriptors (Stephen and Mooney, 2018). The retrosynthetic space presents variations across its population. It seems that the retrosynthetic feasibility of a molecule has a low correlation with atom composition, structure, or reactivity and is related instead to the target molecule's ability to the retrosynthetic rules originally stated by Corey (Coley et al., 2017). The retrosynthetic feasibility has a natural distribution across a big data set with few molecules with high retrosynthetic feasibility and few which are theoretically unfeasible to retrosynthesis. Another finding is that the predictive modeling of retrosynthetic behavior is best achieved when the discussed property is considered as an active biological property of a molecule (David et al., 2003). In other words, retrosynthetic feasibility should be considered like bioactivity, toxicity, lipophilicity of a molecule and predicted as a binary variable. Molecular descriptors are used successfully in giving a one-dimensional, two-dimensional, and three-dimensional representation of the retrosynthetic space, respectively. In a congeneric series, the retrosynthetic space has the same shape and behavior. As stated above ~~before~~, molecular descriptors are effective in exploring big data. When retrosynthetic rules (retrosynthetic analysis) are used with molecular descriptors, the dimensionality of retrosynthetic space can be reduced significantly (Nair et al., 2019). As shown here, potentially, 15 molecules can be used to synthesize the entire big data used (over 1.900.000 compounds).

Lastly, retrosynthetic space exploration can be performed in silica using molecular descriptors. Retrosynthetic space and its properties can be applied to big data. The retrosynthetic feasibility of a molecule is to be found inside the chemical retrosynthetic space.

References

Akhil, K., Wang, L., Ng, C.Y., Maranas, C.D., 2018. Pathway design using de novo steps through uncharted biochemical spaces. Nat. Commun. 9 (1), 184.

Anne-Gaëlle, P., Carbonell, P., Grigoras, I., Faulon, J.-L., 2012. A retrosynthetic biology approach to therapeutics: from conception to delivery. Curr. Opin. Biotechnol. 23 (6), 948–956.

Bade, R., Chan, H.F., Reynisson, J., 2010. Characteristics of known drug space. Natural products, their derivatives and synthetic drugs. Eur. J. Med. Chem 45 (12), 5646–5652.

Basak, S.C., Niemi, C.J., Regal, R.R., Veith, C.D., 1987. Topological indices: their nature, mutual relatedness and applications. Math. Modeling 8, 300–305.

Bøgevig, A., Federsel, H.-J., Huerta, F., Hutchings, M.G., Kraut, H., Langer, T., et al., 2015. Route design in the 21st century: the ICSYNTH software tool as an idea generator for synthesis prediction. Org. Process. Res. Dev 19 (2), 357–368.

Christos, A.N., Watson, I.A., LeMasters, M., Masquelin, T., Wang, J., 2020. Context aware data-driven retrosynthetic analysis. J. Chem. Inf. Model 60 (6), 2728–2738.

Coley, C.W., Rogers, L., Green, W.H., Jensen, K.F., 2017. Computer-assisted retrosynthesis based on molecular similarity. ACS Cent. Sci. 3 (12), 1237–1245.

Corey, E.J., Howe, W.J., Pensak, D.A.J., 1974. Am Chem Soc 96 (25), 7724–7737.

Corey, E.J., Long, A.K., Rubenstain, S.D., 1985. Science 228, 408–418.

Daria, E.K., Zweig, J.E., Newhouse, T.R., 2019. Total synthesis of paspaline a and emindole pb enabled by computational augmentation of a transform-guided retrosynthetic strategy. J. Am. Chem. Soc 141 (4), 1479–1483.

David, C.S., Aynechi, T., Voelz, V.A., Kuntz, I.D., 2003. Information content of molecular structures. Biophys. J. 85 (1), 174–190.

Diudea, M.V., Lungu, C.N., Nagy, C.L., 2018. Cube-rhombellane related structures: a drug perspective. Molecules 23 (10), 2533.

Feng, F., Lai, L., Pei, J., 2018. Computational chemical synthesis analysis and pathway design. Front. Chem 6, 199.

Gasteiger, J., Ihlenfeldt, W.-D., Fick, R., Rose, J.R., 1992. Similarity concepts for the planning of organic reactions and syntheses. J. Chem. Inf. Comput. Sci 32.

Jeffrey, I.S., Woodward, R.B., 2017. A larger-than-life chemistry rock star. Angew. Chem. Int. Ed. Engl 56 (34), 10228–10245.

Jesús Naveja, J., Pilón-Jiménez, B.A., Bajorath, J., Medina-Franco, J.L., 2019. A general approach for retrosynthetic. Mol. Core Anal 11 (1), 61.

Joel, M.S., Harwood, S.J., Baran, P.S., 2018. Radic. Retrosynthesis 51 (8), 1807–1817.

John, R., 2017. Proudfoot. Molecular complexity and retrosynthesis. J. Org. Chem 82 (13), 6968–6971.

John, S.S., Coley, C.W., Bishop, K.J.M., 2019. Learning retrosynthetic, planning through simulated experience. ACS Cent. Sci. 5 (6), 970–981.

Kirkpatrick, P., Ellis, C., 2009. Chemical space. Nature 432 (7019), 823–865.

Laura, K.G.A.-B., Arias-Rotondo, D.M., Biegasiewicz, K.F., Elacqua, E., Golder, M.R., Kayser, L.V., et al., 2020. Organic chemistry: a retrosynthetic approach to a diverse field. ACS Cent. Sci 6 (1), 1845–1850.

Linda, W., 2017. ChemPlanner to integrate with SciFinder. C&EN 95 (25), 35–37.

Mahendra, A., Visini, R., Probst, D., Arús-Pous, J., Reymond, J.-L., 2017. Chemical space: big data challenge for molecular diversity. Chimia. 71 (10), 661–666.

Mirza, A., Desai, R., Reynisson, J., 2009. Known drug space as a metric in exploring the boundaries of drug-like chemical space. Eur. J. Med. Chem 44 (12), 5006–5011.

Molecular Operating Environment (MOE), 2015 Chemical Computing Group Inc., 1010 Sherbooke St. West, Suite #910, Montreal, QC, Canada, H3A 2R7.

Nair, V.H., Schwaller, P., Laino, T., 2019. Data-driven chemical reaction prediction and retrosynthesis. Chimia (Aarau) 73 (12), 997–1000.

Nair, V.H., Schwaller, P., Laino, T., 2019. Data-driven chemical reaction. Predict. Retrosynth. 18 (12), 997–1000.

Nathanyal, J.T., Ayinde, S., Van, K., Liu, J.O., Romo, D., 2019. Pharmacophore-directed retrosynthesis applied to rameswaralide: synthesis and bioactivity of sinularia natural product tricyclic cores. Org. Lett 21 (18), 7394–7399.

Ramachandran, S.1, Kota, P., Ding, F., Dokholyan, N.V., 2011. Automated minimization of steric clashes in protein structures. Proteins 79 (1), 261–270.

Rodrigo, O.M.A.D.S., Miranda, L.S.M., Bornscheuer, U.T., 2017. A retrosynthesis approach for biocatalysis in organic synthesis. Chemistry 23 (50), 12040–12063.

Ruddigkeit, L., van Deursen, R., Blum, L.C., Reymond, J.-L., 2012. Enumeration of 166 billion organic small molecules in the chemical universe database GDB-17. J. Chem. Inf. Model 52 (11), 2864–2875.

Shelby, V.M., Doering, N.A., Sarpong, R., 2020. Retrosynthetic strategies and their impact on synthesis of arcutane natural products. Chem. Sci 11 (29), 7538. 7522.

Stephen, J., Mooney, V., 2018. Pejaver big data in public health: terminology, machine learning, and privacy. Annu. Rev. Public. Health 39, 95–112.

Tomasz, K., Barbara, M.-K., McCormack, M.P., Lima, H., Szymkuć, S., Bhowmick, M., et al., 2018. Efficient syntheses of diverse, medicinally relevant targets planned by computer and executed in the laboratory. Chem 4 (3), 522–532.

Wang, L., Ng, C.Y., Dash, S., Maranas, C.D., 2018. Exploring the combinatorial space of complete pathways to chemicals. Biochem. Soc. Trans 45 (3), 513–522.

Weininger, D., 1988. SMILES: a chemical language and information system. 1. Introduction to methodology and encoding rules. J. Chem. Inf. Comput. Sci 28, 31–36.

Yuan, S., Qin, J.-S., Li, J., Huang, L., Feng, L., Fang, Y., et al., 2018. Retrosynthesis of multi-component metal-organic frameworks. Nat. Commun. 9 (1), 808.

Zhongliang, G., Wu, S., Ohno, M., Yoshida, R., 2020. Bayesian Algorithm for Retrosynthesis 60 (10), 4474–4486.

Approaching history of chemistry through big data on chemical reactions and compounds

7

Guillermo Restrepo[1,2]
[1]Max Planck Institute for Mathematics in the Sciences, Leipzig, Germany,
[2]Interdisciplinary Center for Bioinformatics, Leipzig University, Leipzig, Germany

7.1 Introduction

Although chemoinformatics, bioinformatics, computer-aided drug design, cancer biology, emerging pathogens, and computational toxicology are all subjects of the book containing this chapter, there is another tie, namely chemistry. This science is at the heart of the aforementioned research areas, providing methods, traditions, and practices, as well as social structures and a background language, to advance knowledge and build each of these research areas' specific jargon. There would be no chemoinformatics without substances, molecules, and materials to represent and elaborate on, biological sequences would not exist without knowledge of amino acids or nucleotides, cancer studies would be severely limited without chemical concepts such as ligand-receptor, and toxicological studies would not exist without information on the effects of substances on living systems.

The aforementioned scientific fields are active research areas because there is a chemical tradition that generates a large amount of data and knowledge. Chemistry doubles its material output every 16 years due to the publication of new substances (Llanos et al., 2019), and it is the most productive science in terms of a number of publications, only being surpassed by Engineering (Restrepo and Jost, 2022). This trend was observed in the 1960s by de Solla (1963), confirmed in the 1990s by Schummer (2006) and once more confirmed by us recently (Restrepo and Jost, 2022).

This chapter's central claim is that understanding the workings of chemistry can benefit chemistry and its allied disciplines to a large extent. And we claim that this can be accomplished by utilizing a large amount of chemical data with mathematical and computational tools. Understanding the workings of chemistry entails, for example, detecting historical trends in the expansion of the chemical space, which encompasses all reported chemical substances; or regularities in the discovery of reaction conditions or chemical substance properties; or trends in the production of new signs for communicating chemical results and concepts (Restrepo and Jost, 2022). As a result, this method differs from static studies of large datasets in that it includes a temporal dimension.

Big Data Analytics in Chemoinformatics and Bioinformatics. DOI: https://doi.org/10.1016/B978-0-323-85713-0.00033-5

Traditionally, temporality in chemistry, in its broadest sense, is part of chemistry's history, where narratives and cause-and-effect relationships are intertwined to make sense of chemistry's current state based on past data (Restrepo and Jost, 2022). Our thesis is that, in addition to being a case of big data, the history of chemistry is called to merge its methods and traditions with those of chemoinformatics and other disciplines in order to better understand not only the past of chemistry but also its possible futures.

In the following section, we summarize a formal setting for computationally treating chemical knowledge, which serves as a setting for computational history of chemistry (Restrepo, 2022a). We conclude with a case study of the computational history of chemistry, namely the analysis of the chemical space expansion.

7.2 Computational history of chemistry

Chemical knowledge and its evolution are the broad subjects of history of chemistry. We demonstrate how chemical knowledge can be modeled as a complex dynamical system formed by the interaction of three chemistry subsystems: the material, social, and semiotic systems (Restrepo and Jost, 2022; Fig. 7.1). As any other system, these systems consist of objects and their relationships (Bertalanffy, 1950).

People, academic and scientific societies, committees, enterprises, industries, and other forms of social organization are among the actors in the social system, as are computational objects such as robots and artificial intelligence technologies.

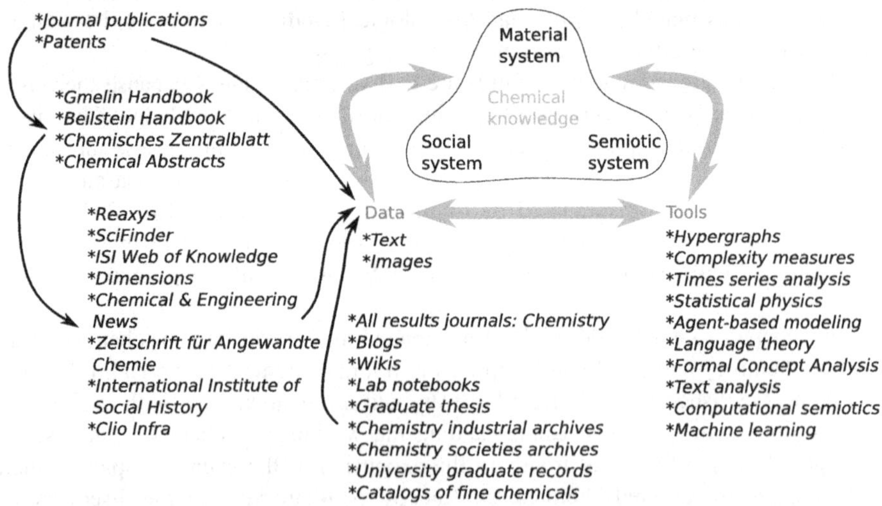

Figure 7.1 Data ecosystem for the computational history of chemistry. The top diagram represents chemical knowledge as a dynamical system resulting from the interaction of the material, semiotic and social systems of chemistry. Different data sources and their formats are depicted as well as some mathematical and computational tools for their treatment.

The objects of this system, held together by economic, political, cultural, academic, and other relationships, comprise the chemistry social system (Restrepo and Jost, 2022).

According to Peirce, a semiotic system involves signs, objects and interpretants (Peirce, 1998; Short, 2007). For example, the object we (interpretants or sign users) associate with the sign "water" in English is a colorless liquid that wets materials, quenches our thirst, flows in rivers, dissolves table salt, and has several other characteristics. This sign was not chosen at random; the interpretant had long thought that the wetting capacity of this liquid was relevant to the intended signification. Chemistry's *semiotic system* includes Peirce's objects, signs, and interpretants, as well as their relationships. Semiotic objects are thus substances and reactions. Interpretants (chemists) have devised various symbols for them, such as empirical formulae, chemical, physical, and biological properties, composition, substance names, atoms, spectra, affinity tables, reaction classes, reaction mechanisms, potential energy surfaces, and others (Restrepo and Jost, 2022). New signs emerge from the interaction of reactions and substances, such as those for chemical bonds and periodic systems as indicators of abstract relationships between substances.

The material system includes substances and reactions, as well as some semiotic objects that later became material objects, such as molecules and atoms. Substances, reactions, atoms, bonds, molecular species, materials, reaction conditions, and the technologies used to interact with those objects range from chemistry glassware to automatized spectroscopic devices and robots for chemical synthesis. Chemical reactions, measurements on substances, and associations between substances or molecular species and materials where they are extracted from published work hold material system objects together (Restrepo and Jost, 2022).

Exploring the history of chemistry computationally through the dynamical system of its knowledge necessitates data and tools for analysis. The general framework we propose is depicted in Fig. 7.1, which discusses various sources of information for the three chemistry systems. Similarly, some methodological tools for investigating the evolution of chemical knowledge are mentioned.

7.2.1 Data and tools

The core data for historical studies is derived from primary sources, the majority of which have traditionally been reported in nonborn digital media such as printed scientific publications and patents. Chemistry began to record its vast amount of data in Handbooks as early as 1817, with the first being *Gmelins Handbuch der anorganischen Chemie*, which specialized in inorganic chemistry over time. Beilstein began publishing the *Handbuch der organischen Chemie*, a book specializing in organic chemistry, in 1881. Beilstein began publishing the *Handbuch der organischen Chemie*, a book specializing in organic chemistry, in 1881. These handbooks were digitized over time and are now part of Reaxys, which is owned by Elsevier and is a large database of chemical information that includes weekly updates from 16,400 journals and patents. Other sources of chemical information include the *Chemisches Zentralblatt* and Chemical Abstracts, which have been digitized and merged in SciFinder, another large database of chemical information owned by the American Chemical Society and updated on a daily basis (Fig. 7.1).

The computational era has made it easier to create and store born-digital chemical information, such as The All Results Journals: Chemistry and various blogs, wikis, and electronic lab notebooks. Other sources of information, though not always in electronic form, include graduate theses from around the world, chemical industry archives, chemical society archives, university graduate records, and fine chemical catalogs (Restrepo and Jost, 2022).

Although Reaxys and SciFinder collect a lot of chemical information, they are far from complete. Despite the fact that they contain information relevant to the social and semiotic system of chemistry, much more data remains stored in industrial and university archives, for example (Fig. 7.1), necessitating digitization and incorporation into a large corpus of chemical information. Collecting, curating, and annotating this data necessitates collaboration between chemists and historians.

The ISI Web of Knowledge, Dimensions, and Chemical & Engineering News archives, as well as those of the *Zeitschrift für Angewandte Chemie*, the International Institute of Social History, and Clio Infra are other sources of information for the social and semiotic systems of chemistry (Restrepo and Jost, 2022). They have become crucial in determining the size of the chemical community over time, as well as the relationship between academia and industry. Paper titles and abstracts provide useful information for assessing changes in the evolution of the semiotic system of chemistry. Some of the tools for dealing with this data will be discussed further below.

Although the majority of the aforementioned data is in text format, the computational approach to the history of chemistry necessitates the use of other formats to store and process information. Images of apparatuses and experimental settings, for example, as well as undigitized spectra and other types of visual representations. With the advent of connection tables, SMILEs, and InChIs, images of molecular structures have been converted into text. However, for the evolution of chemistry's semiotics, those images must also be saved as such. As a result, molecular structures necessitate dual storage for computational studies of the history of chemistry, namely text and images. A further unique information token in chemistry is its various tables, which range from affinity tables to current ones, as well as periodic tables. Although many of them can be stored as text, computational studies of the history of chemistry necessitate the storage and treatment of tables as images.

Some of the computational tools and mathematical settings used to study the history of chemistry and the evolution of chemical knowledge are shown in Fig. 7.1. A central mathematical structure for chemistry is that of hypergraphs (Bretto, 2013), which not only model chemical knowledge but also constitute the best model for chemical reactions (Bernal and Daza, 2011; Klamt et al., 2009; Restrepo and Jost, 2022), which along with their meta-information constitute the core of chemical knowledge (Restrepo and Jost, 2022). Hypergraphs represent relationships between sets that may or may not have internal structure. They are a generalization of graphs that allow only binary relationships. For example, molecular structures are represented as graphs, with atoms serving as the graph's vertices and bonds serving as its edges. Social networks are typically represented as graphs, with social actors as vertices and their relationships, such as collaboration or friendship, as edges. Edges in a hypergraph go beyond binary relationships between vertices and, in general, connect sets of vertices (Restrepo, 2022b). Chemical knowledge, for example, is the result of the interaction of three systems, each with its own internal structure. Chemical knowledge, represented as a hypergraph, is composed of the various relationships between chemistry's social, semiotic, and material systems (Fig. 7.1). This implies that the hypergraph connects, for instance chemists with institutions, publications, substances, reactions and political systems (Restrepo and Jost, 2022). As chemical reactions are at the core of chemical knowledge (Schummer, 1998) and the history of chemistry (Restrepo and Jost, 2022), here we provide some details of this representation by using hypergraphs. Fig. 7.2 depicts a labeled directed hypergraph of a small network of chemical reactions. Substrates from a reaction are grouped together in the set of substrates, which is linked to the set of products from the respective reaction. The hypergraph is used to represent the initial and final sets of substances in a chemical reaction.

The possibility of labeling these hypergraphs with relevant information for further historical studies increases the relevance of the hypergraph of chemical reactions. As previously stated, the material system of chemistry, which contains hypergraphs of chemical reactions, is insufficient to encompass the complexities of chemistry. Because chemical knowledge emerges from the interaction of the material system with the social and semiotic systems, hypergraphs of chemical reactions can be labeled appropriately to meet these additional relationships. In Fig. 7.2, we show two examples of this additional labeling. Substances are linked to their

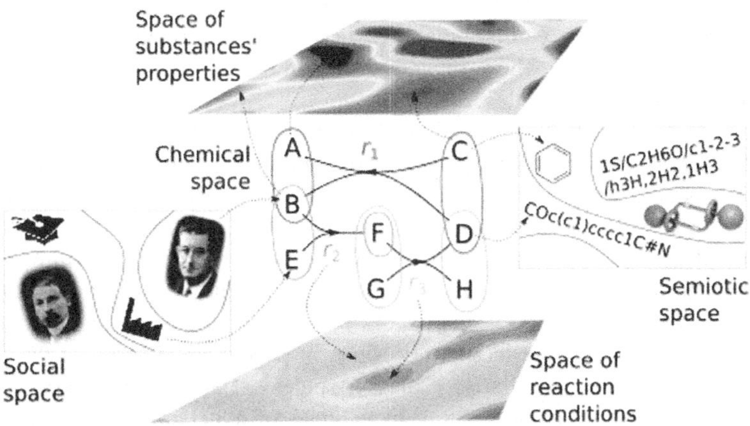

Figure 7.2 Interacting spaces for the computational study of history of chemistry. The chemical space of this depiction is based on four reactions: r_1: C + DA + B, r_2: B + EF, r_3: F + GD + H. These reactions are represented as hypergraphs, where sets of substrates are connected with sets of products. Substances and reactions of the chemical space are connected with other spaces of relevance for the history of chemistry and the evolution of chemical knowledge.

properties, which can be biological, physical, chemical, or ecological (Schummer, 1998). Those properties constitute a space, represented at the top of Fig. 7.2. Similarly, the space of reaction conditions can be associated with the hypergraph of chemical reactions, which contains information on temperatures, reaction times, solvents, catalysts, pHs, pressures, and other important variables informing on how the set of substrates is transformed into the set of products.

Other mathematical theories and computational tools suitable to treat data of relevance for the history of chemistry are discussed (Restrepo and Jost, 2022) and shown in Fig. 7.1. Some of them are complexity measures (Ay et al., 2017, 2011; Jost, 2004), of relevance to quantify the complexity of the different system of chemistry. This quantification may shed light on the effects of social complexity on chemical production, or the relationship between chemistry's semiotic complexity and the expansion of chemical space (Restrepo and Jost, 2022).

Time series analysis (Fulcher et al., 2013; Kantz and Schreiber, 1997) is another useful tool for historical studies because it allows for the study of historical trends as well as forecasting the future of chemical knowledge. This second point is extremely important because it demonstrates that computational approaches to the history of chemistry can not only discuss the past but also shed light on future chemistry trends. However, a word of caution is in order here. The existence of time series analysis, and its recent attention from historians of science (Laubichler et al., 2019), does not imply that it is always possible to make estimates. They are based on the statistical regularity of previous events. For example, we recently discovered that estimates of the future number of new substances are highly inaccurate, despite the fact that these substances have grown at a steady 4.4%/year over the last 200 years (Llanos et al., 2019). It was inaccurate due to the high variability of the time series signal, which is a historical fact (Llanos et al., 2019). We are confident that such an error can be greatly reduced if other driving forces of chemical substance growth, such as those related to the size of the chemical community, are identified (Restrepo and Jost, 2022).

In Fig. 7.2, we depicted the chemical system as a directed hypergraph and left the representation of chemistry's social and semiotic systems open. If these spaces are also modeled with hypergraphs, then interacting hypergraphs provide a formal setting for the evolution of chemical knowledge and the history of chemistry, which is regarded as a generalization of multilayer networks (Aleta and Moreno, 2019; Chodrow and Mellor, 2020; Danziger et al., 2019). In this configuration, each space is a layer. The semiotic and social spaces can be represented as hypergraphs. For example, the co-occurrence of semiotic signs in papers could serve as the foundation for a hypergraph. Similarly, professor-student relationships or co-authorship may serve as criteria for constructing hyperedges of hypergraphs in the social system. As a result, formal tools for dealing with multilayer networks find use in computational studies of the history of chemistry. The statistical physics tools include the mathematics used to analyze these networks (Aleta and Moreno, 2019; Chodrow and Mellor, 2020; Danziger et al., 2019), which are referred in Fig. 7.2. Other statistical physics approaches relevant to the history of chemistry include those developed to study nonequilibrium phase transitions and synchronization and other network coordination patterns (Atay and Jost, 2004; Kaneko, 1984; Lu et al., 2007; Pikovsky et al., 2001). They are useful instruments for, for example, analyzing the specifics of changes in the regime of chemical substance production.

Because one of the goals of computational history of chemistry is to understand and model the emergence of phenomena caused by the interaction of chemistry systems, another computational tool relevant to history of chemistry is agent-based modeling (Restrepo and Jost, 2022). This tool is based on simple local interaction rules among the system's agents, which lead to an emerging pattern after a set number of iterations (Bonabeau, 2002; Eberlen et al., 2017). Agents in the history of chemistry can be scientists, academies, industries, reaction classes, or specific semiotic items. The challenge in applying this technique is determining the simple rules of agent interaction as well as the appropriate size of the interacting agents. These settings are ultimately determined by the available data. Simple rules include the binary decision of whether a new substance will be used with another in a chemical reaction based on chemical similarity.

Language theory can formally treat the evolution of historical changes in chemical language (Matín-Vide, 2003). The basic idea is that any language is built on an alphabet, and that combinations of alphabet characters produce sentences. The grammar of a language encodes the rules for producing formally correct sentences. Stadler's (Andersen et al., 2016, 2017) application of language theory to chemistry is an intriguing one, in which a set of reacting molecules form a graph that is transformed by a chemical rule, typically a name reaction, into a specific set of products.

The evolution of concepts, which can be formally treated using formal concept analysis, is a topic of interest in studies on the history of chemistry (Ganter and Wille, 1999). A context involving objects, attributes, and their relationships is defined using this mathematical technique. The formalism allows for the detection of concepts in a given context, which are characterized by gathering objects with a specific subset of attributes in a closed fashion. For example, in Lavoisian times, the concept of chemical element was viewed as a set of substances with no decomposition reactions, which has evolved to a set of material species with chemical equivalence, atomic number, and lifetimes of at least 10^{-14} seconds (Restrepo, 2020; Restrepo and Jost, 2022). Because contexts can be defined for specific periods of time, the evolution of concepts boils down to analyzing the concepts associated with those contexts, as illustrated by the evolution of the concept of chemical element (Restrepo, 2020; Restrepo and Jost, 2022). Some chemistry-related applications of this mathematical tool have already been reported (Quintero and Restrepo, 2017; Restrepo, 2020).

As previously stated, the various sources of chemical information for historical studies primarily consist of text formats. Text analysis tools will thus be useful in analyzing these large corpora of chemical information. Summarization (Cachola et al., 2020) is one of these techniques that, given an input document, provides a summary of its content using machine learning algorithms. The use of computational semiotic methods is another intriguing text analysis technique with semiotic ramifications (Assaf et al., 2015; Shackell and Sitbon, 2019). This is an appropriate technique for quantifying how ideas and concepts spread over time.

Machine learning is another popular technique nowadays. These methods rely on a large dataset to train an algorithm that is then used to estimate, classify, or make decisions. These algorithms, when combined with the aforementioned methods, form an appropriate set of tools for analyzing the large corpus of chemical information. Machine learning approaches, for example, can be used in conjunction with time series analysis methods to estimate values of specific signals in the material, social, or semiotic system of chemistry. In fact, they have begun to be used in chemistry to estimate synthesis plans (Segler et al., 2018). They can also be used to classify chemical knowledge based on the content of publications (Hook et al., 2018) and even to write reports on parts of the chemical literature that have not been reviewed by humans (Writer, 2019).

The aforementioned tools represent a promising set of methods for processing and extracting knowledge from chemical data. Nothing prevents this set from being expanded to include other methodologies and algorithms.

The following section discusses some preliminary results of applying this computational setting to the study of the history of chemistry, where some of the tools discussed here are used.

7.3 The expanding chemical space, a case study for computational history of chemistry

Despite the novelty of computational approaches to the history of chemistry, there have already been a few studies in this area, such as the investigation of the evolution of chemical space (Llanos et al., 2019) and the influence of this space on the formulation of the periodic system in the 1860s (Leal et al., 2019). In this section, we will summarize the main findings of the first study and discuss the open questions, while the reader is directed to the literature on the second study for more information (Leal et al., 2019).

The study reported by Llanos et al. (2019) sought to determine the rate of expansion of the chemical space as well as the general workings used by chemists to expand the space. We examined 16,356,012 Reaxys reactions and 14,341,955 substances published in scientific journals between 1800 and 2015. We analyzed the annual outcome of new chemicals over time by considering the publication year of a substance as its first report in the database, which led us to the conclusion that the chemical space expands exponentially. In fact, with an annual growth rate of 4.4%, which has not been significantly affected throughout the history of chemistry. This implies that about every 16 years chemists double the number of new substances (Fig. 7.3A).

One advantage of computationally addressing the history of chemistry is that new questions can be formulated and answered that would not be possible to analyze using traditional historical methods. One of these concerns is the effect of World Wars (WWs) on chemical production. Chemistry historians have traditionally examined the role of chemistry in the two World Wars. The effect of wars on

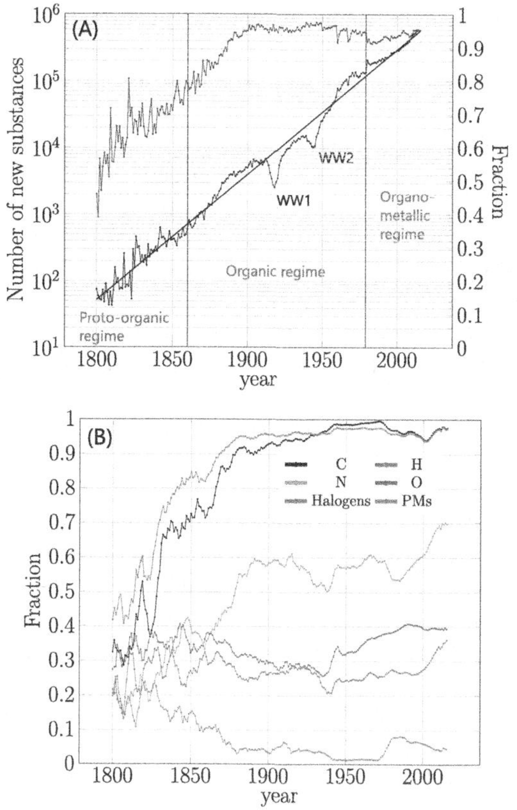

Figure 7.3 The expanding chemical space. (A) Annual growth of number of new substances between 1800 and 2015 (black, left axis). The effects of the World Wars (WWs) are indicated. The three statistical regimes resulting from the annual variability of the number of new substances are indicated, with transitions occurring in 1860 and in 1980 (dotted vertical lines). The exponential equation fitting the growth is indicated as a straight line, with equation: $s_t = 51.85e^{0.04324(t-1800)}$. Fraction of new synthesized compounds to the total of new ones (blue, right axis). (B) Annual fraction of new compounds containing C, H, N, O, halogens, and platinum metals (PMs). These latter corresponding to Fe, Co, Ni, Ru, Rh, Pd, Os, It and Pt. Distributions are convoluted using the moving average method with 5-year window.

chemistry, on the other hand, had never been addressed. Our research enabled us to conclude, and more importantly, quantify, the devastating effect of WW1 on chemistry and the mild effect of WW2. Chemistry was sent back 37 years in WW1 and 16 years in WW2 (Fig. 7.3A). WW1 also resulted in a threefold decrease in the rate of chemical production compared to WW2 (Llanos et al., 2019). The reason for WWI's devastation can be found in the social system of chemistry, which concentrated the chemical industry and research around Germany prior to WWI (Friedman, 2001). After WWI, the chemistry social system was adjusted in such a way that WW2 did not significantly disrupt chemical production. Following WWI, chemistry decentralized from Germany, and other nations, such as the United States, adapted their research and production infrastructures to this new scheme (Friedman, 2001).

Interestingly, WWs had no long-term impact on the expansion of the chemical space; following these events, chemical production recovered and resumed its 4.4% annual growth rate (Fig. 7.3A). This catching-up recovery phenomenon contrasts with other types of production delays, such as the publication of abstracts in other disciplines (de Solla, 1963). Schummer discusses some preliminary research on the possible causes of these phenomena, which are the subject of further investigation in our research group (Schummer, 1997). Our goal is to understand the forces driving this behavior and to develop models that can shed light on chemistry's future material output.

Although WWs have not had a long-term impact on chemical production, they have prompted changes in chemical research. For example, during WWI, the number of As, Sb, and Bi compounds increased while Al, Ga, In, and Tl decreased. During WWII, N and alkali metals declined, but S, B, P, and Si benefited (Llanos et al., 2019). The arsenic warfare agents developed during WW1 may provide an explanation for the rise in As compounds (Radke et al., 2014). Phosphorus compounds became more commonly reported after WWII, when the biological role of phosphorus was established, as well as the use of its compounds in everyday applications and as novel insecticides and other industrial materials (Corbridge, 2013).

From a statistical standpoint, chemical production has undergone two major transitions that have resulted in three distinct production regimes (Fig. 7.3A). We discovered three periods in the history of new chemical production where the variability of the annual output of new substances followed a normal distribution by analyzing the variability of the time series signal depicted in Fig. 7.3A. The first regime, from 1800 to 1860, corresponds to the greatest variability in annual production of new substances. This could be due to the small size of the chemical community, where local setbacks in production of specific research groups could have a large impact on global production. This hypothesis should be investigated further by comparing Fig. 7.3A to annual data on the number of active chemists, which is a topic of current research in our group. Despite the fact that this was the period with the highest percentage of metal compounds reported, C and H compounds dominated throughout the entire period (Fig. 7.3B). Indeed, the second half of the regime was dominated by C, H, N, O, and halogen-based compounds (Fig. 7.3B). According to chemistry historians, this period saw the rise of organic chemistry and, more specifically, the changing role of this type of chemistry, from an analytic approach to a markedly synthetic one (Brock, 1993; Klein, 2003). We called this period the proto-organic regime because of these characteristics (Llanos et al., 2019).

After 1860, a new era of chemical production began, fueled primarily by organic chemistry synthesis. The importance of organic chemical compounds can be seen in the large percentage of carbon and hydrogen compounds that spanned the space during this time period (by 1880, C and H compounds constituted 90% of the new substances) (Fig. 7.3A). This predominance of organic substances has persisted since. In fact, most compounds were made of CHNO as early as 1870, and the same composition is still the most common today (Llanos et al., 2019). The rise of organic chemistry contrasts with a decrease in the percentage of metal-containing compounds (Fig. 7.3B). This was known as the organic regime (Llanos et al., 2019).

The question is about the event or series of events that precipitated the regime change around 1860. Historians have identified the years around 1860 as a period of dramatic changes in chemistry, with significant ramifications in chemistry's semiotics and the way of expanding the chemical space (Klein, 2003). We recently discussed how, prior to 1860, particularly after 1820, the years were dominated by a chemical semiotics based on empirical formulas (Restrepo and Jost, 2022), a semiotic primarily championed by Berzelius (Klein, 2003). Chemists used semiotics to understand and predict new substances in what Klein refers to as a powerful paper tool (Klein, 2003). However, in the 1840s and 1850s, chemists developed more sophisticated analytical techniques that allowed for a more controlled analysis of plant and animal extracts (Brock, 1993; Klein, 2003). This resulted in a flood of organic substances, with previously rare cases of different substances having the same empirical formula becoming the norm, resulting in a semiotic crisis (Restrepo and Jost, 2022). The problem was solved by the introduction and adoption of molecular structural theory (Rocke, 1993), which constituted a more powerful paper tool used by chemists to explore the chemical space in a more controlled manner (Restrepo and Jost, 2022). The structural theory was to chemistry what a tourist guide is to a newcomer eager to explore a city. In addition to the newcomer, chemists could explore the space at random, following different streets and discovering interesting spots from time to time. A tourist guide, on the other hand, allows you to skip the lines and go straight to the city's most interesting sights. This was the situation with structural theory in chemistry; it not only doubled the dimensions of the chemical representation, from unidimensional strings of text (empirical formula) to bi-dimensional drawings of molecules, but it also acted as a tool to discover new substances by applying similar methods to already extracted or synthesized compounds.

The third regime began around 1980 (Fig. 7.3A), and, unlike the transition that occurred by 1860, the event(s) that triggered this transition are still unknown. Some possible causes include chemistry's computerization, but this hypothesis needs to be tested with data from chemistry's social and semiotic systems. This period, which corresponds to our current regime, has been dominated by organic compounds, some of which contain metals. Platinum metal compounds, as well as silicon compounds, increased in popularity during this period (Fig. 7.3B; Llanos et al., 2019). The variability of annual new chemical production is the lowest of the three regimes here, indicating that chemists have more than ever regularized the year-to-year output of new chemical compounds. This is known as the organometallic regime (Llanos et al., 2019).

Chemical substances can come from two sources: extractions or synthesis. The synthesis of urea by Wöhler's in 1828 is widely regarded as the historical event that revealed synthesis as a powerful tool for expanding chemical space (Nicolaou,

2013; Partington, 1964). We discovered that synthesis produced more than half of all new chemicals synthesized and extracted throughout history, with the exception of the first 4 years of the 19th century, when the percentage was slightly lower than 50% (Fig. 7.3A). In particular, new substances containing C, H, N, and O were already about 50% at the time of Wöhler's synthesis, indicating that organic synthesis was well established prior to that. By the turn of the twentieth century, the percentage of chemically synthesized compounds had risen to 90% and has remained there ever since.

Because we already know how chemists populated the chemical space with a conservative set of organogenic elements, the question that arises is how they did so. That is how chemists combined their starting materials to create the ever-expanding chemical space we have today.

We examined the out-degree of substrates, which indicates how frequently substrates are used in reactions, using the hypergraph model for chemical reactions (Fig. 7.2). We discovered that chemists have always preferred certain compounds as starting materials, and that most chemicals are only used as substrates once (Llanos et al., 2019). Strong acids and bases were the most commonly used substrates throughout history, particularly at the beginning of the nineteenth century, before giving way to more organic chemistry-oriented substrates. Acetic anhydride is the most commonly used substrate today, and has been since 1940. Methyl iodide is another important substrate. The question is what is driving these substances to become important substrates in the synthetic toolkit of chemistry. One could argue from a chemical standpoint that acetic anhydride was chosen because of the chemical versatility its two carbonyl groups provide for a variety of chemical reactions; or that methyl iodide is a central substrate for methylation reactions, which are also very popular in synthetic chemistry. Similar arguments could be made for other substrates. To determine the reason(s) for the selection of these toolkit substrates, we must investigate not only the chemistry material system, but also the social and semiotic systems. We need to understand the chemical, social, and semiotic mechanisms that lead to this substrate preference. It's possible that technological progress, combined with economic constraints, propelled acetic anhydride to prominence for reasons other than its chemistry.

To learn more about how substrate selection works in chemistry, we looked at how chemists combined their substrates in chemical reactions. Chemists explicitly described two substrates in roughly half of the reported chemical reactions. Approximately 95% of these two-substrate reactions combine a rarely used substrate with another with a broader range of applications. Acetic anhydride, methanol, and methyl iodide are examples of recurrent substrates. In Llanos et al. (2019), we call this approach to explore the chemical space as the fixed-substrate approach.

Another question raised by modeling the chemical space as a directed hypergraph is whether it is possible to find a hypergraph growth model that leads to a predominance of reactions involving specific substrates such as acetic anhydride and methyl iodide. In this regard, we are developing hypergraph growth models similar to those developed for graphs, such as the Erdős–Rényi, Barabási–Albert and small world models, among others (Newman et al., 2006).

7.4 Conclusions

Chemoinformatics, bioinformatics, computer-aided drug design, cancer biology, emerging pathogens, and computational toxicology are all subjects that benefit from chemistry and the interaction with computation. In this chapter, we propose that by using chemical information (big data) and computation, we can better understand the past, history, and potential future of chemistry. As a result, the aforementioned disciplines benefit from chemistry not only as a provider of experimental settings and a common language to advance their knowledge. These disciplines may learn about their own evolution by studying the evolution of chemistry.

We've talked about how chemical knowledge can be modeled as a complex dynamical system resulting from the interaction of chemistry's material, social, and semiotic systems. The same framework can be used to comprehend the evolution of each of the aforementioned chemistry-related disciplines. Although chemistry has a much larger corpus of information, dating back more than 200 years, the methods described here, particularly the importance of social and semiotic aspects in advancing knowledge, can be used to speculate on the future of the disciplines in question.

We have presented a computational approach to the history of chemistry in addition to motivating a data-driven approach to the evolution of chemical knowledge. In doing so, we presented various sources of digital information for carrying out these studies, as well as mathematical and computational tools for analyzing chemical data from various provenances. Although there are many sources of chemical information that have yet to be digitized, we are confident that interdisciplinary work will fill this gap, with historians, chemists, mathematicians, and computer scientists, among others, contributing to the development of chemical information repositories, primarily for the social and semiotic system of chemistry.

We have shown preliminary results of the exploration of this system by analyzing the growth of the chemical space because the material system of chemistry is the best documented of the three constitutive systems of chemical knowledge. We discovered that chemists have exponentially expanded the set of known substances, unaffected by dramatic events such as World Wars. We've shown how computational studies of this space's expansion can help quantify the impact of setbacks like World Wars on chemical production. Similarly, we demonstrated how data-driven approaches can help debunk accepted historical chemistry ideas, such as the rise of organic chemistry after famous Wöhler' synthesis of urea. Our results indicate that chemical synthesis as early as the dawn of the 19th century is the major chemistry power-force to discover new substances.

We discovered a traditional approach to discovering new substances by analyzing the workings of the chemical space, which we dubbed the fixed-substrate approach, in which chemists traditionally take a well-known chemical such as acetic anhydride and make it react with other less well-studied substrates. The question is whether this traditional approach is motivated by the chemical space, i.e. the chemical nature of the substances, or by social or semiotic factors, such as the influence of chemistry luminaries or the critical role of communication channels in spreading knowledge. Another question that arises is whether similar trends are observed for each of the allied chemistry disciplines covered in this book. So far, how has the space of biological sequences been explored? Are some chemoinformatic algorithms more widely used than others, and does their success depend on the algorithm's efficiency or on social constraints? How has the field of oral-drug-like substances been investigated? Does it follow the same patterns as the rest of the chemical space? We hope that the methods discussed in this chapter will help to answer these questions and inspire further computational research on the history of chemistry.

Acknowledgments

GR thanks Jürgen Jost, Peter F. Stadler, Duc H. Luu, Eugenio J. Llanos and Wilmer Leal for insightful discussions and computations mentioned in this document.

References

Aleta, A., Moreno, Y., 2019. Multilayer networks in a nutshell. Annu. Rev. Condens. Matter Phys. 10 (1), 45–62.

Andersen, J.L., Flamm, C., Merkle, D., Stadler, P.F., 2016. A software package for chemically inspired graph transformationpages In: Echahed, R., Minas, M. (Eds.), Graph Transformation. Springer International Publishing, Cham, pp. 73–88. .

Andersen, J.L., Flamm, C., Merkle, D., Stadler, P.F., 2017. Chemical transformation motifs - modelling pathways as integer hyperflows.

Assaf, D., Yochai, C., Marcel, D., Yair, N., 2015. Opposition theory and computational semiotics. Sign. Syst. Stud. 43, 159–172.

Atay, F.M., Jost, J., 2004. On the emergence of complex systems on the basis of the coordination of complex behaviors of their elements. Complexity 10 (1), 17–22.

Ay, N., Jost, J., Lê, H.V., Schwachhöfer, L.J., 2017. Information geometry, volume 64 of Ergebnisse der Mathematik und ihrer Grenzgebiete. 3. Folge/A Series of Modern Surveys in Mathematics. Springer, Cham.

Ay, N., Olbrich, E., Bertschinger, N., Jost, J., 2011. A geometric approach to complexity. Chaos: An. Interdiscip. J. Nonlin. Sci. 21 (3), 037103.

Bernal, A., Daza, E., 2011. Metabolic networks: beyond the graph. Curr. Comput. Drug. Des. 7 (2), 122–132.

Bertalanffy, L.V., 1950. An outline of general system theory. Br. J. Philos. Sci. 08 (2), 134–165. I.

Bonabeau, E., 2002. Agent-based modeling: methods and techniques for simu- lating human systems. Proc. Natl Acad. Sci. 99 (suppl 3), 7280–7287.

Bretto, A., 2013. Hypergraph Theory. Springer.

Brock, W.H., 1993. The Norton History of Chemistry. W. W. Norton & Company.

I. Cachola, K. Lo, A. Cohan, and D.S. Weld. Tldr: Extreme summarization of scientific documents, 2020.

Chodrow, P., Mellor, A., 2020. Annotated hypergraphs: models and appli- cations. Appl. Netw. Sci. 5 (1), 9.

Corbridge, D.E.C., 2013. Phosphorus: Chemistry, Biochemistry and Technology. CRC Press.

Danziger, M.M., Bonamassa, I., Boccaletti, S., Havlin, S., 2019. Dynamic interdependence and competition in multilayer networks. Nat. Phys. 15 (2), 178–185.

de Solla, D.J., 1963. Price. Little Science, Big Science. Columbia University Press.

Eberlen, J., Scholz, G., Gagliolo, M., 2017. Simulate this! an introduction to agent-based models and their power to improve your research practice. Int. Rev. Soc. Psychol. 30, 149–160.

Friedman, R.M., 2001. The Politics of Excellence. Times Books.

Fulcher, B.D., Little, M.A., Jones, N.S., 2013. Highly comparative time- series analysis: the empirical structure of time series and their methods. J. R. Soc. Interface 10 (83), 20130048.

Ganter, B., Wille, R., 1999. Formal Concept Analysis. Springer.

Hook, D.W., Porter, S.J., Herzog, C., 2018. Dimensions: building context for search and evaluation. Front. Res. Metr. Anal. 3, 23.

Jost, J., 2004. External and internal complexity of complex adaptive systems. Theory Biosci. 123 (1), 69–88.

Kaneko, K., 1984. Period-doubling of kink-antikink patterns, quasiperiodicity in antiferro-like structures and spatial intermittency in coupled logistic lattice: Towards a prelude of a "field theory of chaos". Prog. Theor. Phys. 72 (3), 480–486.

Kantz, H., Schreiber, T., 1997. Nonlinear Time Series Analysis. Cambridge Univ. Press.

Klamt, S., Haus, U.-U., Theis, F., 2009. Hypergraphs and cellular networks. PLOS Comput. Biol. 5 (5), 1–6. 05.

Klein, U., 2003. Experiments, Models, Paper Tools: Cultures of Organic Chemistry in the Nineteenth Century. Stanford University Press.

Laubichler, M.D., Maienschein, J., Renn, J., 2019. Computational history of knowledge: challenges and opportunities. Isis 110 (3), 502–512.

Leal, W., Llanos, E.J., Stadler, P.F., Jost, J., Restrepo, G., 2019. The chemical space from which the periodic system arose. ChemRxiv 8.

Llanos, E.J., Leal, W., Luu, D.H., Jost, J., Stadler, P.F., Restrepo, G., 2019. Exploration of the chemical space and its three historical regimes. Proc. Natl Acad. Sci. 116 (26), 12660–12665.

Lu, W., Atay, F.M., Jost, J., 2007. Synchronization of discrete-time dynamical networks with time-varying couplings. SIAM J. Math. Anal. 39 (4), 1231–1259.

Matín-Vide, C., 2003. Formal grammars and languages. In: Mitkov, R. (Ed.), The Oxford Handbook of Computational Linguistics. Oxford Univ. Press, pp. 157–177.

Newman, M., Barabási, A., Watts, D.J., 2006. The Structure and Dynamics of Networks. Princeton University Press.

Nicolaou, K.C., 2013. The emergence of the structure of the molecule and the art of its synthesis. Angew. Chem. Int. Ed. Eng. 52 (1), 131–146.

Partington, J.R., 1964. A History of Chemistry. Macmillan.

Peirce, C.S., 1998. The Essential Peirce, 2. Indiana University Press.

Pikovsky, A., Rosenblum, M., Kurths, J., 2001. Synchronization. Cambridge Univ. Press.

Quintero, N.Y., Restrepo, G., 2017. Formal Concept Analysis Applications in Chemistry: From Radionuclides and Molecular Structure to Toxicity and Diagnosis. Springer International Publishing, Cham, pp. 207–217.

Radke, B., Jewell, L., Piketh, S., Namieśnik, J., 2014. Arsenic-based warfare agents: Production, use, and destruction. Crit. Rev. Environ. Sci. Technol. 44, 1525–1576.

Restrepo, G., 2020. A Formal Approach to the Conceptual Development of Chemical Element. Oxford University Press, New York, pp. 225–240.

Restrepo, G., 2022a. Computational history of chemistry. Bull. Hist. Chem. 47, 91–106.

Restrepo, G., 2022b. Chemical space: limits, evolution and modelling of an object bigger than our universal library. Digital Discovery. Available from: https://doi.org/10.1039/D2DD00030J.

Restrepo, G., Jost, J., 2022. The Evolution of Chemical Knowledge: A formal Setting for Its Analysis. Springer, Berlin, 2022.

Rocke, A.J., 1993. The Quiet Revolution. University of California Press.

Schummer, J., 1997. Scientometric studies on chemistry I: the exponential growth of chemical substances, 1800–1995. Scientometrics 39 (1), 107–123.

Schummer, J., 1998. The chemical core of chemistry I: a conceptual approach. Hyle 4 (2), 129–162.

Schummer, J., 2006. The philosophy of chemistry: from infancy toward maturity. Boston Stud. Philosophy Sci. .

Segler, M.H.S., Preuss, M., Waller, M.P., 2018. Planning chemical syntheses with deep neural networks and symbolic ai. Nature 555 (7698), 604–610.

Shackell, C., Sitbon, L., 2019. Computational opposition analysis using word embeddings: a method for strategising resonant informal argument. Argum. & Comput. 10 (3), 301–317.

Short, T.L., 2007. Peirce's Theory of Signs. Cambridge University Press.

Writer, B., 2019. Lithium-Ion Batteries. Springer.

Combinatorial and quantum techniques for large data sets: hypercubes and halocarbons

8

Krishnan Balasubramanian
School of Molecular Sciences, Arizona State University, Tempe, AZ, United States

8.1 Introduction

In order to delineate the properties of large data sets and classify the members of the sets as per chosen criterion or properties, one needs computational tools that can both combinatorialy generate and classify large data sets for machine learning, artificial intelligence (AI), and statistical analysis. We review combinatorial techniques that generate a library of molecules that are related to each other as stereoisomers or position isomers, and in general, equivalence classes for large data sets. We have chosen equivalence classes of colorings of hyperplanes of hypercubes due to their applications in biochemical imaging and visualization, neural networks, genetic regulatory networks, Cayley trees, periodic table of elements and AI, etc. (Balasubramanian, 2016a, 2018a,b, 2019, 2020a,b; Banks et al., 2004; Bhaniramka et al., 2000; Carbó-Dorca and Chakraborty, 2019a,b; Carbó-Dorca, 2018; Gowen et al., 2008; Stanley et al., 2009). Furthermore, hypercubes and neural networks find applications in quantum chemistry for the representations of potential energy surfaces of nonrigid molecules, shape analysis, electron density profiles, etc. (Balasubramanian, 2004, 2020d; Mezey, 2012, 2014). Hypercubes find applications in recursive logic, evolving large-scale neural networks, genetic regulatory networks, phylogenetic and other recursive networks, and dynamics of intrinsically ordered proteins (Balasubramanian, 2020a,d; Barthélemy et al., 2005; Forster et al., 2020; Liu and Bassler, 2010; Nandini et al., 2020; Stanley et al., 2009; Wallace, 2011, 2012, 2017). Vertex colorings of hypercubes is a topic of intense scrutiny since the 1800s as some of the earlier results were subsequently shown to be incorrect (Balasubramanian, 2018a). Phylogenetic network analysis of SARS-CoV-2 genomes has been carried out recently (Forster et al., 2020; Nandini et al., 2020) in order to understand the genetic evolution of the corona virus-2 as a function of geographical regions and its relationship to the (assumed) bat origin. These networks and their generalizations have several applications in statistical analysis of the dynamics and the spread of the pandemic through contact tracing, statistical clustering, and other mathematical/AI techniques. Furthermore, AI techniques are becoming increasingly important in the areas of drug discovery and predictive toxicology (Balasubramanian, 2021; Yang et al., 2019; Zhu, 2020). The present author has

Big Data Analytics in Chemoinformatics and Bioinformatics. DOI: https://doi.org/10.1016/B978-0-323-85713-0.00023-2

demonstrated the usefulness of hypercubes for dynamics, spectroscopy, nuclear spin statistics, and rovibronic tunneling splittings of water clusters (Balasubramanian, 2020d). Combinatorial enumeration techniques are also applicable to semiconductor materials and clusters such as Ga_xAs_y (Balasubramanian, 1988a,b, 1990) and fullerene nanospheres (Balasubramanian, 2020e).

Polysubstituted halogenated hydrocarbons, or briefly halocarbons and related heavy metal halides constitute important large data sets as they are perennial environmental pollutants and due to their potential toxicity (Benavides-Garcia and Balasubramanian, 1994; Illenberger and Momigny, 1992; Kaufman et al., 1996; Keng et al., 2020; Koski et al., 1997; Manibusan et al., 2007; Modelli et al., 1992; Punitha et al., 2018; Ravina et al., 2020; Roszak et al., 1993a,b, 1994, 1997, 2001) and toxicology (Balasubramanian and Basak, 2016; Balasubramanian et al., 1986; Basak et al., 2000, 2003; Blair et al., 2004; Costa et al., 2019; Coveney et al. 2016; Crebelli et al., 1992, 1995; Denk et al., 2019; Fang et al., 2008; Friedman, 2000; Fujii et al., 2010; Goldman and Huang, 2018; Kellner et al., 1997; Li et al., 2014, 2016; Liu et al., 2013; Luke et al., 1986, 1987, 1988, 1990; Mclean et al., 2011; Nastainczyk et al., 1978; National Toxicology,1992; Ortiz de Montellano 2010; Trohalaki and Pachter, 2003; Weber et al., 2003; Woo et al., 1985). Quantum chemical tools can provide not only insights into their geometric and electronic parameters but also aid in elucidating the detailed mechanisms of toxic action. Hence the present chapter includes combinatorial techniques for the enumeration and construction of combinatorial libraries of isomers of halocarbons to generate a large data set which can then be studied by a variety of techniques of graph theory, quantum chemistry, genomics and proteomics, neural networks and statistical methods (Balasubramanian and Basak, 2016; Balasubramanian et al., 1986; Blair et al., 2004; Basak et al., 2000, 2003).

Halocarbons have received the attention of several researchers because they are ubiquitous industrial and household chemicals, such as refrigerants, aerosol sprays, pesticides, haloforms as anesthetics, combustion inhibitors, industrial degreasers, etc. These compounds are not only of environmental concern, as they deplete the ozone layer in the stratosphere but they are also of toxicological concern as they cause hepatotoxicity to nephrotoxicity (Balasubramanian and Basak, 2016; Crebelli et al., 1992, 1995; Fang et al., 2008; Fujii et al., 2010; Liu et al., 2013). Even after the ban on some of the halocarbons, they continue to be present in the environment, as they are emitted by seaweeds and marine plants (Keng et al., 2020; Punitha et al., 2018; Ravina et al., 2020). Prolonged exposure to CCl_4 causes not only hepatotoxicity and cirrhosis of the liver but also hepatocellular carcinoma (Edwards, 1941; Fujii et al., 2010). Chloroform causes nephrotoxicity (Fang et al., 2008; Liu et al., 2013) and consequently, the toxicity of halocarbons continues to be a topic of intense research. Hence the toxicity of these species varies strongly in that even the replacement of one halogen by H as in CCl_4 to $CHCl_3$ produces dramatically differing toxicity, and thus there is a compeling need to generate a combinatorial library of these molecules of differing complexity and pursue a comprehensive study of their properties systematically through quantum and other studies.

Experimental studies on halocarbons have been carried out employing anion photoelectron spectroscopy, electron-induced autodissociation of the C−X halide bond, and through a detailed experimental elucidation of the various metabolic products and their interactions with cytochrome P450 enzymes of the liver (Blair et al., 2004; Li et al., 2014; Liu et al., 2013; McLean et al., 2011; Ortiz de Montellano, 2010; Osanai et al., 2010). Thus it is well known that a primary starting process both in the atmosphere and metabolic pathways is electron attachment (Balasubramanian and Basak, 2016; Kaufman et al., 1996; Koski et al., 1997; Luke et al., 1990; Roszak et al., 1993a,b, 1994, 1997, 2001). Electron transfer occurs from an electron-donating enzyme of cytochrome P450 to the halocarbon which then causes the elongation of the C−X bond and subsequent autodissociation of X^- as demonstrated for CCl_4 with CASSCF/MRSDCI studies (Roszak et al., 1994). The potential energy surfaces of the neutral and anionic species undergo crossing near the repulsive well of the neutral, and hence the vertical attachment of an electron to the neutral halocarbon leads to a substantial elongation of a C−Cl bond. This in turn results in the autodissociation of Cl^- due to the PES crossing, thus reactive $\cdot CCl_3$ free radicals are produced. In biological environments, under anaerobic conditions, the halocarbon is reduced and dehalogenated by cytochrome P-450 (Luke et al., 1987). Although this can be inhibited by the presence of O_2 and or NADPH, the halocarbon free radical can abstract protons from the lipid membranes resulting first in the fibrosis of the liver membrane and over a longer period of exposure of halocarbon it causes liver cirrhosis. Consequently, some of the halocarbons metabolize into reactive free radicals and cause damage to the membrane tissues resulting in liver cirrhosis and overlong exposure hepatocellular carcinoma in the case of CCl_4 (Fujii et al., 2010). The halocarbon free radicals bind to the oxidase enzyme in hepatocytes through CYP2E1 in cytochrome P-450; the electron donor enzyme transfers an electron to CCl_4 leading to auto detachment and thus metabolite free radicals. The $\cdot CCl_3$ free radical impairs the hepatocytes by proton extraction changing the permeability of the plasma membranes, lysosome, and mitochondria, thereby causing damage to the liver membranes.

Stimulated by a plethora of experimental and theoretical studies, in this chapter we review combinatorial tools for the study of large data sets with applications to hypercubes and halocarbons. Section 8.2 outlines various combinatorial techniques for creating and partitioning a large library of data sets while Section 8.3 discusses techniques pertinent to halocarbons and Section 8.4 considers hypercubes for large data sets.

8.2 Combinatorial techniques for isomer enumerations to generate large datasets

8.2.1 Combinatorial techniques for large data structures

Combinatorial generation of large data structures is demonstrated using the enumeration of stereoisomers of polysubstituted halocarbons. The technique also demonstrates the complexity arising from the nonrigidity of saturated halocarbons as a consequence of the low-barrier C–C single-bond internal rotations. The generalized wreath product group technique developed for this purpose (Balasubramanian, 1979a,b, 1983, 2016a) for the enumeration of stereo-position isomers including chiral pairs is utilized. This is generalized to other characters with the adaptation of Pólya's theorem (Sheehan, 1967) for the enumeration of both chiral, achiral isomers including meso compounds and position isomers. The combinatorial technique starts with enumerating all 4-trees on n vertices, that is, trees with vertex degrees less than or equal to 4. Alkane trees can be divided into centered and bicentered trees on the basis of the parity of the diameter of a tree (Rains and Sloane, 1999). A tree with even diameter $2m$ must have a node at the midpoint of the tree called the center while a tree of odd diameter $2m + 1$ has a pair of vertices placed at the middle of a path of length $2m + 1$. The generating functions for the two sets of 4-valent trees are obtained as (Rains and Sloane, 1999)

$$C(z) = z + z^3 + z^4 + 2z^5 + 2z^6 + 6z^7 + 9z^8 + 20z^9 + 37z^{10} + 86z^{11} + 181z^{12} + 422z^{13}$$
$$+ 943z^{14} + 2223z^{15} + 5225z^{16} + 12613z^{17} + 30513z^{18} + 74883z^{19} + 184484z^{20} + \ldots \tag{8.1}$$

$$BC(z) = z^2 + z^4 + z^5 + 3z^6 + 3z^7 + 9z^8 + 15z^9 + 38z^{10} + 73z^{11} + 174z^{12} + 380z^{13} + 915z^{14}$$
$$+ 2124z^{15} + 5134z^{16} + 12281z^{17} + 30010z^{18} + 73401z^{19} + 181835z^{20} + \ldots \tag{8.2}$$

Consequently, the total position isomer generating function for alkanes without chirality or stereochemistry is given by

$$GF(C_nH_{2n+2}; \text{position}) = z + z^2 + z^3 + 2z^4 + 3z^5 + 5z^6 + 9z^7 + 18z^8 + 35z^9 + 75z^{10}$$
$$+ 159z^{11} + 355z^{12} + 802z^{13} + 1858z^{14} + 4347z^{15} + 10359z^{16} + 24894z^{17} + 60523z^{18}$$
$$+ 148284z^{19} + 366319z^{20} + \ldots \tag{8.3}$$

For example, the coefficient of z^7 is 9, which gives nine heptane position isomers that are shown in Fig. 8.1. As this generating function scheme does not account for stereochemistry, and accounts for only the position isomers, we include in Fig. 8.1 under each position isomer a * label to designate each chiral center, followed by the number of chiral pairs for each isomer in case there are multiple chiral centers. Thus trees labeled 3 and 4 for heptane are chiral with a single chiral center, thus giving rise to a total of 11 stereo-position isomers for heptane. The corresponding numbers for an octane are 18 and 25, respectively, where 18 corresponds to the position isomers while there are 25 stereo-position isomers which include both chiral pairs and meso isomers. For an octane there exists a position isomer with two

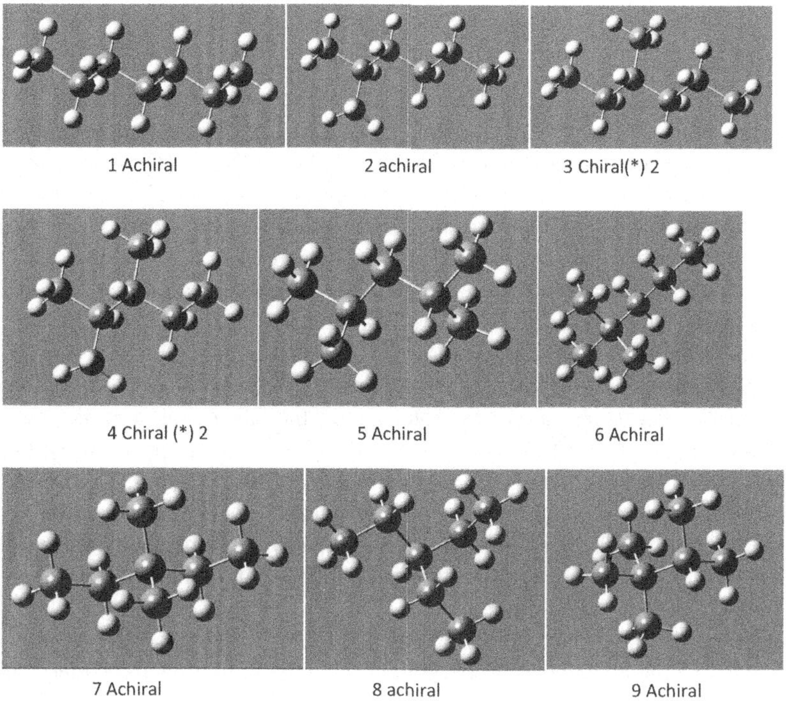

Figure 8.1 Stereo isomers of heptane including chirality that were combinatorially enumerated.

chiral centers such that a mirror plane passes through the center of the molecule, thus resulting in three stereoisomers for this case with one chiral pair and a meso compound. Algorithms have been developed for unique representations of chemical structures with cyclic/acyclic functionalized achiral molecules (Prabhakar and Balasubramanian, 2006).

The combinatorial enumeration of polysubstituted halocarbons starts with the stereo projection of the trees thus enumerated above considering that there are three substituents for the terminal vertices of the tree, two substituents for vertices with degree 2, one substituent for a tertiary carbon or a center with degree 3, and none for vertices of degree 4 (Balasubramanian, 1979a). The vertices of the 4-valent tree can be divided into automorphic equivalence classes on the basis of the automorphism group of the tree which in most general cases are wreath product groups. Suppose D is a set of the vertices of an alkane tree such as the one shown in Fig. 8.1 where we suppress the hydrogens when we refer to quotient trees. The carbon vertices of alkanes are partitioned into automorphic equivalence classes under the action of the automorphism group of the carbon-only quotient tree graph. Let the partitions of the set D of vertices thus generated for alkanes be denoted as Y_1, $Y_2, \ldots Y_m$, where Y_i is an equivalence class partition of the vertices of the alkane under study. Consequently,

$$\bigcup_{i=1}^{m} Y_i = D, Y_i \cap Y_j = \varnothing, \text{ for } i \neq j,$$

where m is the number of equivalence classes of the vertices of the alkane.

Let the nonrigid alkane's point group be G, which includes symmetry operations arising from the rotation around the C–C single bond. Hence G is a wreath product or even a generalized wreath product depending on the alkane tree. For example, for the labeled 5 in Fig. 8.1 the group G is a wreath product $S_2[S_2]$, where S_n is the permutation group containing $n!$ operations. The group G acts on the set D of all carbon vertices, which results in a permutation of vertices within each equivalence class Y_i such that every cycle of the permutation induced by G must be contained within the set Y_i because there cannot be any permutation in i that can mix the members of different Y-sets, as D is partitioned into equivalence classes upon its action on G. Therefore, a vertex coloring of the alkane or a projected three-dimensional-graph such as the ones shown in Fig. 8.1, where the white vertices are replaced with atoms such as F, Cl, Br, and I to form halocarbons. Consequently, a polysubstituted stereo-position isomer of an alkane tree is a map from D to R, where R is the set of colors but because D is partitioned into equivalence classes Y_is, the coloring set can also be partitioned which can provide a tool for stereoisomer enumeration with restrictions.

Let the generalized character cycle index (GCCI) for any character χ of an IR of the group G of the alkane carbon-only tree (i.e., without hydrogens) under consideration be defined as

$$P_G^{\chi} = \frac{1}{|G|}\sum_{g \varepsilon G'} \chi(g) \prod_i \prod_j s_{ij}^{c_{ij}(g)} \tag{8.4}$$

where the sum is over all $g \in G$ and $c_{ij}(g)$ is the number of j-cycles of $g \in G$ upon its action on the set Y_i. The second index j in $c_{ij}(g)$ stands for the cycle length (a j-cycle) generated within the class Y_i for the action of $g \in G$ on Y_i. Eq. (8.4) differs from Pólya's enumeration theorem as we incorporate the character for each irreducible representation of the point group of the alkane and moreover the cycle type products are partitioned into equivalence class of classes on the basis of the partitioning of the set D. The overall cycle index for the enumeration of stereo-position isomers of the polysubstituted alkane is constructed by projecting the tree into a stereograph as described above. Thus for each cycle type s_{ij} in the cycle index (8.4) we associated a Z_{ij} depending on the degrees of the vertices in Y_i. Hence we obtain three types of cycle indices allowing for the internal rotations around the carbon–carbon bonds:

$$\begin{aligned} &Z_{i1} = 1/3\left(s_{i1}{}^3 + 2s_{i3}\right); Z_{i2} = 1/3\left(s_{i2}{}^3 + 2s_{i6}\right); \ldots \\ &Z_{ij} = 1/3\left(s_{ij}{}^3 + 2s_{i3j}\right); \text{if } a \text{ vertex in } Y_i \text{ is } a \text{ terminal vertex.} \end{aligned} \tag{8.5}$$

$$Z_{i1} = s_{i1}{}^2;\; Z_{i2} = s_{i2} \ldots \text{ if } a \text{ vertex in } Y_i \text{ is } a \text{ secondary vertex.} \tag{8.6}$$

$$Z_{i1} = s_{i1}; Z_{i2} = s_{i2} \text{ if } a \text{ vertex in } Y_i \text{ is } a \text{ tertiary vertex.} \quad (8.7)$$

Then the GCCI of the overall nonrigid molecular group of the alkane is obtained by

$$T^{\chi}(s_1, s_2, \ldots s_n) = P_G^{\chi}(s_{ij} \to Z_{ij}) \quad (8.8)$$

The cycle index T^{χ} yields a multinomial generating function for the stereo-position isomers and chiral pairs for polysubstituted halocarbons through the Pólya process. Let $[n]$ be an ordered partition of an integer n into p parts such that $n_1 \geq 0$, $n_2 \geq 0, \ldots, n_p \geq 0$, $\sum_{i=1}^{p} n_i = n$. Define a multinomial function in λs with n_1 colors of the type λ_1, n_2, colors of the type $\lambda_{2\ldots}$ n_p colors of the type λ_p as

$$(\lambda_1 + \lambda_2 + \ldots + \lambda_p)^n = \sum_{[n]}^{p} \begin{pmatrix} & n & \\ n_1 & n_2 & . \; . \; n_p \end{pmatrix} \lambda_1^{n_1} \lambda_2^{n_2} \ldots\ldots \lambda_{p-1}^{n_{p-1}} \lambda_p^{n_p} \quad (8.9)$$

where $\begin{pmatrix} & \mathrm{n} & \\ \mathrm{n}_1 & \mathrm{n}_2 & . \; . \; \mathrm{n}_\mathrm{p} \end{pmatrix}$ are multinomials given by

$$\begin{pmatrix} & n & \\ n_1 & n_2 & . \; . \; n_p \end{pmatrix} = \frac{n!}{n_1! n_2! \ldots\ldots n_{p-1}! n_p!} \quad (8.10)$$

Let the set R of colors is partitioned into sets $R_1, R_2, \ldots, R_m$ with the same number of partitions as the Y-sets such that $R = R_i \cup R_j, R_i \cap R_j = \varnothing$. Assign a weight w_{ij} for each color r_j in the set R_i. Hence the generating function for each irreducible representation for the colorings of vertices of any alkane is obtained by the Pólya substitution as

$$\mathrm{GF}^{\chi} = P_G^{\chi}\left\{ s_{ik} \to \left(w_{i1}^k + w_{i2}^k + \ldots + w_{i,p_i-1}^k + w_{i,p_i}^k \right) \right\} \quad (8.11)$$

where $p_i = | R_i |$. Hence the GFs for each IR and with different colors generate the number of vertex colorings such that each class of the vertices of the alkane tree can be colored independently of the other classes. The GF of the chiral IR generates the number of chiral pairs while the GF of the totally symmetric IR-chiral IR gives the number of achiral isomers and the sum of the numbers obtained from these two IRs enumerates the total number of stereo-position isomers.

8.2.2 Möbius inversion

The Möbius inversion technique is useful in large data sets because it facilitates obtaining various functions for the large data sets in terms of the corresponding functions for the smaller data sets. This can be readily demonstrated with hyperplanes of hypercubes where the generating functions for the cycle types of the permutations for a larger set of hyperplanes can be deduced from the smaller set using the Möbius inversion technique. For an integer q ($q = 1,n$), the computation of the generating functions for the cycle types of various $(n-q)$-hyperplanes of the n-dimensioanl-hypercube involve the Möbius function, a fundamental enumerative combinatorial technique that encompasses generalization of the combinatorial principle of inclusion and exclusion; it has been applied to several areas including music theory (Balasubramanian, 2002), combinatorial isomers with nearest neighbor exclusions (Balasubramanian, 1979b), and graph edge colorings (Balasubramanian, 1988a,b). The Möbius functions provide for various cycle types for the $(n-q)$-hyperplanes of the n-cube using the divisors of the set of all hyperplanes via the simplest cycle types of $q = 1$. Thus the technique involves computing the polynomial generating functions via Möbius sums. One can illustrate this with the case of 5-cube which contains 10 tesseracts ($q = 1$), 40 cells ($q = 2$), 80 faces ($q = 3$), 80 edges ($q = 4$) and 32 vertices ($q = 5$). The generating functions for all cycle types for all values of q representing $(n-q)$-hyperplanes are generated as coefficient of x^q in the polynomial generating function $Q_p(x)$ obtained using the Möbius functions shown below:

$$Q_p(x) = \frac{1}{p}\sum_{d/p} \mu(p/d)F_d(x) \tag{8.12}$$

where the sum is strictly over all divisors d of p, and $\mu(p/d)$ is the Möbius function which takes values 1, −1, −1, 0, −1, 1, −1, 0, 0, 1 ...for arguments 1 to 10; in general, the Möbius function is obtained as follows for any number:

- $\mu(m) = 1$ if *one of m's prime factors is not a perfect square and m contains even number of prime factors,*
- $\mu(m) = -1$ if *m satisfies the same perfect-square condition as before but m contains odd number of prime factors,*
- $\mu(m) = 0$ if *m has a perfect square as one of its factors.*

$F_d(x)$ in Eq. (8.12) is defined as a polynomial in x constructed from the matrix cycle types for $q = 1$ which are readily obtained (Balasubramanian, 2018a). The first row of these elements is represented by a_{1k} while the second rows are denoted by a_{2k} ($k = 1, n$) of the $2 \times n$ matrix cycle type. Then if p is the period of the matrix type shown in the first column of a cycle type for $q = 1$, and define, $g = gcd(k; p)$, $p' = k/g$, $h = gcd(2k; p)$; $p'' = 2k/h$, and define the polynomial $F_p(x)$ in terms of these divisors of the cycle type as

$$Fp(x) = \prod_{k}^{nc} \left(1+2x^{p'}\right)^{ga_{1k}} \left(1+2x^{p''}\right)^{\frac{ha_{2k}}{2}}, \text{ if } h \text{ does not divide } k;$$

$$Fp(x) = \prod_k^{nc} (1+2xp')^{(ga_1k)}, \text{ if } h \text{ divides } k, \tag{8.13}$$

where the product is taken *only* over nc, nonzero columns of the $2 \times n$ matrix cycle types for $q = 1$.

The coefficient of x^q in $Q_p(x)$ obtained from the Möbius sums of various F_d polynomials where d's are strictly divisors of p generate the various cycle types for $(n-q)$-hyperplanes of the n-dimensioanl-hypercube. Consider the following matrix type for the 5-cube $q = 1$:

$$\begin{pmatrix} 0 & 1 & 1 & 0 & 0 \\ 0 & 0 & 0 & 0 & 0 \end{pmatrix} \tag{8.14}$$

As only second and third columns contain nonzero values, we need to consider only these two columns. Thus the maximum period to consider is 6 and hence the possible F polynomials are F_6, F_3, F_2, and F_1 as divisors of 6 are 1, 2, 3, and 6. Applying the GCD followed by the use of Eq. (8.13), we obtain each of these polynomials as

$$F_1(x) = \left(1 + 2x^2\right)\left(1 + 2x^3\right) \tag{8.15}$$

$$F_2(x) = (1+2x)^2\left(1 + 2x^3\right) \tag{8.16}$$

$$F_3(x) = \left(1 + 2x^2\right)(1+2x)^3 \tag{8.17}$$

$$F_6(x) = (1+2x)^5 \tag{8.18}$$

From the F_d polynomials thus constructed above, we obtain the Q_p polynomials using the Möbius sum, shown in Eq. (8.12). Thus we obtain

$$Q_1 = F_1 = 1 + 2x^2 + 2x^3 + 4x^5 \tag{8.19}$$

$$Q_2 = ½\{\mu(2)F_1 + \mu(1)F_2\} = ½\{F_2 - F_1\} = ½\left\{(1+2x)^2\left(1 + 2x^3\right) - \left(1 + 2x^2\right)\left(1 + 2x^3\right)\right\}$$

$$= 2x + x^2 + 4x^4 + 2x^5 \tag{8.20}$$

$$Q_3 = 1/3\{\mu(1)F_3 + \mu(3)F_1\} = 1/3\{F_3 - F_1\} = 1/3\left\{\left(1 + 2x^2\right)(1+2x)^3 - \left(1 + 2x^2\right)\left(1 + 2x^3\right)\right\}$$

$$= 2x + 4x^2 + 6x^3 + 8x^4 + 4x^5 \tag{8.21}$$

$$Q_6 = 1/6\{\mu(1)F_6 + \mu(2)F_3 + \mu(3)F_2 + \mu(6)F_1\} = 1/6\{F_6 - F_3 - F_2 + F_1\}$$

$$= 4x^2 + 10x^3 + 8x^4 + 2x^5. \tag{8.22}$$

The coefficients of x^qs are tabulated below for all possible Q_p polynomials which yield the cycle types for various $(n-q)$-hyperplanes as shown below:

Q_p	x	x^2	x^3	x^4	x^5
Q_1		2	2		4
Q_2	2	1		4	2
Q_3	2	4	6	8	4
Q_6		4	10	8	2
Cycle type	$2^2 3^2$	$1^2 2^1 3^4 6^4$	$1^2 3^6 6^{10}$	$2^4 3^8 6^8$	$1^4 2^2 3^4 6^2$
Hyperplane	$q = 1$ (tesseracts)	$q = 2$ (cells)	$q = 3$ (faces)	$q = 2$ (edges)	$q = 5$ (vertices)

8.2.3 Combinatorial results

We have applied the combinatorial techniques for the enumeration of various super-giant fullerenes, hypercube colorings, trees, and molecules of interest in drug discovery and protein–protein interactions (Balasubramanian, 2004, 2016a, 2018a,b, 2019, 2020a,b,c,d,e; Balasubramanian and Gupta 2019). We consider the polysubstituted isomers of unbranched alkanes for which GCCIs are obtained using wreath product groups depending on the number of carbons, n is odd or even shown below:

n: even

$$T^{A1} = \frac{1}{36}\{s_1^{2n+2} + 4s_1^{2n-1}s_3 + 4s_1^{2n-4}s_3^2 + 3s_2^{n+1} + 6s_2^{n-2}s_6 + 3s_2^{n+1} + 6s_2^{n-2}s_6 + 9s_1^2 s_2^n\} \tag{8.23}$$

$$T^{A2} = \frac{1}{36}\{s_1^{2n+2} + 4s_1^{2n-1}s_3 + 4s_1^{2n-4}s_3^2 - 9s_1^2 s_2^n\} \tag{8.24}$$

n: odd

$$T^{A1} = \frac{1}{36}\{s_1^{2n+2} + 4s_1^{2n-1}s_3 + 4s_1^{2n-4}s_3^2 + 3s_2^{n+1} + 6s_2^{n-2}s_6 + 6s_1^2 s_2^{n-3}s_6 + 12s_1^2 s_2^n\} \tag{8.25}$$

$$T^{A2} = \frac{1}{36}\{s_1^{2n+2} + 4s_1^{2n-1}s_3 + 4s_1^{2n-4}s_3^2 + 3s_2^{n+1} + 6s_2^{n-2}s_6 + 6s_1^2 s_2^{n-3}s_6 + 12s_1^2 s_2^n\} \tag{8.26}$$

The order of the group is 36 because of two nonrigid terminal methyl rotors, and a plane of symmetry is included for the enumeration of both stereo-position, achiral isomers, and chiral pairs including meso compounds for polysubstituted unbranched alkanes. The corresponding multinomial GFs by the Pólya substitution are shown below as a function of the weights λ_i.

n: even

$$\mathrm{GF}^{A1} = \frac{1}{36}\{(\lambda_1+\lambda_2+\cdots+\lambda_k)^{2n+2} + 4(\lambda_1+\lambda_2+\cdots+\lambda_k)^{2n-1}(\lambda_1^3+\lambda_2^3+\cdots.+\lambda_k^3)$$

$$+4(\lambda_1+\lambda_2+\cdots+\lambda_k)^{2n-4}(\lambda_1^3+\lambda_2^3+\cdots.+\lambda_k^3)^2 + 3(\lambda_1^2+\lambda_2^2+\cdots.+\lambda_k^2)^{n+1}$$

$$+6(\lambda_1^2+\lambda_2^2+\cdots.+\lambda_k^2)^{n-2}(\lambda_1^6+\lambda_2^6+\cdots.+\lambda_k^6) + 3(\lambda_1^2+\lambda_2^2+\cdots+\lambda_k^2)^{n+1}$$

$$+6(\lambda_1^2+\lambda_2^2+\cdots.+\lambda_k^2)^{n-2}(\lambda_1^6+\lambda_2^6+\cdots.+\lambda_k^6)^2 + 9(\lambda_1+\lambda_2+\cdots.+\lambda_k)^2$$

$$(\lambda_1^2+\lambda_2^2+\cdots.+\lambda_k^2)^n\} \tag{8.27}$$

$$\mathrm{GF}^{A1} = \frac{1}{36}\{(\lambda_1+\lambda_2+\cdots+\lambda_k)^{2n+2} + 4(\lambda_1+\lambda_2+\cdots+\lambda_k)^{2n-1}(\lambda_1^3+\lambda_2^3+\cdots.+\lambda_k^3)$$

$$+4(\lambda_1+\lambda_2+\cdots+\lambda_k)^{2n-4}(\lambda_1^3+\lambda_2^3+\cdots.+\lambda_k^3)^2 - 9(\lambda_1+\lambda_2+\cdots+\lambda_k)^2(\lambda_1^2+\lambda_2^2+\cdots.+\lambda_k^2)^n\} \tag{8.28}$$

n: odd

$$\mathrm{GF}^{A1} = \frac{1}{36}\{(\lambda_1+\lambda_2+\cdots+\lambda_k)^{2n+2} + 4(\lambda_1+\lambda_2+\cdots+\lambda_k)^{2n-1}(\lambda_1^3+\lambda_2^3+\cdots.+\lambda_k^3)$$

$$+4(\lambda_1+\lambda_2+\cdots+\lambda_k)^{2n-4}(\lambda_1^3+\lambda_2^3+\cdots.+\lambda_k^3)^2 + 3(\lambda_1^2+\lambda_2^2+\cdots.+\lambda_k^2)^{n+1}$$

$$+6(\lambda_1^2+\lambda_2^2+\cdots.+\lambda_k^2)^{n-2}(\lambda_1^6+\lambda_2^6+\cdots.+\lambda_k^6) + 6(\lambda_1+\lambda_2+\cdots+\lambda_k)^2(\lambda_1^2+\lambda_2^2+\cdots.+\lambda_k^2)^n$$

$$+12(\lambda_1+\lambda_2+\cdots+\lambda_k)^2(\lambda_1^2+\lambda_2^2+\cdots.+\lambda_k^2)^{n-3}(\lambda_1^6+\lambda_2^6+\cdots.+\lambda_k^6)\Big\} \tag{8.29}$$

$$\mathrm{GF}^{A2} = \frac{1}{36}\{(\lambda_1+\lambda_2+\cdots+\lambda_k)^{2n+2} + 4(\lambda_1+\lambda_2+\cdots+\lambda_k)^{2n-1}(\lambda_1^3+\lambda_2^3+\cdots.+\lambda_k^3)$$

$$+4(\lambda_1+\lambda_2+\cdots+\lambda_k)^{2n-4}(\lambda_1^3+\lambda_2^3+\cdots.+\lambda_k^3)^2 + 3(\lambda_1^2+\lambda_2^2+\cdots+\lambda_k^2)^{n+1}$$

$$+6(\lambda_1^2+\lambda_2^2+\cdots.+\lambda_k^2)^{n-2}(\lambda_1^6+\lambda_2^6+\cdots.+\lambda_k^6) - 6(\lambda_1+\lambda_2+\cdots+\lambda_k)^2(\lambda_1^2+\lambda_2^2+\cdots.+\lambda_k^2)^n$$

$$-12(\lambda_1+\lambda_2+\cdots.+\lambda_k)^2(\lambda_1^2+\lambda_2^2+\cdots.+\lambda_k^2)^{n-3}(\lambda_1^6+\lambda_2^6+\cdots.+\lambda_k^6)\Big\}. \tag{8.30}$$

The coefficient of a given term, $\lambda_1^i\lambda_2^j\lambda_3^k\lambda_4^l\lambda_5^m$, generates the number of chiral pairs for the A_2 IR; likewise the sum of the coefficients of $\lambda_1^i\lambda_2^j\lambda_3^k\lambda_4^l\lambda_5^m$ of GFs A_1 and A_2 enumerates the stereo-position isomers. The difference in the coefficients of A_1 and A_2 for the term $\lambda_1^i\lambda_2^j\lambda_3^k\lambda_4^l\lambda_5^m$ enumerates the number of achiral isomers for the halocarbon $C_nH_iF_jCl_kBr_lI_m$ when the quotient alkane is unbranched. Table 8.1 shows the combinatorial enumeration of $C_{10}H_iF_jCl_kBr_lI_m$. Fig. 8.2 lists both the stereo and position isomers of $C_6H_{12}Cl_2$. The first 12 isomers correspond to unbranched trees which generate 26 stereoisomers with 10 chiral pairs (Fig. 8.2). These isomers correspond to the partition 12 2 0 0 0 among the ones shown in Table 8.1 which contain $N(A_1) = 16$ and $N(A_2) = 10$ yielding 10 chiral pairs and a total of 26 isomers. The method enumerates all stereo-position isomers including the meso compounds for all possible partitions for halocarbons of fluoro-chloro-bromo-iodo compounds.

The combinatorial numbers enumerated for A_1 and A_2 form interesting integer sequences which are displayed below for the various cases for binomial colorings for alkanes with n carbons:

$n = 6$: $\{1, 3, 16, 38, 86, 133, 185, 196\}, \{0, 2, 10, 32, 70, 118, 160, 176\}$

$n = 7$: $\{1, 4, 22, 67, 172, 327, 522, 674, 742\}, \{0, 2, 15, 55, 150, 295, 481, 624, 692\}$

$n = 8$: $\{1, 4, 29, 102, 316, 700, 1316, 1978, 2562, 2752\}, \{0, 3, 21, 94, 287, 672, 1253, 1922, 2471, 2682$

$n = 10$: $\{1, 5, 46, 218, 874, 2577, 6292, 12504, 21150, 30314, 37680, 40348\}, \{0, 4, 36, 208, 828, 2532, 6168, 12384, 20904, 30104, 37344, 40096\}$

8.3 Quantum chemical techniques for large data sets

8.3.1 Computational techniques for halocarbons

We seek optimal quantum chemical techniques and basis sets for a large combinatorial library of molecules for which electronic properties are needed. Halocarbons offer a challenging platform in that the electron affinities are quite sensitive to the basis sets, as diffuse functions and polarization functions can influence the electron affinities. An augmented correlation consistent basis set including up to 4 f functions with the coupled cluster singles + doubles (CCSD) method can be employed to study a given halocarbon (Balasubramanian and Basak, 2016). However, such methods are very CPU and disk intensive and hence they are not desirable for a large combinatorial library. We find that 6–311 G+ +(2d,p) basis sets in conjunction with B3LYP/DFT techniques appear to offer a promising method for the study of a larger combinatorial library of halocarbons. A smaller 6–311 G + (2d,p) basis set with B3LYP provides very good optimized geometries for the neutral halocarbons and their spectra such as IR and Raman spectra. However, the vertical electron affinities (VEA) values are sensitive to the basis sets and we recommend the 6–311 G+ +(2d,p) basis set for this purpose. Moreover, relativistic effects need to be included for halocarbons containing Br, I, and other heavy atoms (Balasubramanian, 1989, 1990, 1997a,b, 2016b; Balasubramanian and Liao, 1989,

Table 8.1 Few terms in the combinatorics of stereo isomers of unbranched $C_6H_iF_jCl_kBr_lI_m$.[a]

i j k l m	$N(A_1)$	$N(A_2)$ dl-pairs
14 0 0 0 0	1	0
13 1 0 0 0	3	2
12 2 0 0 0	16	10
11 3 0 0 0	38	32
10 4 0 0 0	86	70
9 5 0 0 0	133	118
8 6 0 0 0	185	160
7 7 0 0 0	196	176
12 1 1 0 0	24	23
11 2 1 0 0	107	101
10 3 1 0 0	295	289
9 4 1 0 0	598	583
8 5 1 0 0	927	912
7 6 1 0 0	1150	1130
10 2 2 0 0	455	423
10 1 1 1 1	1734	1734
9 2 1 1 1	6790	6790
8 3 1 1 1	16728	16728
7 4 1 1 1	29160	29160
6 5 1 1 1	38052	38052
8 2 2 1 1	24830	24800
7 3 2 1 1	56440	56440
6 4 2 1 1	89420	89360
5 5 2 1 1	103684	103684
6 3 3 1 1	116648	116648
5 4 3 1 1	164248	164248
4 4 4 1 1	199440	199350
7 2 2 2 1	83780	83660
6 3 2 2 1	173100	172980
5 4 2 2 1	243760	243580
5 3 3 2 1	318024	318024
4 4 3 2 1	386210	386030
4 3 3 3 1	504000	504000
6 2 2 2 2	257048	256480
5 3 2 2 2	471980	471620
4 4 2 2 2	573370	572530
4 3 3 2 2	747920	747560
3 3 3 3 2	976080	976080

[a] $N(A_1) + N(A_2)$ is the total no of stereo-position isomers. $N(A_1)-N(A_2)$ is the total no of achiral isomers. $N(A_2)$ is the total no of chiral pairs.

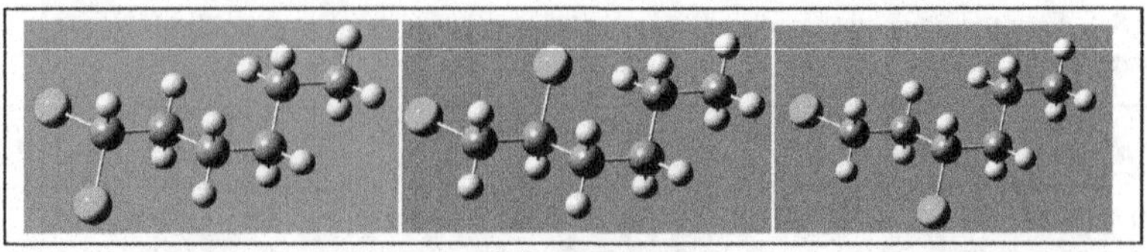

1 Achiral 2 Chiral (*) 2 3 Chiral (*) 2

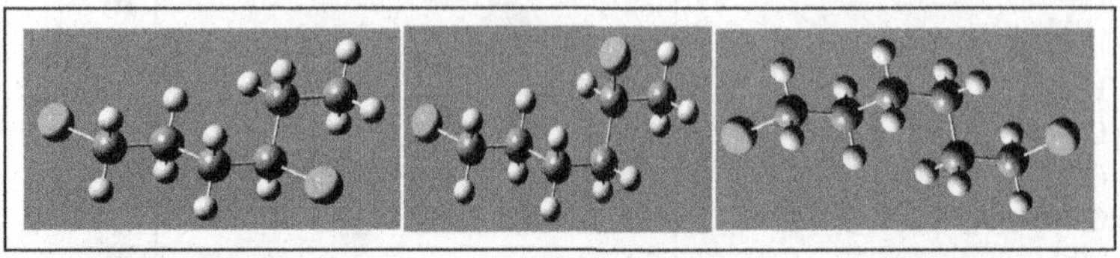

4 Chiral (*) 2 5 Chiral(*) 2 6 Achiral

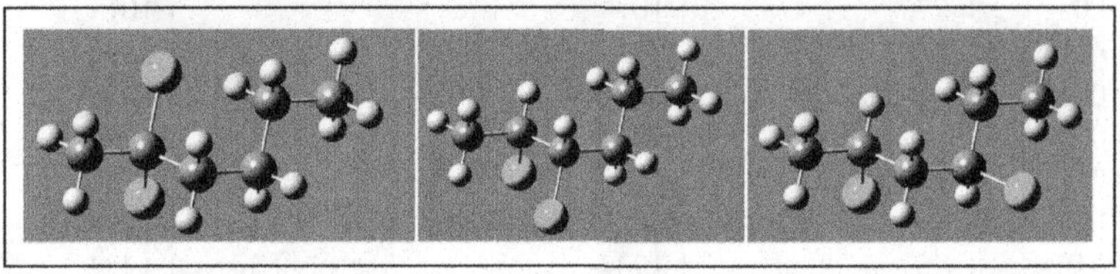

7 AChiral 8-Chiral (*,*) 4 9-Chiral (*,*) 4

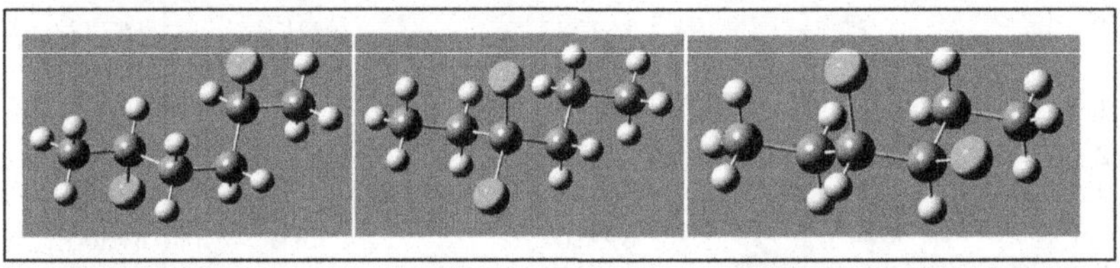

10 Chiral:2 (meso)Achiral1 11-Achiral 12-Chiral:2 (meso)Achiral1

13-Chiral (2) 14-3m-1Cl,3F-Pentane Chiral:2 15-3m-1F,3Cl-Pentane

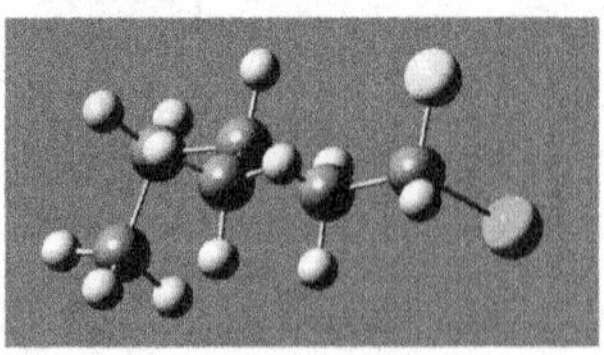

16-3m-1F,1Cl-Pentane

Figure 8.2 Various halocarbon isomers containing six carbons that were combinatorially enumerated.

1991; Balasubramanian et al., 1991; Benavides-Garcia and Balasubramanian, 1991, 1994; Kim and Balasubramanian, 1992; Majumdar et al., 2002; Roszak et al., 1994, 1997). All DFT/B3LYP, MP2 and CCSD computations were carried out with Gaussian codes (Frisch et al., 2009). As a case study, we have previously considered 55 halocarbons from the Crebelli set (Crebelli et al., 1992, 1995). The toxicities of halocarbons are correlated with energetics involved in electron attachment (VEAs) followed by autodetachment and proton extraction potentials from the lipid. Specifically, statistical studies based on hierarchical QSAR (H-QSAR) have been carried out for two toxicity endpoints namely, ARR and D_{37} for the Crebelli set (Basak et al., 2003). Although quantum chemical techniques provide significant insights into elucidating the mechanisms of toxic action, the HQSAR techniques that include topological, topochemical, and quantum chemical parameters offered a greater degree of statistic correlation to toxicity endpoints than quantum chemical parameters by themselves, as toxicity endpoints depend on a number of factors such as the volatility of the chemicals, dermal penetrations, lipophilicity, proton extraction potentials of toxins from the epithelial lipid layers of the liver, electron affinities, and so on. Hence an integrated H-QSAR approach appears to be a better route for machine learning of large data sets to predict end point toxicity parameters (Basak et al., 2003).

8.3.2 Results and discussions of quantum computations and toxicity of halocarbons

As an example to demonstrate the utility of both combinatorial and quantum techniques, in Table 8.2 we display the VEAs of 55 halocarbons obtained by a variety of techniques and basis sets. The VEA is defined as the total energy of the neutral halocarbon minus the total energy of the anion at the equilibrium geometry of the neutral. Hence negative values of VEA mean the anion is higher than the neutral molecule at the neutral geometry. The VEA is sensitive to the basis set and level of theory, as it changes dramatically by the inclusion of diffuse and polarization functions. The B3LYP results for the VEAs are quite comparable to the CCSD results within the same basis set. Table 8.2 shows the results obtained at the MP2/CCSD and B3LYP/6−311++G(2d,p) levels for comparison. As seen from Table 8.2, among the halocarbons in the Crebelli set, CCl_4, CBr_4, $CHBr_3$, CCl_3Br...have positive VEAs. For most of the other molecules in the Crebellli set such as dichloro or dichlorofluoro or difluoro hydrocarbons, VEAs are negative. Among 55 halocarbons (Table 8.2) CF_4 has the most negative VEA (−7.9 eV) in dramatic contrast to CCl_4 (+0.44 eV). It is known that CF_4 is not toxic while CCl_4 is both toxic and carcinogenic in rats and mice (Fujii et al., 2010). $CFCl_3$ which has a VEA of −1.2 eV is on the borderline of toxicity while CF_3Br and CF_3I with VEAs of −0.69 and 0.1 eV, are both toxic. The vertical ionization potentials and the energies of HOMOs of all halocarbons are not impacted much by the inclusion of diffuse or additional polarization functions in contrast to the energies of lowest unocuupied molecular orbitals (LUMOs).

Table 8.2 Vertical electron affinities of 55 compounds in Fig. 8.4 at different levels of theory.

Crebelli		VEA (eV)	VEA (eV)	VEA (eV)
Species	Set	MP2	CCSD	CCSD(T)
CH_2Cl_2	01	−0.674	−0.693	−0.67
$CHCl_3$	02	−0.763	−0.750	−0.722
CCl_4	03	−0.094(0.39)[a]	0.44	0.31
$C_2H_4Cl_2$	04	−0.70	−0.69	−0.66
$C_2H_4Cl_2$	05	−0.77	−0.77	−0.74
$C_2H_3Cl_3$	06	−0.73	−0.70	−0.66
$C_2H_3Cl_3$	07	−0.76	−0.74	−0.704
$C_2H_2Cl_4$	08	−0.74	−0.70	−0.66
$C_2H_2Cl_4$	09	−0.845	−0.854	−0.81
C_2HCl_5	10	−0.79	−0.710	−0.70
C_2HCl_5	11	−0.79	−0.711	−0.697
C_2HCl_3	12	−0.79	−0.76	−0.71
C_2Cl_4	13	−0.99	−1.03	−0.95
$C_2H_2Cl_2$	14	−0.85	−0.86	−0.825
$C_2H_2Cl_2$	15	−0.81	−0.77	−0.74
$C_3H_6Cl_2$	16	−0.61	−0.59	−0.56
$C_3H_6Cl_2$	17	−0.67	−0.66	−0.62
$C_3H_6Cl_2$	18	−0.582	−0.561	−0.532
$C_3H_5Cl_3$	19	−0.51	−0.487	−0.46
1-Cl-butane	20	−0.706−1.11(DFT)[a]−1.49[a]	−0.679 (−1.45[a])	−0.650
2-Cl-butane	21	−0.717−1.12(DFT)[a]-1.48[a]	−0.693−1.46[a]	−0.663
$C_4H_8Cl_2$	22	−0.613		
$C_4H_8Cl_2$	23	−0.781		
1-Cl-2methyl-propane	24	−0.569−0.36-DFT-1.03[a]-0.53DFT[b] -1.03MP2[a]-0.8MP2[b]	−0.529	−0.547
1-Cl-2 methyl-propane	25	−0.55−0.35-DFT-1.04DFT[a] -0.52DFT[b]-1.40MP2[a]-0.76MP2[b]		
$C_5H_{11}Cl$	26	−0.689		
$C_6H_{13}Cl$	27	−0.537		
$C_3H_4Cl_2$	29	−0.702	−0.663	−0.636

$C_3H_4Cl_2$	30	−0.620	−0.593	−0.565
$C_3H_4Cl_2$	31	−0.637	−0.625	−0.599
$C_3H_3Cl_3$	32	−0.715		
$C_4H_6Cl_2$	33	−0.540		
C_4H_7Cl	34	−0.690	−0.663	−0.631
2methylClpropane	35	(−0.52)[b]		
CBr_2ClF	36	(0.71)[b]		
$CHBr_3$	37	(0.62)[b]		
CH_2BrCl	38	−0.64 (−0.61)[b]	−0.629	−0.600
$CBrCl_3$	39	0.085 (0.707)[b]	0.00082	0.071
$CHBrCl_2$	40	−0.717		
$CHBr_2Cl$	41	−0.545		
C_2H_4BrCl	42	−0.737	−0.744	−0.695
1Br-butane	43	(−0.50)[b]		
2Br-butane	44	(−0.53)[b]		
1Cl-3Br-propane	45	(−0.36)[b]		
1,2-dibromo-propane	46	(−0.32)[b]		
1Br-2methyl propane	47	(−0.46)[b]		
2Br-2methyl propane	48	(−0.46)[b]		
1Cl4Br-butane	49	(−0.38)[b]		
4Br-2methylbutane	50	(−0.45)[b]		
CBr_2Cl_2	51	0.439 (0.979)[b]		
CH_2Br_2	52	−0.570 (−0.13)[b]	−0.524	−0.506
CBr_4	53	(1.41)[b]		
$C_2H_2Br_4$	54	(0.49)[b]		
$C_2H_2Br_2$	55	−0.797 (−0.34)[b]	−0.806	−0.770

Result without * designate aug-cc-pvTZ basis.
[a]Results designate 6−311 + G(2d,p) basis.
[b]Results designate 6−311++G(2d,p) basis; all numbers in parentheses are B3LYP results.

Among 55 compounds in Table 8.2 and Fig. 8.3, CCl_4 is very interesting from the standpoint of toxicity as I has a VEA of +0.44 eV, and moreover, the C–Cl bond length elongates from 1.76 to 2.25 Å upon vertical electron attachment (Roszak et al., 1994). This is followed by the spontaneous dissociation of one of the C–Cl bonds, causing the production of reactive $\cdot CCl_3X$(doublet) radical which causes damage to the epithelial lipid membranes of the liver, culminating into fibrosis followed by the cirrhosis of the liver. Proton abstraction by the free radical from the lipid layers of the liver tissues plays a critical role. The toxicity is dependent on the VEA of the halocarbon and C–H bond energy for proton abstraction. The criterion for potential toxicity of a halocarbon as its VEA >-1.4 eV and the C–H bond Energy >3.9 eV. A positive VEA with the C–H bond Energy >3.9 eV would imply that the compound is not only toxic but can also be a potential carcinogen.

Fig. 8.4 shows the computed molecular electrostatic potentials of the first isomer enumerated in Fig. 8.2 (1) or 1,1dichloro-hexane. The left side of Fig. 8.4 depicts the MEP of the neutral halocarbon while the right side of Fig. 8.4 corresponds to the MEP of the anion. The red regions in MEPS correspond to hot regions or

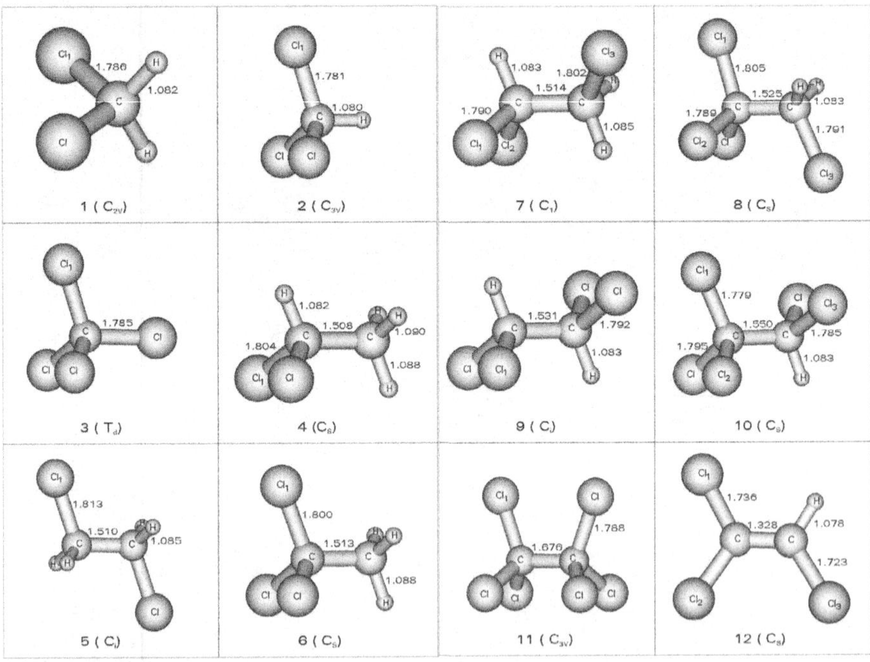

Figure 8.3 Computed geometries of 55 halocarbons in the Crebelli toxicity data.
Source: From Balasubramanian, K. and Basak, S. C., 2016. Metabolic electron attachment as a primary mechanism for toxicity potentials of halocarbons. Curr. Comput. Drug. Des., 12 (1), 62–72.

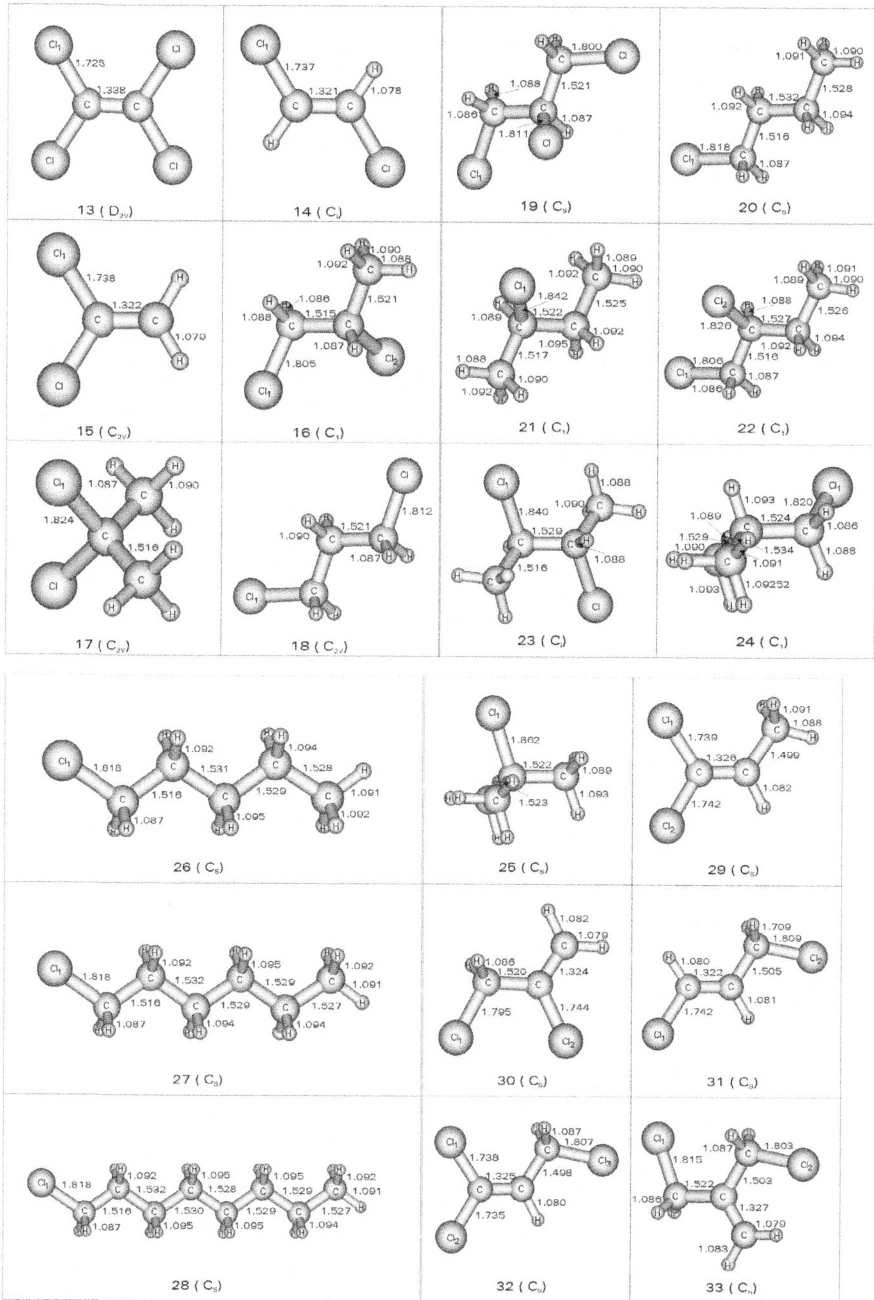

Figure 8.3 (Contiunucd)

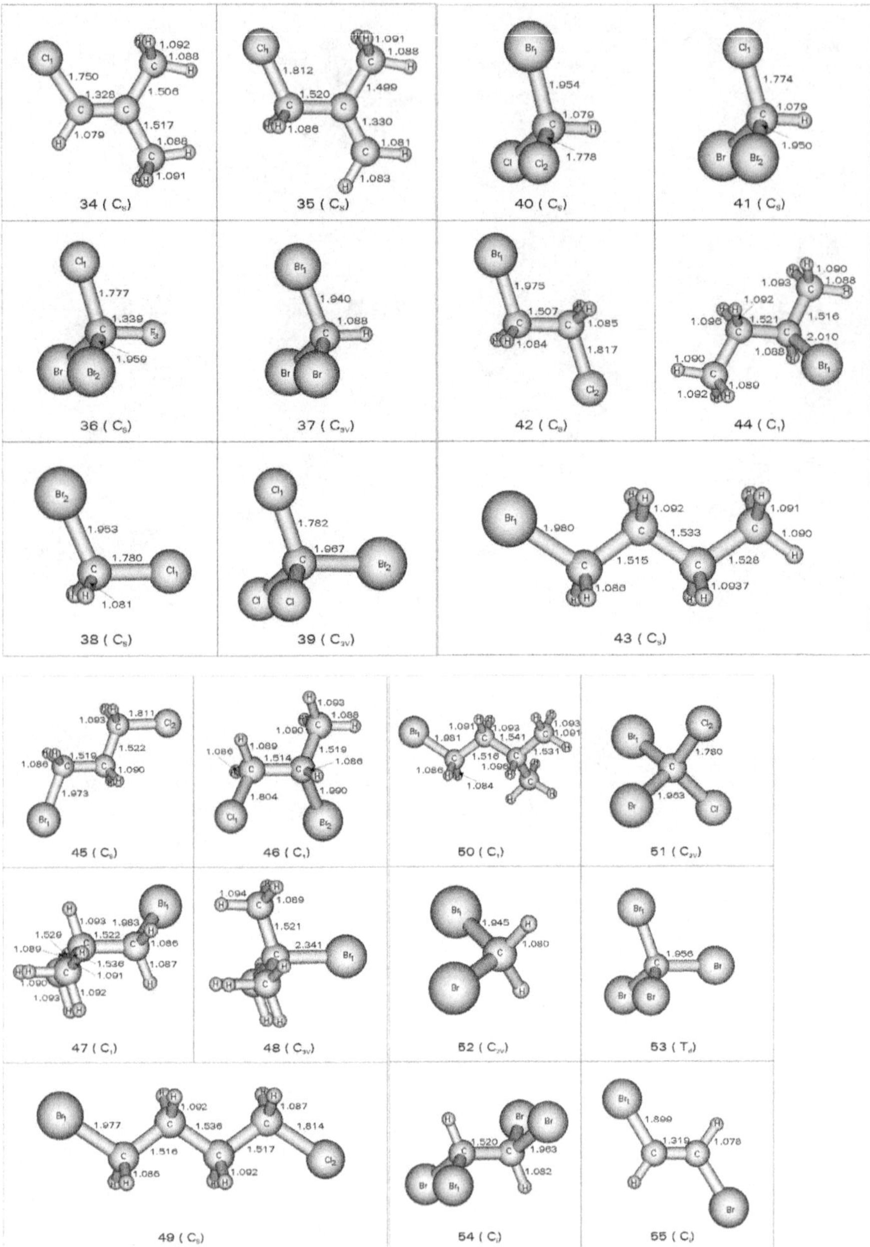

Figure 8.3 (Conitnued)

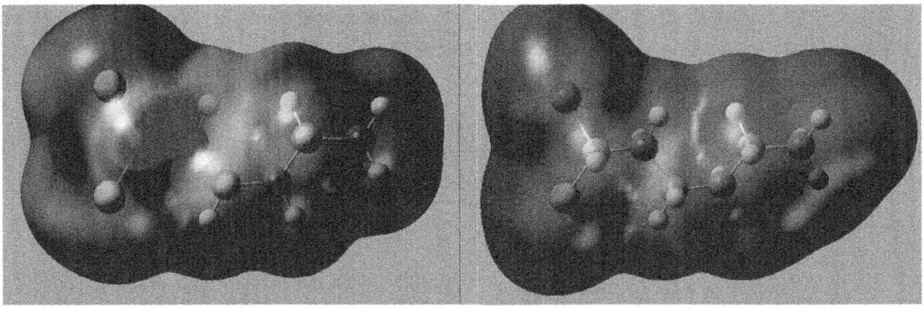

MEP- 1,1dichlorohexane;Fig 2(1);neutral MEP- 1,1dichlorohexane;Fig 2(1);anion

Figure 8.4 Molecular electrostatic potentials of the neutral first isomer in Fig. 8.2 (left) and the corresponding MEP of the anion. The MEPS reveal that upon electron attachment, the regions near both Cl become considerably hot with all cold regions (blue) of the neutral turning yellow or hotter than blue. The extremely hot regions near Cls cause bond elongation followed by autodissociation resulting in the production of toxic free radicals.

attractive regions toward a proton or a positive charge while the blue region is the cold region toward a positive charge. Alternatively, the blue regions are acidic and the red regions are basic. Consequently, a critical comparison of the left or right sides of Fig. 8.4 reveals that upon electron attachment, the regions near both Cl become extremely hot with all cold regions (blue) of the neutral turning yellow or hotter than blue. Therefore upon electron attachment, the extremely hot regions near Cls cause C–Cl bond elongation as most of the anionic charge is carried by Cl atoms (see Fig. 8.4, right). The C–Cl bond elongation thus leads to autodissociation resulting in the production of toxic free radicals. Depending on the halocarbon, the free radicals thus produced cause damage to the lipid membranes of the liver through proton extraction from the lipid at the hot regions in Fig. 8.4, right.

In order to seek optimal techniques for the toxicity predictions for a large data set, several benchmark computations on compounds 24 and 25 (Table 8.2) were considered. The least accurate method, namely, B3LYP/6–311 + G(2d,p) (Table 8.2), gives unrealistically low VEA values primarily because of the lack of diffuse functions that lower the energy of LUMO. Thus the VEAs of compounds 24 and 25 are computed as −1.03, −1.04 eV at the B3LYP level and −1.40, −1.40 eV at the MP2 level employing the 6–311 + G(2d,p) basis set. The AUG-cc-pvTZ basis set yielded −0.36 eV, −0.35 eV at the B3LYP level while the corresponding MP2 results are −0.569 and −0.551 eV. The CCSD and CCSD(T) techniques yield −0.529 and −0.547 eV close to the MP2 results. The results obtained using the 6–311 G++(2d,p) basis set are much improved, and quite closer to the result obtained from a much larger AUG-cc-pvTZ. The computed VEAs for the compounds 24 and 25 are −0.53 and −0.52 eV at the B3LYP/6–311 G++(2d,p) level which compares well with the AUG-cc-pvTZ values of −0.36 and −0.35 eV obtained also at the B3LYP level. The MP2 results are −0.80 and −0.76 eV using the 6–311 G++(2d,p) basis compared to −0.57 and −0.55 eV obtained with the

AUG-cc-pvTZ basis set. Hence the B3LYP method with a relatively large basis set containing diffuse and polarization functions such as the 6−311 G++(2d,p) basis set yields very reasonable VEAs as it yields the correct trend and compares well with more accurate methods such as the CCSD/AUG-cc-pvTZ level which consumes $O(N^3)$ disk space and is significantly more CPU-intensive than B3LYP. The related gallium halides have also been studied using high level quantum chemical methods (Dai and Balasubramanian, 1993).

8.4 Hypercubes and large datasets

In the context of large data sets, we pointed out several applications of hypercubes in the introduction section. As an example, Fig. 8.5 shows a graph-theoretical

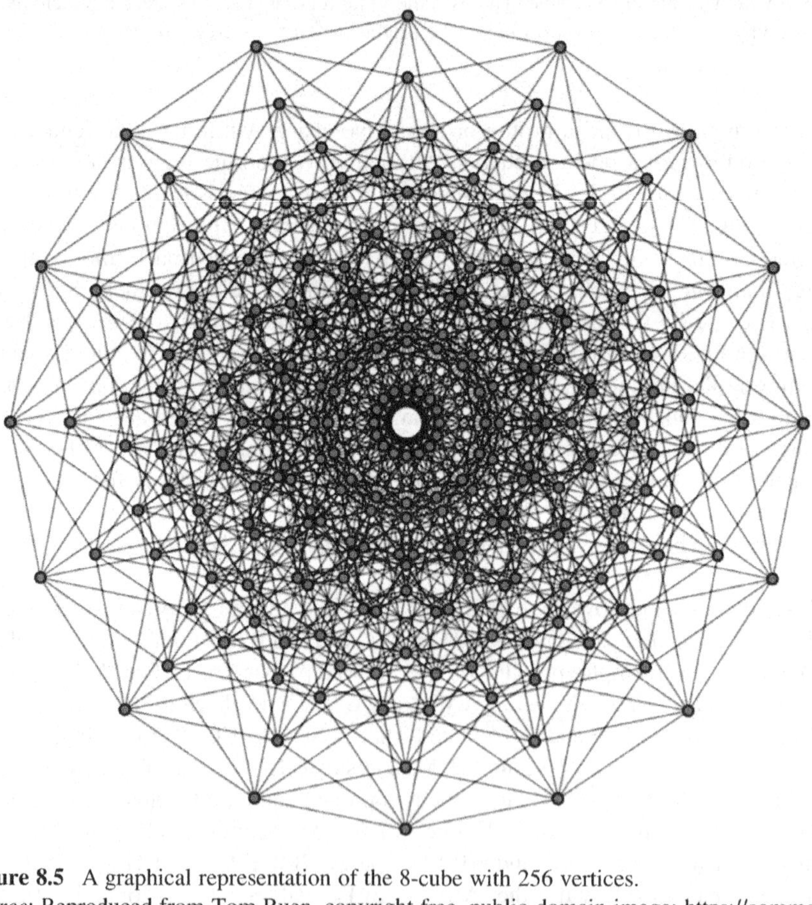

Figure 8.5 A graphical representation of the 8-cube with 256 vertices.
Source: Reproduced from Tom Ruen, copyright free, public-domain image: https://commons.wikimedia.org/wiki/File:8-cube.svg.

depiction of the 8-cube; 256 vertices of the hypercube correspond to the 256 minima in the potential energy surfaces of the water octamer separated by surmountable energy barriers. The automorphism groups of hypercube graphs are wreath product groups, for example, the group of the 8-cube contains 10,321,920 operations that are partitioned into 185 conjugacy classes. The tunneling splitting tables and nuclear spin statistics for the tunneling splittings of the rotational levels of $(H_2O)_8$ can be obtained by employing a computational matrix polynomial generating function technique in conjunction with the Möbius inversion (Balasubramanian, 2020d).

There are 256 vertices, 1024 edges, 1792 faces, 1792 cubical cells, 1120 tesseracts 448 penteracts, 112 hexeracts, and 16 hepteracts in the 8-cube; the action of wreath product $S_8[S_2]$ on the hyperplanes generates cycle types for each set of

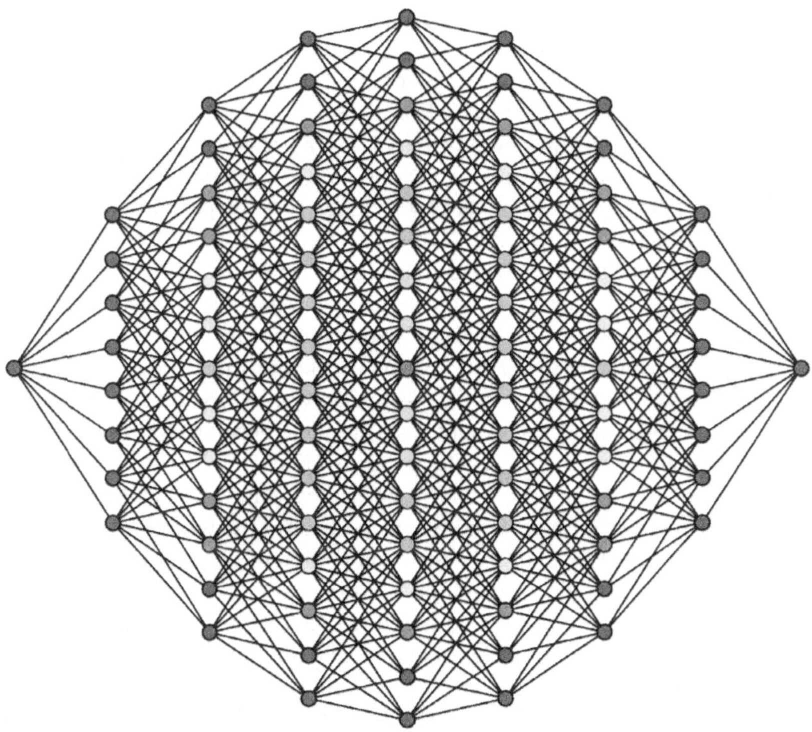

Figure 8.6 Computations of conjugacy classes and the character table generating function through the binomial matrix generating function technique where each term is replaced by 2×8 matrices for the cycle types of the conjugacy classes of the 8-cube for various hyperplane with the Möbius inverter. Each column represents the binomial distribution 1,8,28,56,70,56,28,8,1 of vertices.
Source: Reproduced from Geoff Richards, copy-right free, image: https://commons.wikimedia.org/wiki/File:8-cube_column_graph.svg.

hyperplanes of the 8-cube. For example, the automorphism group acts on cubical cells of the 8-cube to generate permutations of 1792 cells. Hence the generation of the cycle types of each conjugacy class of the 8-cube for the action on 1792 cells is challenging. Fig. 8.6 shows graphical computations of conjugacy classes and the character table generating function through the binomial matrix generating function technique where each term is replaced by 2×8 matrices for the cycle types of the conjugacy classes of the 8-cube for various hyperplane with the Möbius inverter. Thus we have demonstrated the power of the Möbius inversion technique for the generation of cycle types of all hyperplanes of an n-cube from polynomial matrix generators. It is evident that the large data sets require powerful combinatorial techniques that combine group theory, Möbius inversion and multinomial techniques for clustering, partitioning, and other important applications to chemical and biochemical problems including chemical toxicology, drug discovery, protein–protein interactions, and genetic networks.

The equivalence classes of colorings of nD-hypercubes have several applications including genetic regulatory networks (Liu, and Bassler, 2010). The equivalence classes for the colorings of various hyperplanes facilitate the partitioning of complex large data sets into equivalence classes wherein a member of the class represents the properties of all members of the class. In genetics canalization or control of one genetic trait by another trait in the network is important in evolutionary processes. Such networks are represented by n-dimensioanl-hypercubes where the vertices of the n-dimensioanl-hypercube represent the 2^n possible Boolean functions for n traits. The connection between 2-colorings of an n-dimensioanl-hypercube and genetic regulatory pathways was explored earlier (Reichhardt and Bassler, 2007). The necessity to classify the 2-colorings of the vertices into equivalence classes in order to generate a smaller clustering subsets on the basis of equivalence classes for the 2-colorings of the vertices of the n-dimensioanl-hypercube has been demonstrated. Two members in a class would have the same genetic expression, and hence needed computations are reduced. The question of if chirality in colorings would have any implication in the probability of producing chiral traits and thus a biological evolutionary implication of chirality has not been visited thus far. In Fig. 8.7 we have shown the equivalence class for the enumeration of the 2-coloring of the vertices of the cube. The numbers of chiral colorings for face-colorings of the five-dimensional-hypercube are given by 14, 326, 5722, 74973, 811,527, 7,477,975, and 60,113,621 for 3, 4, 5, 6, 7, 8, and 9 greens (remaining reds), respectively. The corresponding results for the edge 2-colorings of the 5-cube are 12,330, 5782, 75,369, 815,762, 60,219,494, and 428,191,237, for 3, 4, 5, 6, 7, 8, and 9 greens (remaining reds), respectively. The 2-colorings of the vertices of the five-dimensional-hypercube produce 2, 26, 148, 653, 2218, 6300, 14,972, 30,730, and 54,528 chiral colorings for 4, 5, 6, 7, 8, 9, 10, 11, and 12 green colors (remaining red), respectively. This suffices to demonstrate the power combinatorics in reducing the complex problem of large data sets into partitions of equivalence classes where a representative in the class is sufficient for the study of the properties of the entire class.

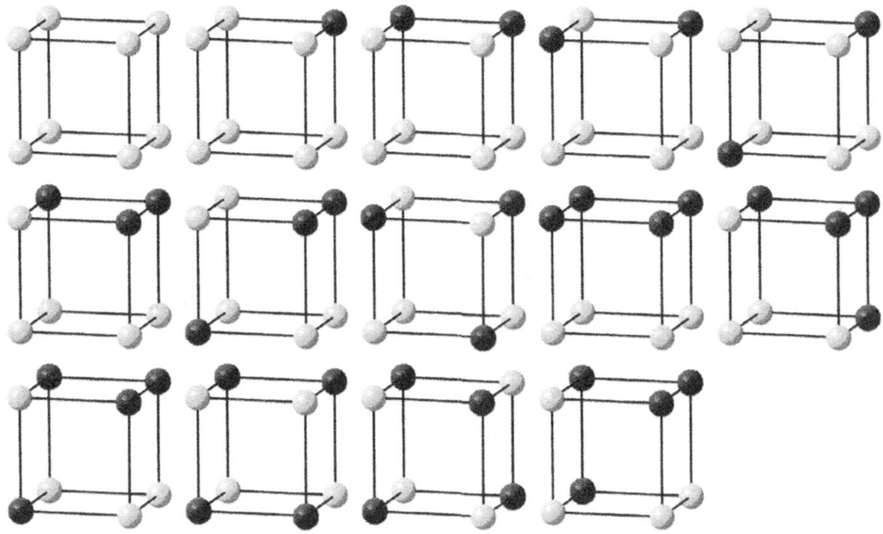

Figure 8.7 Among 14 equivalence classes of 2-colorings of vertices of a cube, the last one is chiral. This is enumerated as the number of A_{1u} irreducible representations for the 2-colorings. For the five-dimensional-hypercube the first chiral coloring appears for 4 greens and 28 reds. There are 2, 26, 148, 653, 2218, 6300, 14972, 30,730, and 54,528 chiral colorings for 4, 5, 6, 7, 8, 9, 10, 11, and 12 green colors (remaining red), respectively for the 2-colorings of the vertices of the five-dimensional-hypercube as enumerated by the A_2 chiral representation of the five-dimensional-hypercube. There are 65,664 such equivalence classes of colorings with 22 green and 10 red colorings; there are 11,875,913,272 equivalence classes for the five-dimensional-hypercube for 11 green, 11 red, and 11 blues.
Source: Reproduced with permission from Balasubramanian, K., 2019. Computational multinomial combinatorics for colorings of 5D-hypercubes for all irreducible representations and applications. J. Math. Chem., 57(2), 655–689.

8.5 Conclusion

In this chapter, we reviewed combinatorial techniques for large data sets with hypercubes and halocarbons as case studies. We also provided an overview of quantum studies for halocarbons which are of both environmental and toxicological concerns. Powerful combinatorial techniques for all irreducible representations can enumerate the polysubstituted halocarbons arising from both branched and unbranched isomers. From the quantum chemical computations and the comparisons with various levels of theory, the B3LYP method in conjunction with the 6−311++G(2d,p) basis set was found to be an optimal and yet feasible choice for pursuing toxicological studies on a large combinatorial library of compounds. Although the CCSD method with AUG-cc-pvTZ basis provides more accurate quantitative results, it is not the ideal choice for a larger library of halocarbons containing several halogen atoms. On the other hand, for a small set of compounds superior complete active space self-consistent field/multi-reference singles + doubles configuration interaction (CASSCF/MRSDCI) techniques are preferred as exemplified by previous studies (Balasubramanian, 1987). We conclude that combinatorial techniques based on Möbius inversion and the variants of Sheehan's theorem provide attractive choices for the partitioning of large data sets into equivalence classes, and these techniques can have far-reaching applications in other areas, for example, spin physics of two-dimensional cavity QED arrays and lattices (Minář et al., 2017), AI, DNA computing and so on. We anticipate the AI techniques to exploit the powerful GPUs for three-dimensional image processing, computer-assisted image perceptions, and in three-dimensional image shape similarity analysis and evolution, for example, the MRI radiomic image dynamics for the prognosis and diagnostics of ovarian cancer (Balasubramanian 2020a). Many such machine learning and AI tools would become an integral part of medicine, drug administration, and pharmacology in the future. Combinatorial tools would form the basis for algorithms in AI and machine learning in toxicology, drug discovery, protein–protein interactions, metal–DNA, metal–protein interactions, genetic regulatory networks, pandemic networks, genomics, proteomics metabolomics, and radiomics.

References

Balasubramanian, K., 1979a. A generalized wreath product method for the enumeration of stereo and position isomers of polysubstituted organic compounds. Theoretica Chim. Acta 51 (1), 37–54.

Balasubramanian, K., 1979b. Enumeration of stable stereo and position isomers of polysubstituted alcohols. Ann. N. Y. Acad. Sci. 319, 33–36.

Balasubramanian, K., 1983. Computer generation of the symmetry elements of nonrigid molecules. J. Comput. Chem. 4 (3), 302–307.

Balasubramanian, K., 1987. Cas scf/ci calculations on Si_4 and Si_4^+. Chem. Phys. Lett. 135 (3), 283–287. Available from: https://doi.org/10.1016/0009-2614(87)85157-6.

Balasubramanian, K., 1988a. Enumeration of the isomers of the gallium arsenide clusters (GamAsn). Chem. Phys. Lett. 150 (1-2), 71–77.

Balasubramanian, K., 1988b. Graph edge colorings and their chemical applications. Theoretica Chim. acta 74 (2), 111–122.

Balasubramanian, K., 1989. Ten low-lying electronic states of Pd3. J. Chem. Phys. 91 (1), 307–313.

Balasubramanian, K., 1990. Electronic structure of (GaAs) 2. Chem. Phys. Lett. 171 (1-2), 58–62.

Balasubramanian, K., 1997a. Relativistic Effects in Chemistry, Parts A. Wiley-Interscience, New York.

Balasubramanian, K., 1997b. Relativistic Effects in Chemistry, Parts B. Wiley-Interscience, New York.

Balasubramanian, K., 2002. Combinatorial enumeration of ragas (scales of integer sequences) of Indian music. J. Integer Seq. 5, 6. A02.2.6.

Balasubramanian, K., 2004. Nonrigid group theory, tunneling splittings, and nuclear spin statistics of water pentamer: $(H_2O)_5$. J. Phys. Chem. A 108 (26), 5527–5536.

Balasubramanian, K., 2016a. Character tables of n-dimensional hyperoctahedral groups and their applications. Mol. Phys. 114 (10), 1619–1633.

Balasubramanian, K., 2016b. Quantum chemical insights into Alzheimer's disease: Curcumin's chelation with Cu (II), Zn (II), and Pd (II) as a mechanism for its prevention. Int. J. Quant. Chem. 116 (14), 1107–1119.

Balasubramanian, K., 2018a. Computational enumeration of colorings of hyperplanes of hypercubes for all irreducible representations and applications. J. Math. Sci. Model. 1 (3), 158–180.

Balasubramanian, K., 2018b. Mathematical and computational techniques for drug discovery: promises and developments. Curr. Top. Med. Chem. 18 (32), 2774–2799.

Balasubramanian, K., 2019. Computational multinomial combinatorics for colorings of 5D-hypercubes for all irreducible representations and applications. J. Math. Chem. 57 (2), 655–689.

Balasubramanian, K., 2020a. Computational and artificial intelligence techniques for drug discovery and administration. Comprehensive Pharmacol. Elsevier, Amsterdam, 2020.

Balasubramanian, K., 2020b. Computational combinatorics of hyperplane colorings of 6D-hypercube for all irreducible representations and applications. J. Math. Chem. 58 (1), 204–272.

Balasubramanian, K., 2020c. Computations of colorings 7D-hypercube's hyperplanes for all irreducible representations. J. Comput. Chem. 41 (7), 653–686.

Balasubramanian, K., 2020d. Nonrigid water octamer: computations with the 8-cube. J. Comput. Chem. 41 (29), 2469–2484.

Balasubramanian, K., 2020e. Combinatorics of supergiant fullerenes: enumeration of polysubstituted isomers, chirality, nuclear magnetic resonance, electron spin resonance patterns, and vibrational modes from C70 to C150000. J. Phys. Chem. A 124 (49), 10359–10383.

Balasubramanian, K., Liao, D.W., 1989. Spectroscopic properties of low-lying electronic states of rhodium dimer. J. Phys. Chem. 93 (10), 3989–3992.

Balasubramanian, K., Liao, D.W., 1991. Spectroscopic constants and potential energy curves of Bi2 and Bi2-. J. Chem. Phys. 95 (5), 3064–3073.

Balasubramanian, K., 2021. Relativistic quantum chemical and molecular dynamics techniques for medicinal chemistry of bioinorganic compounds. Top. Med. Chem. 37, 133–193. Available from: https://doi.org/10.1007/7355_2020_109.

Balasubramanian, K., Basak, S.C., 2016. Metabolic electron attachment as a primary mechanism for toxicity potentials of halocarbons. Curr. Comput. drug. Des. 12 (1), 62–72.

Balasubramanian, K., Gupta, S.P., 2019. Quantum molecular dynamics, topological, group theoretical and graph theoretical studies of protein-protein interactions. Curr. Top. Med. Chem. 19 (6), 426–443.

Balasubramanian, K., Kaufman, J.J., Hariharan, P.C., Koski, W.S., 1986. Energy transfer in Br + -Kr collisions. Chem. Phys. Lett. 129 (2), 165–171.

Balasubramanian, K., Sumathi, K., Dai, D., 1991. Group V trimers and their positive ions: the electronic structure and potential energy surfaces. J. Chem. Phys. 95 (5), 3494–3505.

Banks, D.C., Linton, S.A., Stockmeyer, P.K., 2004. Counting cases in substitope algorithms. IEEE Trans. Vis. Comput. Graph. 10 (4), 371–384.

Barthélemy, M., Barrat, A., Pastor-Satorras, R., Vespignani, A., 2005. Dynamical patterns of epidemic outbreaks in complex heterogeneous networks. J. Theor. Biol. 235 (2), 275–288.

Basak, S.C., Grunwald, G.D., Gute, B.D., Balasubramanian, K., Opitz, D., 2000. Use of statistical and neural net approaches in predicting toxicity of chemicals. J. Chem. Inf. Comput. Sci. 40 (4), 885–890.

Basak, S.C., Balasubramanian, K., Gute, B.D., Mills, D., Gorczynska, A., Roszak, S., 2003. Prediction of cellular toxicity of halocarbons from computed chemodescriptors: a hierarchical QSAR approach. J. Chem. Inf. Comput. Sci. 43 (4), 1103–1109.

Benavides-Garcia, M., Balasubramanian, K., 1991. Spectroscopic constants and potential energy curves for OsH. J. Mol. Spectrosc. 150 (1), 271–279.

Benavides-Garcia, M., Balasubramanian, K., 1994. Bond energies, ionization potentials, and the singlet–triplet energy separations of $SnCl_2$, $SnBr_2$, SnI_2, $PbCl_2$, $PbBr_2$, PbI_2, and their positive ions. J. Chem. Phys. 100 (4), 2821–2830.

Bhaniramka, P., Wenger, R., Crawfis, R., 2000. October. Isosurfacing in higher dimensions. In: Proceedings Visualization 2000. VIS 2000 (Cat. No. 00CH37145), IEEE, pp. 267–273.

Blair, E., Greaves, J., Farmer, P.J., 2004. High-temperature electrocatalysis using thermophilic P450 CYP119: dehalogenation of CCl_4 to CH_4. J. Am. Chem. Soc. 126 (28), 8632–8633.

Carbó-Dorca, R., 2018. DNA, unnatural base pairs and hypercubes. J. Math. Chem. 56 (5), 1353–1356. Available from: https://doi.org/10.1007/s10910-018-0866-9.

Carbó-Dorca, R., Chakraborty, T., 2019a. Divagations about the periodic table: Boolean hypercube and quantum similarity connections. J. Comput. Chem. 40 (30), 2653–2663.

Carbó-Dorca, R., Chakraborty, T., 2019b. Hypercubes defined on n-ary sets, the Erdös–Faber–Lovász conjecture on graph coloring, and the description spaces of polypeptides and RNA. J. Math. Chem. 57 (10), 2182–2194.

Costa, P.J., Nunes, R., Vila-Viçosa, D., 2019. Halogen bonding in halocarbon-protein complexes and computational tools for rational drug design. Expert. Opin. Drug. Discov. 14 (8), 805–820.

Coveney, P.V., Dougherty, E.R., Highfield, R.R., 2016. Big data need big theory too. Philos. Trans. R. Soc. A: Math., Phys. Eng. Sci. 374 (2080), 20160153.

Crebelli, R., Andreoli, C., Carere, A., Conti, G., Conti, L., Ramusino, M.C., et al., 1992. The induction of mitotic chromosome malsegregation in *Aspergillus nidulans* quantitative structure activity relationship (QSAR) analysis with chlorinated aliphatic hydrocarbons. Mutat. Res./Fund. Mol. Mech. Mutagen. 266 (2), 117–134.

Crebelli, R., Andreoli, C., Carere, A., Conti, L., Crochi, B., Cotta-Ramusino, M., et al., 1995. Toxicology of halogenated aliphatic hydrocarbons: structural and molecular determinants for the disturbance of chromosome segregation and the induction of lipid peroxidation. Chemico-biol. Interact. 98 (2), 113–129.

Dai, D., Balasubramanian, K., 1993. Geometries and potential energies of electronic states of GaX_2 and GaX_3 ($X = Cl$, Br, and I). J. Chem. Phys. 99 (1), 293–301.

Denk, M.K., Milutinović, N.S., Dereviankin, M.Y., 2019. Reduction of halocarbons to hydrocarbons by NADH models and NADH. Chemosphere 233, 890–895.

Edwards, J.E., 1941. Hepatomas in mice induced with carbon tetrachloride. J. Natl Cancer Inst. 2 (2), 197–199.

Fang, C., Behr, M., Xie, F., Lu, S., Doret, M., Luo, H., et al., 2008. Mechanism of chloroform-induced renal toxicity: non-involvement of hepatic cytochrome P450-dependent metabolism. Toxicol. Appl. Pharmacol. 227 (1), 48–55.

Forster, P., Forster, L., Renfrew, C., Forster, M., 2020. Phylogenetic network analysis of SARS-CoV-2 genomes. Proc. Natl Acad. Sci. 117 (17), 9241–9243.

Friedman, S.L., 2000. Molecular regulation of hepatic fibrosis, an integrated cellular response to tissue injury. J. Biol. Chem. 275 (4), 2247–2250.

Frisch, M.J.E.A., et al., 2009. Gaussian 09, revision d. 01. Gaussian. Inc., Wallingford CT, p. 201.

Fujii, T., Fuchs, B.C., Yamada, S., Lauwers, G.Y., Kulu, Y., Goodwin, J.M., et al., 2010. Mouse model of carbon tetrachloride induced liver fibrosis: histopathological changes and expression of CD133 and epidermal growth factor. BMC Gastroenterol. 10 (1), 79.

Goldman, M., Huang, Y., 2018. Conformational analysis of 1, 2-dichloroethane adsorbed in metal-organic frameworks. Vibr. Spectrosc. 95, 68–74.

Gowen, A.A., O'donnell, C.P., Cullen, P.J., Bell, S.E.J., 2008. Recent applications of chemical imaging to pharmaceutical process monitoring and quality control. Eur. J. Pharma. Biopharma. 69 (1), 10–22.

Illenberger, E., Momigny, J., 1992. In: Baumgartel, H., Frank, E.U., Grunbein, W. (Eds.), Gaseous Molecular Ions: An Introduction To Elementary Processes Induced By Ionization, Vol. 2. Springer Science & Business Media, New York, pp. 286–287.

Kaufman, J.J., Koski, W.S., Roszak, S., Balasubramanian, K., 1996. Correlation between energetics and toxicities of single-carbon halides. Chem. Phys. 204 (2-3), 233–237.

Kellner, D.G., Maves, S.A., Sligar, S.G., 1997. Engineering cytochrome P450s for bioremediation. Curr. Opin. Biotechnol. 8 (3), 274–278.

Keng, F.S.L., Phang, S.M., Abd Rahman, N., Elvidge, E.C.L., Malin, G., Sturges, W.T., 2020. The emission of volatile halocarbons by seaweeds and their response towards environmental changes. J. Appl. Phycol. 1–18.

Kim, G.B., Balasubramanian, K., 1992. Spectroscopic constants and potential energy curves of GaBr. J. Mol. Spectrosc. 152 (1), 192–198.

Koski, W.S., Roszak, S., Kaufman, J.J., Balasubramanian, K., 1997. Potential toxicity of CF3X halocarbons. Vitro Toxicol. 10 (4), 455–457.

Li, X.X., Zheng, Q.C., Wang, Y., Zhang, H.X., 2014. Theoretical insights into the reductive metabolism of CCl_4 by cytochrome P450 enzymes and the CCl_4-dependent suicidal inactivation of P450. Dalton Trans. 43 (39), 14833–14840.

Li, X.X., Wang, Y., Zheng, Q.C., Zhang, H.X., 2016. Detoxification of 1-methyl-4-phenyl-1, 2, 3, 6-tetrahydropyridine (MPTP) by cytochrome P450 enzymes: a theoretical investigation. J. Inorg. Biochem. 154, 21–28.

Liu, M., Bassler, K.E., 2010. Finite size effects and symmetry breaking in the evolution of networks of competing Boolean nodes. J. Phys. A: Math. Theor. 44 (4), 045101.

Liu, S., Yao, Y., Lu, S., Aldous, K., Ding, X., Mei, C., et al., 2013. The role of renal proximal tubule P450 enzymes in chloroform-induced nephrotoxicity: utility of renal specific P450 reductase knockout mouse models. Toxicol. Appl. Pharmacol. 272 (1), 230–237.

Luke, B.T., Collins, J.R., Loew, G.H., McLean, A.D., 1990. Theoretical investigations of terminal alkenes as putative suicide substrates of cytochrome P-450. J. Am. Chem. Soc. 112 (24), 8686–8691.

Luke, B.T., Loew, G.H., McLean, A.D., 1986. A theoretical examination of substituent effects on the detoxification reaction between glutathione and halogenated methanes. Int. J. Quant. Chem. 29 (4), 883–896.

Luke, B.T., Loew, G.H., McLean, A.D., 1987. Theoretical investigations of the anaerobic reduction of halogenated alkanes by cytochrome P 450. 1. Structures and inverse barriers of heats formation halomethyl radicals. J. Am. Chem. Soc. 109 (5), 1307–1317.

Luke, B.T., Loew, G.H., McLean, A.D., 1988. Theoretical investigation of the anaerobic reduction of halogenated alkanes by cytochrome P-450. 2. Vertical electron affinities of chlorofluoromethanes as a measure of their activity. J. Am. Chem. Soc. 110 (11), 3396–3400.

Majumdar, D., Balasubramanian, K., Nitsche, H., 2002. A comparative theoretical study of bonding in UO2++, UO2+, UO2, UO2− UOmdar, 2U(CO)2 and UO2CO3. Chem. Phys. Lett. 361 (1-2), 143–151.

Manibusan, M.K., Odin, M., Eastmond, D.A., 2007. Postulated carbon tetrachloride mode of action: a review. J. Environ. Sci. Health Part. C. 25 (3), 185–209.

Mclean, K.J., Girvan, H.M., Mason, A.E., Dunford, A.J., Munro, A.W., 2011. Structure, mechanism and function of cytochrome P450 enzymes. Iron-Containing Enzymes 255–280.

Mezey, P.G., 2012. Natural molecular fragments, functional groups, and holographic constraints on electron densities. Phys. Chem. Chem. Phys. 14 (24), 8516–8522.

Mezey, P.G., 2014. Fuzzy electron density fragments in macromolecular quantum chemistry, combinatorial quantum chemistry, functional group analysis, and shape–activity relations. Acc. Chem. Res. 47 (9), 2821–2827.

Minář, J., Söyler, Ş.G., Rotondo, P., Lesanovsky, I., 2017. Effective spin physics in two-dimensional cavity QED arrays. N. J. Phys. 19 (6), 063033.

Modelli, A., Scagnolari, F., Distefano, G., Jones, D., Guerra, M., 1992. Electron attachment to the fluoro-, bromo-, and iodomethanes studied by means of electron transmission spectroscopy and Xα calculations. J. Chem. Phys. 96 (3), 2061–2070.

Nandini, G.K., Rajan, R.S., Shantrinal, A.A., Rajalaxmi, T.M., Rajasingh, I., Balasubramanian, K., 2020. Topological and thermodynamic entropy measures for COVID-19 pandemic through graph theory. Symmetry 12 (12), 1992.

Nastainczyk, W., Ullrich, V., Sies, H., 1978. Effect of oxygen concentration on the reaction of halothane with cytochrome P450 in liver microsomes and isolated perfused rat liver. Biochem. Pharmacol. 27 (4), 387–392.

National Toxicology Program, 1992. Division of toxicology research and testing, management status report (accessed 18.10.92).

Ortiz de Montellano, P.R., 2010. Hydrocarbon hydroxylation by cytochrome P450 enzymes. Chem. Rev. 110 (2), 932–948.

Osanai, M., Sawada, N., Lee, G.H., 2010. Oncogenic and cell survival properties of the retinoic acid metabolizing enzyme, CYP26A1. Oncogene 29 (8), 1135–1144.

Prabhakar, Y.S., Balasubramanian, K., 2006. A simple algorithm for unique representation of chemical structures cyclic/acyclic functionalized achiral molecules. J. Chem. Inf. Model. 46 (1), 52–56.

Punitha, T., Phang, S.M., Juan, J.C., Beardall, J., 2018. Environmental control of vanadium haloperoxidases and halocarbon emissions in macroalgae. Mar. Biotechnol. 20 (3), 282–303.

Rains, E.M., Sloane, N.J., 1999. On Cayley's enumeration of alkanes (or 4-valent trees). J. Integer Seq. 2, Article 99.1.1.

Ravina, M., Facelli, A., Zanetti, M., 2020. Halocarbon emissions from hazardous waste landfills: analysis of sources and risks. Atmosphere 11 (4), 375.

Reichhardt, C.J.O., Bassler, K.E., 2007. Canalization and symmetry in Boolean models for genetic regulatory networks. J. Phys. A: Math. Theor. 40, p4339.

Roszak, S., Balasubramanian, K., Kaufman, J.J., Koski, W.S., 1993a. Multireference configuration interaction study of temporary anion states in haloforms. Chem. Phys. Lett. 215 (5), 427–432.

Roszak, S., Vijayakumar, M., Balasubramanian, K., Koski, W.S., 1993b. A multireference configuration interaction study of photoelectron spectra of carbon tetrahalides. Chem. Phys. Lett. 208 (3-4), 225–231.

Roszak, S., Kaufman, J.J., Koski, W.S., Vijayakumar, M., Balasubramanian, K., 1994. Potential energy curves of ground and excited states of tetra halomethanes and the negative ions. J. Chem. Phys. 101 (4), 2978–2985.

Roszak, S., Koski, W.S., Kaufman, J.J., Balasubramanian, K., 1997. Structure and energetics of CF_3Cl-, CF_3Br-, and CF_3I- radical anions. J. Chem. Phys. 106 (18), 7709–7713.

Roszak, S., Koski, W.S., Kaufman, J.J., Balasubramanian, K., 2001. Structures and electron attachment properties of halomethanes (CXnY m, X = H, F; Y = Cl, Br, I; n = 0, 4; m = 4 n). SAR. QSAR Environ. Res. 11 (5-6), 383–396.

Sheehan, J., 1967. On Pólya's theorem. Can. J. Math. 19, 792–799.

Stanley, K.O., D'Ambrosio, D.B., Gauci, J., 2009. A hypercube-based encoding for evolving large-scale neural networks. Artif. life 15 (2), 185–212.

Trohalaki, S., Pachter, R., 2003. Quantum descriptors for predictive toxicology of halogenated aliphatic hydrocarbons. SAR. QSAR Environ. Res. 14 (2), 131–143.

Wallace, R., 2011. Multifunction moonlighting intrinsically disordered proteins: information catalysis, non-rigid molecule symmetries and the 'logic gate' spectrum. Comptes Rendus Chimie 14 (12), 1117–1121.

Wallace, R., 2012. Spontaneous symmetry breaking in a non-rigid molecule approach to intrinsically disordered proteins. Mol. Biosyst. 8 (1), 374–377.

Wallace, R., 2017. Tools for the future: hidden symmetries. Computational Psychiatry. Springer, Cham, pp. 153–165.

Weber, L.W., Boll, M., Stampfl, A., 2003. Hepatotoxicity and mechanism of action of haloalkanes: carbon tetrachloride as a toxicological model. Crit. Rev. Toxicol. 33 (2), 105–136.

Woo, Y.T., Lai, D.Y., Arcos, J.C., Argus, M.F., 1985. Chemical induction of cancer: structural bases and biological mechanisms, Aliphatic and Polyhalogenated Carcinogens, Volume IIIB. Academic Press, NY.

Yang, X., Wang, Y., Byrne, R., Schneider, G., Yang, S., 2019. Concepts of artificial intelligence for computer-assisted drug discovery. Chem. Rev. 119 (18), 10520–10594.

Zhu, H., 2020. Big data and artificial intelligence modeling for drug discovery. Ann. Rev. Pharmacol. Toxicol. 60, 573–589.

Development of quantitative structure–activity relationship models based on electrophilicity index: a conceptual DFT-based descriptor

9

Ranita Pal[1] *and Pratim Kumar Chattaraj*[2]

[1]Advanced Technology Development Centre, Indian Institute of Technology Kharagpur, Kharagpur, West Bengal, India, [2]Department of Chemistry, Indian Institute of Technology Kharagpur, Kharagpur, West Bengal, India

9.1 Introduction

Quantitative structure–activity relationship (QSAR) deeply explores the possibility of solidifying the link between a compound's knowledge and various aspects of chemistry, biology, statistics and toxicology. It postulates that physicochemical properties (electronic, hydrophobic, steric, etc.) of a molecule strongly correlate with the activity it shows, and can be quantified as

Activity $= f$ (chemical attributes/physicochemical properties).

With the development of new techniques, algorithms and software for calculation and selection of molecular descriptors, drug design and toxicity/activity prediction have undergone huge modifications for the better, since they have shown a significant cutback in animal experiments.

The mid-19th century witnessed the dawn of the quantification of biological activity in terms of structural features and physicochemical properties. Cros (1863), revealed the link between alcohol's water solubility and its toxicities on mammals. Brown and Fraser (1868) later generalized this type of correlation by finding the dependence of the physiological action of a substance on its chemical composition. Subsequently, Richardson (1869) proved that the toxic activity of ethers and alcohols is a function of their water solubility. About two decades later, Richet (1893) claimed the existence of an inverse relationship between the toxic nature of simple organic molecules along with their corresponding solubilities in water, whereas the partition coefficients of similar types of molecules were used as descriptors by Meyer (1899) and Overton (1901) explaining their narcotic behavior. Hammett's idea of incorporating electronic substituent (σ) and reaction (ρ) constants to represent the electronic effects on reaction mechanisms (Hammett, 1935, 1937) took the

Big Data Analytics in Chemoinformatics and Bioinformatics. DOI: https://doi.org/10.1016/B978-0-323-85713-0.00020-7

field of QSAR to a completely new level. Throughout the next few years, important contributions in this field were made by Ferguson (1939), Albert et al. (1945), Albert (1985), Roblin Jr and Bell (1942), and Taft (1952, 1956). Finally, Hansch et al. (1962) introduced the *n*-octanol/water partition coefficient which has, since then, been widely used in various QSAR studies. The famous linear Hansch equations came in 1964 when Hansch and Fujita (1964) combined log *P* with Hammett's electronic constants, which began the era of modern QSAR. Among various quantum chemical global and local descriptors, electrophilicity index (ω) proved to act as an efficient global descriptor toward the prediction of activity (Parthasarathi et al., 2004) and toxicity (Padmanabhan et al., 2006).

The correlation between biological response and chemical attributes are mainly based on (i) molecular geometry (substructural fragments, spatial orientation, dipole distribution, etc.), (ii) physical properties (partition coefficient between lipophilic and aqueous phases), and (iii) reaction chemistry properties (skin sensitization potency, mutagenicity, etc.) (Cronin and Madden, 2010). Quantum chemical concepts such as orbital energies (E_{LUMO} or E_{HOMO}), electronegativity (χ), chemical potential (μ), chemical hardness (η), dipole moment (D), chemical softness (S), polarizability (α), electrophilicity index (ω), etc., have, time and again, proved to act as excellent descriptors toward building effective QSAR models. Global electrophilicity index (ω), along with its square dependence, is explored as a molecular descriptor regarding its effectiveness compared to the widely used hydrophobic parameter, that is, the logarithm of *n*-octanol/water partition coefficient (log *P*).

While lipophilicity alone is equipped to establish the narcotic action of nonpolar chemicals, modeling polar narcosis must include the polarizability of the molecule's electronegative center. Hence, the electrophile-nucleophile interaction has a direct relation with the magnitude of the compound's toxicity. To that end, exploring the electronic behavior of the molecules becomes extremely important while quantifying its biological/ecotoxicological activities. The datasets for this study are procured from the literature Chattaraj et al. (2007a) and Masand et al. (2016), respectively.

9.2 Theoretical background

Conceptual density functional theory (CDFT) (Parr and Yang, 1989) has successfully provided chemists with a hierarchy of well-defined chemical reactivity indices like electronegativity (χ) (Pauling, 1960) and chemical hardness (η) (Parr and Pearson, 1983; Pearson, 1997; Sen and Mingos, 1993). These chemical concepts have significant contributions toward the qualitative understanding and quantitative prediction of chemical reactivity. Pauling's electronegativity displays the tendency of an atom in a molecule to attract electrons to itself. To that effect, he introduced an electronegativity scale in an attempt to quantify the chemical reactivity of a system. His definition opened up the concept of electronegativity to a plethora of new interpretations including those of Mulliken (1934, 1935), Allred-Rochow (1958),

Allred (1961), and Iczkowski and Margrave (1961). The density functional definition of χ was finally established by equating it to the negative value of chemical potential (μ), satisfying the formula

$$\mu = \left(\frac{\partial E}{\partial N}\right)_{v(\vec{r})} = -\chi \tag{9.1}$$

where the first equality provides Parr et al. (1978) definition of chemical potential as the response of total energy (E) with change in the number of electrons (N), external potential $v(\vec{r})$ being constant. The second equality, on the other hand, reflects the work of Iczkowski and Margrave (1961), and a generalization of Mulliken's (1934) definition.

The concept of chemical hardness (η) was identified by Parr and Pearson (1983) as the second derivative of E with respect to N at constant $v(\vec{r})$.

$$\eta = \left(\frac{\partial^2 E}{\partial N^2}\right)_{v(\vec{r})} = \left(\frac{\partial \mu}{\partial N}\right)_{v(\vec{r})} \tag{9.2}$$

where the negative average of ionization potential (IP) and electron affinity (EA) is the definition of μ, and the difference between them is defined as η. Eq. (9.2) is in keeping with the concept of the hard–soft acid–base principle (Ayers, 2005, 2007; Ayers et al., 2006; Cedillo et al., 2000; Chattaraj et al., 1991, 2007b; Chattaraj and Ayers, 2005; Gázquez, 1997)

Applying Koopmans' theorem for closed-shell molecules, we get μ and η as

$$\mu = \frac{E_{\text{LUMO}} + E_{\text{HOMO}}}{2} \tag{9.3}$$

$$\eta = E_{\text{LUMO}} - E_{\text{HOMO}} \tag{9.4}$$

E_{HOMO} and E_{LUMO} are the highest occupied and lowest unoccupied molecular orbital energies, respectively. The ω can then be calculated using Parr's definition (Parr et al., 1999)

$$\omega = \frac{\mu^2}{2\eta} = \frac{\chi^2}{2\eta} \tag{9.5}$$

9.3 Computational details

All the studied compounds are subjected to geometrical optimization at a certain level of theories on the basis of the elements constituting them. Frequency analysis performed on the optimized geometries at the same level revealed no imaginary frequencies ensuring their existence at the minima of their respective potential energy surfaces. While the above computations for the study against *T. brucei* are performed at HF/6−31 G(d) levels (Becke, 1988; Fock, 1930; Hartree, 1928, 1929; Lee et al., 1988; Petersson et al., 1988), using Gaussian 09 package (Frisch et al., 2009), those against *T. pyriformis* are done at HF/6−311 G* in Gaussian 03 program package (Frisch et al., 2004).

Modeling QSARs using linear regressions is performed in Origin 6.0 software (Deschenes and Vanden Bout, 2000). Further, a comparison of ranks by random numbers (CRRNs)-based sum of ranking difference (SRD) (Héberger, 2010; Kollár-Hunek and Héberger, 2013) is employed to determine the relative efficiency of the models constructed.

9.4 Methodology

The central motivation of the present work is to focus on employing electrophilicity index (ω), and it's square and cubic dependence, as descriptors in developing QSAR models. To that end, we have considered two different datasets, (i) $pIGC_{50}$ of 169 aliphatic compounds toward *T. pyriformis*, and (ii) pIC_{50} of 32 pyridyl benzamide derivatives toward *T. brucei*. Among these, the first study encompasses a comparative analysis of the predictive ability of hydrophobicity (log*P*) and electrophilcity (ω) acting as descriptors. Each of the compound datasets considered for the QSAR studies against *T. pyriformis* and *T. brucei* are treated in two different ways. One, where the entire undivided dataset is used and two, where the dataset is divided into three equal groups (say, A, B, and C) followed by clubbing two of the groups to form a training set while the third acts as a test set. To ensure uniformity in the group division, it is done in such a way that the average response values ($pLC_{50}/pIGC_{50}/pIC_{50}$) in each group remain approximately similar.

Multiple linear regression (MLR) analysis is performed on the training set with the experimental toxicity ($pLC_{50}/pIGC_{50}/pIC_{50}$) as the dependent and various descriptor combinations as independent variables. The regression equation so obtained is then employed on the test set to calculate their toxicity values. These theoretically calculated values are then compared with the known experimental $pLC_{50}/pIGC_{50}/pIC_{50}$ values to obtain the error range and hence the efficacy of the descriptors is reflected from the regression coefficient (R^2) and standard deviation (*SD*) values.

To validate the prediction performance of the models constructed by simple MLR, an alternate machine learning technique, Multilayer perceptron (MLP) neural networks, is employed. For the present work, after an extensive empirical study, the hidden layer is set to one with four nonlinear units. The number of output units is set to one corresponding to the toxicity value, and the number of descriptors determines the number of input units. For training the weights of the MLP, backpropagation is used here.

Further, using the CRRN-based SRD method as a tool to compare the generated QSAR models, all the data are organized in matrix form containing rows as objects (R^2 and *SD*) and columns as variables (QSAR models), respectively. The results of all the models are assigned with ranks, each of which is subtracted from the rank of the standard/reference/ideal result. In our study, the ideal (golden standard) in each row corresponds to either the minimum *SD* or the maximum R^2 values. Statistically efficient model corresponds to those with lower SRD values.

9.5 Results and discussion

In the early stages of drug design, QSAR modeling and analysis are practiced to make an estimate of the extent to which a set of molecules will interact with the target/receptor site, to filter databases based on certain desired properties, to predict potential toxic behavior, etc. These help in the decision-making process of drug discovery without actually having to undergo lengthy experiments first. With increasing utilization and advancement of QSAR techniques, huge databases are also becoming easier to access. Handling these, however, becomes challenging since a notable portion of the available data is either not accurate, or is not required for the particular study at hand. A step-by-step data curation thus becomes important while handling big datasets in SAR studies. In the quantum chemical domain, another challenge is choosing the right level of theory for the calculation of various quantum descriptors. Coupled cluster (CC) theory in quantum chemistry offers varying levels of high accuracy depending on the amount of excitation terms included in it, like, singles (S), doubles (D), triples (T), quadruples (Q), etc. While this level can be efficiently used for smaller or perhaps even medium-sized molecules (Balasubramanian and Basak, 2016; Basak et al., 2003), they are way too time-consuming for moderately large compounds even with large computational power. Prioritizing higher accuracy is not always the answer, since the computational cost for such calculations would be too high. A trade-off between these two becomes important to obtain an efficient result with respect to both time and accuracy. For bigger molecules, or for datasets containing a large number of compounds, density functional theory (DFT) can be efficiently used to get reasonably accurate results with low computational cost.

Here we discuss the following case studies where the toxicity of a set of aliphatic compounds against *T. pyriformis* and the human African trypanosomiasis (HAT) healing activity of pyridyl benzamides.

9.5.1 *Tetrahymena pyriformis*

The efficacy of the developed models is determined based on R^2, adjusted R^2 and *SD* values (a regression model is statistically relevant only when its $R^2 > 0.60$) (Golbraikh and Tropsha, 2002; Tropsha, 2010). The dataset used toward the study against *T. pyriformis* contains 169 aliphatic compounds with diverse functionality, namely, saturated alcohols, monoesters, diesters, carboxylic acids, ketones, amino alcohols and unsaturated alcohols. From the MLR analysis on the undivided dataset, we find that the R^2 corresponding to the one-parameter models based on ω, ω^2, and ω^3 ranges from 0.703 to 0.779, that is, on average, it is capable of describing approximately 74% of the data (Table 9.1). Hydrophobicity, on the other hand, is capable of describing a higher percentage in most cases except that in amino alcohols. The descriptor log *P* has direct control over the toxic behavior of a compound owing to its ability to measure the toxicodynamic process of the molecule reaching the active site. This is evident from the positive value of the coefficients of log *P* and $(\log P)^2$ in all the developed SAR models. However, $(\log P)^2$ contains a negative coefficient in the models for amino alcohols suggesting that in this case hydrophobicity is not important in toxic doses and this also hints toward the polar nature of the receptor site (Hansch et al., 2001, 2002). Among all the descriptor combinations used, dual parameter QSARs $\{\omega, \log P\}$, $\{\omega^2, \log P\}$ and $\{\log P, (\log P)^2\}$ provide the best correlations.

9.5.2 *Tryphanosoma brucei*

The HAT healing activity of a dataset of 32 pyridyl benzamides studied by Masand et al. in terms of the following two QSAR models are:

$$pIC50 = 4.8523(\pm 0.3676) - 0.9802(\pm 0.4082) * \mathrm{GATS8}c + 0.1240(\pm 0.0734) * \mathrm{RDF40}p + 0.0552(\pm 0.0099) * \mathrm{RDF55}s \quad [9.1]$$

$$pIC50 = 6.4521(\pm 1.2282) - 3.1455(\pm 2.3630) * E1s + 0.0545(\pm 0.0492) * \mathrm{RDF40}m + 0.0489(\pm 0.0111) * \mathrm{RDF55}s \quad [9.2]$$

where: GATS8c = Geary autocorrelation of lag-8/weighted by atomic charges, RDF40p = radial distribution function-040/weighted by relative polarizabilities, RDF55s = radial distribution function-055/weighted by relative I-state, E1s = 1st component accessibility directional WHIM index/weighted by relative I-state, RDF40m = radial distribution function-040/weighted by relative mass

Models 1.1 showed a correlation coefficient of 0.828. Our incorporation of ω and ω^2 by replacing one or two of the above descriptors have produced pretty interesting results. While replacing GATS8c and RDF40p, both one-by-one and simultaneously, by ω and ω^2 produced an R^2 in the range 0.683–0.760, replacing RDF55s drastically lowers the R^2 value suggesting its importance as a descriptor for this dataset. Fig. 9.1 depicts the plots between experimental and predicted pIC_{50} for the test sets of model 1.1. Model 1.2, for the undivided dataset, does not provide statistically relevant results, however, some descriptor combinations do produce models with R^2 values closer to the threshold. The CRRN-based SRD method also confirms the results obtained from MLR analysis. Model 1.1 shows the best predictive ability (SRD = 0), closely followed by models singly-substituted descriptor combinations {GATS8c, ω, RDF55s}, {ω, RDF40p, RDF55s}, and {ω^2, RDF40p, RDF55s} with SRD value 6.25%.

Table 9.1 R^2 values obtained from MLR analysis on complete sets for *Tetrahymena pyriformis* using ω, ω^2, ω^3, $\log P$, $(\log P)^2$, and their combinations separately as descriptors.

R^2 *w.r.t.*	Saturated alcohols	Carboxylic acids	Monoesters	Diesters	Ketones	Amino acids	Unsaturated alcohols
ω	0.715	0.750	0.756	0.739	0.779	0.748	0.301
ω^2	0.709	0.734	0.758	0.733	0.771	0.746	0.296
ω^3	0.703	0.728	0.760	0.725	0.762	0.743	0.288
$\log P$	0.981	0.919	0.930	0.910	0.975	0.340	0.868
$(\log P)^2$	0.895	0.882	0.889	0.814	0.881	0.156	0.790
ω, $\log P$	0.981	0.919	0.932	0.958	0.975	0.879	0.868
ω^2, $\log P$	0.982	0.919	0.932	0.957	0.975	0.878	0.868
ω, $(\log P)^2$	0.905	0.917	0.889	0.911	0.959	0.857	0.790
ω^2, $(\log P)^2$	0.906	0.918	0.900	0.912	0.959	0.853	0.799
ω, ω^2	0.732	0.785	0.763	0.748	0.876	0.748	0.302
$\log P$, $(\log P)^2$	0.983	0.937	0.933	0.912	0.975	0.387	0.890

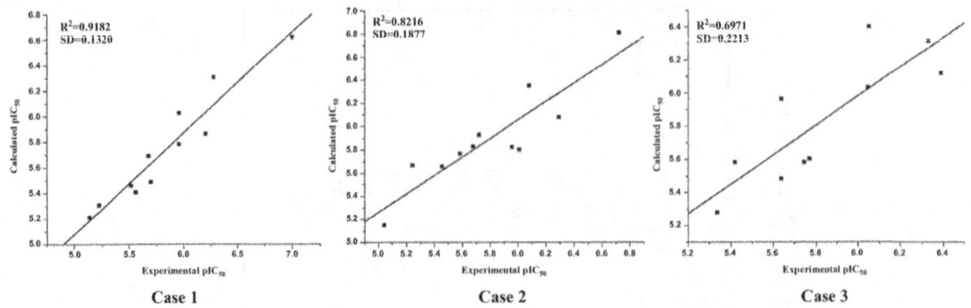

Figure 9.1 Plots of experimental versus calculated pIC_{50} along with their respective R^2 and SD values for the test sets of model 1.1.

9.6 Conclusion

The present work encompasses the role of global electrophilicity index as a descriptor in predicting biological and ecotoxicological activities. Single parameter quantitative structure–toxicity relationship (QSTR) analysis for the benzene derivatives and 169 aliphatic compounds revealed that ω and ω^2 are sufficiently capable of predicting their toxicities toward *T. pyriformis*. When compared with the results obtained from the most widely used log *P*-based models, it is found that they produce marginally close correlation coefficients. The fact that the values of the hydrophobic parameter used here are experimentally obtained, whereas electrophilicity index is easily computed without conducting any costly and time-consuming experiments, makes ω much more affordable to use. Having said that, it is to be noted that the efficiency of any descriptor depends on the homogeneity of the dataset used.

The incorporation of ω and ω^2 in conjunction with some of the descriptors screened and selected by Masand et al., to develop new QSAR models showed satisfactory predicting power (R^2 above 0.7 in most of the cases). The R^2 did not cross the threshold value when RDF55s descriptor is removed from model 1.1, revealing its significance in explaining the HAT healing activity for the dataset considered.

In conclusion, it is safe to claim that toxicity/activity prediction using QSAR models with electrophilicity index as a descriptor can be used as a preliminary step to provide statistically sound results. Further improvement on the coefficient of determination can be achieved by adding relevant descriptors, if and when needed. Electrophilicity index quantifies the electronic environment of a compound and thus is heavily dependent on the molecular geometry and its elemental composition. Being a quantum chemical descriptor, it correlates the molecular structure to the biological/toxicological activity being studied, which essentially is the main purpose of QSAR.

Acknowledgments

PKC would like to thank Dr. Subhash Basak for kindly inviting him to contribute a chapter in the book "Big data analytics in Chemoinformatics and Bioinformatics." He also thanks the

DST, New Delhi, for the J. C. Bose National Fellowship, grant number SR/S2/JCB-09/2009. RP thanks CSIR for her fellowship.

Conflict of interest

The authors declare that they have no conflict of interest regarding the publication of this article, financial, and/or otherwise.

References

Albert, A., Rubbo, S.D., Goldacre, R.J., Davey, M.E., Stone, J.D., 1945. The influence of chemical constitution on antibacterial activity. Part II: a general survey of the acridine series. Br. J. Exp. Pathol. 26 (3), 160.

Albert, A., 1985. Selective Toxicity, 7th ed. Chapman & Hall, London.

Allred, A.L., Rochow, E.G., 1958. A scale of electronegativity based on electrostatic force. J. Inorg. Nucl. Chem. 5 (4), 264–268.

Allred, A.L., 1961. Electronegativity values from thermochemical data. J. Inorg. Nucl. Chem. 17 (3–4), 215–221.

Ayers, P.W., 2005. An elementary derivation of the hard/soft-acid/base principle. J. Chem. Phys. 122, 141102.

Ayers, P.W., Parr, R.G., Pearson, R.G., 2006. Elucidating the hard/soft acid/base principle: a perspective based on half-reactions. J. Chem. Phys. 124 (19), 194107.

Ayers, P.W., 2007. The physical basis of the hard/soft acid/base principle. Faraday Discuss. 135, 161–190.

Balasubramanian, K., Basak, S., C., 2016. Metabolic electron attachment as a primary mechanism for toxicity potentials of halocarbons. Curr. Comput. drug. Des. 12 (1), 62–72.

Basak, S.C., Balasubramanian, K., Gute, B.D., Mills, D., Gorczynska, A., Roszak, S., 2003. Prediction of cellular toxicity of halocarbons from computed chemodescriptors: a hierarchical QSAR approach. J. Chem. Inf. computer Sci. 43 (4), 1103–1109.

Becke, A.D., 1988. Density-functional exchange-energy approximation with correct asymptotic behavior. Phys. Rev. A 38 (6), 3098–3100.

Brown, A.C., Fraser, T.R., 1868. V—On the connection between chemical constitution and physiological action. Part. I—on the physiological action of the salts of the ammonium bases, derived from strychnia, brucia, thebaia, codeia, morphia, and nicotia. Earth Environ. Sci. Trans. R. Soc. Edinb. 25 (1), 151–203.

Cedillo, A., Chattaraj, P.K., Parr, R.G., 2000. Atoms-in-molecules partitioning of a molecular density. Int. J. Quantum Chem. 77 (1), 403–407.

Chattaraj, P.K., Ayers, P.W., Melin, J., 2007b. Further links between the maximum hardness principle and the hard/soft acid/base principle: insights from hard/soft exchange reactions. Phys. Chem. Chem. Phys. 9 (29), 3853–3856.

Chattaraj, P.K., Lee, H., Parr, R.G., 1991. HSAB principle. J. Am. Chem. Soc. 113 (5), 1855–1856.

Chattaraj, P.K., Ayers, P.W., 2005. The maximum hardness principle implies the hard/soft acid/base rule. J. Chem. Phys. 123 (8), 086101.

Chattaraj, P.K., Roy, D.R., Giri, S., Mukherjee, S., Subramanian, V., Parthasarathi, R., et al., 2007a. An atom counting and electrophilicity based QSTR approach. J. Chem. Sci. 119 (5), 475–488.

Cronin, T.D., Madden, J.C., 2010. In Silco Toxicology - Principles and Applications. Royal Society of Chemistry, Cambridge, UK.

Cros, A., 1863. Action de l'alcool amylique sur l'organisme. Faculté de. médecine de Strasbg .

Deschenes, L.A., Vanden Bout, D.A., 2000. Origin 6.0: Scientific Data Analysis and Graphing SoftwareVanden BoutUniversity of Texas, Austin Origin Lab Corporation (formerly Microcal Software, Inc). http://www.originlab.com. Commercial price: 595. Academicprice: 446.

Ferguson, J., 1939. The use of chemical potentials as indices of toxicity. Proc. R. Soc. London. Ser. B-Biological Sci. 127 (848), 387–404.

Fock, V., 1930. Näherungsmethode zur Lösung des quantenmechanischen Mehrkörperproblems. Z. für Phys. 61 (1–2), 126–148.

Frisch, M.J., Trucks, G.W., Schlegel, H.B., Scuseria, G.E., Robb, M.A., Cheeseman, J.R., et al., 2004. Gaussian 03, Revision C. Gaussian, Wallingford, CT, 02.

Frisch, M.J., Trucks, G.W., Schlegel, H.B., Scuseria, G.E., Robb, M.A., Cheeseman, J.R., et al., 2009. Gaussian 09, Version D. Gaussian, Wallingford, CT, 01.

Gázquez, J.L., 1997. The hard and soft acids and bases principle. J. Phys. Chem. A 101 (26), 4657–4659.

Golbraikh, A., Tropsha, A., 2002. Beware of q2!. J. Mol. Graph. Model. 20 (4), 269–276.

Hammett, L.P., 1935. Some relations between reaction rates and equilibrium constants. Chem. Rev. 17 (1), 125–136.

Hammett, L.P., 1937. The effect of structure upon the reactions of organic compounds. Benzene derivatives. J. Am. Chem. Soc. 59 (1), 96–103.

Hansch, C., Maloney, P.P., Fujita, T., Muir, R.M., 1962. Correlation of biological activity of phenoxyacetic acids with Hammett substituent constants and partition coefficients. Nature 194 (4824), 178–180.

Hansch, C., Fujita, T., 1964. p-σ-π Analysis. A method for the correlation of biological activity and chemical structure. J. Am. Chem. Soc. 86 (8), 1616–1626.

Hansch, C., Kurup, A., Garg, R., Gao, H., 2001. Chem-bioinformatics and QSAR: a review of QSAR lacking positive hydrophobic terms. Chem. Rev. 101 (3), 619–672.

Hansch, C., Hoekman, D., Leo, A., Weininger, D., Selassie, C.D., 2002. Chem-bioinformatics: comparative QSAR at the interface between chemistry and biology. Chem. Rev. 102 (3), 783–812.

Hartree, D.R., 1928. The wave mechanics of an atom with a non-coulomb central field. Part iii. Term values and intensities in series in optical spectra, Mathematical Proceedings of the Cambridge Philosophical Society, 24. Cambridge University Press, pp. 426–437, July.

Hartree, D.R., 1929. The distribution of charge and current in an atom consisting of many electrons obeying Dirac's equations, Mathematical Proceedings of the Cambridge Philosophical Society, 25. Cambridge University Press, pp. 225–236, April.

Héberger, K., 2010. Sum of ranking differences compares methods or models fairly. TrAC. Trends Anal. Chem. 29 (1), 101–109.

Iczkowski, R.P., Margrave, J.L., 1961. Electronegativity. J. Am. Chem. Soc. 83 (17), 3547–3551.

Kollár-Hunek, K., Héberger, K., 2013. Method and model comparison by sum of ranking differences in cases of repeated observations (ties). Chemometrics Intell. Laboratory Syst. 127, 139–146.

Lee, C., Yang, W., Parr, R.G., 1988. Development of the Colle-Salvetti correlation-energy formula into a functional of the electron density. Phys. Rev. B 37 (2), 785–789.

Masand, V.H., Mahajan, D.T., Maldhure, A.K., Rastija, V., 2016. Quantitative structure−activity relationships (QSARs) and pharmacophore modeling for human African trypanosomiasis (HAT) activity of pyridyl benzamides and 3-(oxazolo [4, 5-b] pyridin-2-yl) anilides. Medicinal Chem. Res. 25 (10), 2324−2334.

Meyer, H., 1899. Welche Eigenschaft der Anaesthetica bedingt ihre narkotische Wirkung?. Naunyn-Schmiedeberg's. Arch. Exp. Pathol. Pharmakol. 42, 109−118.

Mulliken, R.S., 1934. A new electroaffinity scale; together with data on valence states and on valence ionization potentials and electron affinities. J. Chem. Phys. 2 (11), 782−793.

Mulliken, R.S., 1935. Electronic structures of molecules XI. Electroaffinity, molecular orbitals and dipole moments. J. Chem. Phys. 3 (9), 573−585.

Overton, C.E., 1901. Studien über die Narkose zugleich ein Beitrag zur allgemeinen Pharmakologie. Fischer, Jena, Germany.

Padmanabhan, J., Parthasarathi, R., Subramanian, V., Chattaraj, P.K., 2006. Group philicity and electrophilicity as possible descriptors for modeling ecotoxicity applied to chlorophenols. Chem. Res. Toxicol. 19 (3), 356−364.

Parr, R.G., Donnelly, R.A., Levy, M., Palke, W.E., 1978. Electronegativity: the density functional viewpoint. J. Chem. Phys. 68 (8), 3801−3807.

Parr, R.G., Pearson, R.G., 1983. Absolute hardness: companion parameter to absolute electronegativity. J. Am. Chem. Soc. 105 (26), 7512−7516.

Parr, R.G., Yang, W., 1989. Density functional theory of atoms and molecules. Oxford. University Press, New York.

Parr, R.G., Szentpály, L.V., Liu, S., 1999. Electrophilicity index. J. Am. Chem. Soc. 121 (9), 1922−1924.

Parthasarathi, R., Subramanian, V., Roy, D.R., Chattaraj, P.K., 2004. Electrophilicity index as a possible descriptor of biological activity. Bioorganic & medicinal Chem. 12 (21), 5533−5543.

Pauling, L., 1960. The Nature of the Chemical Bond, 260. Cornell university press, Ithaca, NY, pp. 3175−3187.

Pearson, R.G., 1997. Chemical Hardness: Applications from Molecules to Solids. Wiley-VCH, Weinheim.

Petersson, A., Bennett, A., Tensfeldt, T.G., Al-Laham, M.A., Shirley, W.A., Mantzaris, J., 1988. A complete basis set model chemistry. I. The total energies of closed-shell atoms and hydrides of the first-row elements. J. Chem. Phys. 89 (4), 2193−2218.

Richardson, B., 1869. Physiological research on alcohols. Med. Gazzette 2, 703−706.

Richet, C., 1893. Reports of the sessions of the society of biology and its subsidiaries. Soc. Biol. Ses. Fil. 9, 775−776.

Roblin Jr, R.O., Bell, P.H., 1942. Structure and reactivity of sulphanilamide type compounds. J. Am. Chem. Soc. 64, 2905−2917.

Sen, K.D., Mingos, D.M.P., 1993. Structure and Bonding: Chemical Hardness, 80. Springer, Berlin.

Taft, R.W., 1952. Polar and steric substituent constants for aliphatic and o-Benzoate groups from rates of esterification and hydrolysis of esters1. J. Am. Chem. Soc. 74 (12), 3120−3128.

Taft, R.W., 1956. Separation of polar, steric and resonance effects in reactivity. Steric Eff. Org. Chem. 556−675.

Tropsha, A., 2010. Best practices for QSAR model development, validation, and exploitation. Mol. Inform. 29 (6-7), 476−488.

Pharmacophore-based virtual screening of large compound databases can aid "big data" problems in drug discovery

10

Apurba K. Bhattacharjee
Department of Microbiology and Immunology, Biomedical Graduate Research Organization, School of Medicine, Georgetown University, Washington, DC, United States

10.1 Introduction

An estimated 2–3 quintillion bytes ($2-3\times10^{17}$) of data are generated every day in the world (Bulao, 2020). Every major sector of the economy and scientific organizations should consider how to handle "big data" efficiently for solving important problems across disciplines to obtain useful interpretations. About 30% of the above amount of data comes from the health care systems, mostly from laboratories, medical records, cell-phone health apps and wearable technologies. Although drug discovery programs generate relatively a smaller percentage of this data, it is still a challenging problem. To shortlist compounds from about a billion available drug-like compounds for high-throughput screening (HTS) against a specific disease target from biological space of about 10^4-10^5 human proteins is a monumental task (Blum and Reymond., 2009). Extensive computational power is required to screen millions of compounds in the shortest possible time and simultaneously interpret them for rapid decision-making. In order to address this "big data" problem, several major virtual screening efforts have been initiated utilizing grid and distributed computing networks (UN Global Pulse, 2012). However, more efficient tools are constantly necessary for quick interpretation to reduce clinical attrition for an early decision. Big data analytics using machine learning (ML) and artificial intelligence (AI) tools are providing significant support to the R&D of the pharmaceutical industry, however, many more challenges are yet to be resolved. Intelligent augmentation (IA) of AI tools is being developed to improve the efficiency of the process. AI generally refers to replacing people with machines but AI tools with intelligence augmentation (IA) are actually assisting ML technologies to further efficiency. IA may now set to take over from AI when it comes to the progress of such efforts (Doraiswamy. 2017). Problems associated with available data include general reproducibility as well as the context associated with the experiment for success. Apart from them, AI with IA can also generate new hypotheses prospectively by linking data together (Doraiswamy. 2017). Big data

Big Data Analytics in Chemoinformatics and Bioinformatics. DOI: https://doi.org/10.1016/B978-0-323-85713-0.00014-1

from the context of biology, chemistry, drug discovery and clinical trials are complex processes and thus require more validation and scrutiny from time to time.

Drug discovery itself has become more complex than ever before with rapidly changing technologies. The process may require ten to fifteen years for taking a new compound from the bench of discovery to the shelve of a pharmacy for human use costing approximately US$2–3 billions. The procedure before approval has to undergo several rounds of stringent animal testing, toxicity evaluations and human clinical trials. Despite this huge cost involved in the process, fewer drugs are approved by the FDA (Food and Drug Administration, USA). Only nineteen drugs were approved recently. Consequently, the investment in research and development (R&D) took a steep downward trend to about 3.5% making more challenges to drug discovery programs (Blum and Reymond., 2009; Bleicher et al., 2003). Nevertheless, with many odds and challenges, the pharmaceutical industry will have to remain an integral part of our healthcare systems. The overall goal of a pharmaceutical company is to find a lead compound that can be optimized to give a drug candidate. Chemical synthesis to modify the lead molecule to improve its chances for a successful drug is usually the process of optimization. However, there are many challenges to achieving the goal. Despite a billion drug-like compounds being there only about 30 million compounds are actually available to target a disease-specific protein out of 10^4–10^5 human proteins (Fig. 10.1) (Blum and Reymond., 2009).

A pharmaceutical company can usually handle about one million compounds in its HTS for testing. Obviously, the number is a small proportion of the compounds actually "available." Thus, more HTS screening and large synthesis efforts would be necessary. Again, large-scale screening is expensive and not all targets are suitable for HTS. Moreover, current therapeutic targets are of seven main classes, (receptors 45%, enzymes 28%, hormones and factors 11%, nucleic acids 2%, ion channels 5%, nuclear receptors 2%, and unknown 7%). Enzymes and receptors represent a major part of them (Bleicher et al., 2003).

Therefore, new technologies are necessary and valuable for improving efficiency and cost-effectiveness (Kubinyi, 2006). Virtual screening of large compound databases using a range of in silico techniques facilitate the selection of a smaller number of

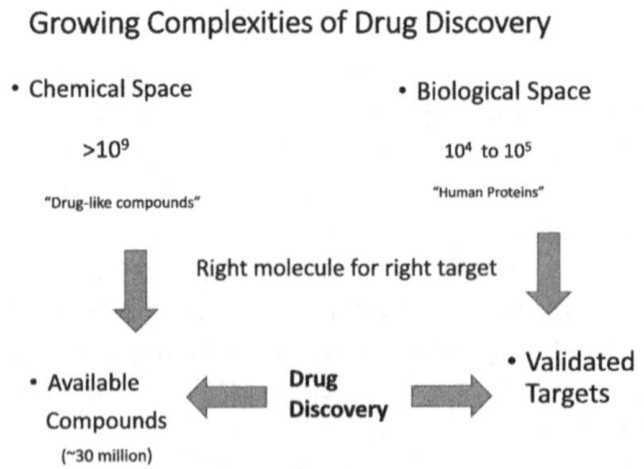

Figure 10.1 Schematic illustration showing complexities of compound selection in drug discovery.

compounds to better serve the HTS. ML methods in recent years along with IA enabled further improvement of the process. Several in silico methods have provided IA to the ML methods (Bleicher et al., 2003). Since ML methods with IA can provide significant efficiency in the downselection of compounds from virtual screening, the focus of the chapter is on the potential use of pharmacophore models in the augmentation process with the goal of rapid drug discovery.

10.2 Background of data analytics, machine learning, intelligent augmentation methods and applications in drug discovery

10.2.1 Applications of data analytics in drug discovery

Data analytics usually carry out the collection, manipulation, and analysis of big and diverse data sets. These tools are capable of providing (a) predictive insights to identify targets of a drug that could be considered in the pipeline; (b) increased monitoring capacity and statistical evaluations for better patient recruitments; and (c) data collection from public domain information network for early identification of any previous report of adverse drug reactions (Stupokevitch et al., 2020). Although these approaches were often used in clinical studies through mining different clinical databases, fewer uses were available in the field of drug discovery. Even with molecular information or sequence information being available, clinical trials were not enough for a conclusive study for paucity of number of patients. However, with lowering of costs for genomic sequencing and possibility of multiple initiatives for connecting patient profiles and clinical information, large-scale databases are now available for better clinical studies (Brown et al., 2018). Thus, AI and ML technologies are becoming more important and useful in the fight against diseases. However, upgrading skills to make use of latest technologies for speeding up the design and development of a novel drug-like compound requires multiple coordinations. The approach should include previous data applications from preclinical studies, clinical trials, and postnew drug application surveillance to enable a fairly accurate prediction of FDA approval and patient outcomes. Effective collaboration with different partners would be necessary to accelerate data mining while maintaining each other's intellectual property. By mining different databases of partners, compounds can be obtained to find closely matched pairs and by analyzing them, rules can be created for virtual molecules to predict the impacts of chemical structural changes. Automation needs to be applied for big data to allow monitoring trials in real-time. The process should be able to identify safety and operational signals in order to avoid costly issues, such as adverse events and delays.

10.2.2 Machine learning in drug discovery

ML is often considered as tool to replace bench chemists from certain routine practices in research. Thus, for example, ML can be used to predict new reagents capable of making the desired material helping chemists to use these algorithms instead of searching literatures or using intuition. However, since researchers can still attempt to train these algorithms by introducing biases, the approach can be counterproductive for ML and can ultimately degrade its performance (Jia et al., 2019). Therefore, careful consideration should be required for useful applications for ML techniques.

Background of ML indicates that substructure analysis was first used on large datasets in 1973. The idea was that each fragment substructure should be able to contribute in the same way to activity of a specific disease continuously (Varnek and Baskin, 2012). Substructure analysis overcomes the problem of dimensionality when it comes to analyzing structures in drug design. These methods were used by the pharmaceutical industry to explore large activity datasets quite vigorously. However, with the advent of HTS methods to the industry, efforts from ML thinned out (Varnek and Baskin, 2012). At that time, the ML idea was primarily based on the hypothesis that each fragment substructure of a molecule contributes for a particular type of activity constantly, regardless of other substituents. Thus, in general, the methods used only fragment-based fingerprints. Each fragment had a weight assigned to it reflecting its differential occurrence in the training set of active and inactive compounds. Different types of weight schemes emerged in this way to assess the quality of the models. For example, (a) an unknown compound is scored by summing the weights for all fragments it contains; (b) scores are used to rank the test-set compounds in decreasing probability order of activity.

The weight for a fragment substructure is considered to be some or all of the followings:

(a) Active (ACT) and inactive (INACT), the numbers of active and inactive compounds in a training set.
(b) ACT(i) and INACT(i), the numbers of active and inactive molecules in the training set that contain the *i*th fragment.

A typical example for operating the weights can be formulated in the form of the following equation which is also widely known as the Bayesian classifier:

$ACT(i)/ACT(i) + INACT(i)$

Furthermore, recursive partitioning (RP) strategies can also be introduced. A classification decision tree can be constructed from qualitative data. For example,

– active/inactive, soluble/insoluble, toxic/nontoxic

A drug-likeness rule based approach (Lipinski et al., 2001; Rishton, 2003) as shown in Fig. 10.2 can be applied to provide the best statistical classification. For example,

– drug/nondrug: $MW < 500 | MW > 500$

Repeat each set arising from the previous classifications iteratively till no more reasonable classification can be achieved.

However, the above procedure may generate good models but caution has to be taken since often poor predictive power was observed if not performed carefully. Better results could be possible by the following strategies: (a) use of leave-many-out strategies to validate, and (b) easy to interpret/drive what-next decisions

- ➢MW < 500
- ➢clogP < 5
- ➢H-bond donors (sum of OHs and NHs) < 5
- ➢H-bond acceptors (sum of Ns and Os) <10
- ➢No. of rotatable bonds <10
- ➢Polar surface area (PSA) < 140 A^2
- ➢Peptides unsuitable

Figure 10.2 Illustration of "drug-likeness" characteristics proposed by C.A. Lipinski et al. *Sources*: From Lipinski, C.A., Lombardo, F., Dominy, B.W., Feeney, P.J. (2001). Adv. Drug. Deliv. Rev., 1; 46(1–3): 3–26. doi: 10.1016/s0169-409x(00)00129-0. PMID: 11259830. Rishton, G.M. (2003). Drug. Discov. Today, 8: 86–96. doi: 10.1016/s1359644602025722. PMID: 12565011.

(Hamman et al., 2010; Wang et al., 2008). Fig. 10.2 provides some few insights on the classification strategy of a training set.

10.2.3 Application of other computational approaches in drug discovery

Over the years, computer-aided drug design and discovery (CADDD) initiatives have encountered many challenges starting from the availability of large-scale genome and proteome data for diseases, to pathogens and hosts as well as accessibility for analysis and interpretation (Kapetanovic, 2008; Devillers, 2018; Bhattacharjee, 2018). Computer science methods rapidly came forward to assist several initiatives by developing accurate software for solving many challenges of chemical and biological problems. More and more software are being created utilizing and streamlining the drug discovery approaches. Due to the escalating costs in the development phase of a drug molecule, objective quality assessment is necessary much before the clinical trial phase within pharmaceutical research. Instead of the linear process for compound optimization, a parallel strategy is adopted iteratively so that the profile of chemical entities can rapidly be shaped allowing properties to be multidimensionally balanced.

It is now becoming possible to identify potential proteins and metabolic pathways for pathogens as well as microorganisms and parasites more accurately for discovery of many novel biomolecular therapeutics. Life sciences, computer science, and data science are converging in helping each other in the process. Although several challenges still exist, advanced data analytic tools can help the industry enormously to overcome unfavorable R&D expenses. However, further complementary efforts within organizations would be necessary to maximize skills with data science resources that should include the emerging advances of AI and ML tools (Bleicher et al., 2003).

10.2.4 Predictive drug discovery using molecular modeling

In earlier days, drug discovery primarily involved natural products or compounds of animal origin for trials to identify drug candidates. The process used to be high-throughput screening (HTS) for in vitro evaluations with combinatorial chemistry to test a few thousand compounds iteratively for achieving about 1% hit rate. However, with the paradigm shift toward designing molecular property-based combinatorial compound libraries, increasing quality-based product information started becoming available (Folkers et al., 2004). Libraries are being generated more on specific ligand or bio-structural information, augmented with modern integrated computational and synthetic methods. However, HTS continues to remain the main source of chemistry initiatives in pharmaceutical research. Thus, time taken for assay development, logistical hurdles and issues related to compound acquisition require alternative approaches to complement the lead discovery. Development of new algorithms now allows annotation and grouping of biological targets as well as chemical structures creating establishments of new multi disciplines approaches. For example, efforts for joining chemical genomics where chemical topology space can be seamlessly generated with the biological target space. This would allow systematic screening of targeted chemical libraries of small molecules against individual drug targets to achieve the goal of identification of novel drugs and drug targets and the design of customized ligands (Folkers et al., 2004; Jacoby, 2009; Weill and Curr, 2011).

Since the investment goal of pharmaceutical companies is to recover initial R&D costs, the focus is primarily centered on discovery of compounds that are likely to succeed in clinical trials and the market. As mentioned earlier that even though there are about a billion drug-like compounds in the chemical space, only a small proportion of about 30 millions are actually available for drug discovery (Blum and Reymond., 2009). Therefore, choosing the right molecule for the right target continues to be a challenging task using the conventional HTS methods. Alternative approaches are necessary for rapid downselection of available compounds against validated disease-specific targets. Several in silico technologies have emerged to take up the challenges over past several decades and have performed remarkably well in many frontiers of drug discovery including virtual screening (Brust et al., 2009; Ren et al., 2009; Chen et al., 2009). Virtual screening essentially is an in silico approach used for search of large databases to downselect a smaller number of compounds for biological testing. In silico predictive modeling of ADME/toxicity properties has become more reliable and accurate over the years. These tools can now predict the pharmacokinetic profiles of potential drug molecules far more accuracy. Advanced ML mathematical models and simulations too have assisted to predict pharmacokinetic profiles more accurately and enabled a better understanding of how the potential drug molecule will act in the body.

Virtual screening can be performed for both in-house and commercial databases. Choosing compounds to purchase from external sources or finding compounds from in-house can both help to decide which compounds to be synthesized for optimal success. One key advantage of database searching over de novo design is that it

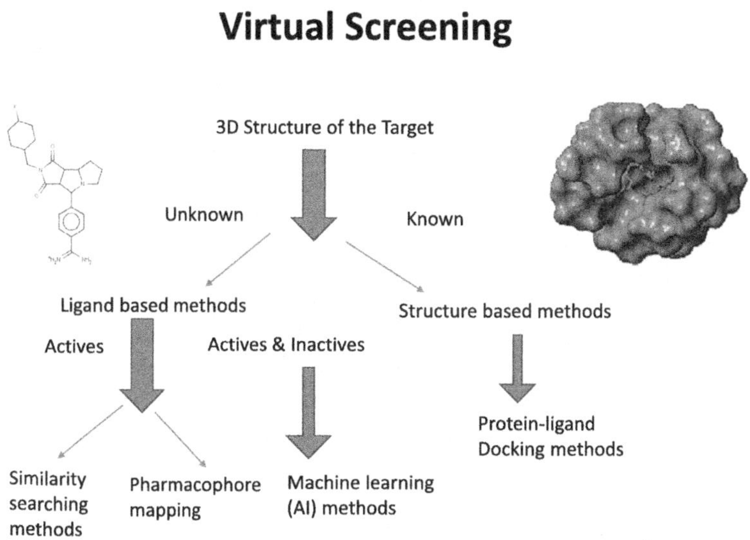

Figure 10.3 Schematic illustration of virtual screening methods.

allows the identification of existing compounds that are either readily available or have a known synthetic procedure. However, it is important to note that the technique applied in the virtual screening process depends on the amount of information available about the particular disease target. A schematic diagram of the virtual screening process is shown in Fig. 10.3.

Although following many such techniques shown in Fig. 10.3, virtual novel compounds can be created in computers (de novo design), there is no guarantee that the virtual compounds will be active unless tested. Thus, custom synthesis may be necessary only when tested and showed promise to be in a pipeline for clinical trials.

Without the information on X-ray crystallographic protein structures, virtual screening of compound libraries is now possible to identify thousands of potential drug candidates and shortlist them to a handful of 10–15 compounds for experimental validation using intelligent modeling techniques. ML tools for training sets of disease-specific potent substructures have been used for virtual screening of large compound databases to identify new potentially active compounds for in vitro testing. However, further IA of the ML methods may bring better efficiency of the virtual screening process (Brown et al., 2018).

10.3 Pharmacophore modeling

Pharmacophore modeling is one such technique that can perhaps augment the intelligence (IA) of the ML methods. Pharmacophore modeling will be discussed in more details here to comment on how big data analytics can be better augmented by this in silico technique for drug discovery and design process. Visual imaging is often used for facial recognition of millions of people with the help of data analytics by intelligence departments of countries (Nilsson, 2009; Vracko et al., 2015). Pharmacophore imaging may similarly be used by interacting with ML tools to further improve the quality of virtual screening of millions of compounds to identify potential disease-specific compounds. However, quality algorithms are to be developed in the area to avoid false positives and statistical components are to be properly controlled.

Pharmacophore modeling has been used for the discovery of compounds against many diseases including cancer, diabetes, and HIV (Brust et al., 2009; Chen et al., 2009). Although in the past, combinatorial chemistry coupled with HTS was widely used in most drug discovery programs, in recent years large-scale genome mappings and availability of crystallographic protein structures have led a shift toward structure-based drug discovery. Importantly, the shift has led to opening of new chapters for drug discovery starting from X-ray crystallographic protein structure determination, active site analysis to design and identification of new ligands through virtual screening of databases, docking of ligands at the active site and pharmacophore modeling. Pharmacophores from specific disease-active ligands have the additional advantage that it is useful even when crystallographic protein structures are unknown (Nilsson, 2009; Vracko et al., 2015; Bhattacharjee et al., 2012). Pharmacophores not only provide useful templates for virtual screening to identify new compounds but also provide useful insights for designing novel ligands known as de novo design of active compounds for an unknown target protein. Since pharmacophore transcends chemical structural class and captures only features responsible for activity, the use of pharmacophore for virtual screening has additional advantage of identification of different active chemical classes having the potential for opening up new chapters of chemotherapeutic studies. The concept of pharmacophore is outlined as follows. When does a molecule become potentially a drug molecule? A molecule when able to optimally interact with the disease-specific target structure to trigger or inhibit its biological response, the molecule will have the potential to be a drug against the disease. At the molecular level, it is a combination of steric and electronic features only allows a molecule to optimally interact with the target to trigger or inhibit its biological response. Therefore, a combination of steric and electronic properties will constitute the pharmacophore of a drug molecule against a disease. The modern definition of pharmacophore as provided by the Union of Pure and Applied Chemistry (IUPAC) in 1998 (Leach et al., 2010) states similarly that a pharmacophore is "an ensemble of steric and electronic features that are necessary for optimal interaction with a specific receptor target structure (a protein or an enzyme) to trigger or inhibit its biological response" (Seidel et al., 2010). Although pharmacophore definition indicates it to be an abstract idea, it is possible to represent a pharmacophore in three-dimensional space as a statistical distribution of chemical features such as hydrogen bond acceptors and donors, aliphatic and aromatic hydrophobic sites, ring aromaticity, and ionizable sites required for interaction with complementary sites of the target structure. Features of a pharmacophore are represented as vectors and geometrical spheres.

The vector directionality features, namely those associated with hydrogen bonding are derived from symmetry, a number of localized lone pairs, and the environment around the atom in a molecule (ligand atom) (Güner, 2000). More importantly, the concept of pharmacophore becomes important to a medicinal chemist when it is demonstrated that if two structurally different compounds share the same features, the compounds are likely to have the same biological activity. For example, two structurally dissimilar compounds sharing the same pharmacophore features containing two hydrogen bond acceptors and two hydrophobic sites as shown in Fig. 10.4 should have similar biological activities. Indeed, both compounds were found experimentally to have similar antimalarial activities (Bhattacharjee et al., 2004a).

Pharmacophores are frequently used by medicinal chemists in drug discovery programs. The use of pharmacophore for virtual screening of compound databases is an intelligent and knowledge-based approach for efficient discovery of potentially bioactive compounds including new chemical classes. A pharmacophore is used in two ways to identify new compounds that share its features, and thus may exhibit the same biologic response. In the first approach, de novo design can be performed by linking the disjointed parts of a pharmacophore together with chemical fragments to generate a hypothetical structure which is chemically reasonable but completely novel required to be synthesized. The second approach is the virtual screening of a compound database with the pharmacophore model to identify new potential hit compounds. Therefore, virtual screening with pharmacophore can be complementary to HTS and together with ML could be very efficient for rapid drug

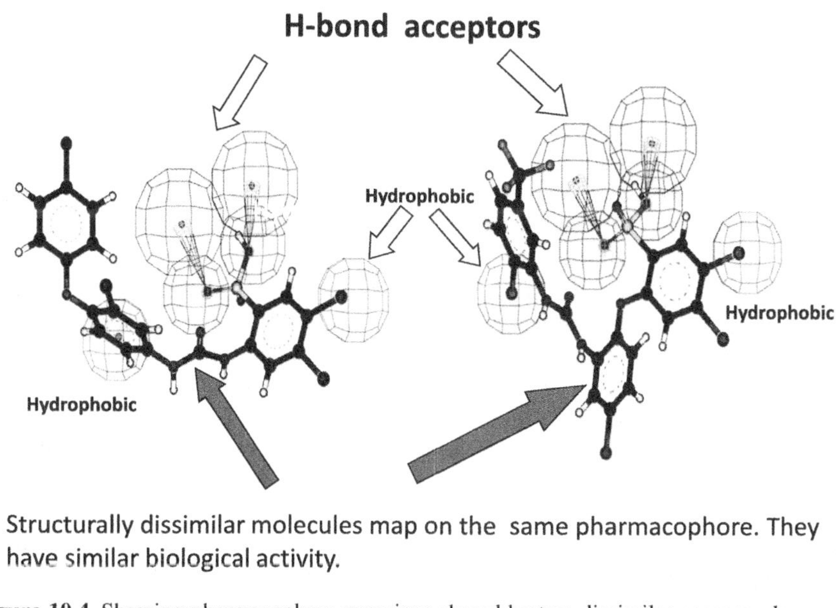

Figure 10.4 Showing pharmacophore mappings shared by two dissimilar compounds indicating similar interactions and biological activity.

discovery (Kapetanovic, 2008). Pharmacophore-based virtual screening can help to enrich active molecules in the hit list compared to a random selection of test compounds.

Pharmacophore models can be generated from a set of known active compounds using typical three-dimensional QSAR pharmacophore generation methods. The two widely used pharmacophore modeling and screening software are Ligandscout (Temml et al., 2014) and Discovery Studio (Discovery Studio, 2007). Both software are widely used for the generation, validation, and application for virtual screening of compound databases to identify new compounds. In addition, phase from Schrödinger is also used frequently (Dixon et al., 2006).

Validation of a quantitative pharmacophore model is extremely important for reliability and usefulness of the model. It can be performed by generating models that were able to identify active structures and predict the activity accurately. Mapping of pharmacophore had been used as a very useful tool for rapid discovery of effective drugs (Dror et al., 2004; Wolber et al., 2008). A test set of active compounds could also be generated for determination of correlation (R) and compared with that of the training set. However, for ultimate validation, the compounds identified by virtual screening of databases should be tested in vitro first and if show promise, in vivo testing for successful outcomes.

The best generated and validated pharmacophore model may be employed as a search query to retrieve molecules from different compound databases to discover new compounds. The identified compounds are to be mapped onto the pharmacophore model and estimated (predicted) activities EC_{50} or IC_{50} values are to be noted. Estimated activities of less than 500 nM are usually considered for further study.

Pharmacophore can be particularly useful for efficient search of databases when the mapped model onto the most active compound is converted to a three-dimensional shape-based template. It is because the template now will not only reflect the complementarity for the binding site but also will account for steric factors associated with the target structure. Since both pharmacophore features and steric requirements being embedded now in the screening procedure, it would make a powerful tool for rapid discovery of potentially bioactive compounds (Bhattacharjee, 2018, 2014).

Software combining ML with pharmacophores may be developed to construct classification models to distinguish between active and inactive compounds from the diverse data set. A validated shape-based three-dimensional template of a pharmacophore model in conjunction with ML should be an ideal one for virtual screening of large databases. The shape-based pharmacophore features can be used to train the ML tools for virtual screening of compound databases. Classification of new molecules may be carried out on the basis of potent in vitro activity data. The procedure should be able to obtain an optimal ensemble of pharmacophore models having a validated capacity to differentiate between active and inactive compounds. This could be identified by the RP approach. Normal Bayesian classification and support vector machine methodologies could then be applied to generate classification models by integrating multiple validated pharmacophores. In this way, improved predictability and specificity for a particular disease-specific target protein could be achieved rather than a single pharmacophore model. The improved model should then be able to perform a better search for more potent and efficacious compounds through virtual screening of large compound databases. Significant improvements in finding efficacious compounds following the procedure have been reported (Wang et al., 2016). Two case studies from the author's own laboratory are provided below which could be useful for possible augmentation of ML by the pharmacophore model in developing training sets for discovery of new potent compounds.

10.3.1 Case studies

The first example relates to the pharmacophore model used for discovery of new antimalarials targeting cyclin-dependent kinases (CDKs). A robust pharmacophore model was developed by us from a training set of 15 structurally diverse CDK inhibitors with a wide range of activity (Bhattacharjee et al., 2004b). The model allowed us to discover several potent new malarial kinase (Pfmrk) inhibitors through an in-house database search of about 300,000 compounds. The model was found to contain two hydrogen bond acceptor functions and two hydrophobic sites including one aromatic ring hydrophobic function (Fig. 10.5) (Bhattacharjee et al., 2004b). Although the model was solely developed from structure–activity considerations, it was found to be consistent with a large number of crystallographic structures of CDKs with bound inhibitors. The model was also found to be consistent with the structure-functional requirements for binding of kinase inhibitors at the ATP binding site.

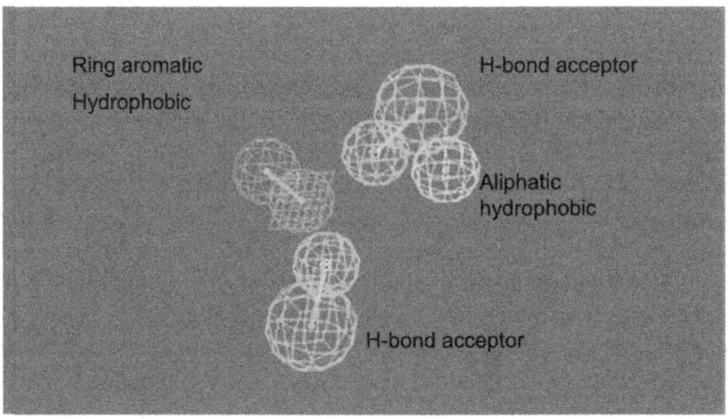

Figure 10.5 Pharmacophore model for malarial cyclin-dependent kinases (pfmrk) developed from known CDKs.
Source: From Bhattacharjee, A.K., Geyer, J.A., Woodard, C.L., Kathcart, A.K., Nichols, D. A., Prigge, S.T., et al. (2004b). J. Med. Chem. 47: 5418–5426.

Since CDK inhibitors are important compounds for exploring efficacy against cancer, the model could be useful for possible augmentation in ML methods (Wang et al., 2016) to allow rapid discovery of kinase inhibitors targeting CDKs.

The second example is related to a pharmacophore model used for the discovery of potential nontoxic compounds against pesticides originating from organophosphorus (OP) compounds. The commonly used pesticides in the US include paraoxon, chlorpyrifos, and TEPP-tetraethyl pyrophosphate. The acute toxicity of these OP compounds is due to inhibition of the enzyme acetylcholinesterase (AChE, EC 3.1.1.7) resulting in the accumulation and overstimulation of postsynaptic cholinergic receptors of the neurotransmitter acetylcholine (ACh) in synapses of the central and peripheral nervous systems (Bakshi et al., 2000; Marrs, 1993; Taylor, 1990). AChE, a serine hydrolase functions by hydrolyzing (removing) the neurotransmitter acetylcholine in the synaptic junction of nerve terminals in humans and animals. Discovery of AChE inhibitors and novel reactivators for OP-inhibited AChE are both important goals for neurologic therapeutics. Since a serious loss of cholinergic function in the central nervous system contributes significantly to the cognitive symptoms associated with AD and advanced age, the focus on the development of AChE inhibitors as new AD therapeutics too got much attention in recent years. Although the discovery of AChE inhibitors is an important goal for the development of neurologic therapeutics, the focus of this study was to identify novel reactivators for OP-inhibited AChE and binding affinities of identified compounds at the active site for possible potential AChE inhibitors. Nucleophiles, such as oximes are used for decades to displace the phosphate moiety of the OP which reacted with the enzyme's active site serine hydroxyl group, thereby reactivating AChE (Bajgar, 2004). However, oximes have several disadvantages including toxicity, CNS penetrability, and limited reactivation efficacy against the nerve agent tabun (Kassa et al., 2005; Musilek et al., 2007). In fact, none of the oximes is regarded as a broad-spectrum antidote for all nerve agents. Compounds from nonoxime chemical classes having good blood–brain barrier (BBB) penetrability have not been widely explored (Bedford et al., 1986; Okuno et al., 2008). In pursuit of those objectives, we adopted an in silico pharmacophore modeling strategy to identify potential nonoxime inhibitors and reactivators through virtual screening of compound databases. Initiating the approach, an in silico pharmacophore model for reactivation of OP (tabun)-inhibited AChE from binding affinity data of oximes was developed (Bhattacharjee et al., 2010). The pharmacophore model (Seidel et al., 2010) had one hydrogen bond acceptor function, one hydrogen bond donor function, and one aromatic hydrophobic (aromatic ring) function located in specific geometric regions surrounding the molecular space (Fig. 10.6).

The model allowed discovery of several potent nonoxime reactivators of OP (DFP)-inhibited AChE by evaluation of in vitro efficacy (Bhattacharjee et al., 2012; 2015). A few of the shortlisted nonoxime compounds were found to have significant in vivo efficacy against DFP-induced (OP-agent-induced) neuropathology in guinea pigs. One compound showed efficacy comparable to 2-PAM (used by the US Army in warfare) against brain symptoms for OP (DFP)-inhibited AChE (Bhattacharjee et al., 2015). Although the model was solely developed from structure–activity

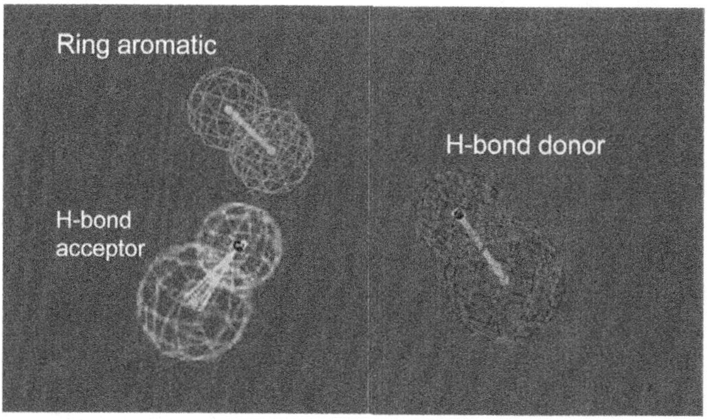

Figure 10.6 Pharmacophore model for reactivators against organo-phosphate poisoning agents. Developed from the binding affinity of oximes to tabun-inhibited AChE.
Source: From Bhattacharjee, A.K., Kuca, K., Musilek, K., Gordon, R.K. (2010). Chem. Res. Toxicol. 23: 26–36.

relationships, it is consistent with the observations of X-ray crystal structures of AChE with bound reactivators. It proved to be quite predictive and instrumental in conducting a targeted in silico screening of thousands of compounds from the in-house WRAIR-CIS database, Maybridge and ChemNavigator databases followed by in silico downselection through in silico evaluations for favorable BBB penetrability, octanol–water partition (Clog P), and toxicity (rat oral LD50) (Bhattacharjee et al., 2012, 2015). The model should have the potential to augment (IA) ML in virtual screening to further enhance efficiency of discovery of specific compounds against ACh hazards from OP-pesticides.

Both the above two case studies provide useful training sets (Bhattacharjee et al., 2004b, 2012, 2015) with known active and known inactive compounds which can be used combining substructure analysis models from ML methods and pharmacophore models for augmenting efficiency of virtual screening of large compound databases. The combination is likely to have a stronger potential to improve the drug discovery approach (Bandyopadhyay and Agrafiotis, 2008; Muthas et al., 2008; Gurujee and Deshpande, 1978).

10.4 Concluding remarks

Pharmacophore models augmented with ML could further improve the efficacy of large-scale virtual screening of millions of commercially available compounds or more complete sets of FDA-approved older drugs. In silico ADME/toxicity computational efforts can then prioritize to downselect a few compounds for testing. Thus, the combined approach should be more useful to leverage the HTS data that has already been developed at a great cost. In this study, we have focused on pharmacophore modeling in conjunction with ML but the approach may also be combined with the data from a variety of high-throughput screens to provide much larger training sets. There is also the scope to apply different computational approaches beyond those described here to identify disease-specific active compounds. Drug discovery excels by improving collaboration of groups by streamlining communication between discovery, clinical development, and medical affairs as well as external academics. This may provide insights over different research results and clinical identification leading to potential opportunities in personalized medicine. Academic research may provide the latest scientific breakthroughs and initiatives to other pharmaceutical companies. Combined expertise of structure-based knowledge, molecular biology, computational chemistry, pharmacophore modeling, virtual screening, AI, ML, and medicinal chemistry should be ideal for a successful drug discovery and development program.

References

Bhattacharjee, A.K., 2014. Role of in silico stereoelectronic properties and pharmacophores in aid of discovery of novel antimalarials, antileishmanials, and insect repellents (E-book) In: Basak, S.C., Restrepo, G., Villaveces, J.L. (Eds.), Advances in Mathematical Chemistry and Applications, 1. Bentham Science Publishers, Amsterdam, pp. 273–305.

Bhattacharjee, A.K., 2018. Pharmacophore modeling applied to mosquito-borne diseases. In: Devillers, J. (Ed.), Computational Design of Chemicals for the Control of Mosquitoes and Their Diseases. CRC Press, Taylor & Francis Group, Boca Raton, FL, USA, pp. 139–169.

Bhattacharjee, A.K., Hartell, M.G., Nichols, D.A., Hicks, R.P., Stanton, B., van Hamont, J. E., et al., 2004a. Eur. J. Med. Chem. 39, 59–67.

Bhattacharjee, A.K., Geyer, J.A., Woodard, C.L., Kathcart, A.K., Nichols, D.A., Prigge, S.T., et al., 2004b. J. Med. Chem. 47, 5418–5426.

Bajgar, J., 2004. Adv. Clin. Chem. 38, 151–216.

Bakshi, K.S., Pang, S.N.J., Snyder, R., 2000. J. Toxicol. Environ. Health A 59, 282–283.

Bandyopadhyay, D., Agrafiotis, D.K., 2008. A self-organizing algorithm for molecular alignment and pharmacophore development. J. Comput. Chem. 29, 965–982.

Bedford, C.D., Howd, R.A., Dailey, O.D., Miller, A., Nolen, H.W., Kenley, R.A., et al., 1986. J. Med. Chem. 29, 2174–2183.

Bhattacharjee, A.K., Kuca, K., Musilek, K., Gordon, R.K., 2010. Chem. Res. Toxicol. 23, 26–36.

Bhattacharjee, A.K., Marek, E., Le, H.T., Gordon, R.K., 2012. Eur. J. Med. Chem. 49, 229–238.

Bhattacharjee, A.K., Marek, E., Le, H.T., Ratcliffe, R., DeMar, J.C., Pervitsky, D., et al., 2015. Eur. J. Med. Chem. 90, 209–220.

Bleicher, K.H., Böhm, H.-J., Müller, K., Alanine, A.I., 2003. Hit and lead generation: beyond high-throughput screening. Nat. Rev. Drug. Discov 2, 369. Available from: https://doi.org/10.1038/nrd1086.

Blum, L.C., Reymond., J.L., 2009. J. Am. Chem. Soc. 1 131 (25), 8732. Available from: https://doi.org/10.1021/ja902302h. PMID: 19505099.

Brown, N., Cambruzzi, J., Cox, P.J., Davies, M., Dunbar, J., Plumbley, D., et al., 2018. (Chapter 5)—Big Data in Drug Discovery 57, 277–356.

Brust, A., Palant, E., Croker, D.E., Colless, B., Drinkwater, R., Patterson, B., et al., 2009. J. Med. Chem, 52. pp. 6991–7002.

Bulao, J., 2020. How much data is created everyday in 2020 <https://techjury.net/blog/how-much-data-is-created-every-day/#gref> (accessed 10.10.20).

Chen, J.J., Liu, T.L., Yang, L.J., Li, L.L., Wei, Y.Q., Yang, S.Y., 2009. Chem. Pharm. Bull. 57, 704–709.

Discovery Studio, 2007. DS Version 2.5. Accelrys Inc., San Diego, CA <http://accelrys.com/products/discovery-studio/>.

Dixon, S.L., Smondyrev, A.M., Knoll, E.H., Rao, S.N., Shaw, D.E., Friesner, R.A., 2006. J. Comput. Aided Mol. Des. 20, 647–671. http://www.Schrodinger.com.

Dror, O., Shulman-Peleg, A., Nussinov, R., Wolfso, H.J., 2004. Curr. Med. Chem. 11, 71–90.

Folkers, G., Kubinyi, H., Müller, G., Mannhold, R., 2004. Chemogenomics in Drug Discovery: A Medicinal Chemistry Perspective. Wiley-VCH, Weinheim, ISBN 978-3-527-30987-0.

Gurujee, C.S., Deshpande, V.L., 1978. An improved method of substructure analysis. Comput. Struct 8 (1), 147–152. Available from: https://doi.org/10.1016/0045-7949(78)90171-2.

Kubinyi, H., 2006. Success stories of computer-aided design. In: Ekins, S., Wang., B. (Eds.), Computer Applications in Pharmaceutical Research and Development. Wiley-Interscience, p. 377.

Hamman, F., Gutmann, H.N., Voigt, H.C., Drewe, J., 2010. Clin. Pharmacol. Ther 88, 52–59.

Devillers, J. (Ed.), 2018. Computational Design of Chemicals for the Control of Mosquitoes and Their Diseases. CRC Press, Taylor & Francis Group, Boca Raton, FL, USA.

Jacoby, E., 2009. Chemogenomics: Methods and Applications. Humana Press, N.J. Totowa, ISBN 978-1-60761-273-5.

Jia, X., Lynch, A., Huang, Y., et al., 2019. Nature 573, 251–255. Available from: https://doi.org/10.1038/s41586-019-1540-5.

Kapetanovic, I.M., 2008. Chem. Bio. Interact. 171, 165–176.

Kassa, J., Kuca, K., Cabal, J., 2005. Biomed. Pap. 149, 419–423.

Leach, A.R., Gillet, V.J., Lewis, R.A., Taylor, R., 2010. J. Med. Chem. 53, 539–558. Available from: https://doi.org/10.1021/jm900817u.

Lipinski, C.A., Lombardo, F., Dominy, B.W., Feeney, P.J., 2001. Adv. Drug. Deliv. Rev. 46 (1–3), 3–26. Available from: https://doi.org/10.1016/s0169-409x(00)00129-0. PMID: 11259830.

Marrs, T.C., 1993. Organophosphate poisoning. Pharmacol. Ther. 58, 51–66.

Musilek, K., Kuca, K., Jun, D., Dolezal, M., 2007. Curr. Org. Chem. 11, 229–238.

Muthas, D., Sabnis, Y.A., Lundborg, M., Karlen, A., 2008. Is it possible to increase hit rates in structure-based virtual screening by pharmacophore filtering? An investigation of the advantages and pitfalls of post-filtering. J. Mol. Graph. 26, 1237–1251.

Nilsson, N.J., 2009. The Quest for Artificial Intelligence. Cambridge University Press, ISBN 9781139642828.

Güner, O.F. (Ed.), 2000. Pharmacophore, perception, development, and use in drug design. University International Line (IUL) Biotechnology Series, San Diego,.

Okuno, S., Sakurada, K., Ohta, H., Ikegaya, H., Kazui, Y., Akutsu, T., et al., 2008. Toxicol. Appl. Pharmacol. Pharmacol 227, 8–15.

Taylor, P., 1990. In: Gilman, A.G., Rall, T.W., Neis, A.S., Taylor, P. (Eds.), Goodman & Gilman's the Pharmacological Basis of Therapeutics, eighth ed. pp. 131–149.

Doraiswamy, P.M., 2017. Forget AI. The real revolution could be IA. World Economic Forum.

Ren, J.X., Li, L.L., Zou, J., Yang, L., Yang, J.L., Yang, S.Y., 2009. Eur. J. Med. Chem. 44, 4259–4265.

Rishton, G.M., 2003. Drug. Discov. Today 8, 86–96. Available from: https://doi.org/10.1016/s1359644602025722. PMID: 12565011.

Seidel, T., Ibis, G., Bendix, F., Wolber, G., 2010. Drug. Disc. Today: Technol. 7, 221–228.

Stupokevitch, B., Sweenor, D., Swiderek, S., 2020. Reporting, predictive analytics, & everything in between, a guide to selecting the right analytics for you. O'Reilly (Ed) <https://www.investopedia.com/terms/d/data-analytics.asp>.

Temml, V., Kaserer, T., Kutil, Z., Landa, P., Vanek, T., Schuster, D., 2014. Future Med. Chem 6, 1869–1881.

UN Global Pulse, 2012. Big data for development: challenges and opportunities <http://www.unglobalpulse.org/projects/> (accessed 16.11.20).

Varnek, A., Baskin, I., 2012. J. Chem. Inf. Model. 52, 1413–1437.

Vracko, M., Basak, S.C., Bhattacharjee, A.K., 2015. Curr. Comput. Aided Drug. Des 11, 197.

Wang, H., Duffy, R.A., Boykow, G.C., Chackalamannil, S., Madison, V.S., 2008. J. Med. Chem. 51, 2439–2446. Available from: https://doi.org/10.1021/jm701519h.

Wang, S., Sun, H., Liu, H., Li, D., Li, Y., Hou, T., 2016. Mol. Pharmaceut. 13 (8), 2855–2866. Available from: https://doi.org/10.1021/acs.molpharmaceut.6b00471.

Weill, N., Curr, N., 2011. Top. Med. Chem. 11, 1944–195555. Available from: https://doi.org/10.2174/156802611796391212.

Wolber, G., Seidel, T., Bendix, F., Langer, T., 2008. Drug. Disc. Today 13, 23–29.

A new robust classifier to detect hot-spots and null-spots in protein–protein interface: validation of binding pocket and identification of inhibitors in in vitro and in vivo models

11

Yanrong Ji[1], Xin Tong[2], DanDan Xu[2], Jie Liao[2], Ramana V. Davuluri[1], Guang-Yu Yang[2] and Rama K. Mishra[3]

[1]Department of Preventive Medicine, Division of Health and Biomedical Informatics, Northwestern University Feinberg School of Medicine, Chicago, IL, United States, [2]Department of Pathology, Northwestern University Feinberg School of Medicine, Chicago, IL, United States, [3]Department of Biochemistry and Molecular Genetics, Feinberg School of Medicine, Northwestern University, Chicago, IL, United States

11.1 Introduction

Protein–protein interactions (PPIs) comprise a multitude of promising therapeutic targets, yet the development of inhibitors is challenging, particularly for protein pairs that lack extensive structural characterization. To circumvent time-consuming mutagenesis and co-crystallization studies, we developed a new computational framework for the efficient identification of PPI hotspots (HSs) by considering the interacting conjoint triads of amino acids from the two interacting proteins (Chang et al., 2010). We developed a machine learning (ML)-based classifier to identify HSs and nullspots (NS) in the PP interface, to assess the druggability of the binding pocket/groove containing the HSs and demonstrated the utility of the approach to find small molecule inhibitors.

It is well-known that conjoint triads of residues are most important while studying PPIs (Chang et al., 2010). Conjoint triad, which treats three continuous amino acids as a single unit, has been shown to be useful set of features for studying PPIs. In a conjoint triad, the rotameric state of the middle residue is very much influenced by its two neighboring residues. While identifying the HSs, earlier workers (Chang et al., 2010) have considered only the conjoint triad residues of a particular amino acid of interest for its HS evaluation using physicochemical properties of the concerned amino acid by examining the sequence only rather than the three-dimensional

Big Data Analytics in Chemoinformatics and Bioinformatics. DOI: https://doi.org/10.1016/B978-0-323-85713-0.00022-0

conformation of the conjoint residues. It is clear that when only using the primary sequence, one can get the location of a residue of interest without any information about its conformation or of the influence of its neighboring residues. Again, the previous studies did not consider the conjoint triad of its binding partner residues. To obtain an accurate picture of the interactions, we considered 50 crystal poses of the two interacting proteins. To determine HS and NS residues in these 50 complexes, 505 alanine mutations from experimental data have been considered. The residues where $\Delta\Delta G > 2.0$ kcal/mol have been defined as HS. If $\Delta\Delta G < 2.0$ kcal/mol, it is considered an NS (Moreira et al., 2007). We then choose the conjoint interacting partners based on the intermolecular atomic contacts within 3 Å. After defining 505 conjoint triad pairs of interacting partners, we computed 252 novel interacting descriptors such as the Jurs surface-based descriptors (Stanton et al., 2004), three-dimensional solvent accessibility descriptors (Shukla et al., 2014), E-state interacting pharmacophore (Kier and Hall, 2001) and interacting two-dimensional fingerprint descriptors (Vellay et al., 2009). Importantly our classification method for predicting HS and NS residues extended and improved upon the existing methods in a number of key respects. First, these descriptors along with some earlier descriptors (Munteanu et al., 2015) were used as feature variables to build the classifiers considering the experimental HS and NS residues as training and testing data. The combined data consist of 120 HSs and 385 NSs as observations for the two classes. In order to objectively and unbiasedly evaluate the classifier's performance, we randomly partitioned our dataset (505 residues/observations) into ¾ training and ¼ testing sets, preserving the proportions of HS/NS to be approximately equal. The training data therefore contain 379 observations consisting of 90 HS and 289 NS cases, while testing data contain 126 observations with 30 HS and 96 NS.

11.2 Training and testing of the classifier

As a proof of principle, we have evaluated seven commonly used and proven ML algorithms—random forest (RF) (Breiman, 2001), radial basis function kernel-based support vector machine (svmRadial) (Wang et al., 2004), adaptive boosting (AdaBoost.M1) (Freund, 2001), extreme gradient boosting (XGBoost/xgbTree) (Friedman, 2001), boosted logistic regression (LogitBoost) (Friedman et al., 2000), C5.0 decision tree (Quinlan, 1993) and penalized discriminant analysis (PDA) (Hastie et al., 1995). Some of these were previously found to be among the best performing ones in a study by Moreira et al. (2017) and Kuhn (2008). To deal with the class imbalance problem, we applied the Synthetic Minority Over-sampling TEchnique (SMOTE), which creates extra synthetic examples in the minority class while downsampling the majority class (Chawla et al., 2002). To overcome overfitting of the classifiers, we repeated 10-fold cross-validation for 10 iterations. Briefly, the training set was further randomly partitioned into 10 equal-sized subsets by resampling, with each subset being hold out once for validation/testing while classifier trained on the remaining nine subsets, and the entire process was repeated for 10 times. The final training performance of the classifier is reflected by averaged classification accuracy of the total $10 \times 10 = 100$ validations. The order of performing SMOTE and 10-fold CV also matters, as applying SMOTE prior to CV eliminates the true imbalanced nature of the validation subset and limits the entire CV on one subsampling pattern (i.e., all cross-validations were performed on the same set of observations being upsampled/downsampled), resulting again in inflated model performance. To avoid this problem, we integrated the 10-fold CV with SMOTE by doing an independent SMOTE on each training subset, as recommended in the *caret* package. In this case, each training set has distinct observations being subsampled, while the class imbalance is preserved for each holdout testing set during CV. This results in CV accuracy on training data more reflective of actual performance metrics on an independent testing dataset, minimalizing the potential overfitting problem.

11.2.1 Variable selection using recursive feature elimination

To find important features that contribute most to HS and NS discrimination, we incorporated a feature/variable selection step in the model-building process using the RF-based recursive feature elimination (RF-RFE) algorithm (Diaz-Uriarte and Andres, 2006). Briefly, descriptors ranked in terms of importance were recursively removed and the model was refit on the retained variables at different iterations, after which the model performance was evaluated. While the removal of the least important variables is expected to have a low impact on model performance, the removal of highly important variables will result in a significant drop in model performance. This is often considered a cutoff for isolating important features contributing most to classification. We again incorporated a feature selection step within resampling-based repeated 10-fold CV as mentioned previously in order to prevent overfitting and to make sure that the variables being selected are generalizable to an independent new dataset (Ambroise and Mclachlan, 2002).

11.2.2 Random forest performed best using both published and combined datasets

We started with the partial dataset (250 out of 881 total variables reported) used in building the SpotOn HSs identification server, and evaluated seven different classifiers (Moreira et al., 2017). Most models worked fairly well in classifying HS and NS except for PDA and LogitBoost. Importantly, using only 250 available out of 881 total reported variables, RF and EGB were able to achieve an AUROC of 0.848 and 0.846 as well as testing accuracy of 0.811 and 0.833, respectively. In variable importance, the top important predictors mostly include solvent-accessible surface area (SASA)-related features, the number of nearby residues at certain distance considered as well as hydrophobicity of these residues, as they were repeatedly ranked top by different classifiers.

In order to evaluate the predictive value of our new computed features and build a better-performing classification model, we combined our 252 new features with the original 250 published features and retrained the classifiers using the new combined matrix. Most classifiers, except PDA and svmRadial, showed constant or slightly decreased accuracy and AUROC. PDA and svmRadial failed to do effective classification in this case. RF showed the best performance with an AUROC of 0.891 and testing accuracy of 0.841, respectively (Table 11.1). Moreover, variable importance plot shows two newly calculated descriptors, Jurs_RPSA (relative polar and apolar surface area) and Jurs_FPSA_3 (fractional charge surface area), among the top 30 most important variables.

To robustly identify important variables, we further performed the RF-RFE-based feature selection method with repeated 10-fold cross-validation. For the RF classifier trained with a combined dataset, we found that the model performance generally stabilizes at around 110 variables, as the repeated CV accuracy does not tend to further improve much after this number (Fig. 11.1). We, therefore, selected these 110 top ranking variables as the most important discriminating features. While previously published SASA-related properties again were ranked top in terms of their importance, we identified 20 new features (from the feature set) among the top 110 variables. Some of the new features, Jurs_FNSA_1, Jurs_FPSA_1, Jurs_FNSA_2, ES_Sum_sNH3, Jurs_PNSA_2 and Jurs_PNSA_1 were ranked in the top 30.

Finally, in order to demonstrate that the 20 newly calculated features are essential and responsible for the improvements in model performance, we repeated variable selection followed by model fitting on both original feature sets, as well as the 90 remaining variables after the removal of 20 new features from the $N = 110$ top variables on the new combined set. For the original dataset, repeated CV accuracy stabilized after $N = 45$ variables which were selected as the cutoff. The model was refitted using $N = 30$, 40, 45, 50, 60, and 100 variables to observe the model performance difference. Using optimal $N = 45$ variables, an AUROC of 0.837 and an accuracy of 0.803 was achieved. Importantly, using a similar number of variables

Table 11.1 Model performance metrics based on 1/4th testing data for seven classifiers using the combined dataset.

Metric	C5.0	Logit-boost	PDA	Svm-Radial	Ada-boost. M1	rf	Xgb-tree
AUROC	0.82	0.71	0.59	0.57	0.80	0.89	0.81
Accuracy	0.83	0.75	0.56	0.61	0.79	0.84	0.80
Kappa	0.47	0.30	0.06	0.09	0.39	0.53	0.42
Sensitivity	0.50	0.43	0.50	0.43	0.50	0.57	0.50
Specificity	0.93	0.85	0.58	0.67	0.88	0.93	0.90
PPV	0.68	0.48	0.27	0.29	0.56	0.71	0.60
NPV	0.86	0.83	0.79	0.79	0.85	0.87	0.85
Precision	0.68	0.48	0.27	0.29	0.56	0.71	0.60
Recall	0.50	0.43	0.50	0.43	0.50	0.57	0.50
F1	0.58	0.46	0.35	0.35	0.53	0.63	0.55

Notes: Training set performance reflected by accuracy from resampling-based 10-fold cross-validation coupled with SMOTE upsampling. Testing set performance was indicated by 10 metrics. *AUROC*, area under receiver operating curve; *PPV*, positive predicative value; *NPV*, negative predicative value.

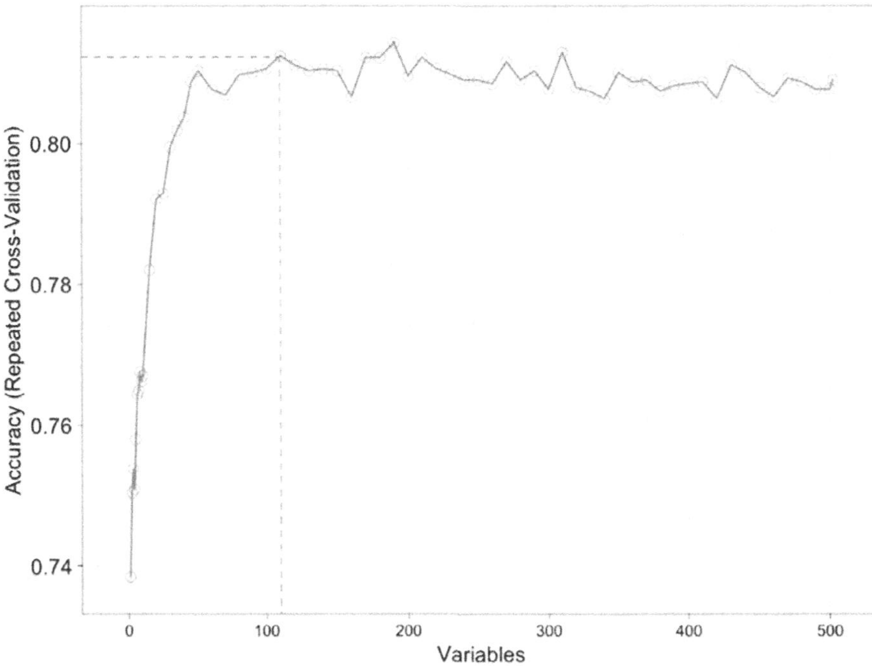

Figure 11.1 Random forest (RF)-RFE variable selection picked 20 new features among the top 110 combined variables responsible for improvements in the accuracy of the RF classifier.

($N = 100$) from the original set as training on the combined dataset did not improve the performance (AUROC = 0.844, accuracy = 0.818). To rule out the possibility that differences between datasets other than the 20 new features are responsible for the improved accuracy, we also removed the 20 new features from the previous $N = 110$ top variables and trained the classifier only using the remaining 90 on a combined dataset, and the model performance dropped to a level comparable to the training of the original set (AUROC = 0.841, accuracy = 0.810). **Therefore, our results demonstrate that the improvement in the classification model performance was largely due to the 20 newly computed features, which highlights the value of our new features in HS identification.**

11.3 Technical details to develop novel protein–protein interaction hotspot prediction program

11.3.1 Training data

The combined dataset has 120 HSs and 385 NSs with 250 published features (Moreira et al., 2017) and 252 new features. While computing the new features we did not incorporate the time-consuming quantum chemical features and built a robust classifier.

11.3.2 Building and validating a novel classifier by evaluating state-of-the-art feature selection and machine learning algorithms

We obtained a PPI HSs classification accuracy of 0.841 and an area under ROC (AUROC) of 0.891 using RF, while we were able to achieve a similar level of classification performance (testing accuracy = 0.825, AUROC = 0.882) only using top 110 variables in terms of ranked importance picked by variable selection. We further demonstrated that 20 of the 110 variables were from our newly calculated feature set, while removing the variables nullifies the significant improvement in model performance. Although the classifier has achieved a reasonably good accuracy, many further considerations and steps offer additional room for improvements for future model.

The central factor determining the performance of a ML classifier is quality of the features in successfully aiding in distinguishing the two classes. We have demonstrated that only a small subset of variables used ($\sim$20%) were discriminative enough to accurately predict the classes. Highly sparse (i.e., many observations are zeros) and noisy features tend to negatively affect the model performance. Therefore, feature selection is a critical step of the model building technique. Currently, many feature selection techniques were developed, which could be roughly categorized into three types: filter, wrapper and embedded (Saeys et al., 2007). Filter methods refer to the independent statistical checking or screening of individual features for their relevance to the outcomes, or class labels. Wrapper methods instead often associate the addition/deletion of features with model performance evaluation and use that as a central criterion to determine whether a variable should be included/removed from the final model. Recursive feature elimination is one commonly used type of wrapper methods while there are other methods available as well, including Genetic Algorithms (Jones et al., 1994), Simulated annealing (Kirkpatrick et al., 1983) and others. In addition, many ML algorithms embed a feature selection step within its model training, meaning that the model fitting will only be attempted upon an optimized selection of variables instead of the full set, and such inbuilt feature selection was often shown to be more effective than selecting variables from outside of the model. Specifically in the context of RF, common algorithms other than recursive feature elimination (RFE) include Boruta (Kursa and Rudnicki, 2010), Vita (Degenhardt et al., 2017) and permutation-based Altmann (Altmann et al., 2010). These methods were evaluated and compared with RFE to extract truly relevant features so as to improve model performance. Note that in a previous study more than 800 variables were used to obtain an AUROC of 0.91 while we only used 110 variables to obtain a comparable AUROC of 0.88 (Moreira et al., 2017). It is expected that addition of more refined features will further improve the model performance significantly. We also evaluated the model performance by adding hyperparameter tuning as one of the steps during model training. It is known that hyperparameter tuning prevents the model from converging to the local minima and therefore leads to significant model performance improvements (Richardson et al., 2015).

Finally, model meta ensembling, or model stacking, was evaluated to improve the classification performance. Model stacking involves training multiple best-performing base learners and then combining them to obtain an even stronger second-level meta- or super-learner, which in practice nearly always outperforms individual base learners (van der Laan et al., 2007). First, we performed an unsupervised clustering to determine homogenous clusters within the data, and train different algorithms as base learners that perform best in each cluster. To ensure adequate diversity of initial models in building the level I ensemble, we first trained all base learners, eliminate those performing worse than others and leave all the rest in constructing the next level of meta-learner. Base models resulting from the same or similar classes of ML algorithms could be further diversified by using different subsamples of observations, features or hyperparameter combinations. Also, instead of having only one stage of meta-learner, multiple layers of super-learners were trained and eventually summarized into a final stage stacked model. In summary, with a more refined feature set, more careful model training and application of the model stacking method, we were able to build a highly accurate classifier for the identification of HSs in the PPI interface.

11.4 A case study

11.4.1 Identification of a druggable protein–protein interaction site between mutant p53 and its stabilizing chaperone DNAJA1 using our machine learning-based classifier

A survey of the literature reveals that the TP53 gene is the most frequently mutated gene in human cancer which encodes tumor suppressor p53 protein (Vogelstein et al., 2000; Ciriello et al., 2013). When p53 functioning is altered the cell proliferation is not regulated effectively and DNA damage accumulates in the cells leading to uncontrolled cell division and ultimately generating tumor growth. Hence, targeting the degradation of mutant p53 (mutp53) might serve as an excellent strategy for cancer prevention.

It is well known that protein folding and homeostasis do depend on a very complex network of molecular chaperones (Jeng et al., 2015). Chaperone proteins generally make "triage" decisions about whether the substrate proteins will undergo folding or degradation, particularly for mutp53-like proteins (Kim et al., 2013; Parrales et al., 2016; Tracz-Gaszewska et al., 2017; Wawrzynow et al., 2018).

It has been reported that DNAJA1 (DNAJ homolog subfamily A member 1) is an important co-chaperone of heat shock protein 70 (Hsp70) (Terada and Oike, 2010; Qiu et al., 2006) which has been suggested to be a vital player for mutp53 stability and oncogenic function (Parrales et al., 2016; Xu et al., 2019). Until now the DNAJA1 crystal structure has not yet been solved but the p53 structure is available in the protein databank. We used the in-silico tools to build a high-resolution type homology model of DNAJA1. Then we considered the p53 having protein databank code 6FF9.pdb and mutated it (R175H) to study the PPIs in order to identify the HS using our RF generated classification model.

11.4.2 Building the homology model of DNAJA1 and optimizing the mutp53 (R175H) structure

To build a homology model structure of DNAJA1 we considered the primary sequence of human DNAJA1 having the NCBI accession code NP_001539.1 which has 397 residues. We tried to build a comparative homology model using a few template structures. To find out the templates we used blast/psi-blast to search the templates using the *p*-sequence of DNAJA1 as the query. In the search, we did not get any single structure having a sequence similarity $> 60\%$ with DNAJA1 sequence. In the absence of a good single template, we used a multitemplate-based algorithm to build a three-dimensional model of DNAJA1. We used the prime module implemented in the Schrodinger platform to build the model (Farid et al., 2006). Prime 3.1 is a well-validated protein structure prediction suite of programs that integrates comparative modeling and fold recognition into a single interface. The comparative modeling path incorporates complete protein structure prediction including template identification, alignment, and model building. Furthermore, Prime allows for refinement of the side-chain prediction, loop prediction, and energy minimization. The alignment steps were then used to align the templates with the query. Unfortunately, the alignment was not optimal and had several gaps. Secondary structure prediction (SSP) tools were used to obtain better alignment. Multiple templates have been used after matching the different parts of the query sequences for making comparative models for DNAJA1. The homology model was further validated using MolProbity guidelines (Chen et al., 2010). The MolProbity score for our model was found to be 98% percentile and the model has a $<3\%$ clash score, and $<5\%$ poor rotamers. A MolProbity score of more than 95 percentile predicts that the homology model structure is a quality structure for in-silico studies. The model structure is shown in Fig. 11.2A.

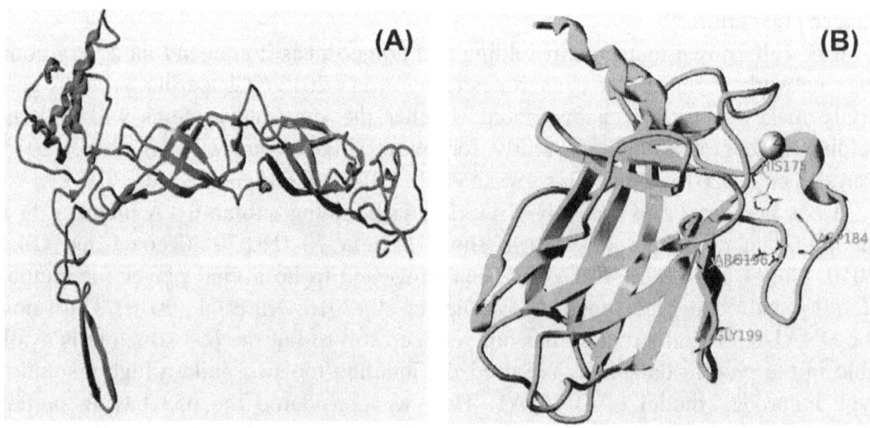

Figure 11.2 Homology model of DNAJA1 (A) and energy minimized mutp53^{R175H} (B).

The other binding partner mutp53^{R175H} structure has also not yet been crystallized. However, another mutp53^{R180K} structure has been solved and is available in the protein databank with the accession code 6FF9.pdb. We changed the K180 to R180, R175 to H175 and subjected the structure to energy minimization in OPLS3 forcefield (Harder et al., 2016). The energy-minimized structure is shown in Fig. 11.2B.

11.4.3 Protein–protein docking

We used protein–protein docking (PPD) as the tool to assemble separate protein components into protein–protein complexes or assemblies using computational methods. Typically, PPD starts with the three-dimensional structures of the individual components, often referred to as a protein receptor and a protein ligand. The Z-dock coupled with R-dock docking suites implemented in Discovery Studio (Accelrys, 2010) was used to carry out the PPD of DNAJA1 and mutp53^{R175H}. In order to execute the PPD, we need to assign one protein as the receptor and the other one as the ligand. The designated ligand was rotated in every 6 degrees in the space of Euler angles around the receptor and interacting energy between the two proteins was computed for 2000 events. Then we reverse the assignment of receptor and ligand and carried out the same rotations and generated another 2000 poses. These poses were then clustered based on the energy bins and we looked for similar energy bins and picked 10 similar bins from both the runs. These bins are then analyzed and 50 best conformers of the PP complexes were obtained based on the lowest energy scale. Using our in-house SITE-ID tool (Zhu et al., 2015) we examined the existence of putative small molecule binding pockets in these 10 conformers of the complex. Then we identified the interacting conjoint triads from both DNAJA1 and mutp53^{R175H} and computed all the 110 features (descriptors) as outlined in variable selection in ML section. These features are fed into the classifier and HS and NS are predicted and are shown in Table 11.2 for various conformers of the complex.

Table 11.2 The hotspot and nullspot predicted using random forest classifier.

DNAJA1-mutp53^{R175H}complex	**HS**	**NS**
Conformer 1	3	7
Conformer 2	5	3
Conformer 3	4	8
Conformer 4	7	4
Conformer 5	9	4
Conformer 6	2	7
Conformer 7	0	4
Conformer 8	3	8
Conformer 9	2	4
Conformer 10	1	6

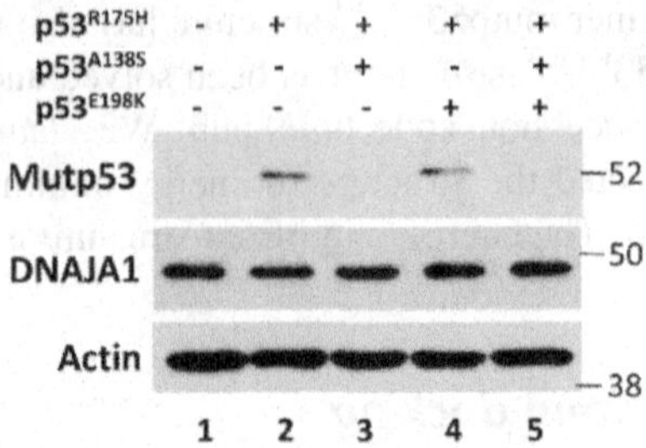

Figure 11.3 Effects of different mutations of p53 and DNAJA1 on their interaction and mutp53 stability.

Analyzing the HS/NS data obtained from the classifier we considered the conformer 5 having the most HS and low NS sites. However, we wanted to validate these in-silico findings using site-directed mutagenesis experiments. Out of the 9 HSs sites in this conformer, 4 more prominent HSs are observed and they are Ala138 and Glu198 from mutp53^{R175H} and other two are Pro84 and Lys125 from DNAJA1 co-chaperone protein.

Using site-directed mutagenesis approach we generated multiple mutants of p53 constructs such as mutp53$^{R175H/E198K}$, mutp53$^{R175H/A138S}$, mutp53$^{R175H/E198K/A138S}$ and DNAJA1 mutants are DNAJA1^{K125Q}, DNAJA1^{P84S}, and DNAJA1$^{K125Q\ /\ P84S}$. We considered AsPC-1, a p53 null-human pancreatic cancer cell line (Rodicker and Putzer, 2003) and transfected it with different p53 constructs. We observed that p53$^{R175H/A13S}$ plasmid and p53$^{R175H/E198K/A138S}$ plasmid led to a loss of mutp53^{R175H} expression, and transfection with p53 $^{R175H/E198K}$ plasmid resulted in a decrease in mutp53 expression but the DNAJA1 protein level remained unchanged (Fig. 11.3). The other biological validations can be found in our previously published paper (Tong et al., 2020).

Human pancreatic cancer AsPC-1 cells (p53-null cells) were transfected with the control plasmid (lane 1), mutp53R175H plasmid (lane 2), mutp53R175H/A138S plasmid (lane 3), mutp53R175H/E198K plasmid (lane 4) or (5) mutp53R175H/A138S/E198K plasmid (lane 5) for 24 hours, and the expression of mutp53 and DNAJA1 were determined by Western blotting (Mutp53 antibody: SC-99 from Santa Cruz Biotechnology).

11.4.4 Small molecules inhibitors identification through drug-like library screening against the DNAJA1- mutp53^{R175H} interacting pocket

In order to further explore the druggable characteristics of the binding pocket identified through ML classifier and validated through the site-directed mutagenesis, we performed a virtual high throughput screening (vHTS) of a large drug-like library to get an inhibitor binding to this site. The goal of the vHTS campaign is to find out drug-like compounds which are amenable to chemical modifications to generate hits to lead optimization. Very often, the hits find out by the conventional high throughput method generate hits which are nondrug-like, difficult for chemical modifications and loaded with different liabilities. Hence, we created a small molecule library using multilayers filterings such as Lipinski (Lipinski, 2004), Veber (Wan-Mamat et al., 2009), and 239 PAINs filters (Baell and Holloway, 2010) to ZINC database (Sterling and Irwin, 2015) containing approximately 50 million small molecules. After applying all these filters, we obtained a curated library containing approximately 10 million compound sets. This proprietary library has been previously used by us for many different disease targets to generate hit molecules (Mishra et al., 2015b, 2016; Mishra and Singh, 2015a; Villa et al., 2017; Zhu et al., 2015). We used the 3-tier screening platform implemented in Schrodinger suite (Sherman et al., 2006) to screen 10 million compounds. This screening generated 121 hits with docking scores <−6.0. We then

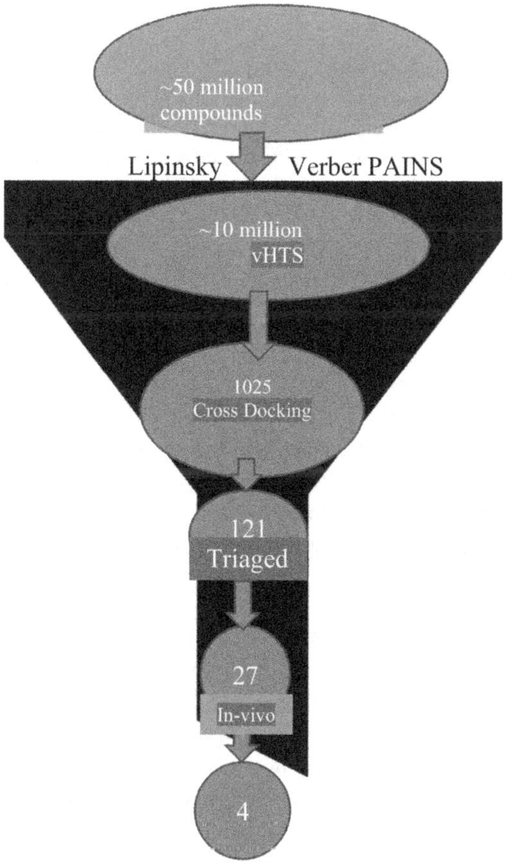

Figure 11.4 Schematic representation of the screening cascade.

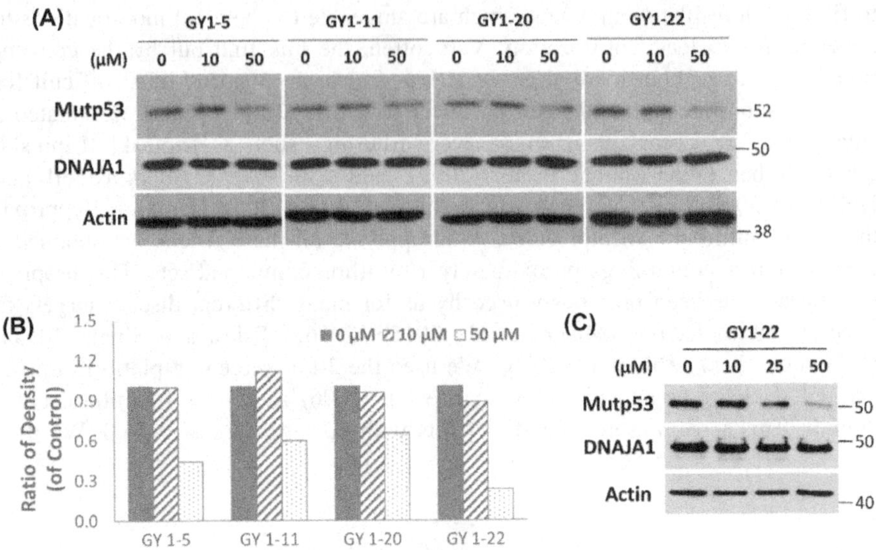

Figure 11.5 Screening with small molecule drug-like library against DNAJA1-mutP53R175H interacting pocket. (A) Four of the 27 top hits showed the effect of reducing mutp53 protein level in mouse pancreatic cancer P03 cells. Cells were treated with different concentrations of the compounds for 24 h and the protein levels of mutp53 and DNAJA1 were determined by Western blotting. (B) The mutp53 band intensities were determined by densitometry and were normalized to loading control Actin. (C) Human colon cancer LS123 cells (containing the p53R175H mutation) were treated with different concentrations of GY1-22 for 24 h and the protein levels of mutp53 and DNAJA1 were detected by Western blotting.

cross-validate these hits with Gold docking tools (Verdonk et al., 2005) to see the similar docking poses of the hits. Based on the consensus docked poses and scores we selected 27 compounds for further validation of biological testing (Fig. 11.4). To validate the hits, we used a murine pancreatic carcinoma cell model (PO3 cells), containing a $p53^{R175H}$ mutation which equals human $p53^{R175H}$ mutation. Out of the 27 in-silico hits four compounds showed a significant reduction of $mutp53^{R175H}$ expression at a dose range of 10–50 μM, where DNAJA1 protein levels remained unchanged Fig. 11.5. Out of the four hits, the GY-22 compound is the most promising hit. Furthermore, we also tested GY-22 compound in LS123 a human colon cancer cell line which also contains $mutp53^{R175H}$ and observed that $mutp53^{R175H}$ expression has reduced Fig. 11.5C.

Analyzing the binding pose of the best hit GY1-27 we observed that one of the benzimidazoles of this compound had two potential hydrogen bonds with Ala138 backbone and the side chain of Glu198 of $mut53^{R175H}$. The phenyl ring of GY1-27 showed a strong hydrophobic interaction with Pro184 of DNAJA1 druggable site (Fig. 11.6C).

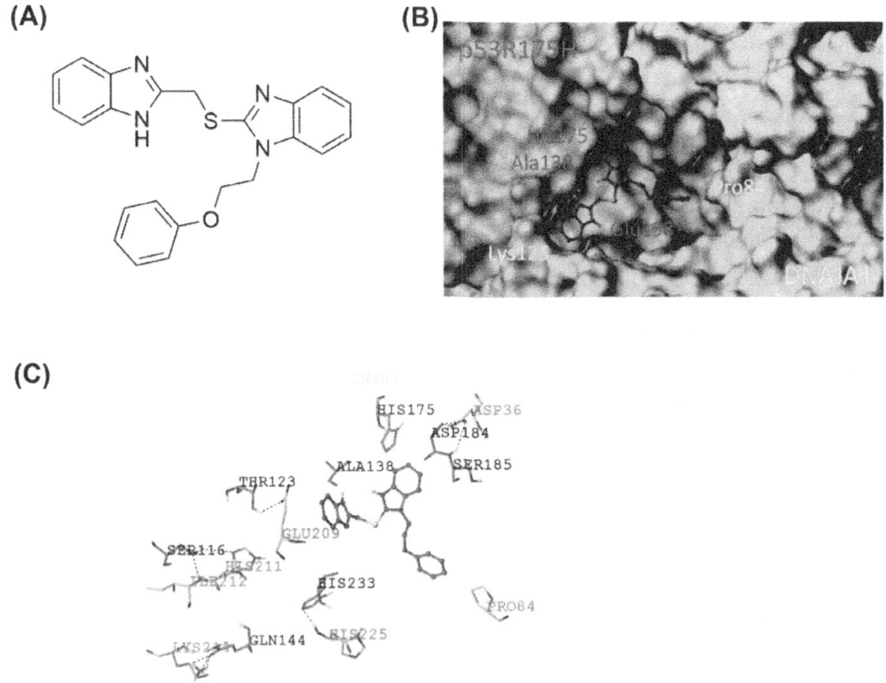

Figure 11.6 Validation of the most promising hit GY1-22 on its ability to reduced mutp53 expression. (A) Chemical structure of GY1-22. (B) GY1-22 docked pose on DNAJA1-mutp53R175H complex. (C) Binding pocket residues interacting with GY1-22.

11.5 Discussion

The inhibition of PPIs continues to be a major area of investigation given the importance of these systems in both normal and adverse biological processes. Interest in this area continues to grow rapidly and perceptions regarding the feasibility and success of PPI modulation have changed given high-profile clinical successes with drugs like navitoclax and venetoclax. Interestingly, these PPI success stories are based on the principle that most of the free energy of the binding interactions has been mediated by binding "HSs" that can be targeted with small molecules. Historically, identifying PPI HSs and evaluating their druggability have been carried out at the bench, one PPI pair at a time, over a period of months or years. While computational methods such as PPD have been developed to assist, there is currently a major gap in the knowledge around PPIs since the current approaches depend on three-dimensional structural data and often suggest hundreds of potential PPI interfaces. To address current limitations in locating druggable PPI interfaces, in this article a novel platform that uses experimentally determined HSs and NSs considering 51 crystallized protein complexes. With these data we have built a robust classifier using ML approaches to predict new HSs in an interface between

DNAJA1 and mutp53^{R175H}. In this chapter, we used seven well-known ML approaches to build a robust HS/NS classifier. Finally, based on our novel descriptor set, the RF (rt) has produced a robust classifier.

Carrying out the PPD of DNAJA1 and mutp53^{R175H} we obtained 2000 complexes and then filtering out we left with energetically similar 50 complexes. Applying site-ID only 10 of them showed potential small molecule binding sites. We then computed 110 features for each of the conjoint triad and fed into the classifier to obtained HS/NS which gave us the best ligand binding pocket.

It may be noted that so far only in-silico experiments were performed by many researchers but here in this chapter, we validated the in-silico findings using biological experiments. Hence, the site-directed mutagenesis was performed and the in-silico identified HS was validated. Not only we validated the binding site in the interface of DNAJA1-mutp53^{R175H} but also showed that a low double-digit inhibitor was obtained to decrease the mutp53^{R175H} expression by carrying out a vHTS of huge compound database containing 10 million molecules. Furthermore, we carried out in-vitro experiments along with animal experiments (Tong et al., 2020) to validate the inhibitor GY1-22.

Author contribution

RD, GY, and RM conceived this work and designed the experiments. YJ, RM, and RD performed the feature selection and build classifier. XT, DX, JL, and GY performed all biological testing. RD, GY, and RM wrote the chapter.

Acknowledgment

A part of this work was performed by the Northwestern University ChemCore, which is partially funded by Cancer Center Support Grant P30CA060553 from the National Cancer Institute awarded to the Robert H. Lurie Comprehensive Cancer Center. All the biological studies were supported by NIH R01 DK10776, CA172431, and CA164041 grants to Dr. Guang-Yu Yang.

Conflicts of interest

The authors declare that they have no conflicts of interest with the contents of this article.

References

Accelrys, 2010. Discovery Studio. San Diego, CA.

Altmann, A., Tolosi, L., Sander, O., Lengauer, T., 2010. Permutation importance: a corrected feature importance measure. Bioinformatics 26, 1340–1347.

Ambroise, C., Mclachlan, G.J., 2002. Selection bias in gene extraction on the basis of microarray gene-expression data. Proc. Natl Acad. Sci. USA. 99, 6562–6566.

Baell, J.B., Holloway, G.A., 2010. New substructure filters for removal of pan assay interference compounds (PAINS) from screening libraries and for their exclusion in bioassays. J. Med. Chem. 53, 2719–2740.

Breiman, L., 2001. Random forests. Mach. Learn. 45, 5–32.

Chang, D.T., Syu, Y.T., Lin, P.C., 2010. Predicting the protein-protein interactions using primary structures with predicted protein surface. BMC Bioinform. 11 (Suppl 1), S3.

Chawla, N.V., Bowyer, K.W., Hall, L.O., Kegelmeyer, W.P., 2002. SMOTE: synthetic minority over-sampling technique. J. Artif. Intell. Res. 16, 321–357.

Chen, V.B., Arendall 3rd, W.B., Headd, J.J., Keedy, D.A., Immormino, R.M., Kapral, G.J., et al., 2010. MolProbity: all-atom structure validation for macromolecular crystallography. Acta Crystallogr. D. Biol. Crystallogr. 66, 12–21.

Ciriello, G., Miller, M.L., Aksoy, B.A., Senbabaoglu, Y., Schultz, N., Sander, C., 2013. Emerging landscape of oncogenic signatures across human cancers. Nat. Genet. 45, 1127–1133.

Degenhardt, F., Seifert, S., Szymczak, S., 2017. Evaluation of variable selection methods for random forests and omics data sets. Brief Bioinform.

Diaz-Uriarte, R., de Andres, S.A., 2006. Gene selection and classification of microarray data using random forest. Bmc Bioinforma. 7.

Freund, Y., 2001. An adaptive version of the boost by majority algorithm. Mach. Learn. 43, 293–318.

Friedman, J.H., 2001. Greedy function approximation: a gradient boosting machine. Ann. Stat. 29, 1189–1232.

Farid, R., Day, T., Friesner, R.A., Pearlstein, R.A., 2006. New insights about HERG blockade obtained from protein modeling, potential energy mapping, and docking studies. Bioorg Med. Chem. 14, 3160–3173.

Friedman, J., Hastie, T., Tibshirani, R., 2000. Additive logistic regression: a statistical view of boosting. Ann. Stat. 28, 337–374.

Harder, E., Damm, W., Maple, J., Wu, C., Reboul, M., Xiang, J.Y., et al., 2016. OPLS3: a force field providing broad coverage of drug-like small molecules and proteins. J. Chem. Theory Comput. 12, 281–296.

Hastie, T., Buja, A., Tibshirani, R., 1995. Penalized discriminant-analysis. Ann. Stat. 23, 73–102.

Jeng, W., Lee, S., Sung, N., Lee, J., Tsai, F.T., 2015. Molecular chaperones: guardians of the proteome in normal and disease states. F1000Res 4.

Jones, G., Robertson, A.M., Willett, P., 1994. An introduction to genetic algorithms and to their use in information-retrieval. Online & Cdrom Rev. 18, 3–13.

Kier, L.B., Hall, L.H., 2001. Database organization and searching with E-state indices. SAR. QSAR Environ. Res. 12, 55–74.

Kim, Y.E., Hipp, M.S., Bracher, A., Hayer-Hartl, M., Hartl, F.U., 2013. Molecular chaperone functions in protein folding and proteostasis. Annu. Rev. Biochem. 82, 323–355.

Kirkpatrick, S., Gelatt, C.D., Vecchi, M.P., 1983. Optimization by simulated annealing. Science 220, 671–680.

Kuhn, M., 2008. Building predictive models in R using the caret package. J. Stat. Softw. 28, 1–26.

Kursa, M.B., Rudnicki, W.R., 2010. Feature selection with the Boruta package. J. Stat. Softw. 36, 1–13.

Lipinski, C.A., 2004. Lead- and drug-like compounds: the rule-of-five revolution. Drug. Discov. Today Technol. 1, 337–341.

Mishra, R.K., Shum, A.K., Platanias, L.C., Miller, R.J., Schiltz, G.E., 2016. Discovery and characterization of novel small-molecule CXCR4 receptor agonists and antagonists. Sci. Rep. 6, 30155.

Mishra, R.K., Singh, J., 2015a. A structure guided QSAR: a rapid and accurate technique to predict IC50: a case study. Curr. Comput. Aided Drug. Des. 11, 152–163.

Mishra, R.K., Wei, C., Hresko, R.C., Bajpai, R., Heitmeier, M., Matulis, S.M., et al., 2015b. In silico modeling-based identification of glucose transporter 4 (GLUT4)-selective inhibitors for cancer therapy. J. Biol. Chem. 290, 14441–14453.

Moreira, I.S., Fernandes, P.A., Ramos, M.J., 2007. Hot spots—a review of the protein-protein interface determinant amino-acid residues. Proteins 68, 803–812.

Moreira, I.S., Koukos, P.I., Melo, R., Almeida, J.G., Preto, A.J., Schaarschmidt, J., et al., 2017. SpotOn: high accuracy identification of protein-protein interface hot-spots. Sci. Rep. 7, 8007.

Munteanu, C.R., Pimenta, A.C., Fernandez-Lozano, C., Melo, A., Cordeiro, M.N., Moreira, I.S., 2015. Solvent accessible surface area-based hot-spot detection methods for protein-protein and protein-nucleic acid interfaces. J. Chem. Inf. Model. 55, 1077–1086.

Parrales, A., Ranjan, A., Iyer, S.V., Padhye, S., Weir, S.J., Roy, A., et al., 2016. DNAJA1 controls the fate of misfolded mutant p53 through the mevalonate pathway. Nat. Cell Biol. 18, 1233–1243.

Qiu, X.B., Shao, Y.M., Miao, S., Wang, L., 2006. The diversity of the DnaJ/Hsp40 family, the crucial partners for Hsp70 chaperones. Cell Mol. Life Sci. 63, 2560–2570.

Quinlan, J.R., 1993. C4.5: Programs for Machine Learning. Morgan Kaufmann Publishers, San Mateo, Calif.

Richardson, B.G., Jain, A.D., Speltz, T.E., Moore, T.W., 2015. Non-electrophilic modulators of the canonical Keap1/Nrf2 pathway. Bioorg Med. Chem. Lett. 25, 2261–2268.

Rodicker, F., Putzer, B.M., 2003. p73 is effective in p53-null pancreatic cancer cells resistant to wild-type TP53 gene replacement. Cancer Res. 63, 2737–2741.

Saeys, Y., Inza, I., Larranaga, P., 2007. A review of feature selection techniques in bioinformatics. Bioinformatics 23, 2507–2517.

Sherman, W., Day, T., Jacobson, M.P., Friesner, R.A., Farid, R., 2006. Novel procedure for modeling ligand/receptor induced fit effects. J. Med. Chem. 49, 534–553.

Shukla, A., Sharma, P., Prakash, O., Singh, M., Kalani, K., Khan, F., et al., 2014. QSAR and docking studies on capsazepine derivatives for immunomodulatory and anti-inflammatory activity. PLoS One 9, e100797.

Stanton, D.T., Mattioni, B.E., Knittel, J.J., Jurs, P.C., 2004. Development and use of hydrophobic surface area (HSA) descriptors for computer-assisted quantitative structure-activity and structure-property relationship studies. J. Chem. Inf. Comput. Sci. 44, 1010–1023.

Sterling, T., Irwin, J.J., 2015. ZINC 15—ligand discovery for everyone. J. Chem. Inf. Model. 55, 2324–2337.

Terada, K., Oike, Y., 2010. Multiple molecules of Hsc70 and a dimer of DjA1 independently bind to an unfolded protein. J. Biol. Chem. 285, 16789–16797.

Tong, X., Xu, D., Mishra, R.K., Jones, R.D., Sun, L., Schiltz, G.E., et al., 2020. Identification of a druggable protein-protein interaction site between mutant p53 and its stabilizing chaperone DNAJA1. J Biol Chem.

Tracz-Gaszewska, Z., Klimczak, M., Biecek, P., Herok, M., Kosinski, M., Olszewski, M.B., et al., 2017. Molecular chaperones in the acquisition of cancer cell chemoresistance with mutated TP53 and MDM2 up-regulation. Oncotarget 8, 82123–82143.

Van Der Laan, M.J., Polley, E.C., Hubbard, A.E., 2007. Super learner. Stat. Appl. Genet. Mol. Biol. 6, Article25.

Vellay, S.G., Latimer, N.E., Paillard, G., 2009. Interactive text mining with pipeline pilot: a bibliographic web-based tool for PubMed. Infect. Disord. Drug. Targets 9, 366–374.

Verdonk, M.L., Chessari, G., Cole, J.C., Hartshorn, M.J., Murray, C.W., Nissink, J.W., et al., 2005. Modeling water molecules in protein-ligand docking using GOLD. J. Med. Chem. 48, 6504–6515.

Villa, S.R., Mishra, R.K., Zapater, J.L., Priyadarshini, M., Gilchrist, A., Mancebo, H., et al., 2017. Homology modeling of FFA2 identifies novel agonists that potentiate insulin secretion. J. Investig. Med. 65, 1116–1124.

Vogelstein, B., Lane, D., Levine, A.J., 2000. Surfing the p53 network. Nature 408, 307–310.

Wan-Mamat, W.M., Isa, N.A., Wahab, H.A., Wan-Mamat, W.M., 2009. Drug-like and non drug-like pattern classification based on simple topology descriptor using hybrid neural network. Conf. Proc. IEEE Eng. Med. Biol. Soc. 2009, 6424–6427.

Wang, J.P., Chen, Q.S., Chen, Y., 2004. RBF kernel based support vector machine with universal approximation and its application. Adv. Neural Netw. - Isnn 3173 (Pt 1), 512–517. 2004.

Wawrzynow, B., Zylicz, A., Zylicz, M., 2018. Chaperoning the guardian of the genome. Two-faced role molecule chaperones p53 tumor suppressor action. Biochim. Biophys. Acta Rev. Cancer 1869, 161–174.

Xu, D., Tong, X., Sun, L., Li, H., Jones, R.D., Liao, J., et al., 2019. Inhibition of mutant Kras and p53-driven pancreatic carcinogenesis by atorvastatin: mainly via targeting of the farnesylated DNAJA1 in chaperoning mutant p53. Mol. Carcinog. 58, 2052–2064.

Zhu, J., Mishra, R.K., Schiltz, G.E., Makanji, Y., Scheidt, K.A., Mazar, A.P., et al., 2015. Virtual high-throughput screening to identify novel activin antagonists. J. Med. Chem. 58, 5637–5648.

Mining big data in drug discovery—triaging and decision trees

12

Shahul H. Nilar
Global Blood Therapeutics, San Francisco, CA, United States

12.1 Introduction

> *"Start a huge, foolish project, like Noah...it makes absolutely no difference what people think of you."*
>
> *Jalal-uddin Rumi (1207–73)*

The search for particular pieces of an intricate puzzle, the components of which are random and scattered, is always an interesting and challenging endeavor. The components that bring together the overall picture that is being constructed or envisaged can often be likened to searching for a needle in a haystack. The search can follow a standard operating procedure such as: looking for that particular element (like the color of the building block) to initiate the search process or another property or characteristic associated with that particular region of space. The exploration and search for the various pieces can follow an operational procedure or an evolving thought process, driven by learning from the previous pieces chosen and the features associated with these components. This learning guides the search for the next acceptable component of the puzzle along a rule-based thought process that changes as the search progresses, in other words, in a supervised manner. The changes accommodate the learnings as the search evolves. Unlike the scenario described above, a second search method allows for the features that define the next acceptable piece to not necessarily be based on a fixed set of properties such as color, for example. The two scenarios presented above can be likened to constructing a jigsaw puzzle. The first has a picture to guide the selection process of the components while the second has no pictorial guidance and is in fact monochromatic. The latter scenario allows each piece to be used on either face of the piece and can be further complicated if the number of pieces are small and there is a myriad to choose from. This scenario is an example of an unsupervised learning process.

12.2 Big data in drug discovery

There are many examples of the application of Big Data analytics in small molecule drug discovery. The exploration of suitable and focused therapeutic targets from the massive sequence databases generated from genome analysis is one example.

Big Data Analytics in Chemoinformatics and Bioinformatics. DOI: https://doi.org/10.1016/B978-0-323-85713-0.00019-0

The search for suitable starting chemical structures from millions of data points from which a drug candidate can be derivatized is another example. The phrase "data point" in the second example is an oversimplification. Within this "data point" is the information about the underlying chemical structure that cheminformatics techniques seek to uncover from the vast amounts of available data. It is the latter category of search techniques and the methods used in current small molecule drug discovery that will be discussed in this chapter.

Finding a suitable chemical moiety as a starting point to optimize and develop into a clinical candidate is the first step in the small molecule drug discovery paradigm. To find such moieties, a collection of compounds, usually the corporate compound library, is screened against the biological system of interest. The high-throughput screening (HTS) procedure of such libraries of compounds has been successfully applied towards this task and meaningful starting points identified (Broach and Thorner, 1996; Wildey et al., 2017; Inglese et al., 2006; Macarron et al., 2011; O'Brien and Fallah Moghaddam, 2013; Bamborough et al., 2019). The size of such screening libraries can range from a few thousands to millions of compounds and the molecules that comprise the library are chosen to span a wide spectrum of structural diversity (De La Rosa et al., 2006; Arora et al., 2020; Wang et al., 2011; Quartararo et al., 2020; Zhao et al., 2019).

The algorithms and computational methods used in the design of compound libraries that are used in HTS campaigns have been discussed extensively in the literature and will not be repeated here (Eurtivong and Reynisson, 2019; López-Vallejo et al., 2012; Schneider et al., 2009; Follmann et al., 2019; Huggins et al., 2011). Some of the earlier screening libraries have used the Lipinski Rules, a much discussed set of rules, which triage compounds to be included in the screening set based on a set of physicochemical properties. These rules use the molecular weight, the number of hydrogen bond acceptors and donors and the lipophilicity, as calculated by the logarithm of the partition coefficient of the compound between octanol and water, as criteria within a range of numerical values for each property. Molecules that satisfy these criteria are progressed to the next step in the analysis and evolution of the lead optimization process (Lipinski et al., 1997; Lipinski, 2004, 2016; Shultz, 2018).

The Bemis–Murcko method (Bemis and Murcko, 1996) of decomposing molecular structures into rings, linkers, frameworks and side chain moieties can be helpful in the task of data mining as chemically meaningful substructures can be retrieved from the data. Such an analysis illustrates the scaffold composition of the data set in terms of the number of occurrences of each structural moiety that has been identified. In the analysis of large data sets, defining the chemical substructures that one is looking for is a challenging task. Besides the dimension of the screening results data set itself, the lack of familiarity with the chemical structures of the compounds that have been screened in the assay adds another layer of complexity. Furthermore, with the current trend of moving away from aromaticity and into acyclic chemistry space, an updated definition of a concept of a scaffold is necessary. The ideal expectation in the data mining process is the discovery of new information. In the small molecule drug discovery process, this would correspond to the identification of novel chemical scaffolds as starting points for further optimization.

In the past two decades, there has been a focus on fragment-based screening (FBS) where the molecular weight of the entities in the screening pool are less than 350 Daltons. Such libraries are smaller in size, usually in the range of a few thousands (Hajduk and Greer, 2007; Erlanson, 2011; Kirsch et al., 2019; Noble et al., 2016; Yokokawa et al., 2016). The success of FBS approaches in drug discovery has been extensively reviewed in the literature (Erlanson et al., 2016, 2020). A theoretical analysis of the intermolecular interaction profile between a ligand and the protein, both moieties expressed in a one-dimensional representation has led to the concept of molecular complexity (Hann et al., 2001; Leach and Hann, 2011). Although a precise definition of molecular complexity is subjective, a conceptual definition can be briefly described as the optimum number of interaction points between the ligand and the protein that results in a binding interaction that has a measurable outcome. Molecular recognition can be viewed as the composite result of the selectivity between the interacting moieties and the sensitivity of the experimental measuring process. These interactions can be between hydrogen bond donors and acceptors, favorable Coulombic atom-centered partial charges and regions of lipophilicity, to name a few, and have emerged as an underlying criterion in the selection of the screening entities. Extending the work of Hann (Hann et al., 2001) and using a combinatorial algebraic technique in terms of the number of such interaction points on the protein and the ligand within the one-dimensional model description, it has been shown that chemical structures with lower complexity such as fragments have a higher selectivity of interacting with the protein. The sensitivity of the measurement of the intermolecular recognition process was evaluated using the cumulative distribution function of the Boltzmann distribution and is lower for fragments as observed in the experiment (Nilar et al., 2013).

The biology associated with the therapeutic area or disease target is constituted in an assay which most often is enzymatic or phenotypic in nature, resulting in binding or functional data. The measurement obtained from the assay/screening process is an approximation to the in vivo biology studied; however, experience has shown that this process is a reliable first step in drug discovery. The biological response is the result of the compounds in the screening library interacting with the biology as represented in the assay: the larger the screening library, the larger will be the data set of results. Most often, the first screen is done at a single concentration for each of the compounds in the library for cost-effective reasons. Nevertheless, whichever type of library or assay is used for screening against the biological system, the primary challenge at this stage in the drug discovery process, remains in the methodologies used to extract the optimum chemical structures from the enormous data generated by the screening process. Briefly described, how does one mine the data set of assay results to produce a series of chemically meaningful scaffolds which constitute an interesting initial structure−activity relationship (SAR) within each series? The SAR extracted from the screening data should not only be amenable to the optimization of the identified scaffolds towards potent drug-like properties but also span a novel intellectual property space.

12.3 Triaging

The methods used to mine the data from the screening experiment need to be relevant and capable of recovering clusters of molecules that are chemically meaningful. Each cluster comprises a set of molecules that contain a maximum common scaffold (MCS) around which the presence of pendant substituent groups allows for a degree of similarity. Within each cluster, the chemical structures exhibit a SAR that displays a meaningful change in the biological response which is a reflection of the structural variations among the structures in each subset or cluster. It is preferable that among the clusters that have been retrieved, the common scaffold in each cluster, taken together with the counterparts in the other clusters, constitute a set of diverse structural moieties. This allows for different sets of hit molecules to be considered in the first step of SAR exploration, augmentation and analysis.

The results from a typical screening process usually form a skewed distribution as shown in Fig. 12.1. The primary inhibition percentage of nearly 83,000 compounds in the CD40 Signaling pathway in BL2 cells obtained from the PUBChem data set AID1053188 (https://pubchem.ncbi.nlm.nih.gov/bioassay/1053188) have been plotted as a count histogram. There are fewer compounds of higher biological response, with larger numbers of compounds with lower activities. Mining the compounds in the range of the lowest tier of the biological response is not a meaningful exercise given that the response measured could fall within the noise level of the assaying process.

A standard technique used in the pharmaceutical industry is to choose a value "*d*" for the biological response and explore only those molecules above this cut-off value which are considered to be the "active" molecules. The numerical value

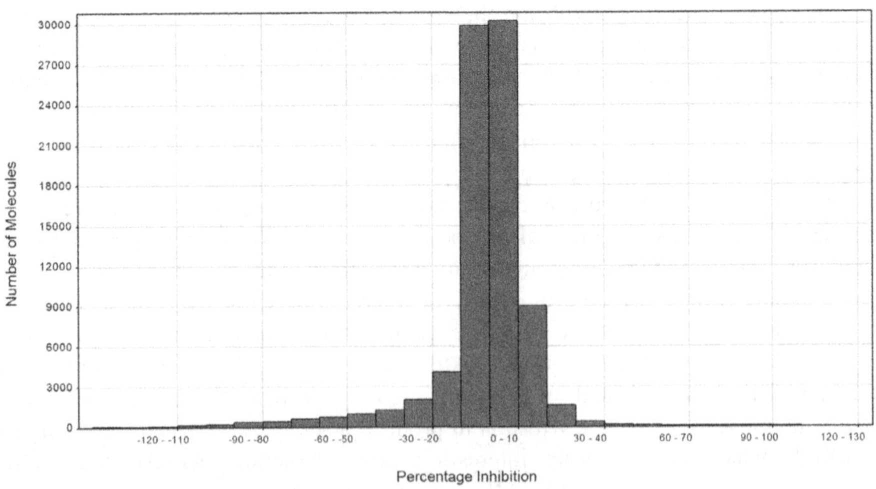

Figure 12.1 Molecule count distribution histogram of activity from a high-throughput screen (ChemBL data set: AID1053188).

chosen, most often, depends on the number of active molecules that can be studied further with the available resources or dictated by the statistics of the assay. Another metric that is used as a cut-off value is three standard deviations from the mean of the activity distribution, which assumes (even though highly unlikely) that the data set is inherently binomial (Yeo, 2012). Further analysis of these molecules in this subset includes reconfirmation studies in the form of dose-response behavior followed by clustering of the molecules to recover SAR trends around MCSs. This process reduces the data set to a size that can be studied by relying on the knowledge of human experts.

Focusing the study on the subset above the cut-off value "*d*", which is expected to contain some false positives, is convenient. However, this process ignores the data below the cut-off point as shown in Fig. 12.2.

As the cut-off value is arbitrarily assigned, there is a possibility of missing interesting chemical scaffold classes which are found in the second set, below the cut-off criterion. The knowledge that is recovered from such a mining process is clearly a subset of the information available in the entire data set of results. As before, the data above the cut-off value is clustered using an algorithm of choice and MCSs identified. It is common practice to mine into the bulk of the screening data below the cut-off value, based on the MCSs identified from the data above the cut-off value. This process helps augment the SAR for scaffold clusters extracted from the "active" data set. The assumption in this process is that the only information that can be retrieved from the data set below the cut-off is that which is related to the clusters identified from the "active" set. The rest of the information in the second data set, which might include novel scaffolds, is ignored.

The question then arises on the methodologies available to explore as much of the data as possible within the available hardware resources. Studying the entire data set would be the ideal solution. However, this approach might not be efficient

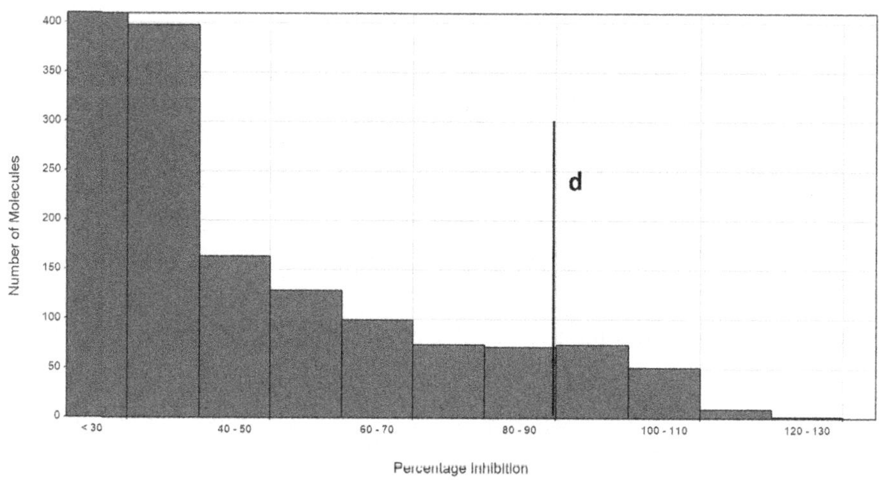

Figure 12.2 The use of a cut-off value for a generic assay.

as the molecules with low activity are really not interesting to explore as hits, even though such molecules can be valuable in exploring and rationalizing the SAR spectrum of the scaffolds and understanding activity cliffs.

Ideally, if the screening set can be optimally separated into two subsets, partitioned by the cut-off criterion, the subset above "*d*" will contain the compounds of interest while the second subset will have a random distribution of activities. The information content in the two subsets is, ideally, very different. Based on the data mining query, the former data set is more informative than the latter. If the first data set were to contain the active compounds that can be classified into SAR series around different scaffolds, then this data set behaves differently from the rest of the screening data. Although this is an idealistic interpretation, it is easier to visualize the partitioning process.

The compound set enrichment method (Varin et al., 2010) has been used to identify SAR series by incorporating a compound classification technique together with the Kolmogorov–Smirnov (KS) test (Birnbaum and Tingey, 1951). Seven screening data sets from PubChem were analyzed using this method. The number of data points across the data sets was in the range of a few thousand to hundreds of thousands. It was shown that the application of the KS statistic in partitioning the data sets followed by clustering analysis, was able to extract SAR series around interesting chemical scaffolds. Some of the activity trends included molecules of low activity.

The KS test, in general, compares two data distributions with the null hypothesis being the assumption that the two subsets have no difference in the distribution of the activity values for the compounds. In the HTS context, the subset above the cut-off value "*d*" can be compared with the data below the cut-off value termed here as upper and lower sets for convenience. The expectation is that the upper data set has more relevant information in terms of active scaffolds than the lower data set. The two data sets are transformed into the corresponding empirical cumulative distribution functions F_{upper}, F_{lower} and the KS statistic $D_{\max}$ between these two functions evaluated as

$$D_{\max} = \sup_x |F_{\text{lower}}(x) - F_{\text{upper}}(x)$$

where $\sup_x$ is the Supremum of the separation distances between the two cumulative distribution functions. The null hypothesis can be rejected if the level of significance of $D_{\max}$ given by Press et al. (1988)

$$\text{Probabilty}(D_{\max} > \alpha) = Q_{\text{KS}}\left(\sqrt{\frac{N_1 N_2}{N_1 + N_2}}\, D_{\max}\right)$$

where N_1 and N_2 are the sample sizes of the two distributions and Q_{KS} is defined by

$$Q_{\text{KS}}(x) = 2\sum_{j=1}^{\infty}(-1)^{(j-1)}\exp(-2j^2x^2)$$

The equation for Q_{KS} is approximate, nevertheless, it is sufficient in the HTS experiment as the sizes of the data sets are large. Critical values for α can be obtained from published tables for the KS statistic in the literature (Massey, 1951).

Iterative analysis of screening data sets using the compound structural information and corresponding biological activity fingerprints has been shown to be another successful method in retrieving active compounds (Paricharak et al., 2016). Starting with small subsets of the screening data and iteratively expanding the compound selection based on similarity, the method has been shown to be successful in retrieving the top 0.5% of the hits across many HTS data sets. By analyzing the data for as little as 15,000 compounds for each of the 34 data sets studied, the iterative screening method was capable of mining diverse chemical moieties from 0.5% of the molecules with the highest activity from the entire screening set.

12.4 Decision trees

The resulting data set from a screening protocol can be decomposed into a set of chemically meaningful frameworks using a hierarchical scaffold classification method by building decision trees. Learning from the chemical moieties that form the leaf nodes of the trees, a set of generic prioritization rules have been proposed and successfully applied to a selection of publicly available data sets (Schuffenhauer et al., 2007).

12.5 Recursive partitioning

Recursive partitioning (RP) is a statistical method that subdivides a data set into decision trees based on a set of descriptors or inputs and this method has been successfully applied across a wide range of problems in diverse fields of study (Rusinko et al., 1999; Ferreira and Andricopulo, 2019; Jia et al., 2020). In the study of chemical moieties, physicochemical properties such as molecular weight, lipophilicity (represented by a calculated log P), the number of hydrogen bond donors and acceptors are some such examples. When the data set is large, as in the case of HTS libraries, such descriptors might not be sufficient to assign molecules to a branch of the decision tree with a high level of confidence. Using many such physicochemical properties can be helpful in constructing meaningful decision trees; however, the results should be closely studied for correlation among the input descriptors and overfitting of the data. The building of decision trees using the RP method can become difficult when the dimension of the active and inactive data sets is imbalanced. In addition, the effect of noise on the descriptors can interfere with the RP process in distinguishing the smaller from the larger molecular classes. One of the advantages of the RP method is that the decision trees can be easily interpreted which is essential in rationalizing SAR trends and behaviors (Nicolaou et al., 2002).

The standard RP method described above generates a single tree in the process of categorizing the molecules in terms of the predefined descriptors. Building many such trees, referred to as ensemble recursive partitioning (ERP), has been shown to generate more accurate models. In the analysis of 232 compounds for the estrogen receptor binding,

It has been shown (Tong et al., 2003) that by generating 1–7 RP trees, the combined accuracy of the ERP model reduces the number of misclassifications from 17 (for a single tree) to 5 (for the ensemble model). As an improvement to the ensemble or consensus model by analyzing a larger data set, it has been shown that the combination of dissimilar trees with different levels of complexity can result in a higher degree of enrichment (van Rhee, 2003). The successful application of the random forest method, which uses multiple such decision trees, in mining assay results for tropical diseases such as Dengue fever and tuberculosis has been reported (Yeo, 2012). The method was shown to be able to successfully extract hit molecular series from the screening data of a few thousand compounds and performed better than using a cut-off criterion described above.

Extracting information from data sets by correlating chemical substructures with the measured biological response is another approach to analyzing screening data (Yeo et al., 2012; Sun et al., 2016). This approach has the potential to establish an understanding between the chemical substructure of interest and the biological data from the assay. The identified substructures, which have a higher occurrence among the active molecules, can be used as the input descriptors in building RP trees. Based on the substructure motifs extracted from the data set according to a set of association rules, a decision tree can be constructed by iteratively filtering the compounds in the data set. At each node of the tree, one of the substructure descriptors that is mostly associated with the active molecules is used to partition the molecules into branches that contain or lack the presence of the particular chemical moiety. The former branch is associated with the active molecules. Repeated splitting of each branch around a node that contains another substructure further partitions the data set. At the end of the splitting process of the data, some of the leaf nodes will contain, mainly, a larger number of active molecules while the other nodes will contain a majority of the inactive molecules.

Analyzing the compounds above the cut-off threshold "*d*" for correlation rules, building decision trees and filtering the compounds in the lower section of the screening data (below the "*d*" value) through these decision trees helps build SAR families that have meaningful information incorporating, both, the chemical and biological information in the classes. RP methods are partitional by definition and using structural features that are correlated with activity as a set of inputs might confuse the construction of the decision tree as some of these features might be present in a few of the inactive molecules as well. This behavior is due to the multidomain character of molecules—a particular chemical scaffold or chemical moiety can be part of multiple chemical classes and consequently reflect a different SAR behavior (MacCuish et al., 2001; Muratov et al., 2020).

12.6 PhyloGenetic-like trees

The complex interplay between the physicochemical properties of the compounds in the screening set and the influence of these properties on the biological response that is measured in the assay is the fundamental knowledge that needs to be mined in the analysis process, usually resulting in the identification of chemically meaningful SAR classes. It is these classes that are further analyzed and progressed in lead optimization studies.

The ideal method should be able to view and analyze all or most of the data from a HTS experiment simultaneously. Furthermore, the method should be able to extract from the data meaningful classes that reflect variations in the biological response measured in the HTS. Another attractive feature of such a method is that the search algorithm is independent of predefined substructures and can generate the substructures "on the fly" and learn from the data along the way. This approach is a challenging task even for smaller data sets in the range of a few thousands of molecules.

A PhyloGenetic-like tree (PGLT) has been described as having a "bushy carriage but a single trunk" (Nock and Nielsen, 2020). In general terms, phylogenetic trees represent the evolution of a set of objects, and in small molecular drug discovery, these objects are the molecules comprising the screening library. In the HTS data mining paradigm, this description of a PGLT can be viewed as a root with many branches. Each branch contains the common scaffold together with the active and the inactive molecules around this scaffold, effectively defining a SAR. An important difference between a PGLT and a decision tree is that the leaf nodes are not connected to the children in PGLT. A detailed description of a successful application of PGLT in data mining has been published (Nicolaou et al., 2002). An advantage of the approach described in this work is that the mining algorithm is not constricted to a set of predefined substructures or keys, and hence, by allowing the substructures to be generated "on the fly" (Tamura et al., 2002; Nicolaou and Pattichis, 2006), dependent on the data set that is being analyzed, the classes of compounds that are mined from the data is a better reflection of the SAR trends within molecular domains.

12.7 Multidomain classification

Molecules in a screening library can be viewed as being comprised of different domains, each substructural moiety being considered a domain. A cluster of molecules around this domain can define a SAR—each molecule in this cluster contains a particular chemical moiety. It becomes quickly apparent that one molecule can belong to multiple domains as each substructure within a molecule can be the basis of a SAR cluster. This observation is the basis of multidomain classification. A tremendous advantage in multidomain classification is that the probability of missing a SAR trend from the data is low, as each molecule has been viewed in multiple data environments, each of which reflects an SAR trend. The SAR for each domain subclass reflects the biological response of that chemical moiety in different chemical environments. A drawback of this approach is that the visual inspection of the results can be cumbersome as molecules are repeated in the output and might not be visually appealing. A solution to this dilemma is to view the biological response within each domain and focus on those subclasses which have a wide and interesting spectrum of biological response. When such a subclass contains a reasonable number of molecules, the information within that domain can be helpful in interpreting a SAR.

Within the multidomain classification scheme, the population of classes (or clusters) by compounds allows for the compounds to be classified when a part of the compound structure is satisfied within the specified class. This allows molecules to occupy more than one class, an advantage that is not allowed if the classification rules were tight as in a decision tree method.

In Fig. 12.3, the compounds that constitute a class from a PGLT/multidomain classification analysis of the data set AID1053188 described above are shown.

Each compound tile contains the PubChem Compound ID on the left and the percentage inhibition, measured at 10 μM, on the right with the MCS for the class highlighted in red. The classification allowed for the definition of fuzzy atoms such that molecules containing thiophene and furan can be included within a class to highlight the variation in the structural diversity with the corresponding biological activity. The distribution of the activities of the compounds is shown in Fig. 12.4. As discussed above, a compound can appear in another class in the MDC approach to data mining. The compound with PubChem Compound ID

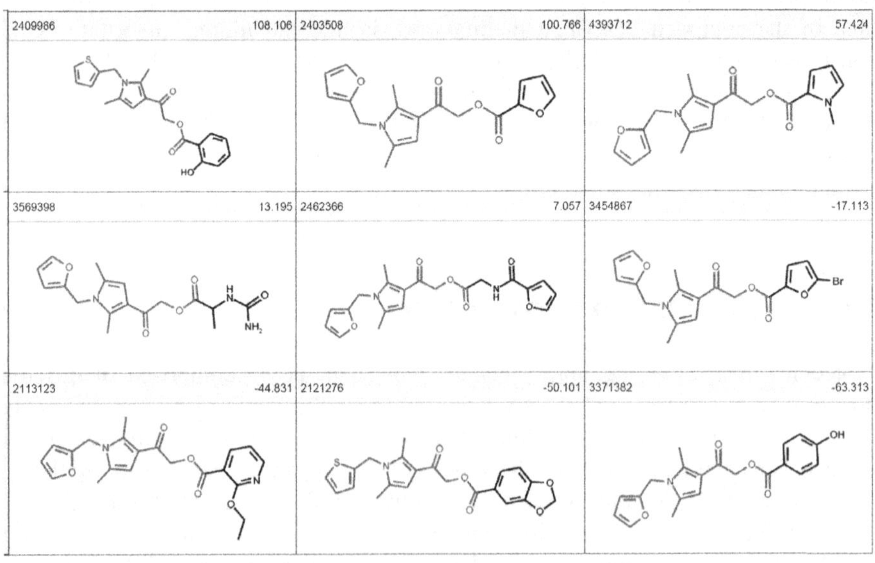

Figure 12.3 Constituent molecules in a class from a PhyloGenetic-like tree classification.

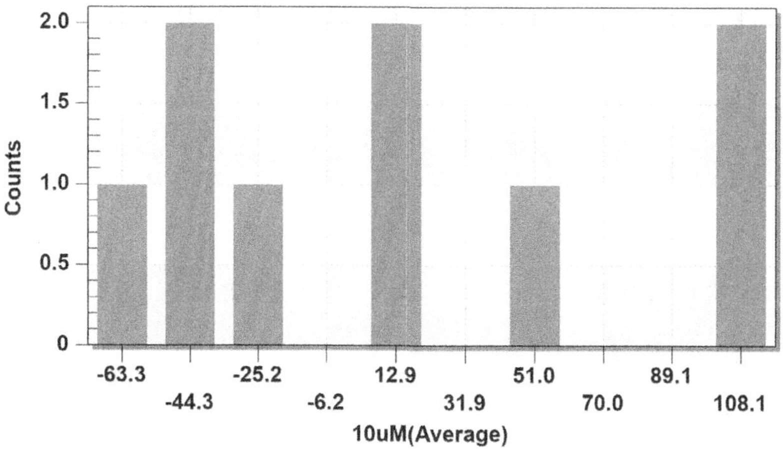

Figure 12.4 Distribution of the activities (percentage inhibition – x-axis) of the nine compounds in Fig. 12.3.

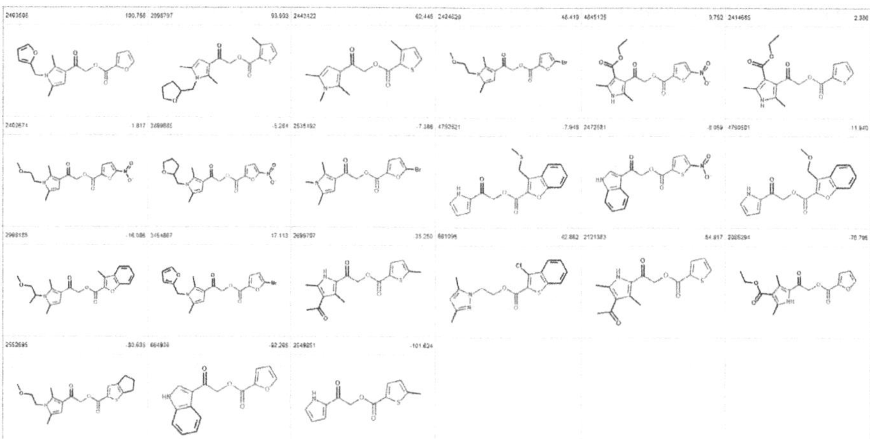

Figure 12.5 A Different structure–activity relationship for Compound ID 2403508.

2403508 belongs to another class within the PGLT classification manifold as shown in Fig. 12.5. The MCS, colored in red, is smaller in scope in this class compared to the scaffold in Fig. 12.3; however, the substitution pattern around this scaffold is more diverse than in the former case. Furthermore, the spectrum of biological activity for the molecules in Fig. 12.5 has a wider range as shown in Fig. 12.6. The results of the clustering calculations of the PubChem data set shown in Figs. 12.3–12.6 was done using the MedChem Studio Module of ADMET Predictor (2020).

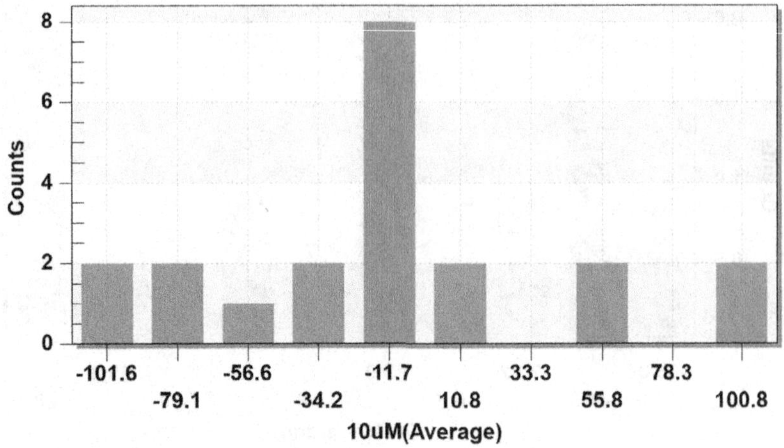

Figure 12.6 Distribution of the activities (percentage inhibition – *x*-axis) of the 21 compounds in Fig. 12.5.

The fuzzy atom approach allows molecules with substructural domains with the similar chemical environment but different atomic arrangements to be included within a single class. Such an approach allows for the inclusion of compounds that contain benzene, pyridine, pyrimidine and such variations, five-membered heterocyclic rings such as the thiophenes, furans, oxazoles and isoxazoles to be included within the same class which can result in a more informative SAR analysis.

12.8 Fuzzy trees and clustering

Is it possible to classify molecules in the screening library into clusters with each molecule associated with an occupancy number? Some molecules belonging to a cluster based on a certain criterion have an occupancy value of 1, others that do not belong to the cluster, an occupancy value of 0, while another subset of molecules have a partial membership depending on a set of predefined criteria. The third set of molecules can be associated with the concept of "intermediate truth" based on probability concepts as described in the theory of fuzzy logic and can have a class occupancy value between 0 and 1.

The hypothesis of the data miner can include some imprecision built into the query such as the mining of the data for a less stringent structural query as discussed above. An unsupervised method such as fuzzy clustering can be used in this context to address multidomain and ambiguous clustering. Fuzzy decision trees can be meaningful in this context as it affords a degree of flexibility to the data miner in defining the queries. These trees are constructed similar to the classical decision trees by choosing a measure such as a physicochemical property, descriptor or any of the keys used to describe the molecules in the data set. The data set is partitioned based on a similarity measure and a metric M that allows the splitting of the data set based on an increasing level of accuracy. The concept of molecular similarity has been discussed extensively in the literature (Willett, 2009; Xu and Hagler, 2002).

The similarity $\sigma(A,B)$ between two molecules (A,B) is defined based on a set of predefined keys such as substructural fragments in the MACCS keys (Durant et al., 2002) or a subset physicochemical properties of the two molecules. The quantification of molecular similarity requires the definition of a metric and many such metrics are available in the literature. Two commonly used metrics are the Tversky measure σ_{Tv} (A,B) and the Tanimoto metric σ_{Tm} (A,B) defined by Eqs. (12.1) and (12.2), respectively.

$$\sigma_{Tv} = \frac{\kappa(A \cap B)}{(\alpha.[\kappa(A) - \kappa(C)] + \beta.[\kappa(B) - \kappa(C)] + \kappa(C))} \tag{12.1}$$

$$\sigma_{Tm} = \frac{\kappa(A \cap B)}{\kappa(A) + \kappa(B) - \kappa(A \cap B)} \tag{12.2}$$

In the above equations, $\kappa(X)$ denotes the number of bits in molecule X that has been set to 1. Eq. (12.1) reduces to the Tanimoto definition of molecular similarity given in Eq. (12.1) when $\alpha = \beta = 1$.

These metrics are more applicable in tight clustering methods such as RP. In the fuzzy clustering approach, such similarity measures need to be redefined in consensus with the query used in the retrieving algorithm (Bouchon-Meunier et al., 2007; Askari, 2020). One of the earliest applications of the concepts of fuzzy logic in drug discovery analyzed two data sets: one a set of HIV-1 reverse transcriptase inhibitors and another data set of antirhinovirus inhibitors (Russo et al., 1998). It was shown that an accurate study and extraction of knowledge is possible if the vocabulary and grammar of the queries are well defined.

Data mining is a challenge as the dimensionality of the computation can scale exponentially depending on the queries and the algorithm used. In the particular case of analyzing HTS data in small molecule drug discovery, within each data point is encapsulated chemical information that has to be retrieved. Interpreting this information in a chemically meaningful context along with the variation in the biological response is the fundamental knowledge that needs to be extracted in the construction of a meaningful SAR.

The successful mining of a HTS data set is crucial to the success of a drug discovery program. Retrieving novel information for different classes (scaffolds) of compounds is a critical first step in lead optimization in the early stages of drug discovery.

"Do not be satisfied with stories, how things have gone with others – Unfold your own Myth." Jalal-uddin Rumi (1207–1273)

Acknowledgments

The author thanks Ms. Lina Q. Setti of Global Blood Therapeutics, South San Francisco, California, USA, Dr. Ranjit Muhandiram, Department of Molecular Genetics, University of Toronto, Toronto, Canada and Dr. Ngai Ling Ma for critically reading the manuscript and many insightful suggestions.

References

ADMET Predictor, 2020. Simulations Plus, Inc, Lancaster, CA <https://www.simulations-plus.com/software/admetpredictor>.

Arora, R., Liew, C.W., Soh, T.S., Otoo, D.A., Seh, C.C., Yue, K., et al., 2020. Two RNA tunnel inhibitors bind in highly conserved sites in Dengue virus NS5 polymerase: structural and functional studies. J. Virol.

Askari, S., 2020. Fuzzy C-means clustering algorithm for data with unequal cluster sizes and contaminated with noise and outliers: review and development. Expert. Syst. Appl. 113856.

Bamborough, P., Chung, C.W., Demont, E.H., Bridges, A.M., Craggs, P.D., Dixon, D.P., et al., 2019. A qualified success: discovery of a new series of ATAD2 bromodomain inhibitors with a novel binding mode using high-throughput screening and hit qualification. J. Med. Chem. 62 (16), 7506–7525.

Bemis, G.W., Murcko, M.A., 1996. The properties of known drugs. 1. Molecular frameworks. J. Med. Chem. 39 (15), 2887–2893.

Birnbaum, Z.W., Tingey, F.H., 1951. One-sided confidence contours for probability distribution functions. Ann. Math. Stat 22 (4), 592–596.

Bouchon-Meunier, B., Detyniecki, M., Lesot, M.J., Marsala, C., Rifqi, M., 2007. Real-world fuzzy logic applications in data mining and information retrieval. Fuzzy Logic. Springer, Berlin, Heidelberg, pp. 219–247.

Broach, J.R., Thorner, J., 1996. High-throughput screening for drug discovery. Nature 384 (6604 Suppl), 14–16.

De La Rosa, M., Kim, H.W., Gunic, E., Jenket, C., Boyle, U., Koh, Y.H., et al., 2006. Tri-substituted triazoles as potent non-nucleoside inhibitors of the HIV-1 reverse transcriptase. Bioorg. Med. Chem. Lett. 16 (17), 4444–4449.

Durant, J.L., Leland, B.A., Henry, D.R., Nourse, J.G., 2002. Reoptimization of MDL keys for use in drug discovery. J. Chem. Inf. Comput. Sci. 42 (6), 1273–1280.

Erlanson, D.A., 2011. Introduction to fragment-based drug discovery. Fragment-Based Drug Discovery and X-ray Crystallography. Springer, Berlin, Heidelberg, pp. 1–32.

Erlanson, D.A., De Esch, I.J., Jahnke, W., Johnson, C.N., Mortenson, P.N., 2020. Fragment-to-lead medicinal chemistry publications in 2018. J. Med. Chem. 63 (9), 4430–4444.

Erlanson, D.A., Fesik, S.W., Hubbard, R.E., Jahnke, W., Jhoti, H., 2016. Twenty years on: the impact of fragments on drug discovery. Nat. Rev. Drug. Discov. 15 (9), 605.

Eurtivong, C., Reynisson, J., 2019. The development of a weighted index to optimise compound libraries for high throughput screening. Mol. Inform. 38 (3), 1800068.

Ferreira, L.L., Andricopulo, A.D., 2019. ADMET modeling approaches in drug discovery. Drug. Discov. Today 24 (5), 1157–1165.

Follmann, M., Briem, H., Steinmeyer, A., Hillisch, A., Schmitt, M.H., Haning, H., et al., 2019. An approach towards enhancement of a screening library: the next generation library initiative (NGLI) at Bayer—against all odds? Drug. Discov. Today 24 (3), 668–672.

Hajduk, P.J., Greer, J., 2007. A decade of fragment-based drug design: strategic advances and lessons learned. Nat. Rev. Drug. Discov. 6 (3), 211–219.

Hann, M.M., Leach, A.R., Harper, G., 2001. Molecular complexity and its impact on the probability of finding leads for drug discovery. J. Chem. Inf. Comput. Sci. 41 (3), 856–864.

Huggins, D.J., Venkitaraman, A.R., Spring, D.R., 2011. Rational methods for the selection of diverse screening compounds. ACS Chem. Biol. 6 (3), 208–217.

Inglese, J., Auld, D.S., Jadhav, A., Johnson, R.L., Simeonov, A., Yasgar, A., et al., 2006. Quantitative high-throughput screening: a titration-based approach that efficiently identifies biological activities in large chemical libraries. Proc. Natl Acad. Sci. 103 (31), 11473–11478.

Jia, C.Y., Li, J.Y., Hao, G.F., Yang, G.F., 2020. A drug-likeness toolbox facilitates ADMET study in drug discovery. Drug. Discov. Today 25 (1), 248–258.

Kirsch, P., Hartman, A.M., Hirsch, A.K., Empting, M., 2019. Concepts and core principles of fragment-based drug design. Molecules 24 (23), 4309.

Leach, A.R., Hann, M.M., 2011. Molecular complexity and fragment-based drug discovery: ten years on. Curr. Opin. Chem. Biol. 15 (4), 489–496.

Lipinski, C.A., 2004. Lead-and drug-like compounds: the rule-of-five revolution. Drug. Discov. Today: Technol. 1 (4), 337–341.

Lipinski, C.A., 2016. Rule of five in 2015 and beyond: target and ligand structural limitations, ligand chemistry structure and drug discovery project decisions. Adv. drug. Deliv. Rev. 101, 34–41.

Lipinski, C.A., Lombardo, F., Dominy, B.W., Feeney, P.J., 1997. Experimental and computational approaches to estimate solubility and permeability in drug discovery and development settings. Adv. Drug. Deliv. Rev. 23 (1–3), 3–25.

López-Vallejo, F., Giulianotti, M.A., Houghten, R.A., Medina-Franco, J.L., 2012. Expanding the medicinally relevant chemical space with compound libraries. Drug. Discov. Today 17 (13–14), 718–726.

Macarron, R., Banks, M.N., Bojanic, D., Burns, D.J., Cirovic, D.A., Garyantes, T., et al., 2011. Impact of high-throughput screening in biomedical research. Nat. Rev. Drug. Discov. 10 (3), 188–195.

MacCuish, J., Nicolaou, C., MacCuish, N.E., 2001. Ties in proximity and clustering compounds. J. Chem. Inf. Comput. Sci. 41 (1), 134–146.

Massey, Jr. F.J., 1951. The Kolmogorov-Smirnov test for goodness of fit. J. Am. Stat. Assoc 46 (253), 68–78.

Muratov, E.N., Bajorath, J., Sheridan, R.P., Tetko, I.V., Filimonov, D., Poroikov, V., et al., 2020. QSAR without borders. Chem. Soc. Rev. .

Nicolaou, C.A., Pattichis, C.S., 2006. Molecular substructure mining approaches for computer-aided drug discovery: a review. Proc. ITAB 26–28.

Nicolaou, C.A., Tamura, S.Y., Kelley, B.P., Bassett, S.I., Nutt, R.F., 2002. Analysis of large screening data sets via adaptively grown phylogenetic-like trees. J. Chem. Inf. Comput. Sci. 42 (5), 1069–1079.

Nilar, S.H., Ma, N.L., Keller, T.H., 2013. The importance of molecular complexity in the design of screening libraries. J. Comput. Mol. Des. 27 (9), 783–792.

Noble, C.G., Lim, S.P., Arora, R., Yokokawa, F., Nilar, S., Seh, C.C., et al., 2016. A conserved pocket in the dengue virus polymerase identified through fragment-based screening. J. Biol. Chem. 291 (16), 8541–8548.

Nock, R., Nielsen, F., 2020. The phylogenetic tree of boosting has a bushy carriage but a single trunk. Proc. Natl Acad. Sci. USA. 117 (16), 8692.

O'Brien, Z., Fallah Moghaddam, M., 2013. Small molecule kinase inhibitors approved by the FDA from 2000 to 2011: a systematic review of preclinical ADME data. Expert. Opin. Drug. Metab. & Toxicol. 9 (12), 1597–1612.

Paricharak, S., IJzerman, A.P., Bender, A., Nigsch, F., 2016. Analysis of iterative screening with stepwise compound selection based on Novartis in-house HTS data. ACS Chem. Biol. 11 (5), 1255–1264.

Press, W.H., Teukolsky, S.A., Vetterling, W.T., Flannery, B.P., 1988. Numerical Recipes in C. Cambridge University Press.

Quartararo, A.J., Gates, Z.P., Somsen, B.A., Hartrampf, N., Ye, X., Shimada, A., et al., 2020. Ultra-large chemical libraries for the discovery of high-affinity peptide binders. Nat. Commun. 11 (1), 1–11.

Rusinko, A., Farmen, M.W., Lambert, C.G., Brown, P.L., Young, S.S., 1999. Analysis of a large structure/biological activity data set using recursive partitioning. J. Chem. Inf. Comput. Sci. 39 (6), 1017–1026.

Russo, M., Santagati, N.A., Pinto, E.L., 1998. Medicinal chemistry and fuzzy logic. Inf. Sci. 105 (1–4), 299–314.

Schneider, P., Tanrikulu, Y., Schneider, G., 2009. Self-organizing maps in drug discovery: compound library design, scaffold-hopping, repurposing. Curr. Med. Chem. 16 (3), 258–266.

Schuffenhauer, A., Ertl, P., Roggo, S., Wetzel, S., Koch, M.A., Waldmann, H., 2007. The scaffold tree – visualization of the scaffold universe by hierarchical scaffold classification. J. Chem. Inf. Model. 47 (1), 47–58.

Shultz, M.D., 2018. Two decades under the influence of the rule of five and the changing properties of approved oral drugs: miniperspective. J. Med. Chem. 62 (4), 1701–1714.

Sun, Y., Zhou, H., Zhu, H., Leung, S.W., 2016. Ligand-based virtual screening and inductive learning for identification of SIRT1 inhibitors in natural products. Sci. Rep. 6, 19312.

Tamura, S.Y., Bacha, P.A., Gruver, H.S., Nutt, R.F., 2002. Data analysis of high-throughput screening results: application of multidomain clustering to the NCI anti-HIV data set. J. Med. Chem. 45 (14), 3082–3093.

Tong, W., Hong, H., Fang, H., Xie, Q., Perkins, R., 2003. Decision forest: combining the predictions of multiple independent decision tree models. J. Chem. Inf. Comput. Sci. 43 (2), 525–531.

van Rhee, A.M., 2003. Use of recursion forests in the sequential screening process: consensus selection by multiple recursion trees. J. Chem. Inf. Comput. Sci. 43 (3), 941–948.

Varin, T., Gubler, H., Parker, C.N., Zhang, J.H., Raman, P., Ertl, P., et al., 2010. Compound set enrichment: a novel approach to analysis of primary HTS data. J. Chem. Inf. Model. 50 (12), 2067–2078.

Wang, Q.Y., Bushell, S., Qing, M., Xu, H.Y., Bonavia, A., Nunes, S., et al., 2011. Inhibition of dengue virus through suppression of host pyrimidine biosynthesis. J. Virol. 85 (13), 6548–6556.

Wildey, M.J., Haunso, A., Tudor, M., Webb, M., Connick, J.H., 2017. High-throughput screening, Annual Reports in Medicinal Chemistry, Vol. 50. Academic Press, pp. 149–195.

Willett, P., 2009. Similarity methods in chemoinformatics. Ann. Rev. Inf. Sci. Technol. 43, 3–71.

Xu, J., Hagler, A., 2002. Chemoinformatics and drug discovery. Molecules 7 (8), 566–600.

Yeo, W.K., Go, M.L., Nilar, S., 2012. Extraction and validation of substructure profiles for enriching compound libraries. J. Comput. Mol. Des. 26 (10), 1127–1141.
Yeo, W.K., 2012. In Silico Methodologies for Selection and Prioritization of Compounds in Drug Discovery (Doctoral dissertation). National University of Singapore.
Yokokawa, F., Nilar, S., Noble, C.G., Lim, S.P., Rao, R., Tania, S., et al., 2016. Discovery of potent non-nucleoside inhibitors of dengue viral RNA-dependent RNA polymerase from a fragment hit using structure-based drug design. J. Med. Chem. 59 (8), 3935–3952.
Zhao, G., Huang, Y., Zhou, Y., Li, Y., Li, X., 2019. Future challenges with DNA-encoded chemical libraries in the drug discovery domain. Expert. Opin. Drug. Discov. 14 (8), 735–753.

Section 3

Bioinformatics and computatioanl toxicology section

Use of proteomics data and proteomics-based biodescriptors in the estimation of bioactivity/toxicity of chemicals and nanosubstances

13

Subhash C. Basak[1] *and Marjan Vracko*[2]
[1]Department of Chemistry and Biochemistry, University of Minnesota, Duluth, MN, United States, [2]Theory Department, Kemijski inštitut/National Institute of Chemistry Ljubljana, Slovenia

13.1 Introduction

The completion of the Human Genome Project (Human Genome Project Timeline of Events, 2022) resulted in a tremendous growth spurt in the omics sciences, such as genomics, proteomics, metabolomics, and so on, as well as some significant challenges for bioinformaticians. A cell contains millions of proteins (Milo, 2013) and only a small fraction of those can be characterized using current technology. Proteomics methods, including two-dimensional gel electrophoresis (2-DE) and mass spectrometry (Anderson et al., 1987; Basak, 2010; Basak and Gute, 2008; Basak et al., 2005, 2006, 2010, 2016, 2016a; Cox and Mann, 2011; Hawkins et al., 2006; Lee et al., 2017; Sinitcyn et al., 2018; Vracko et al., 2006, 2018, 2018a; Witzmann and Grant, 2003; Yeoh et al., 2011) provide us with a wealth of data on the cellular status of a large number of proteins in cells, tissues, and organisms under normal physiological conditions as well as when biological systems are disrupted by drugs, toxins, or disease-causing organisms.

Proteins are the workhorses of the cell, determining the structure, function, and regulatory activities of cells, tissues, and organs. They make up the proteome as a whole (James, 1997). Proteomics technologies can quantify the status of thousands of cellular proteins and their post-translational modifications (Altelaar et al., 2013; Sinitcyn et al., 2018; Witzmann and Grant, 2003).

In this chapter, we will discuss the novel mathematical/computational approaches developed by the authors of this article over the last few decades in the big data realm of proteomics for the characterization of drug and toxicant effects on biological systems, as well as the effect of nanoparticles (NPs) on cellular function. We will look at data from two proteomics technologies: 2-DE and mass spectrometry-based proteomics.

Big Data Analytics in Chemoinformatics and Bioinformatics. DOI: https://doi.org/10.1016/B978-0-323-85713-0.00025-6

13.2 Proteomics technologies and their toxicological applications

13.2.1 Two-dimensional gel electrophoresis

The 2-DE technology gives us a momentary snapshot of the proteome of the cell.

A map derived from 2-DE provides data on the charge, mass, and abundance of approximately 2000 individual proteins or spots consisting of similar proteins (Witzmann and Grant, 2003). Fig. 13.1 presents a proteomics map that provides a snapshot of the protein patterns of a cell derived by the 2-DE method.

The management of such large amount of information is a daunting task (Basak, 2010). Basak and colleagues used graph theory, chemometrics, and information theory methods to develop biodescriptors of proteomics maps from 2-DE data over the last decade (Basak, 2010; Basak and Gute, 2008; Basak et al., 2005, 2010; Randic et al., 2001; Vracko and Basak, 2004; Vracko et al., 2006). They also conducted preliminary comparative studies of such biodescriptors versus individual proteins/protein spots in understanding cellular function perturbations caused by drug and toxicant exposure (Hawkins et al., 2006). Historically, the majority of proteomics biodescriptor approaches arose from an interdisciplinary research effort led by Basak and colleagues

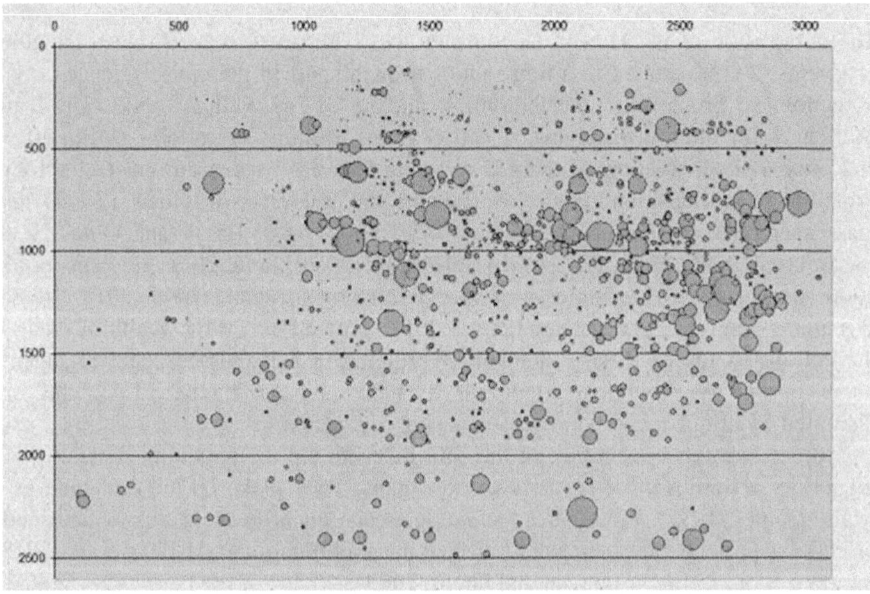

Figure 13.1 Proteomics map showing the distribution of proteins according to their respective charge, mass and abundance. in two-dimensional proteomics methods the characterization of the distribution pattern of the total cellular protein mass is based on charge (*x*-axis) and mass (*y*-axis). The size of the bubble is related to the amount or abundant of a particular protein or closely related proteins accumulated in a single protein spot. *Source*: From Witzmann and Grant (2003).

to characterize DNA sequences and their alterations by chemical and biological processes (Basak, 2010; Basak and Gute, 2008; Nandy et al., 2006) For the sake of brevity, a detailed description of the mathematical basis of the various types of biodescriptors and their applications is not provided here. Here, we briefly mention the four approaches taken for mathematical proteomics analysis of data generated by 2-DE method:

(a) Graph invariants like leading eigenvalues of the D/D matrix derived from graphs of proteomics maps (Randic et al., 2001)
(b) Information-theoretic Shannon type (Shannon, 1948) biodescriptors derived from the information-theory based partitioning of spots on proteomics maps (Basak et al., 2005)
(c) Spectrum-like descriptors derived from the projection of the three-dimension data of 2-DE maps (charge, mass and abundance of spots) into the three (x, y), (y, z) and (x, z) planes (Vracko and Basak, 2004; Vracko et al., 2006)
(d) The protein abundances in treated and control samples were compared to generate a similarity index. It can be used to describe the abundance changes of a single protein, a pattern of proteins, or entire proteomics maps.

We describe two approaches we developed for characterization of 2-DE-based proteomics maps here: (1) An information-theoretic approach to quantifying proteomics maps and (2) A chemometric approach to deriving spectrum-like mathematical proteomics descriptors and their application in characterizing the effects of peroxisome proliferators such as perfluorooctanoic acid (PFOA), perfluorodecanoic acid (PFDA), Clofibrate, and dietheylhexylphthalate (DEHP) on cellular proteomics

13.2.1.1 Information theoretic approach for the quantification of proteomics maps

As previously stated, two-dimensional proteomics methods characterize the distribution pattern of total cellular protein mass based on charge (x) and mass (m) (y). When a biological target is exposed to a toxicant, its transcriptional and translational machinery are disrupted, resulting in protein spot redistribution and the appearance or disappearance of some proteins. For quantifying such complex patterns, information theory is an adequate mathematical tool. We used information theory previously to characterize the neighborhood complexity of atomic bonding patterns within molecules (Basak, 1999, 1987; Basak et al., 1979; Roy et al., 1983). A similar approach based on Shannon's information content (Shannon, 1948) was utilized in characterizing the proteomic patterns of cells exposed to four peroxisome proliferators, viz., perfluoro-octanoic acid (PFOA), perfluorodecanoic acid (PFDA), clofibrate, and diethylhexyl-phthalate (DEHP).

In the information theoretic formalism, a set A of N objects (***proteomics spots in this case***) is partitioned into subsets A_i $(i = 1, 2, \ldots, h)$ with cardinalities N_i; $\Sigma N_i = N$. A probability scheme is then associated to the distribution:

$$\begin{pmatrix} A_1, A_2, \ldots, A_h \\ p_1, p_2, \ldots, p_h \end{pmatrix}$$

In the above formulation, $p_i = N_i/N$.

Subsequently, the information content (IC) or the complexity index (bio-descriptor) of the system consisting of N objects is computed by Shannon's (1948) formula:

$$IC = -\sum_{l=1}^{h} p_l, \log_2 p_l$$

The IC biodescriptors for proteomics maps were calculated using the first 200 and 500 most abundant spots, as well as all 1054 spots that were experimentally characterized by 2-DE. Based on the calculated complexity biodescriptor values shown in Table 13.1, the information theoretic descriptor could differentiate between the control and peroxisome proliferator-exposed biological systems. Fig. 13.2 depicts our general approach to characterization of proteomics maps using mathematical biodescriptors.

13.2.1.2 Chemometric approach for the calculation of spectrum-like mathematical proteomics descriptors

When proteomic maps are numerically encoded, the numerical representations can be analyzed using chemometrical methods, as described above. The goal of this analysis is to cluster (group) objects in proteomics maps so that quantifiers of representations can be used as biodescriptors in quantitative structure-activity relationship (QSAR) models, and to identify proteins that are involved in a specific biological process. Hierarchical clustering, principal component analysis (PCA), and self-organizing maps (SOMs) are some of the methods used artificial neural

Table 13.1 Calculated values of map information content based on five treatments using the most abundant 200, most abundant 500, and the whole set of 1054 proteins from tissue after exposure to perfluorooctanoic acid, perfluorodecanoic acid), clofibrate, and diethyl hexyl phthalate.

Treatments	Map information content (MIC) values derived from different number of spots on the proteomics maps exposed to PFOA, PFDA, Clofibrate and DEHP		
	200	500	1054
Control	3.8390	3.9486	3.9892
PFOA	3.7865	3.9112	3.9519
PFDA	3.7702	3.8861	3.9289
Clofibrate	3.7769	3.9051	3.9584
DEHP	3.7033	3.8387	3.8923

Note: Biochemically all these chemicals are peroxisome proliferators and may lead to various toxic effects including cancer.
Source: From Peraza et al. (2006).

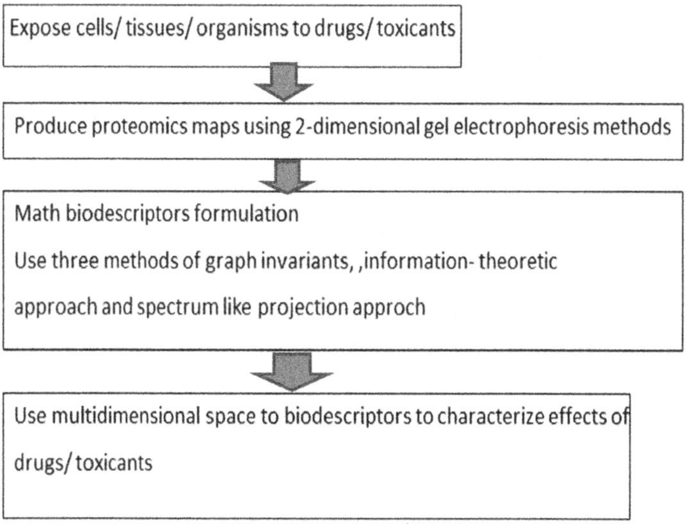

Figure 13.2 Schematic representation of methods from experimental generation of proteomics data using two-dimensional gel electrophoresis, calculation of mathematical proteomics descriptors and characterization of effects of drugs/toxicants using multidimensional proteomics bio-descriptors space.

networks alias Kohonen (Vračko and Zupan, 2015; Drgan et al., 2016). On the data set reported by Anderson et al. (1987), a cluster analysis was performed. In that study, mouse liver cells were exposed to five peroxisome proliferators: LY171883, DEHP, clofibric acid, WY14643, and nafenopin, as well as one nonproliferator (LY163443). The proteomics data was compared to the nontreated (control) group. A dose-dependent study was also conducted for one proliferator (LY171883). Fig. 13.3 shows the similarity index between control samples and those treated with different doses. The diagram on the left side of Fig. 13.3 indicates to a hormetic (biphasic) response (Calabrese, 2018). Further proteomics analyses were performed on samples obtained after 5 or 35 days of compound administration. A comparison of proteomics maps using the similarity index reveals that samples exposed for a longer period of time are more perturbed than those exposed for a shorter period of time. Results show that the chemometrical method, hierarchical clustering (Fig. 13.4) and SOM clearly separate proliferators from control and nonproliferators (Vracko and Basak, 2004).

In a study reported by Vracko et al. (2006), the researchers examined data from rat hepatocytes treated with 14 halocarbons. The proteomic maps were created using two-dimensional electrophoresis, and each map contained 1401 proteins. On the other hand, six toxicological endpoints were determined using different in vitro techniques: cell viability assay (MTT), membrane integrity assay, cell thiols, lipid peroxidation, reactive oxygen species production (ROS), and catalase activity. The study's goal was to identify protein sets (patterns) associated with a specific

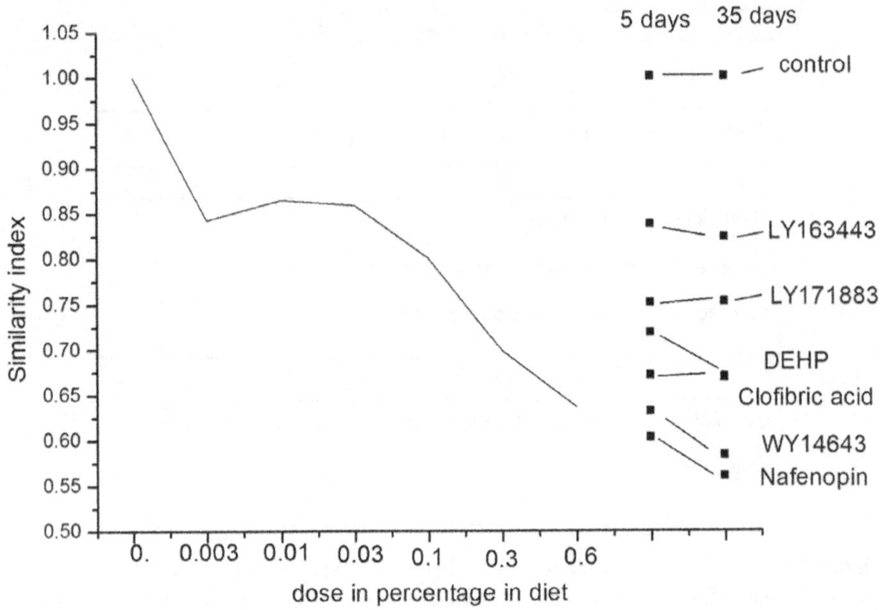

Figure 13.3 The similarity index for different administrative dose of proliferator LY171883. The right side shows the indices for different exposure times.

toxicological endpoint. An algorithm based on the similarity index and the genetic algorithm was created for this purpose. The scheme of the algorithm is shown in Fig. 13.5. Several tens of thousands of different protein patterns, each containing 10–20 proteins, were examined for their relationship with biological activities. After analyzing all patterns with a correlation coefficient greater than 0.75, we chose three proteins that were linked to all biological activities.

13.2.2 Mass spectrometry-based proteomics technology and their applications in mathematical nanotoxicoproteomics

The use of nanomaterials in industrial and consumer goods is rapidly expanding. This raises the likelihood of occupational and consumer exposure to these materials (Musazzi et al., 2017). As a result, there is concern about their potential impact on human health and the environment. NPs' physical, chemical, and biological properties are strongly influenced by their size, shape, and physical state, making them suitable for a variety of applications. Although many efforts have been made to develop nanotoxicology methods, there is currently no unified protocol for assessing the risk of NPs (Musazzi et al., 2017). However, existing data on the diverse biological effects of NPs can be used to conduct chemo- and bioinformatics-based research on these substances (Vracko et al., 2018, 2018a; Winkler et al., 2014).

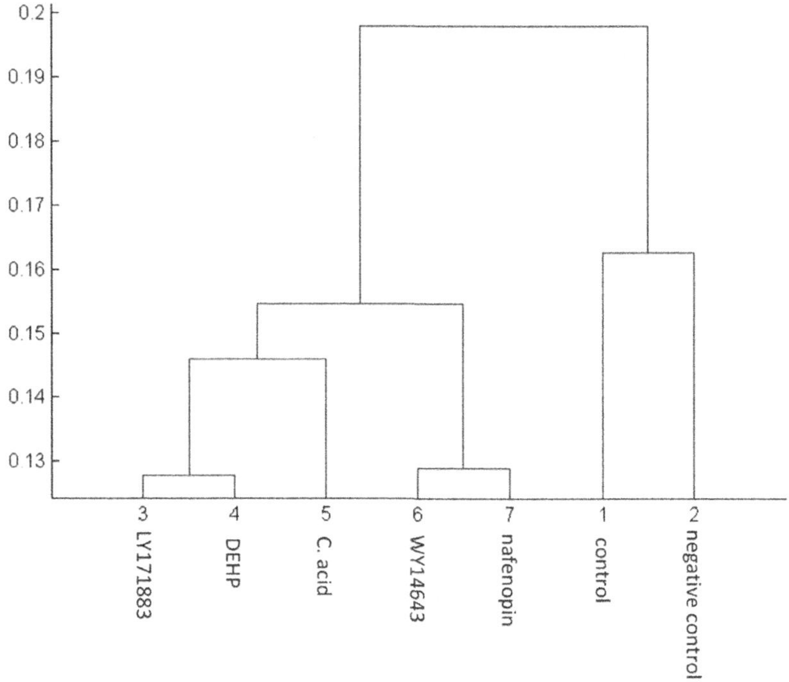

Figure 13.4 Dendrogram shows the separation of five proliferators from control and negative control samples.

The proteomics technique has been used to investigate the biological effects of NPs. Tilton et al. (2014) described the experimental details of cell exposure, proteomics analysis, and toxicity data presented here. In their study, three cell types were considered: a co-culture of Caco-2 and HT29-HTX goblet cells, primary small airway epithelial cell (SAEC), and THP-1 macrophage like cells. These cells were exposed to two kinds of NPs: multiwalled carbon nanotubes (MWCNT) and TiO_2 nanobelts (TiO_2 NB) under different experimental conditions: for two different durations of exposure: 3 and 24 hours and two different concentrations: 10 and 100 μg/mL. Results of chemometrical analysis presented here (Vračko et al., 2018a) is based on using of similarity index, which describes the changes in protein abundances, hierarchical clustering, and PCA. We asked the question: *Which parameter, type of nanoparticles, time of exposure, or concentration, has the largest influence on clustering of samples*? Analyzing the dendrograms (Fig. 13.6) and PCA score plots (Fig. 13.7) for all three cell types the time of exposure emerged as the most influential factor. It was attempted to identify the set of individual proteins that were disrupted by nanomaterials applied under specific conditions. Some evidence suggests that histones may be the target proteins most affected by long-term exposure (Vračko et al., 2018a). In a previous study Basak et al. (2016) reported on the subset of proteins, which were strongly perturbed by specific type of NPs.

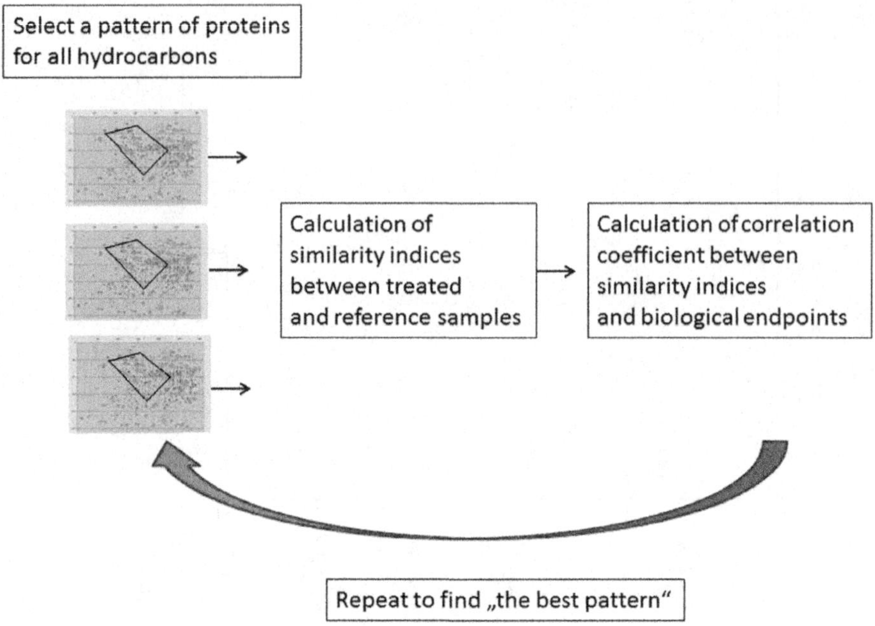

Figure 13.5 Scheme of algorithm to select the patterns of proteins related to biological activity of halocarbons.

13.3 Discussion

In the postgenomic era, the field of proteomics is expanding alongside other related fields such as genomics and metabolomics, all of which generate a large amount of data, also known as big data (Basak, 2010; Basak et al., 2016; Sinitcyn, et al., 2018). We need to understand the role of perturbations in proteomics patterns as a whole or as individual proteins in normal physiological, pathological, or toxicological conditions. Proteomics methods, both 2-DE and mass spectrometry-based, generate data on a large number of proteins (p—number of predictors) for a small number of cases (chemicals or nanomaterials, n—number of cases). In such rank-deficient situations, where $n < p$ ordinary least squares (OLS) type regression models are invalid. (Basak, 2019). As a result, we used robust methods such as ridge regression, PCA, and SOM to analyze proteomics data for drugs, toxins, and NPs (Hawkins, et al., 2006; Vracko et al., 2006; Vracko and Basak, 2004). The results of the analyses presented in this chapter show that these methods have a good chance of being useful in mathematical/computational proteomics approaches to pharmacology and toxicology.

Proteomics approaches are finding useful applications in diverse areas including new drug discovery (dos Santos Vasconcelos and Rezende, 2021; Li et al., 2021; Witzmann and Grant, 2003), toxicology (Basak, 2010; Basak et al., 2010; Vracko et al., 2018, 2018a) and the management of health care (Dash et al., 2019).

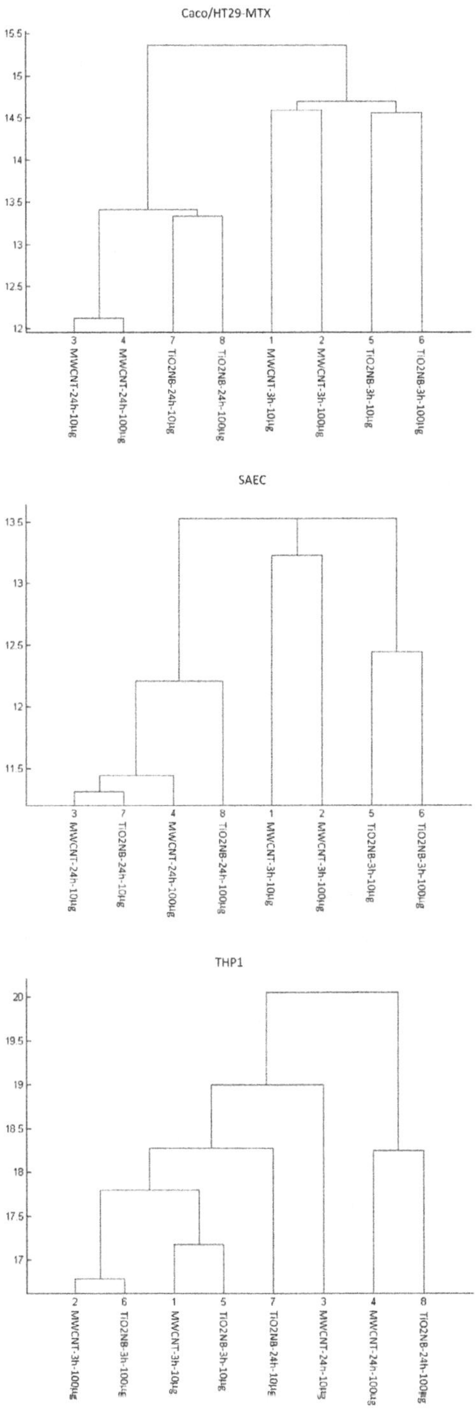

Figure 13.6 Dendrograms show the similarity among different samples. The separation between short term and long-term exposure proteomics data of cells to multiwalled carbon nanotube and TiO_2 nanobelt.

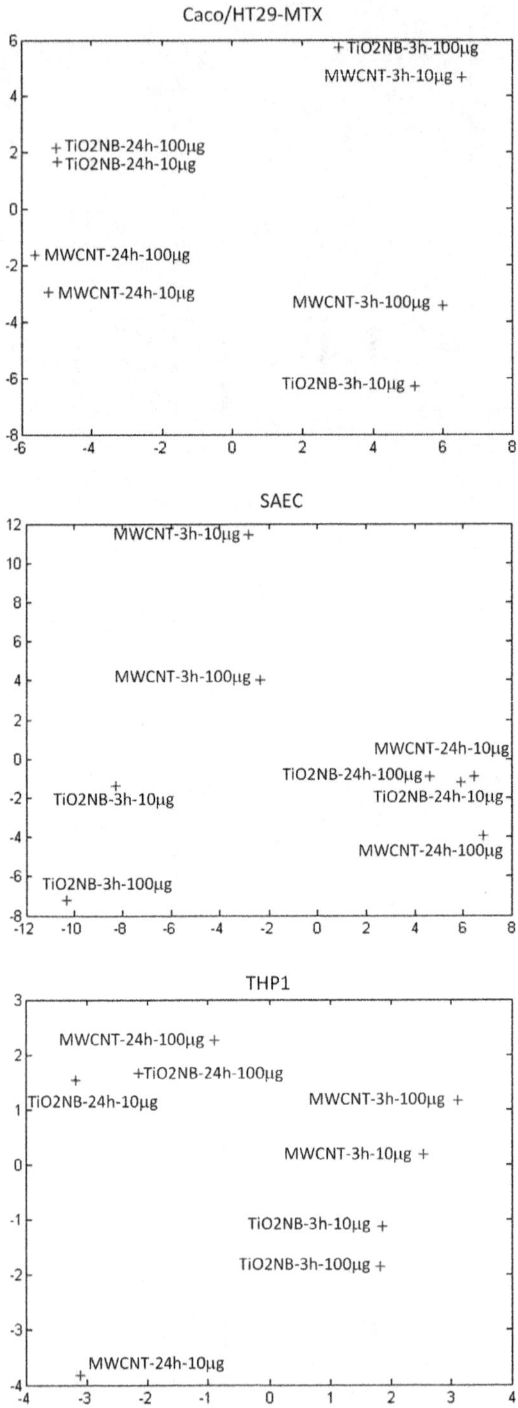

Figure 13.7 Score plots of samples recorded under different conditions for three cell types. (Caco/HT29-MTX, SAEC, and THP1).

We must develop and validate robust and useful methods for analyzing and interpreting such large amounts of data. We hope that the methods developed by us and others will help to meet this critical need in modern biomedical research.

Acknowledgment

The authors are grateful to Frank A. Witzmann for providing two-dimensional gel electrophoresis data on the effects of drugs and toxicants on cells and tissues as well as mass spectrometry data on effects of nanoparticles on cultured cells.

References

Altelaar, A.F., Munoz, J., Heck, A.J., 2013. Next-generation proteomics: towards an integrative view of proteome dynamics. Nat. Rev. Genet. 14 (1), 35–48.

Anderson, N.L., Giere, F.A., Nance, S.L., Gemmell, M.A., Tollaksen, S.L., Anderson, N.G., 1987. Effects of toxic agents at the protein level: quantitative measurement of 213 mouse liver proteins following xenobiotic treatment. Fund. Appl. Toxicol. 8, 39–50.

Basak, S.C., 1987. Use of molecular complexity indices in predictive pharmacology and toxicology: a QSAR approach. Med. Sci. Res. 15, 605–609.

Basak, S.C., 1999. Information theoretic indices of neighborhood complexity and their applications. In: Devillers, J., Balaban, A.T. (Eds.), Topological Indices and Related Descriptors in QSAR and QSPR. Gordon and Breach Science Publishers, The Netherlands, pp. 563–593. , 1999.

Basak, S.C., 2010. Role of mathematical chemodescriptors and proteomics-based biodescriptors in drug discovery. Drug. Dev. Res. 72, 1–9.

Basak, S.C., 2019. Importance of proper statistical practices in the use of chemodescriptors and biodescriptors in the twenty-first century. Future Med. Chem. 11, 1–4.

Basak, S.C., Gute, B.D., 2008. Mathematical descriptors of proteomics maps: background and applications. Curr. Opin. Drug. Discov. Devel. 11, 320–326.

Basak, S.C., Roy, A.B., Ghosh, J.J., 1979. Study of the structure-function relationship of pharmacological and toxicological agents using information theory. In: Avula, X.J.R., Bellman, R., Luke, Y.L., Rigler, A.K. (Eds.), Proceedings of the Second International Conference on Mathematical Modelling. University of Missouri-Rolla, Rolla, Missouri, pp. 851–856.

Basak, S.C., Gute, B.D., Witzmann, F.A., 2005. Information-theoretic biodescriptors for proteomics maps: development and applications in predictive toxicology. Conf. Proc., WSEAS Trans. Inf. Sci. Appl. 7, 996–1001.

Basak, S.C., Mills, D., Gute, B.D., 2006. Quantitative structure-toxicity relationships using chemodescriptors and biodescriptors. In: Riviere, J.E. (Ed.), Biological Concepts and Techniques in Toxicology: An Integrated Approach. Taylor & Francis, New York, pp. 61–82.

Basak, S.C., Gute, B.D., Monteiro-Riviere, N., Witzmann, F.A., 2010. Characterization of toxicoproteomics maps for chemical mixtures using information theoretic approach. In: Mumtaz, M. (Ed.), Principles and Practice of Mixtures Toxicology. Wiley-VCH Verlag GmbH & Co, KGaA, Weinheim, pp. 215–232.

Basak, S.C., Vracko, M., Witzmann, F.A., 2016. Nanotoxicology and big data management: a mathematical/computational approach to characterize complex proteomics patterns derived from exposure of cells to nanomaterials. Curr. Comput. Aided Drug. Des. 12, 253–254.

Basak, S.C., Vracko, M., Witzmann, F.A., 2016a. Mathematical nanotoxicoproteomics: quantitative characterization of effects of multi-walled carbon nanotubes (MWCNT) and TiO_2 nanobelts (TiO_2-NB) on protein expression patterns in human intestinal cells. Curr. Comput. Aided Drug. Des. 12, 259–264.

Calabrese, E., 2018. Hormesis: path and progression to significance. Int. J. Mol. Sci. 19, 2871. Available from: https://doi.org/10.3390/ijms19102871.

Cox, J., Mann, M., 2011. Quantitative, high-resolution proteomics for data-driven systems biology. Annu. Rev. Biochem. 80, 273–299.

Dash, S., Shakyawar, S.K., Sharma, M., Dash, S., 2019. Big data in healthcare: management, analysis and future prospects. J. Big Data 6, 54. Available from: https://doi.org/10.1186/s40537-019-0217-0.

dos Santos Vasconcelos, C.R., Rezende, A.M., 2021. Systematic *in silico* evaluation of Leishmania spp. proteomes for drug discovery. Front. Chem. 9, 607139. Available from: https://doi.org/10.3389/fchem.2021.607139.

Drgan, V., Župerl, Š., Vračko, M., Comob, F., Novič, M., 2016. Robust modelling of acute toxicity towards fathead minnow (*Pimephales promelas*) using counter-propagation artificial neural networks and genetic algorithm. SAR&QSAR Environ. Sci. 27, 501–519.

Hawkins, D.M., Basak, S.C., Kraker, J.J., Geiss, K., Witzmann, F.A., 2006. Combining chemodescriptors and biodescriptors in quantitative structure-activity relationship modeling. J. Chem. Inf. Model. 46, 9–16.

Human Genome Project Timeline of Events <https://www.genome.gov/human-genome-project/Timeline-of-Events> (accessed 17.05.21).

James, P., 1997. Protein identification in the post-genome era: the rapid rise of proteomics. Q. Rev. Biophys. 30 (4), 279–331.

Lee, J.A., Carragher, N.O., Berg, E.L., 2017. Empirical drug discovery: a view from the proteome. Drug. Discov. Today: Technol. 23, 1–5.

Li, J., Liu, S., Shi, J., Wang, X., Xue, Y., Zhu, H.J., 2021. Tissue-specific proteomics analysis of anti-COVID-19 nucleoside and nucleotide prodrug-activating enzymes provides insights into the optimization of prodrug design and pharmacotherapy strategy. ACS Pharmacol. Transl. Sci. 4 (2), 870–887.

Milo, R., 2013. What is the total number of protein molecules per cell volume? A call to rethink some published values. Bioessays 35, 1050–1055.

Musazzi, U.M., Marini, V., Casiraghi, A., Minghetti, P., 2017. Is the European regulatory framework sufficient to assure the safety of citizens using health products containing nanomaterials? Drug. Discov. Today 22, 870–882.

Nandy, A., Harle, M., Basak, S.C., 2006. Mathematical descriptors of DNA sequences: development and application. ARKIVOC 9, 211–238.

Peraza, M.A., Burdick, A.D., Marin, H.E., Gonzalez, F.J., Jeffrey, M., Peters, J.M., 2006. The toxicology of ligands for peroxisome proliferator-activated receptors (PPAR). Toxicol. Sci. 90, 269–295.

Randic, M., Witzmann, F.A., Vracko, M., Basak, S.C., 2001. On characterization of proteomics maps and chemically induced changes in proteomes using matrix invariants: application to peroxisome proliferators. Med. Chem. Res. 10, 456–479.

Roy, A.B., Basak, S.C., Harriss, D.K., et al., 1983. Neighborhood complexities and symmetry of chemical graphs and their biological applications. In: Avula, X.J.R., Kalman, R.E.,

Liapis, A.I., Rodin, E.Y. (Eds.), Mathl. Model. Sci. Tech. Pergamon Press, pp. 745–750.

Shannon, C.E., 1948. A mathematical theory of communication. Bell Syst. Tech. J. 27, 379–423.

Sinitcyn, P., Rudolph, J., Cox, J., 2018. Computational methods for understanding mass spectrometry–based shotgun proteomics data. Annu. Rev. Biomed. Data Sci. 1, 207–234.

Tilton, S.C., Karin, N.J., Tolic, A., Xie, Y., Lai, X., Hamilton Jr, R.F., et al., 2014. Three human cell types respond to multi-walled carbon nanotubes and titanium dioxide nanobelts with cell-specific transcriptomic and proteomic expression patterns. Nanotoxicology 8, 533–548.

Vracko, M., Basak, S.C., 2004. Similarity study of proteomic maps. Chemometr. Intell. Lab. Syst. 70, 33–38.

Vračko, M., Zupan, J., 2015. A non-standard view on artificial neural networks. Chemom. Intell. Lab. Syst. 149, 140–152.

Vracko, M., Basak, S.C., Geiss, K., Witzmann, F.A., 2006. Proteomics maps-toxicity relationship of halocarbons studied with similarity index and genetic algorithm. J. Chem. Inf. Model. 46, 130–136.

Vracko, M., Witzmann, F.A., Basak, S.C., 2018. Editorial. a possible chemo-biodescriptor framework for the prediction of toxicity of nanosubstances: an integrated computational approach. Curr. Comput. Aided Drug. Des. 14, 2–4.

Vračko, M., Basak, S.C., Witzmann, F.A., 2018a. Chemometrical analysis of proteomics data obtained from three cell types treated with multi-walled carbon nanotubes and TiO_2 nanobelts. SAR. QSAR Environ. Res. 28, 567–577. Available from: https://doi.org/10.1080/1062936X.2018.1498015.

Winkler, D.A., Burden, F.R., Yan, B., Wickeder, R., Tassa, C., Shaw, S., et al., 2014. Modelling and predicting the biological effects of nanomaterials. SAR. QSAR Environ. Res. 25, 161–172.

Witzmann, F.A., Grant, R.A., 2003. Pharmacoproteomics in drug development. Pharmacogenomics J. 3, 69–76.

Yeoh, L.C., Dharmaraj, S., Gooi, B.H., Singh, M., Gam, L.H., 2011. Chemometrics of differentially expressed proteins from colorectal cancer patients. World J. Gastroenterol. 17 (16), 2096–2103.

Mapping interaction between big spaces; active space from protein structure and available chemical space

14

Pawan Kumar[1], *Taushif Khan*[2] *and Indira Ghosh*[3]
[1]National Institute of Immunology, Aruna Asaf Ali Marg, New Delhi, Delhi, India,
[2]Immunology and Systems Biology Department, OPC-Sidra Medicine, Ar-Rayyan, Doha, Qatar, [3]School of Computational & Integrative Sciences, Jawaharlal Nehru University, New Delhi, Delhi, India

14.1 Introduction

The availability of multiomics data offers an alluring opportunity to explore and understand numerous biological paradigms. The availability of high throughput, inexpensive and scalable experiments in areas such as genomics, transcriptomics, proteomics, metabolomics, metagenomics, pharmacogenomics, epigenomics, interact-omics and imaging generate a huge volume of data within a small timeframe (D'argenio, 2018; Vamathevan et al., 2019). This deluge of data must be managed with efficient computational algorithms and analytics for extracting usable information which can be further translated into knowledge (Cook et al., 2020). Integration of knowledge from above-mentioned diverse platforms not only enables us to gain insight into complex biological phenomena but also overlays the path for personalized medicine with disease monitoring and targeted therapeutics (D'argenio, 2018; Kasson, 2020; Saylor et al., 2020; Eckhardt et al., 2020).

With the exquisite available space of data provided by multiomics techniques; it will become puzzling to navigate for specific functionality. In the context of cell survival, biomolecules must function to tailor various structural and functional requirements. The diversity of biomolecules like DNA, RNA, proteins, carbohydrates, lipids and many complex macromolecules have been fine-tuned by evolutionary selection over millions of years. DNA and RNA contained in the genome; are the transcript of the organism and its hereditary properties to propagate genetic information. Translated from the genome, proteins are known to be the working force of cells. The information flow from gene to protein is the "backbone" of life, driven by complex networks of essential cellular components. However, large strings of combinations of just four-letter codes (A, T[U], G, C in DNA[RNA]) produce a diverse number of permutations to generate 20 natural amino acids, hence forming a large number of unique proteins by sequence and structures. Evolution in

Big Data Analytics in Chemoinformatics and Bioinformatics. DOI: https://doi.org/10.1016/B978-0-323-85713-0.00029-3

its due course (~3.8 billion years), produced more and more sophisticated life forms through mechanisms like gene duplication, divergence and combination which results in different life forms creating divergence in the genome which is reflected in the organism's protein sequence and structure.

The enormous diversity in protein structure and function space is rather a result of its component, the amino acids. A simple combinatorial exercise of sequence possibility for a 100-residue long protein suggests 20^{100} different possible sequences (considering 20 frequently occurring natural amino acids). Similarly, exponential expansion in structure space can be observed by considering the similar combinatorial approach as shown in Fig. 14.2A (Khan and Ghosh, 2015). However, considering the physico-chemical nature of amino acids only a fraction could be able to fold in thermodynamically stable conformation (Brenner et al., 1997; Chan and Dill, 1990; Dabrowski-Tumanski and Sulkowska, 2017; Davidi et al., 2018). But to be a functional protein, these numbers even become drastically small (Bartlett et al., 2002; Borgwardt et al., 2005; Kahraman et al., 2010; Khan and Ghosh, 2015; Khersonsky et al., 2018). In the context of the genomic landscape, presently, only a limited number of protein sequences are available in literature and databases, and even less data is available for which protein structures are known. While the bottleneck in protein structural space is challenging experimental set-up, over the last decades, we have witnessed a "resolution revolution," with cutting-edge techniques like electron cryo-microscopy (cryo-EM) (Mccafferty et al., 2020; Kühlbrandt, 2014).

With cutting-edge technologies like nuclear magnetic resonance (NMR), three-dimensional electron microscopy (3DEM)/electron tomography (ET) and cryo-electro electron microscopy (EM) more complex structures are now being solved which was not possible by conventional macromolecular crystallography (MX) (Consortium, 2019). Presently, Protein Data Bank (PDB) (https://www.rcsb.org/), one of the members of wwPDB Core Archives (http://www.wwpdb.org/), contains protein structural data solved using the above-mentioned three techniques. Fig. 14.1 shows the availability of the three-dimensional structural data in the current volume of the PDB (December 2020), and it indicates that a significant number of protein

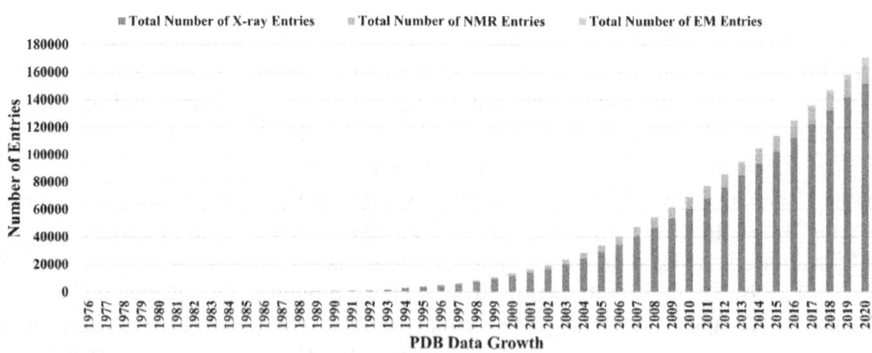

Figure 14.1 Cumulative number of three-dimensional -protein structures solved by three different approaches and available in the Protein Data Bank.

entries ($\sim$88%) are solved using the X-rays technique starting from 1976 to now while NMR driven entries are gaining space from 2000 onwards. Though the number of entries for 3DEM is still meager in number, however, its contribution can be understood from the fact that large protein complexes such as the structure and function of ribosome was studied by V. Ramakrishnan, Thomas A. Steitz and Ada E. Yonath and received the Nobel prize for Chemistry in 2009 (https://www.nobelprize.org/prizes/chemistry/2009/summary/). Further, this technique has enabled us to understand the major infectious component (spike protein) of the severe acute respiratory syndrome-Coronavirus-2 (SARS-CoV-2) virus (Salvatori et al., 2020) which has currently caused a global pandemic. Overall, among the different challenges at the various levels, protein three-dimensional structural space is growing exponentially since its inception and empowering the scientific community to understand the molecular interactions at the atomic level.

Techniques like X-ray, NMR spectroscopy, EM have been explored to target specific questions in structural genomics by purifying and solving macromolecular structures over the years. Although these tools have resulted in some seminal discoveries, we are yet to understand the intricate balance between structure and function. Understanding the arrangements between these elements will enhance our ability to design tailored proteins by mapping required functional components. At this juncture as well as in future it will be paramount to leverage the available big data space to map the crucial interacting components within the structural scaffold. However, it is evident that the flow of information by the protein is conveyed by its interactions with small and large molecules. So, representing proteins in their interacting space is of utmost importance to reduce such large data for functional analysis for designing inhibitors and drugs eventually.

14.2 Background

14.2.1 Navigating protein fold space

Like all life forms (or any intelligent design), the intricate trade-off between functionality and stability is crucial for survival (Lapenta and Jerala, 2020; Graham et al., 2019; Khersonsky et al., 2018). The process of natural selection fine-tunes the different fitness functions to retain or eliminate several components via the process of evolution, through various mechanisms like gene duplication and divergence (Copley, 2020). In the universe of protein fold space, several approaches have been developed to understand this process (Mccafferty et al., 2020; Taylor, 2020). A better insight into the protein fold universe, will not only provide a broad understanding of organizational principles but also highlight designing codes to tailor functionality.

14.2.2 From amino acid string to dynamic structural fold

Proteins are mainly composed of 20 amino acids. The long chain of amino acid, a product of translation linked by peptide bond (from N to C terminal) referred to as protein sequence (or one-dimensional structures). Amino acids occurring in protein are of different sizes (number of atoms) and varied physicochemical properties. Differences in amino acids occur in their respective sidechains that can be classified as polar (uncharged), charged and hydrophobic (more on the standard feature of amino acids can be found in classic biochemistry textbooks). For functional requirements, proteins need to organize or fold to form a three-dimensional structure which is influenced by the sequence of amino acids, their physicochemical properties and environment where it gets expressed. Alpha helix and beta structures are observed recursively as ordered secondary structures, other than turns which provide protein with its geometrical compact definite shape. Many combinations of such secondary elements provide hierarchical complex protein structures for their specific functions in the cell, such as enzymatic, structural, and signaling purposes (Taylor, 2002; Fleming et al., 2006). Different folding rules have been established to monitor the occurrence and architecture of these recurring super-secondary structures (Adamian and Liang, 2001; Chan and Dill, 1990; Przytycka et al., 2002; Taylor, 2002; Zhang et al., 2011). For example, a study on all beta proteins (Przytycka et al., 2002), proposed folding rules like hairpin rules, beta-wind rules, and bridge rules, which investigate general organization principles of beta proteins. These rules conceptualize basic structural properties and describe the folding of the hierarchical organization.

Mimicking the evolutionary processes both experimental and computation studies attempted to explore the sequence-structure relationship (Marks et al., 2012; Panchenko et al., 2005). Characterizing the compatibility of protein sequences for a given structure, also known as an inverse-folding problem (Koehl and Levitt, 2002) provided critical insight into the nature of the folding process. Studies on proteins of different structure families (SH3 family, all-beta proteins, alpha/beta barrel fold proteins), tested across different protein sizes (number of residues) have revealed that the native sequence of a protein is only one among all sequence compatible with the native structure (Koehl and Levitt, 2002; Han et al., 2007). Among many reported examples, mutational studies on T4-lysozymes (Dessailly et al., 2013) and Arc repressor proteins (Stewart et al., 2019; Koehl and Levitt, 2002) are well characterized. These studies illustrated that bacteriophage T4 lysozyme retains its robust scaffold even after mutation of 10 consecutive amino acids, whereas a double mutation in Arc repressor protein resulted in major structural changes (each alpha helix changed to beta-sheet). Similar changes have been observed in proteins G and L, which despite having only approximately 20% sequence identity, shares similar structural folds (Karanicolas and Brooks, 2002).

Analyzing the size of sequence space for a structure by entropy study suggests that the size of sequence space reflects the usage of the observed fold. It has been known that structures that are robust to random mutation are likely to be frequently found in nature (Kolodny et al., 2013). Identifying such frequent structural topology (in the context of protein structures, topology refers to the relative arrangement of structural components in three-dimensional structure, see Section 14.3 for detail) is of great interest for protein designing and engineering. The designable structural scaffolds accommodate high mutational rates (Lapenta and Jerala, 2020;

Panchenko, 2003; Panchenko et al., 2005; Verma and Pandit, 2019). As described by Panchenko et al. (2005), the degree of structural variation with sequence variation (within proteins of the homologous family) might be analogous to stress–strain relationship in materials (from solid mechanics). The physical body undergoes deformations (geometrical) after applying stress. Amino acid substitution or mutation introduces stress in protein structure. Protein responds to such stress either by incorporating small adjustments while maintaining native scaffolds or transforming into different structures. A balance between strong functional constraints and stringent structural scaffolds represents the evolutionary plasticity of structure (Hanson et al., 2014; Skolnick and Gao, 2013). Hence, systematic investigation of protein structure–function is crucial to evaluate the effect of different influencing criteria.

14.2.3 Elements for classification of protein

The three-dimensional scaffold of protein structures provides the frame for the vital sites on the proteins that carry out several functions like catalytic, protein–protein interaction, and ligand binding (Dessailly et al., 2013). As discussed previously, the three-dimensional structure is a result of local regular structural patterns called secondary structure elements and forms the structural components of proteins. They provide the basic scaffold or skeleton that can facilitate functionality (Morrone et al., 2011; Petrey et al., 2009). The alpha-helix and beta structures are the most frequently observed ordered secondary structures. Based on regular backbone H-bond formation, secondary structural patterns are observed (alpha, beta or turns). As there is no involvement of side chains, these structures can be formed with any set of amino acids (with different preferences) (Cao et al., 2016; Fleming et al., 2006; Kabsch and Sander, 1983; Martin et al., 2005). However, every sequence of protein has a unique function to perform, such as enzymes, signaling molecules, structural molecules, transcription factors, etc.

14.2.4 Available methods for classifying proteins

Based on shared characteristics, biological organisms have been grouped into different taxonomy (De Queiroz and Gauthier, 1994). Grouping on shared or similar characteristics has been the main orientation to understand and study the nature of any complex system. For biological classification, structural (or morphological) similarity reflects the evolutionary relationship of species, grouped into different classes such as Kingdom, Phylum, Class, Order, Family, and Species. Each class represented a certain level of common behavior shared among their representing elements. Similarly, proteins have been classified according to their sequence and (or) structural similarity (Tseng and Li, 2012; Andreeva and Murzin, 2010; Marchler-Bauer et al., 2015; Moutevelis and Woolfson, 2009; Schaeffer et al., 2017; Sillitoe et al., 2019). As many numbers of proteins sharing a similar sequence and structure composition, this classification differs based on their respective similarity cutoffs (heuristic criteria). The main advantage of such classification is that it provides an easy (but crude) way to predict the function, evolution and physicochemical nature of any new protein sharing similar structural or sequence properties (De Lima Morais et al., 2011).

The expansion of the protein database calls for more sophisticated measures of annotation. Different approaches provide a plethora of information giving a biological perspective to structural data. The most popular, hierarchical classification system [Class (C), Architecture (A), Topology (T) and Homologous superfamily (H) (CATH) (Sillitoe et al., 2019), Structural Classification Of Proteins (SCOP) (Andreeva and Murzin, 2010)] aims to understand structural data by cataloging the fold space based on similarity measures at sequence, evolution, function and structure levels. Both classification system considers fold-space to be discrete, and classification is based on protein domains (Rackovsky, 2015; Mura et al., 2019; Sadowski and Taylor, 2010). Identification of protein domains remains semiautomatic achieved by different measures of hydrophobic clusters. However, both SCOP and CATH databases could not identify any new folds or topologies in quite some time. It poses serious questions in the nature of topology space and the conventional hierarchical classification schema, used for protein structures.

Both SCOP and CATH identify protein domains in solved structures (reported in RCSB) and classify domains into different groups in a hierarchical fashion. At a lower level of classification, domains are compared based on sequence homology, in SCOP these groups are called "family" and in CATH as "Homology." Depending on the functional and evolutionary relationship, domains in families, are grouped together as superfamily (a SCOP definition absent in CATH). The next level of classification considers the general arrangement of secondary structures (as "fold" in SCOP and "topology, architecture" in CATH). At the highest level of classification, the major composition of secondary structures (alpha, beta or alpha/beta) form "class." Two classification systems have similar grouping criteria except for CATH, which extensively recognizes the overall shape of the fold as "architecture." SCOP provides more functional and evolutionary relationships between protein domains, whereas CATH treats structural features more carefully.

14.3 Protein topology for exploring structure space

Analysis of protein domains evidenced the recurrence of small structural motifs (or super-secondary structures such as α-hairpins, β-hairpins, αβ motifs) (Grainger et al., 2010; Khan and Ghosh, 2015; Przytycka et al., 2002). Using these structural elements, traversing from one structural architecture to another becomes handy and is being reported by different studies described as the "Russian doll effect" (Krishna and Grishin, 2005; Rackovsky, 2015). Continuity in fold space could also be apprehended from the very limited number of distinct folds observed in nature (varying from ~1500 to 2000). With very little to no novel topology or fold found since 2001 in spite of exponential growth in protein structures, kept the community wondering about the nature of fold space (Govindarajan and Goldstein, 1996; Grainger et al., 2010; Sadowski and Taylor, 2010). Even CATH and SCOP in their newer release (SCOP v2) identify that with a hierarchical or discrete way of classification, protein structure organization would be a difficult task because many overlapping structural motifs have increased. In the new prototype of SCOP (v2), an ontology-based approach has been adopted to overcome the limitations of the tree-like hierarchical way of classification (Andreeva et al., 2015, 2020). The conventional similarity measures used to cluster folds and topology have limitations to address the evolutionary process of insertion and deletion of structural fragments (Sadowski and Taylor, 2012; Gordeev and Efimov, 2013). Moreover, the definition of fold and domains are also debatable with ambiguities in defining boundaries.

Over the years, many pioneering works on an alternative view of protein fold space have been proposed, using topological and multiple abstract descriptors (Khan et al., 2018; Gordeev and Efimov, 2013; Schaeffer et al., 2017). Work using the ideal forms in protein structure by Taylor et. al. (Grainger et al., 2010; Sadowski and Taylor, 2010; Taylor, 2020; Taylor, 2002) focused on setting up rules to describe protein structure architecture in layer form. The development of a "periodic table" or rules to address the organization of secondary structure in globular proteins is a good example of the importance and emergence of topological descriptors in protein structure study.

Topology helps to look at subjects qualitatively rather than quantitatively. The qualitative perspective may be useful in the following cases:

- Identifying possibilities or tendencies that can be investigated more closely with quantitative analysis. It is like identifying possibilities, without necessarily showing how, or identifying impossibility, which is critical for protein designing (known as avoiding pool structures by implementing "negative designing" (King and Lai, 2013; Lindorff-Larsen et al., 2005).
- For coarse-grained systems, which are approximate models of the real-world, only qualitative meaningful results are provided.

Topology-based classification helps in understanding the protein folding process (Lindorff-Larsen et al., 2005; Lai et al., 2012; Ramakrishnan et al., 2012; Wang et al., 2012). With topology, protein structural fold can be cataloged, that further used to explore the presence of structural modules and reemergence due to positive selection as shown for the case of alpha-helical proteins (Khan and Ghosh, 2015).

14.3.1 Modularity in protein structure space

The development of robust architecture consists of "distinct processes that while operating in coordination, can be dissociated into separate elements." This concept has been the core for wide applications (Wagner et al., 2007), from software architecture (functions, classes), and network theory (subgroups) to engineering (Lorenz et al., 2011). The concept of modularity is important because, in addition to robust design and evolvable segments, modules save the search space for individual building blocks

Modules could be denoted as pattern connectedness and could be explained by two main characteristics (i) independence and (ii) widespread. Modules could act as independent entities which are observed frequently across the datasets. Modules should also have the capability to evolve separately.

14.3.2 Data-driven approach to extract topological module

Modularity in protein structures could be explained from the point of function and occurrence in the metabolic pathway. Studies on protein functional domains have reported that the evolution of different metabolic pathways has been guided by convergent and divergent evolution of protein domains. De-novo biosynthesis of purine has been one of the well-known examples of convergent evolution (Li et al., 2009). Three separate proteins, GAR synthetase, GAR transformylase, and AIR synthetase, in *Escherichia coli* and *Bacillus subtilis* form a single tri-functional protein in humans and chickens through convergent evolution (Park et al., 2015; Aimi et al., 1990). In biological systems, hierarchical modularity works at several levels. Modules act as a bridge between robustness and resolvability, whose compensation is essential for sustainability. The next big question is: how to define modules in protein structures? The most widely used domain-based approach, which is an evolutionary unit, has been well explained and used extensively. These are considered as structural modules and are also known as autonomous folding units. However, the definition of the domain remains subjective.

One systematic approach to solving this problem has been illustrated by focusing on the tertiary contact between secondary structures (Khan and Ghosh, 2015; Khan et al., 2018). The composition and arrangement of secondary structure elements (or topology) provide the necessary structural scaffold for protein structure and function. Earlier we have reported a component-based approach to protein structure space independent of evolution as addressed by conventional structure classification resources (Sillitoe et al., 2019; Andreeva et al., 2015). We have further illustrated the use of "contact string" in exploring protein topological space using secondary structure elements and Topological Contact/Interaction String, as illustrated in Fig. 14.2C and D. Briefly, we have compared the topology and protein space in "prevalent" (P) and "nonprevalent" (NP) classes. As shown in Fig. 14.2B, comparing the density distribution, a clear distinction in distributions of "P" and "NP" can be observed. For each case, the maximum density of the data can be found around their respective mean and interquartile regions, whose values vary for topology and proteins in both cases. Examining the distributions, it can be observed that the topologies in "P" are only approximately 20% of the total topology space, whereas it caters to approximately 70% of total proteins, which is reverse in the case of "NP." This characteristic of distribution for topology is fairly evident, however, proteins have subtle higher variance, distributed around a mean of approximately 60% for "P" and approximately 40% for "NP." Similar analysis has been performed for different datasets and among structural classes (Khan et al., 2018). Among all studied cases, we have observed the consistent distribution of topology space, with tolerable variance in protein distribution across structure classes.

As the nature of topology space depends on the number of secondary structures, protein structure space has been scanned in three exclusive structural classes (all

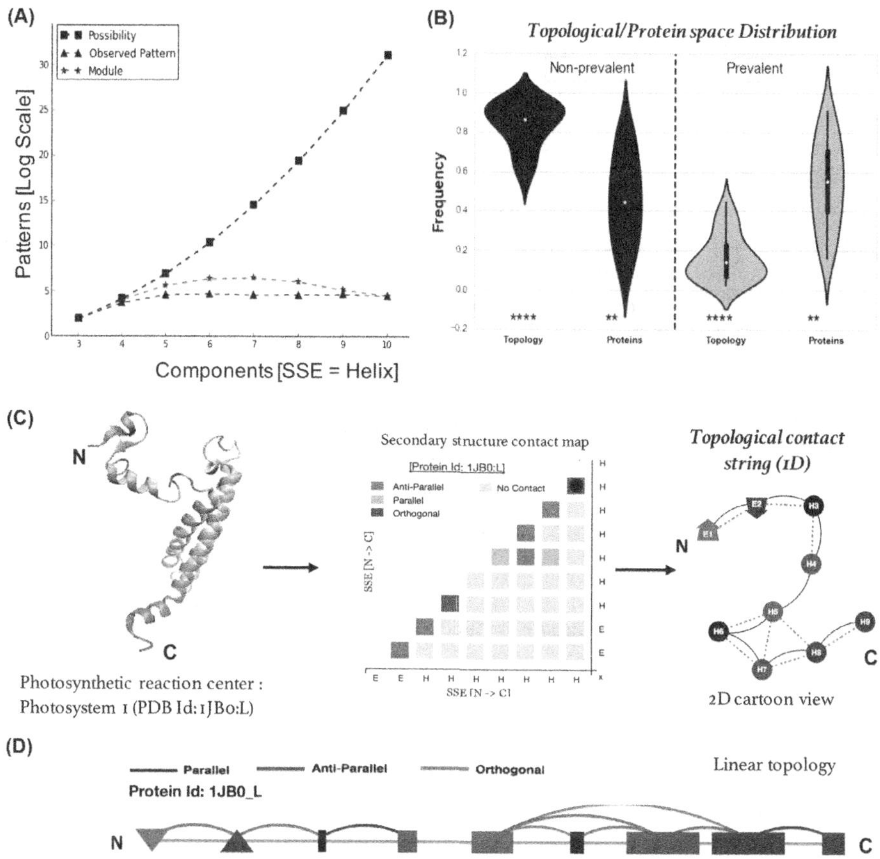

Figure 14.2 (A) Comparing theoretical possible patterns and observed structural patterns and modules in alpha-helical proteins. (B) Distribution of topology and proteins in groups of "nonprevalent" (left to dashed line) and "prevalent" (right to dashed lines) from all protein chains from approximately 82,000 protein chains and domains that have been analyzed from nonredundant datasets (Khan et al., 2018). The shape of the violin plot describes the kernel density estimation of the distribution of data in different topologies and proteins. A summary of statistics is shown in the inner boxplot. The white dot represents the median, thick gray bar shows the interquartile range and the thin line describes the 95% confidence interval. Comparison of distribution with nonparametric Wilcoxon rank-sum test has been performed and P-values are indicated as "*" (**** $P < 0.001$ and ** $P < 0.01$) in the bottom. (C and D) A graphical representation for contact string generation from three-dimensional protein structure, (C) illustration of a protein three-dimensional structure from PDB id:1JB0; chain L followed by constructing of adjacency matrix that maps tertiary contact and relative orientation of major secondary structures for a given protein. The matrix is used to generate topological contact string, which can be used to represent abstract two-dimensional and linear topological (D) representation.

alpha, all beta and alpha-beta). From the extensive analysis of different datasets, we have reported that topology space distribution is biased. Using a component-based approach, we have found that protein topology space is modular and composed of a small number of topological building blocks. These modules could be made an analogy with lego blocks and proteins as an architecture emerging from these lego-blocks.

Domains emerged from the fusion of shorter peptides, often called repetitive units (Fig. 14.3). The repeating units like beta-hairpins (found in β-trfoil, jelly-roll, immunoglobin fold, bladed propeller), alpha–alpha repeats (collagens, TPR repeats, globin like-fold, membrane all-alpha), βαβ-repeats (TIM (named after Triose-

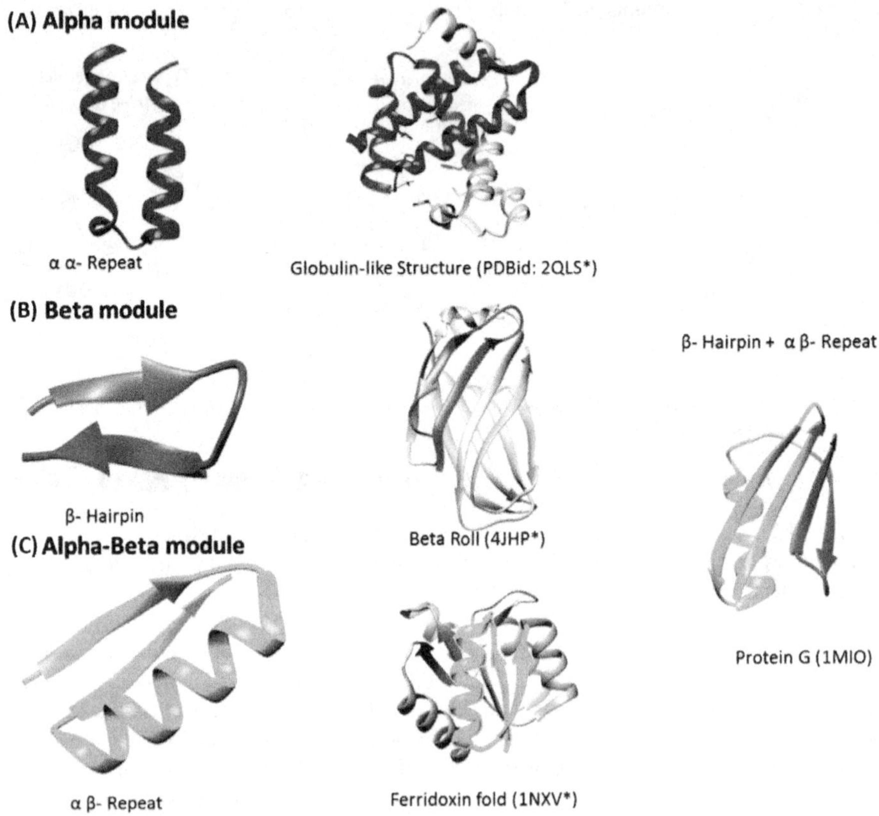

Figure 14.3 Repetitive units in protein structures. Graphical illustration of some of the frequently occurring structural repeats observed in protein structures among different classes. (A) Alpha–alpha repeat is a prominent structural module also found in the globulin-like structure, (B) beta-hairpin is one of the prevalent structural modules forming complex architectures like beta rolls, (C) alpha–beta repeat structural modules are present in ferredoxin fold. These structural modules can also present in a different combination, as in protein G, where both beta-hairpin and alpha–beta repeat modules form the structure. Representative proteins (truncated PDB) have been taken from RCSB (http://www.rcsb.org).

phosphate IsoMerase, a conserved metabolic enzyme with alpha-beta barrel)-barrel, ferridoxin fold, updown bundles, leucine-rich repeats] has been evidently observed. Higher organization of these repeats has the potential to build more complex proteins.

14.4 Scaffolds curve the functional and catalytic sites

Limited fold space indicates a plausible relationship between function and structure. Different folds can perform the same function, sometimes with the same catalytic cluster and mechanism; a result of convergent evolution among different folds for similar enzymatic functions (Bork et al., 1993; Irwin and Tan, 2014).

Extracting the precise biological role of protein from structure or sequence has been a complex task that drives most research interest (Skolnick and Gao, 2013; Tseng and Li, 2012; Smith and Hecht, 2011). The classical paradigm in structural biology is to investigate the structure in order to understand the biological function at a molecular level. But with the rapid increase in protein structure repository and from an applicability aspect, the current challenge is to extract useful biological and biochemical roles in the organism from a given three-dimensional structure. This is also useful in understanding evolution in protein structures and designing proteins of the desired function.

Structural homology has been an important toolkit for identifying or annotating the function of newer proteins, which are otherwise difficult to extract from sequences (Fry et al., 2009; Lee et al., 2010; Sillitoe et al., 2019). As structures have evolved more slowly than sequences, they can be used as structural relics to decipher protein evolution in a functional context. This was supported by the limitation in fold space, which suggests that evolution reuses similar folds to accomplish biological objectives (TIM barrel). But, the interplay between structure and function is not straight as other complex phenomena of biology. There are several glaring examples of different folds which perform the same function (trypsin and subtilisin). However, at this juncture, not a single approach or method is available which can solve the problem at hand. The limitation contributes broadly to the nature of data, that has been used in protein structure analysis, as structural data usually contains only information about the biochemical function of the protein. Its biological role in the cell or organism has multiple levels of complexity, which might be deciphered by combinations of diverse information of omics (like interactome, genome, and proteome) data (Mccafferty et al., 2020).

The interplay of protein structure and function is the reflection of billions of years of evolution, during which organisms have undergone multiple convergent and divergent evolution. This can be reflected by the underlying examples (Dellus-Gur et al., 2013; Martinez Cuesta et al., 2014; Nasir et al., 2014).

- **Same function different folds (convergent evolution):** Carbonic anhydrases (EC: 4.2.1.1) occur in two different folds such as left-handed beta-sheet (PDB id:1THJ) and flat beta-sheet (1DMX), however, performs same function (Fig. 14.4).

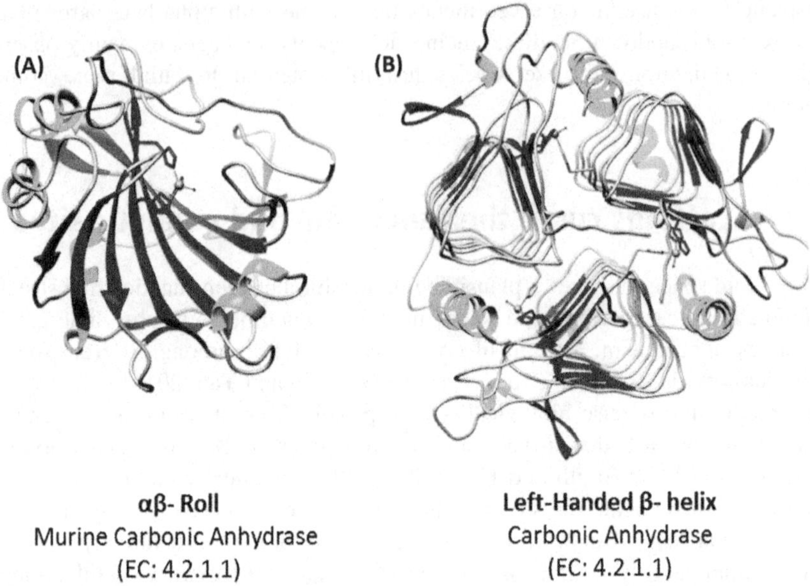

Figure 14.4 Enzymatic activity of carbonic anhydrase (same function) with different structure folds. The figure shows protein folds with (A) alpha–beta roll (PDB Id:1DMX:A) and (B) flat beta-sheet (PDB Id:1THJ) structure with left-handed beta-helix configuration.

- **Different function same fold (divergent evolution):** TIM barrel is a generic scaffold catalyzing 15 different enzymes. TIM barrel protein was reported as xylose isomerase, aldose reductase, enolase, and adenosine deaminase.

It was assumed that molecular recognition is mainly based on geometric and electrostatic complementarities, which are broadly referred to as functional sites. The catalytic site is a subset of the functional site (Dessailly et al., 2013; Redfern et al., 2008). The functional site includes locations (residues) in protein which are directly involved in function, that can be a binding site to other molecules (protein or small ligands like epitopes) and the catalytic site which performs the enzymatic activity. In most of the known proteins, interacting and enzymatic sites are present principally in different locations but nearby. The interacting and ligand binding sites are mainly situated on the protein surface whereas the catalytic site prefers to be in concave surfaces (or cavities) (Borgwardt et al., 2005; Feldman and Labute, 2010).

An extensive analysis of binding site properties suggests that there is a lack of complementarity between the ligand and binding surfaces. A study of the variation in the shape of the binding pocket and their ligands by spherical harmonic shape descriptor (Kahraman et al., 2010; Li et al., 2011) reports the relative size of active sites and substrates. According to the latest results, it can be concluded that binding pockets are on average three times larger than their bound-ligands, providing a "buffer-space" for the water molecule. This buffer zone was known to be complimenting the "entropic" loss of ligand and binding site residues by providing space for vibrational motion and flexibility inside the binding site. Comparing the shapes of binding pockets and their ligands, it has been observed that geometrical complementarity is not enough to drive specific molecular recognition, because of the presence of large variations in hydrophobic and electrostatic environments experienced by the ligands. A study on the binding of the same ligands to nonhomologous proteins suggests that the variation in the physicochemical property, as well as electrostatic, are driven by different functional requirements. For example:

- **Binding of heme type B:** Heme is bound by more than 20 different protein folds, which results in a diverse environment for heme binding. But, variation in electrostatic potential (ESP), describes the function-specific alteration in ligand binding surface environment. For structural complementarity, ESP around the propionate group should be positive, and that of around Fe should be negative. But, in case of bovine endothelial nitric oxide synthetase (eNOS), the ESP environment is observed to be reversed (Kahraman et al., 2010). eNOS has the function of creating nitric oxide, hence it must bind to L-arginine (substrate molecule). The positive ESP around Fe(II) in eNOS is thus advantageous for substrate-binding by increasing the binding affinity towards substrates compared to Fe(III) by 80 folds.
- **Binding of ATP:** Adenosine-5′-triphosphate has three phosphate groups which make ATP highly negatively charged. The surface complementarity for ATP binding in case of ESP is supposed to be positively charged. In most ATP-binding proteins positive ESP (avg 11.33 kcal/mole) has been observed mainly because of the coordination of metal ions (Kahraman et al., 2010). But, this metal ion coordination is absent in the ATP binding of DNA ligase and biotin carboxylase.

This suggests that the promiscuity in protein function is based on the variation in the local binding environment. The lack of physicochemical or shape complementarity of the ligand–protein binding phenomenon raises questions about basic docking methodologies. The convergent nature of evolution provides multiple binding solutions for the same ligand for function-specific roles.

14.4.1 Signature of catalytic site in protein structures

Enzymes accelerate the chemical reactions without undergoing any permanent chemical change, that is, accelerate the biological mechanism while maintaining fine structural specificity (i.e., stereospecific and region-specific, regioselectivity is used if a reaction gives a predominance of one of two products, a minor and major product. The reaction is regiospecific if one product is formed exclusively). The precise nature of enzymes is a result of their distinct catalytic sites which are different from protein regions to which substrates, cofactors, or metals bind. A guideline or set of rules that characterize catalytic site residues has been proposed based on the following points (for details, see Holliday et al., 2007; Holliday et al., 2014; Rahman et al., 2014):

1. Direct involvement in the catalytic mechanism.
2. Exerting effect (electrostatic or acid-base) on another resides (or water molecule) which is directly involved in the catalytic mechanism.

3. Stabilization of intermediates (transition-state).
4. Exerting an effect on a substrate or cofactor which aids in catalysis.

The above set of rules helps in the classification of enzyme function, which describes their role in the catalytic mechanism. Barlette et al. (Bartlett et al., 2002; Furnham et al., 2014) and later on Thornton group reported various aspects of catalytic residues involving their physicochemical, shape (Kahraman et al., 2010), local environment (Holliday et al., 2007; Holliday et al., 2014).

14.4.2 Protein function-based selection of topological space

Only fold is not sufficient, a protein must accomplish its objective, that is, function. For a topology to be selected over others, it must perform the function. We have analyzed protein function for prevalent and nonprevalent topologies using different annotation approaches, as reported by Khan and Ghosh (2015). It has been observed that the functional diversity score for prevalent topologies is comparatively higher than that of nonprevalent topologies. The prevalence of topology could be explained by its functional diversity. Literature evidence suggests that protein scaffolds engineered by natural selection optimizes the specificity and affinity of protein function (Engelhardt et al., 2011; Espinosa-Soto and Wagner, 2010; Rorick, 2012). Investigation of enzymes and protein scaffolds shows that the distribution of active site residues in different protein regions influences the capability of enzyme diversity.

In this section, we analyze the distribution of active site residues in protein scaffold and nonscaffold regions and compare them with topology variation. Enzyme annotation has been extracted from Catalytic Site Atlas (CSA v2) (Furnham et al., 2014) for proteins from the latest nonredundant dataset of all-alpha, all-beta, and alpha–beta proteins. The distribution of amino acids in enzymes with a fraction of nonscaffold residues in the active site has been illustrated in Fig. 14.5. As shown in Fig. 14.5A that for more than 40% (420 out of 964) of proteins, all active site residues are located in regular secondary structure elements or scaffold regions. Contributions of at least one nonscaffold residues in the active site are found in more than 55% of proteins. The fraction of enzymes with active site residues in only nonscaffold regions is observed to be very low (~0.5%). Analyzing enzymes from individual structural classes illustrates that the alpha–beta structure class has the greatest number of proteins as enzymes (63.4%) followed by alpha (28.9%) and beta (7.5%). The observed results are in accordance with the previously reported study of enzyme distribution in structure class (Martinez Cuesta et al., 2015). From Fig. 14.5, it can also be observed that for alpha–beta proteins and beta proteins, more than 60% of residues occur (Table 14.1) as nonscaffold residues in active sites. On the contrary for alpha helices, most of the active sites are placed in scaffold regions (~ >62%).

A comparative analysis of enzyme function diversity of topologies (grouped in different CATH architectures) has been shown in Table 14.1. Functional diversity refers to the number of different enzyme activity a protein performs with same

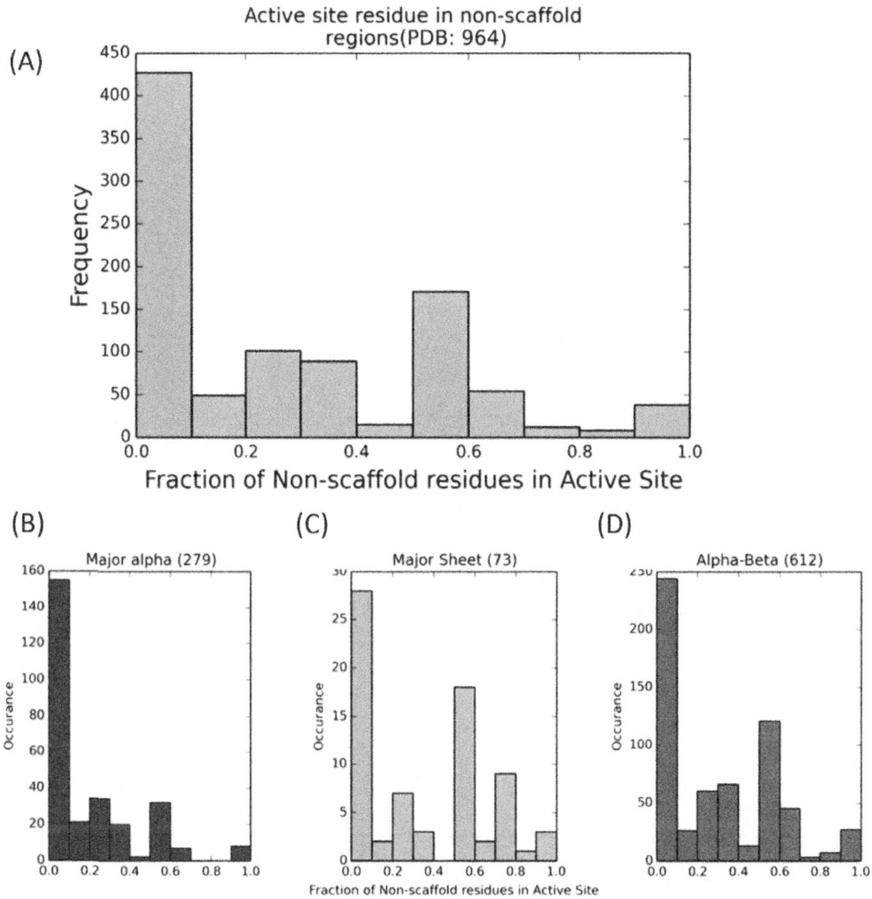

Figure 14.5 Distribution of nonscaffold residues in active site regions in protein structure space. Histogram shows, population of protein structures (y-axis: frequency/occurrence) w.r.t fraction of nonscaffold residues in active site (x-axis) of all structural class (A), the lower panel (B–D) illustrates the same individually in alpha, beta and alpha–beta protein classes.

structural architecture. The percentage of prevalent and nonprevalent topologies (Fig. 14.2B) in each architecture group has been calculated. From Table 14.1, it can be observed that mostly enzymes belong to alpha–beta proteins. A correlation between the fraction of nonscaffold residues in the active site and functional diversity for each structural class can be observed (Table 14.1). At the architecture level, three-layer sandwich and alpha–beta barrel structures have been reported to have maximum functional diversity (log(#EC) > 4) because of topologies like Rossman and TIM-barrel fold. This is in agreement with previously reported results of high functional diversity in TIM-barrels as compared to DHFR fold (Espinosa-Soto and Wagner, 2010).

Table 14.1 Protein functional diversity in topology space.

Class	Architecture (CATH id)	Percentage of Topology by Contact String		Log(#EC)	Fraction of Active site residues in nonscaffold	
		P[a]	Np[b]		Avg.	Std.
Alpha-beta	3-Layer(aba) Sandwich (3.40)	83.21	16.79	4.81	0.79	0.26
	Alpha-beta barrel (3.20)	87.14	12.86	4.07	0.67	0.27
	Alpha-beta complex (3.90)	73.4	26.6	3.63	0.73	0.3
	2-Layer sandwich (3.30)	82.33	17.67	3.25	0.61	0.38
	4-Layer sandwich (3.60)	83.33	16.67	2.7	0.63	0.21
	Roll (3.10)	76	24	2.56	0.75	0.1
	3-Layer (bba) sandwich (3.50)	66.66	33.34	1.09	0.36	0.45
Only beta	Sandwich (2.60)	61.22	38.78	2.83	0.87	0.24
	Beta barrel (2.40)	80.21	19.79	2.48	0.63	0.3
	Distorted sandwich (2.70)	54.1	45.9	1.6	0.77	0.17
	3 Solenoid (2.160)	23.12	76.88	1.6	0.54	0.17
	Beta complex (2.170)	56.2	43.8	1.38	0.68	0.21
	6-Propeller (2.120)	32.78	67.22	1.38	0.46	0.4
	Beta roll (2.30)	48.1	51.9	0.69	1	0
Only-alpha	Orthogonal bundle (1.10)	54.14	45.86	3.13	0.88	0.16
	Up-down bundle (1.20)	43.9	56.1	2.19	0.96	0.12
	Alpha-barrel (1.50)	21.2	78.8	1.6	0.93	0.1

Notes: Enzyme diversity (number of distinct enzymes in log scale Log(#EC)) for different observed (a) prevalent and (b) nonprevalent topology grouped into CATH architecture from the dataset used in the study. Average fraction (Avg.) and standard deviation (Std.) of nonscaffold residues in each protein active site for a CATH architecture has been shown in columns (Avg. and Std.).
Source: From Khan et al. (2018).

Comparing enzyme reaction with the objective of analyzing the local reaction environment of the protein scaffold, it has been observed that proteins with prevalent topology have low similarity scores for substructure similarity, bond changes, and reaction centers. This indicates that proteins with prevalent topology can accommodate diverse enzymatic activity as compared to the same in nonprevalent topology. Analysis of scaffold and nonscaffold regions of enzymes from the dataset indicates that topologies that have a high fraction of active site residues in nonscaffold regions are found to have high functional diversity, measure by number of diverse enzymatic activities. By mapping our topology to CATH architectures, we have found that prevalent topologies have highly represented in functionally diverse architecture. From the functional analysis, it has been shown that the prevalence of topology can be influenced by its ability to accommodate diverse functions.

The use of structural scaffolds for designing enzymes with new and diverse functionality has been reported by mimicking natural evolutionary processes (Richard, 2019; Gerlt and Babbitt, 2009). The reported prevalent topology sets acquired from big data of protein structure and function, can be used as candidate scaffolds to test the evolution of different functional activities as well as an engineer for more efficient or desired functionality.

14.4.3 Protein dynamics and transient sites

For cell survival, both folding and function are vital. It was long known that protein function is a result of dynamics, and the energy landscape, which is attributed to kinetic traps (Fersht, 2000), frustrations (Wensley et al., 2010, 2012), metastability and alternative folds (Gershenson et al., 2014). Attaining native configuration from a wide possibility space has been a perilous path for the protein chain. For a protein to be functional, different design principles encoded by nature have been identified that smooth the energy landscape and protect proteins from eventual misfolding and aggregation. The guided landscape enables proteins to avoid aggregation which is a leading cause of several precarious diseases (Valastyan and Lindquist, 2014). For example, amyloid accumulation can cause toxic effect and leads to Alzheimer's disease, Parkinson's disease, amyloid lateral sclerosis (ALS), spongiform encephalopathies, type 2 diabetes, cataract, etc.

Protein folding reaction has been probed by different experimental (Oliveberg and Wolynes, 2005; Scheraga et al., 2007; Wolynes et al., 2012) and computational approaches (Morris and Searle, 2012). Among experimental techniques, different spectroscopic methods have been used, and folding has been probed by monitoring reaction coordinates like overall secondary structure content by far-UV circular dichroism (Greenfield and Fasman, 1969), energy transfer between dipole−dipole interaction by Forster resonance energy transfer (Schuler and Eaton, 2008), etc. Investigating protein folding reactions by the computational study has been developed rapidly over the years aided by rapid improvement in computational infrastructures. Different dedicated supercomputers have been established (like Anton from D.E Shaw group or Fold@Home from Baker Lab), which deals with the computational complexity caused by the numerous degree of freedom of a polypeptide chain. With extensive computation time, only smaller proteins (less than 80 residues) are observed to be successfully folded (Kmiecik et al., 2012). However, with the use of coarse grain systems (high level of abstraction) protein folding reaction could also be investigated with much less computation and time investment. Over the years, a number of studies suggests that folding reaction can be studied efficiently and can be replicated using abstract models, which has been found to successfully reproduced previously established experimental results (Chen et al., 2006; Wathen and Jia, 2009; Lammert et al., 2009).

A reduced model of protein has been used to understand various aspects of protein structure and function (Banavar et al., 2004; Go, 1984; Dill, 1990). The challenge is to reduce the complexity of the system while maintaining the detail level. From the lattice model and tube model, it has been shown that secondary structures in proteins are substantially stabilized by chain compactness (Li et al., 2004). Based on the theory of "minimum frustration," structure-based models provide an abstract but feasible approach to examining protein folding kinetics (Noel et al., 2016). In this approach contact tracing among residues is the main guiding force to achieving starting structure and has been used to computationally mimic different well-known folding paradigms (Dabrowski-Tumanski and Sulkowska, 2017; Ljubetic et al., 2017; Sponer et al., 2018).

14.4.4 Learning methods for the prediction of proteins and functional sites

Over the last two decades, with an increase in computing power and significant strides in machine learning (ML) framework, the field of protein has also benefited like other fields of research (Shi et al., 2019; Tsuchiya and Tomii, 2020; Shirai and Terada, 2020). With the recent release of the biennial Critical Assessment of protein Structure Prediction (CASP, 2020) result, the artificial intelligent (AI)-based solution from DeepMind, AlphaFold has achieved significant progress in protein structure prediction (Senior et al., 2020). In subsequent competitions (2018 and 2020), the accuracy of de-novo structure prediction [measured in the Global Distance Test (GDT)] has been increased from approximately 59 to 87 GDT. Similarly, leveraging the diverse large amount of data from the functional assay in combination with structural data, several deep-learning applications have been developed to assess changes in protein dynamics with ligand binding as well as for allosteric changes (Tsuchiya and Tomii, 2020; Shimizu and Nakayama, 2020). Using publicly available data from resources like DrugBank (Wishart et al., 2008), Metador and SuperTarget (Gunther et al., 2008), Tsubaki et al. have reported a model combining graph neural network and convolutional neural network to predict protein-compound interaction (Tsubaki et al., 2019). Further, protein regions were weighed (neural attention) to estimate interaction strength and identify compound-binding sites. Using an autoencoder, unsupervised neural network, Tsuchiya et al., (2019), have shown the utility of learning algorithms by identifying allosteric changes in PDZ2 protein domain in PTPN13, induced by the binding of RAPGEF6. Integration of molecular dynamic data to train an autoencoder-based regression method has been used to extract subtle dynamic changes in apo and holo forms of the PDZ2 domain, which can be further developed to extract anomalies in protein dynamics as well as to predict disease states. In parallel, the development of "comparative perturbed-ensembles analysis," which explicitly uses protein dynamics from molecular dynamics (MD) simulations to explore the link between protein dynamics and biological function has been used to identify allosteric pathways in cyclophilin (Yao and Hamelberg, 2019). In summary, in the era of big data with sophisticated data-mining techniques in combination with learning models, complex biological problems are now approachable, even with limited resources. This envisages a new genre in drug discovery.

14.5 Protein interactive sites and designing of inhibitor

As mentioned above, the substrate/inhibitor interaction space can be analyzed thoroughly to design some novel inhibitor molecules to inhibit protein function. In the substrate binding region, incoming small ligand molecule engages with available complementary interactions projected from the protein side (Brooijmans and Kuntz, 2003) and the water-mediated buffer zone (Radoux et al., 2016; Goodford, 1985), this engagement helps the ligand molecule to recognize and stabilize the interaction upon binding. As discussed above, among the available large number of interaction sites in the substrate-binding region, only some subsites encompass high interaction energy and so-called hotspot regions of that binding site (Radoux et al., 2016). So, only targeting the characterized hotspot regions for a novel selective inhibitor designing process can drastically reduce the large interaction space. Hence, the required time and investment can also be reduced enormously. But it is of utmost importance to identify those hotspots from large interactive complementary spaces in the active site.

14.5.1 Interaction space exploration for energetically favorable binding features identification

Molecular interaction fields (MIFs) (Goodford, 1985) present a reliable way of finding the complementary interactions site by griding the protein site under consideration. Depending upon the protein type and binding site of interest, a small molecular probe-type can be employed for this process; however, finding a complementary site cannot be directly suited for inhibitor designing. To use the protein-specific MIF information more systematically, Ghosh and her team developed a Clique-based ab-initio pharmacophore modeling program named "CliquePharm" which can thoroughly analyze the MIFs of three probes for many similar classes of proteins to identify the ensemble of pharmacophore features required for the multi-class specific inhibitor design (Kaalia et al., 2016; Kumar et al., 2018). CliquePharm explores all possible high interaction MIF points across the selected proteins and is further guided by protein binding site-specific anchor residues and minimum and maximum interfeatures distance criteria. Developing ensembles of pharmacophore is a better approach considering the active site spans a large amount of free space and not just a catalytic site. The program can determine the pharmacophore models having the maximum number of features (Fig. 14.6), which can be further employed for novel inhibitor designing.

14.5.2 Protein dynamics guided binding features selection

Among the many factors, protein flexibility is one of the major determinants of its function (Teague, 2003). Many studies are reported in the literature suggesting the importance of protein flexibility in achieving the ligand molecule's selectivity and high affinity against protein (Damm and Carlson, 2007; Hornak and Simmerling, 2007).

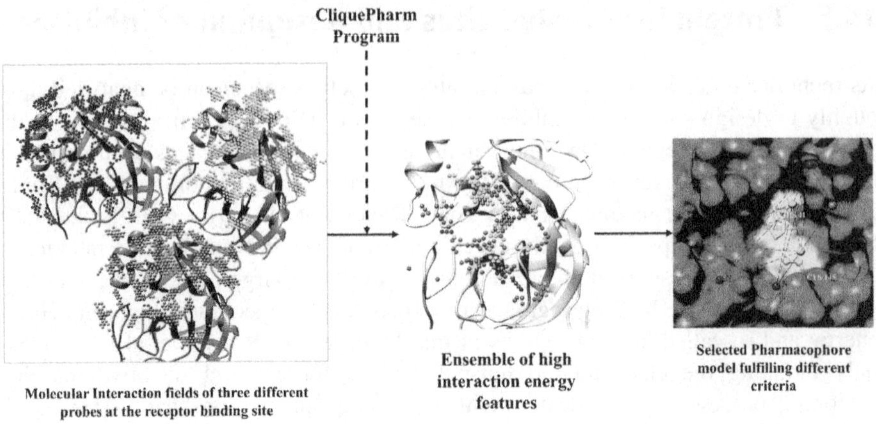

Figure 14.6 CliquePharm approach explores the protein binding site for energetically favorable region determination and pharmacophore feature modeling. SARS-CoV-2 Mpro target protein is illustrated here for receptor binding site-specific feature modeling. By implementing the CliquePharm approach, the seven features pharmacophore model is identified for Mpro protein occupying the major subsite regions required for novel inhibitor designing (Kumar and Ghosh 2020).
Source: From Kumar, P. & Ghosh, I., 2020. Molecular multi-target approach on COVID-19 for designing novel chemicals. In: Roy, K. (ed.) In Silico Modeling of Drugs Against Coronaviruses—Computational Tools and Protocols. Humana Press.

In one case study, *Plasmodium falciparum* ATPase6 (PfATPase6) also known as sarco/endoplasmic reticulum calcium-dependent ATPase (SERCA) of *P. falciparum* is one of the known probable target receptor protein of the antimalarial artemisinin (Haynes and Krishna, 2004; O'neill et al., 2010). In 2003, Krishna and his colleagues reported the possible molecular mechanism for this multidomain receptor protein (Eckstein-Ludwig et al., 2003). Considering the importance of this protein in the parasite developmental cycle (Ghartey-Kwansah et al., 2020; Arnou et al., 2011), a detailed mechanism of artemisinin binding and protein dynamics was elaborated by Shandilya et al. (2013). The study highlighted that only activated artemisinin (Fe-artemisinin) could disrupt the function of PfATPase6 by modulating the receptor state open (active form) to close (inactive form). This work provided the mechanistic view of the receptor dynamics and suggested that small molecules targeting only a specific subset of interactions can inhibit the protein's function among the all-available interactions in the receptor-binding site.

To further analyze the pharmacophore feature-driven interaction profile, an in-house protocol named dynamic pharmacophore (Dynaphore) docking has been developed and used for this purpose (Kumar, 2019). Dynaphore docking includes the flexibility of docking the different sizes and types of pharmacophore feature combinations in the defined receptor binding site and highlights the consistent interactions during the protein dynamics (Fig. 14.7). This method also suggests the minimum set of feature combinations required to achieve the desired effect.

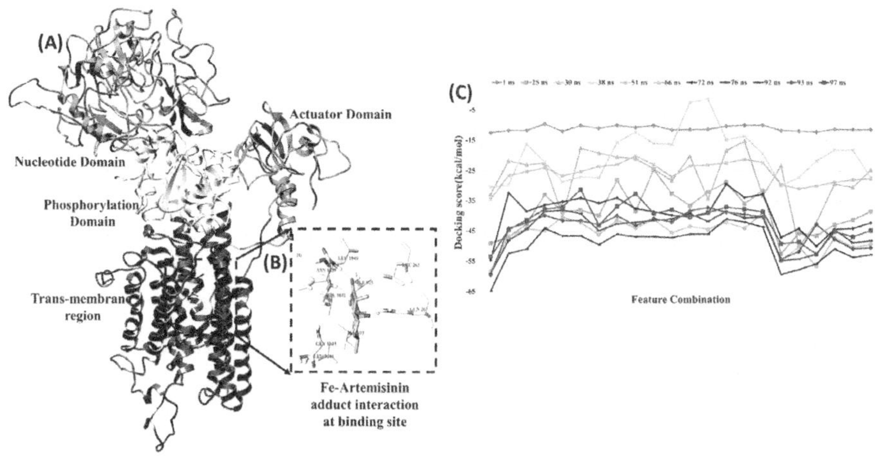

Figure 14.7 Multidomains PfATPase6 receptor protein with Artemisinin binding site at the transmembrane region (A and B). Developed pharmacophore features used for interaction profiling and highly interacting feature mapping (C).

By clustering, 11 representative structures of PfATPase6 are determined and used to understand the dynamic interaction profile by combinatorial docking of the 23 pharmacophore features (Kumar et al., 2017) and Fig. 14.7C shows the docking score for corresponding features set at the y-axis. This analysis emphasizes the role of domain movements and taking care of flexibility in the drug-binding site. The features score selectivity is seen when PfATPase6 changes its state from open to semiclose to finally closed state. In the close state PfATPase6, the six feature combinations have the highest dock score (−64.84 kcal/mol); however, a set of four feature combinations also give a very comparable score. So, from the pharmacophore docking-based interaction profile generation, critical features can be identified that contribute to receptor conformational changes. These recognized elements can be further used for prioritizing the in-silico screening of hits for novel inhibitor designing.

14.5.3 Protein flexibility and exploration of ligand recognition site

Protein flexibility is always a critical component of its function and also a significant player in understanding its interactions with other biomolecules (Amaral et al., 2017). The protein's dynamic nature can be facilitated to adopt different conformational states upon external stimuli such as small ligand binding (Richard, 2019; Teague, 2003; Gershenson et al., 2014) and modification of the protein's residue (Spicer and Davis, 2014). Advances in computational applications have enabled us to understand and examine the different possible protein−ligand interaction sites and the corresponding ligand interaction pathway at the atomic level. In one such study, casein kinase 2 (CK2) is examined to predict the different bound states of ellagic acid (small molecule) with the help of the supervised molecular dynamics (SuMD) technique (Sabbadin and Moro, 2014). Starting from the unbound CK2 receptor state, SuMD has identified the ellagic acid recognition pathways along with stable and metastable binding sites characterized by enthalpic and configurational entropic components (Panday et al., 2019). This study suggests all possible binding states for ellagic acid guided from favorable interactions to the most stable binding site, hence exploring many ligand recognition possibilities.

14.5.4 Artificial intelligence to understand the interactions of protein and chemical

The drug discovery pipeline is very resource-dependent involving large-scale money and time investment (Paul et al., 2010). Many new computational tools/techniques have emerged with time, especially the ML/deep learning (DL)-based approaches to predict highly reliable inhibitor candidates for further experimental validation (Chan et al., 2019; Ou-Yang et al., 2012). The exponential growth of many dimensional experimental data in biology has opened the immense possibility of ML/DL-driven drug repurposing (Aliper et al., 2016; Zhou et al., 2020) or novel candidate search with more accuracy and scalability than classical computational approaches (Ekins et al., 2019; Chan et al., 2019) thus, also improving the required time and cost-efficiency. Due to the applicability advantages of ML-driven methods (Chan et al., 2019), tedious processes can now be automatized and optimized to substantially speed up the different stages of the drug discovery pipeline (Mamoshina et al., 2016).

The speedy growth of the different chemical databases such as PubChem (Kim et al., 2019), ChEMBL (Mendez et al., 2019) has facilitated developing ML models predicting other bioactivities such as large-scale structure−activity relationship (SAR) for various target proteins (Tyrchan and Evertsson, 2017), oral exposure (Leach et al., 2006), distribution coefficient (logD) (Warner et al., 2010; Keefer et al., 2011), ADME and toxicity properties and mode of action (Schonherr and Cernak, 2013). The potential application of the ML models is not only limited to the extensive data-driven bioactivity prediction modeling; many states of art models are now reported in the literature emphasizing the utility of the ML/DL in de-novo drug discovery (Zhavoronkov et al., 2019; Kadurin et al., 2017; Putin et al., 2018; Segler et al., 2018). DL models are expanding and boosting the potential application of computer-aided discovery, for example, finding novel candidates from the synthetic chemical library or de-novo candidate design (Zhavoronkov et al., 2019; Mamoshina et al., 2016), reaction mechanism elucidation (Segler and Waller, 2017), chemical synthesis accessibility planning (Coley et al., 2018), protein three-dimensional -structure prediction (Senior et al., 2020) and drug target identification (Jimenez et al., 20f19). Very recently, with the help of the DL algorithm developed by Google offshoot DeepMind has made a gigantic step in predicting the protein three-dimensional structure with near to the experiment accuracy (https://www.nature.com/articles/d41586-020-03348-4).

Currently, the Coronavirus disease-2019 (COVID-19) pandemic situation has affected more than 226 million lives with 4.66 million death and is still a serious threat to humankind (https://covid19.who.int/ September 17, 2021). To search for novel solutions to fight against this pandemic situation, much computational and experimental research has been conducted worldwide (Chahrour et al., 2020). Along with conventional in silico approaches, ML/DL-based models are also reported in the literature to address the drug repurposing opportunity (Zhou et al., 2020; Batra et al., 2020; Zeng et al., 2020). This suggests that despite the availability of less experimental data, different level of data integration can also be used to develop reliable ML models. Apart from drug repurposing, ML is also used to quickly detect the COVID-19 infection by analyzing chest X-ray images (Jain et al., 2020b, Yoo et al., 2020; Togacar et al., 2020; Jain et al., 2020a).

14.6 Intrinsically unstructured regions and protein function

With the advance in protein research, it has been noted that certain regions of protein do not form any regular known structural patterns but remain unstructured in a normal functional state (Wright and Dyson, 2015; Meszaros et al., 2018; Meszaros et al., 2009). In case of disordered regions, the amino acid composition contributes negatively toward the formation of such interresidue contacts. Mostly, a significant presence of small, hydrophilic and proline residues, introduces a conformational uncertainty in these regions (Dosztanyi et al., 2005; Meszaros et al., 2009). A number of resources explore this compositional bias in amino acid composition [IUPred (Meszaros et al., 2018), ANCHOR]. With the advance in protein research, it has been noted that certain regions of protein do not form any regular known structural patterns but remain unstructured in a normal functional state (Wright and Dyson, 2015; Meszaros et al., 2018; Meszaros et al., 2009). In case of disordered regions (IDRs) or proteins (IDP), the amino acid composition contribute negatively toward the formation of interresidue contacts, which is crucial for retaining a conformational state. Mostly, a significant presence of small, hydrophilic and proline residues, introduces a conformational uncertainty in these regions (Dosztanyi et al., 2005; Meszaros et al., 2009). Indeed, this group of proteins challenges the classic sequence-structure–function relationship, hence numerous approaches have been developed for the functional characterization of these proteins (Atkins et al., 2015; Van Der Lee et al., 2014; Liu et al., 2019; Wallmann and Kesten, 2020). A mechanism like gene duplication is reported to be maintaining the distribution of disorder regions in the genome (Yruela et al., 2018). Studies on plant proteome reveal that amount of paralogous intrinsically disorder regions has a positive correlation with the number of chromosomes (Yruela et al., 2018; Wallmann and Kesten, 2020). IDP is found to be providing a complex framework to deal with stress conditions (Van Der Lee et al., 2014). Due to their rapidly modulating flexibility, IDPs can cater to multiple binding sites to different binding partners and hence serves as a central hub in the overall interaction network (Kim et al., 2008; Tompa and Fuxreiter, 2008; Wright and Dyson, 2015). Although, IDP and IDRs have been successfully identified in a number of functional pathways (Wallmann and Kesten, 2020), like late embryogenesis abundant response to stress conditions (Magwanga et al., 2018), dynamic liquid–liquid phase separation for P-bodies or nucleolus formation (Bergeron-Sandoval et al., 2016), microtubule organization (Mukrasch et al., 2009), cryptochrome signaling (Czarna et al., 2013), regulators in cell cycle progression (Grimmler et al., 2007) as well as transcriptional regulation (Bourbousse et al., 2018), their much functional space remained untapped.

14.7 Conclusions

Protein sequence, length of 300 amino acids (only 20 variants) having many diverse topologies can encompass as estimated theoretically a big data space, however in reality to achieve selective and diverse biological functions whole of the space are not available. Moreover, the shape of soluble proteins has domain structures and prevails to accommodate active sites where major chemical reactions/interactions happen. This smart way of reduction of complex space based on functional application has allowed researchers to map the chemical interaction sites of proteins with their complementary ligands, driven by interactions. The present review describes such diversity in topology and interactions evolving from structural diversity of both protein and ligands. Many membrane bound/associated proteins also take part in major biological functions and provide another large scale of topology to enumerate, not included here. In recent years drug designing for the disease has been explored from the disease point of view, that is, finding the relevant target(s) and pathway(s) causing the disease, then finding the chemicals which may disrupt interactions and decelerate the disease. Recently available combinatorial libraries of molecules/chemicals are also really vast, with 120 million drug-like purchasable molecules (Sterling and Irwin, 2015). In addition, the computationally estimated resource has been generated using up to 17 C, N, O, S, and/or halogen atoms with less than 300 Da and a virtually enumerated library available containing more than 166 billion compounds (Reymond, 2015), this opens up an enormous chemical space available to explore for new compounds using interactive pharmacophore approach as described above. Natural compounds are another exploratory space for searching ligands because most of them are modified/exact metabolites of many biochemical pathways in plants and animals, which are the machinery to function cellular biology in living systems. Diversity and selectivity are often balanced for living systems to perform their functions like growth, sustainment and interaction with environments. In this perspective large data explosion may play a better selection in survival; generating, understanding, and interpreting such a large system at a molecular level will drive drug designing research in future.

Acknowledgments

PK would like to acknowledge the NSM-funded PDB-India project for financial support and the National Institute of Immunology (NII) for the computational facility. Shailesh Pandey, PhD, is acknowledged for sharing his research publication for inclusion in this review.

References

Adamian, L., Liang, J., 2001. Helix-helix packing and interfacial pairwise interactions of residues in membrane proteins. J. Mol. Biol. 311, 891–907.

Aimi, J., Qiu, H., Williams, J., Zalkin, H., Dixon, J.E., 1990. De novo purine nucleotide biosynthesis: cloning of human and avian cDNAs encoding the trifunctional glycinamide ribonucleotide synthetase-aminoimidazole ribonucleotide synthetase-glycinamide ribonucleotide transformylase by functional complementation in *E. coli*. Nucleic Acids Res. 18, 6665–6672.

Aliper, A., Plis, S., Artemov, A., Ulloa, A., Mamoshina, P., Zhavoronkov, A., 2016. Deep learning applications for predicting pharmacological properties of drugs and drug repurposing using transcriptomic data. Mol. Pharm. 13, 2524–2530.

Amaral, M., Kokh, D.B., Bomke, J., Wegener, A., Buchstaller, H.P., Eggenweiler, H.M., et al., 2017. Protein conformational flexibility modulates kinetics and thermodynamics of drug binding. Nat. Commun. 8, 2276.

Andreeva, A., Murzin, A.G., 2010. Structural classification of proteins and structural genomics: new insights into protein folding and evolution. Acta Crystallogr. Sect. F. Struct. Biol. Cryst. Commun. 66, 1190–1197.

Andreeva, A., Howorth, D., Chothia, C., Kulesha, E., Murzin, A.G., 2015. Investigating protein structure and evolution with SCOP2. Curr. Protoc. Bioinforma. 49, 1261–12621.

Andreeva, A., Kulesha, E., Gough, J., Murzin, A.G., 2020. The SCOP database in 2020: expanded classification of representative family and superfamily domains of known protein structures. Nucleic Acids Res. 48, D376–D382.

Arnou, B., Montigny, C., Morth, J.P., Nissen, P., Jaxel, C., Moller, J.V., et al., 2011. The *Plasmodium falciparum* Ca(2 +)-ATPase PfATP6: insensitive to artemisinin, but a potential drug target. Biochem. Soc. Trans. 39, 823–831.

Atkins, J.D., Boateng, S.Y., Sorensen, T., Mcguffin, L.J., 2015. Disorder prediction methods, their applicability to different protein targets and their usefulness for guiding experimental studies. Int. J. Mol. Sci. 16, 19040–19054.

Banavar, J.R., Cieplak, M., Maritan, A., 2004. Lattice tube model of proteins. Phys. Rev. Lett. 93, 238101.

Bartlett, G.J., Porter, C.T., Borkakoti, N., Thornton, J.M., 2002. Analysis of catalytic residues in enzyme active sites. J. Mol. Biol. 324, 105–121.

Batra, R., Chan, H., Kamath, G., Ramprasad, R., Cherukara, M.J., Sankaranarayanan, S., 2020. Screening of therapeutic agents for COVID-19 using machine learning and ensemble docking studies. J. Phys. Chem. Lett. 11, 7058–7065.

Bergeron-Sandoval, L.P., Safaee, N., Michnick, S.W., 2016. Mechanisms and consequences of macromolecular phase separation. Cell 165, 1067–1079.

Borgwardt, K.M., Ong, C.S., Schonauer, S., Vishwanathan, S.V., Smola, A.J., Kriegel, H.P., 2005. Protein function prediction via graph kernels. Bioinformatics 21 (Suppl 1), i47–i56.

Bork, P., Sander, C., Valencia, A., 1993. Convergent evolution of similar enzymatic function on different protein folds: the hexokinase, ribokinase, and galactokinase families of sugar kinases. Protein Sci. 2, 31–40.

Bourbousse, C., Vegesna, N., Law, J.A., 2018. SOG1 activator and MYB3R repressors regulate a complex DNA damage network in Arabidopsis. Proc. Natl Acad. Sci. USA 115, E12453–E12462.

Brenner, S.E., Chothia, C., Hubbard, T.J., 1997. Population statistics of protein structures: lessons from structural classifications. Curr. Opin. Struct. Biol. 7, 369–376.

Brooijmans, N., Kuntz, I.D., 2003. Molecular recognition and docking algorithms. Annu. Rev. Biophys. Biomol. Struct. 32, 335–373.

Cao, C., Wang, G., Liu, A., Xu, S., Wang, L., Zou, S., 2016. A new secondary structure assignment algorithm using calpha backbone fragments. Int. J. Mol. Sci. 17, 333.

Chahrour, M., Assi, S., Bejjani, M., Nasrallah, A.A., Salhab, H., Fares, M., et al., 2020. A bibliometric analysis of COVID-19 research activity: a call for increased output. Cureus 12, e7357.

Chan, H.S., Dill, K.A., 1990. Origins of structure in globular proteins. Proc. Natl Acad. Sci. USA 87, 6388–6392.

Chen, M., Wilson, C.J., Wu, Y., Wittung-Stafshede, P., Ma, J., 2006. Correlation between protein stability cores and protein folding kinetics: a case study on *Pseudomonas aeruginosa* apo-azurin. Structure 14, 1401–1410.

Chan, H.C.S., Shan, H., Dahoun, T., Vogel, H., Yuan, S., 2019. Advancing drug discovery via artificial intelligence. Trends Pharmacol. Sci. 40, 592–604.

Coley, C.W., Green, W.H., Jensen, K.F., 2018. Machine learning in computer-aided synthesis planning. Acc. Chem. Res. 51, 1281–1289.

Consortium, W., 2019. Protein Data Bank: the single global archive for 3D macromolecular structure data. Nucleic Acids Res. 47, D520–D528.

Cook, C.E., Stroe, O., Cochrane, G., Birney, E., Apweiler, R., 2020. The European Bioinformatics Institute in 2020: building a global infrastructure of interconnected data resources for the life sciences. Nucleic Acids Res. 48, D17–D23.

Copley, S.D., 2020. Evolution of new enzymes by gene duplication and divergence. FEBS J. 287, 1262–1283.

Czarna, A., Berndt, A., Singh, H.R., Grudziecki, A., Ladurner, A.G., Timinszky, G., et al., 2013. Structures of Drosophila cryptochrome and mouse cryptochrome1 provide insight into circadian function. Cell 153, 1394–1405.

D'argenio, V., 2018. The High-Throughput analyses era: are we ready for the data struggle? High. Throughput 7.

Dabrowski-Tumanski, P., Sulkowska, J.I., 2017. Topological knots and links in proteins. Proc. Natl Acad. Sci. USA 114, 3415–3420.

Damm, K.L., Carlson, H.A., 2007. Exploring experimental sources of multiple protein conformations in structure-based drug design. J. Am. Chem. Soc. 129, 8225–8235.

Davidi, D., Longo, L.M., Jablonska, J., Milo, R., Tawfik, D.S., 2018. A bird's-eye view of enzyme evolution: chemical, physicochemical, and physiological considerations. Chem. Rev. 118, 8786–8797.

De Lima Morais, D.A., Fang, H., Rackham, O.J., Wilson, D., Pethica, R., Chothia, C., et al., 2011. SUPERFAMILY 1.75 including a domain-centric gene ontology method. Nucleic Acids Res. 39, D427–D434.

De Queiroz, K., Gauthier, J., 1994. Toward a phylogenetic system of biological nomenclature. Trends Ecol. Evol. 9, 27–31.

Dellus-Gur, E., Toth-Petroczy, A., Elias, M., Tawfik, D.S., 2013. What makes a protein fold amenable to functional innovation? Fold polarity and stability trade-offs. J. Mol. Biol. 425, 2609–2621.

Dessailly, B.H., Dawson, N.L., Mizuguchi, K., Orengo, C.A., 2013. Functional site plasticity in domain superfamilies. Biochim. Biophys. Acta 1834, 874–889.

Dill, K.A., 1990. Dominant forces in protein folding. Biochemistry 29, 7133–7155.

Dosztanyi, Z., Csizmok, V., Tompa, P., Simon, I., 2005. The pairwise energy content estimated from amino acid composition discriminates between folded and intrinsically unstructured proteins. J. Mol. Biol. 347, 827–839.

Eckhardt, M., Hultquist, J.F., Kaake, R.M., Huttenhain, R., Krogan, N.J., 2020. A systems approach to infectious disease. Nat. Rev. Genet. 21, 339–354.

Eckstein-Ludwig, U., Webb, R.J., Van Goethem, I.D., East, J.M., Lee, A.G., Kimura, et al., 2003. Artemisinins target the SERCA of *Plasmodium falciparum*. Nature 424, 957–961.

Ekins, S., Puhl, A.C., Zorn, K.M., Lane, T.R., Russo, D.P., Klein, J.J., et al., 2019. Exploiting machine learning for end-to-end drug discovery and development. Nat. Mater. 18, 435–441.

Engelhardt, B.E., Jordan, M.I., Srouji, J.R., Brenner, S.E., 2011. Genome-scale phylogenetic function annotation of large and diverse protein families. Genome Res. 21, 1969–1980.

Espinosa-Soto, C., Wagner, A., 2010. Specialization can drive the evolution of modularity. PLoS Comput. Biol. 6, e1000719.

Feldman, H.J., Labute, P., 2010. Pocket similarity: are alpha carbons enough? J. Chem. Inf. Model. 50, 1466–1475.

Fersht, A.R., 2000. Transition-state structure as a unifying basis in protein-folding mechanisms: contact order, chain topology, stability, and the extended nucleus mechanism. Proc. Natl Acad. Sci. U S A 97, 1525–1529.

Fleming, P.J., Gong, H., Rose, G.D., 2006. Secondary structure determines protein topology. Protein Sci. 15, 1829–1834.

Fry, B.G., Roelants, K., Champagne, D.E., Scheib, H., Tyndall, J.D., King, G.F., et al., 2009. The toxicogenomic multiverse: convergent recruitment of proteins into animal venoms. Annu. Rev. Genomics Hum. Genet. 10, 483–511.

Furnham, N., Holliday, G.L., De Beer, T.A., Jacobsen, J.O., Pearson, W.R., Thornton, J.M., 2014. The Catalytic Site Atlas 2.0: cataloging catalytic sites and residues identified in enzymes. Nucleic Acids Res. 42, D485–D489.

Gerlt, J.A., Babbitt, P.C., 2009. Enzyme (re)design: lessons from natural evolution and computation. Curr. Opin. Chem. Biol. 13, 10–18.

Gershenson, A., Gierasch, L.M., Pastore, A., Radford, S.E., 2014. Energy landscapes of functional proteins are inherently risky. Nat. Chem. Biol. 10, 884–891.

Ghartey-Kwansah, G., Yin, Q., Li, Z., Gumpper, K., Sun, Y., Yang, R., et al., 2020. Calcium-dependent protein kinases in malaria parasite development and infection. Cell Transpl. 29, 963689719884888.

Go, N., 1984. The consistency principle in protein structure and pathways of folding. Adv. Biophys. 18, 149–164.

Goodford, P.J., 1985. A computational procedure for determining energetically favorable binding sites on biologically important macromolecules. J. Med. Chem. 28, 849–857.

Gordeev, A.B., Efimov, A.V., 2013. Modeling of folds and folding pathways for some protein families of (alpha + beta)- and (alpha/beta)-classes. J. Biomol. Struct. Dyn. 31, 4–16.

Govindarajan, S., Goldstein, R.A., 1996. Why are some proteins structures so common? Proc. Natl Acad. Sci. USA 93, 3341–3345.

Graham, B.S., Gilman, M.S.A., Mclellan, J.S., 2019. Structure-based vaccine antigen design. Annu. Rev. Med. 70, 91–104.

Grainger, B., Sadowski, M.I., Taylor, W.R., 2010. Re-evaluating the "rules" of protein topology. J. Comput. Biol. 17, 1371–1384.

Greenfield, N., Fasman, G.D., 1969. Computed circular dichroism spectra for the evaluation of protein conformation. Biochemistry 8, 4108–4116.

Grimmler, M., Wang, Y., Mund, T., Cilensek, Z., Keidel, E.M., Waddell, M.B., et al., 2007. Cdk-inhibitory activity and stability of p27Kip1 are directly regulated by oncogenic tyrosine kinases. Cell 128, 269–280.

Gunther, S., Kuhn, M., Dunkel, M., Campillos, M., Senger, C., Petsalaki, E., et al., 2008. SuperTarget and matador: resources for exploring drug-target relationships. Nucleic Acids Res. 36, D919–D922.

Han, J.H., Batey, S., Nickson, A.A., Teichmann, S.A., Clarke, J., 2007. The folding and evolution of multidomain proteins. Nat. Rev. Mol. Cell Biol. 8, 319–330.

Hanson, B., Westin, C., Rosa, M., Grier, A., Osipovitch, M., Macdonald, M.L., et al., 2014. Estimation of protein function using template-based alignment of enzyme active sites. BMC Bioinforma. 15, 87.

Haynes, R.K., Krishna, S., 2004. Artemisinins: activities and actions. Microbes Infect. 6, 1339–1346.

Holliday, G.L., Almonacid, D.E., Mitchell, J.B., Thornton, J.M., 2007. The chemistry of protein catalysis. J. Mol. Biol. 372, 1261–1277.

Holliday, G.L., Rahman, S.A., Furnham, N., Thornton, J.M., 2014. Exploring the biological and chemical complexity of the ligases. J. Mol. Biol. 426, 2098–2111.

Hornak, V., Simmerling, C., 2007. Targeting structural flexibility in HIV-1 protease inhibitor binding. Drug. Discov. Today 12, 132–138.

Irwin, D.M., Tan, H., 2014. Evolution of glucose utilization: glucokinase and glucokinase regulator protein. Mol. Phylogenet Evol. 70, 195–203.

Jain, G., Mittal, D., Thakur, D., Mittal, M.K., 2020a. A deep learning approach to detect Covid-19 coronavirus with X-Ray images. Biocybern. Biomed. Eng. 40, 1391–1405.

Jain, R., Gupta, M., Taneja, S., Hemanth, D.J., 2020b. Deep learning based detection and analysis of COVID-19 on chest X-ray images. Appl. Intell. .

Jimenez, J., Sabbadin, D., Cuzzolin, A., Martinez-Rosell, G., Gora, J., Manchester, J., et al., 2019. PathwayMap: molecular pathway association with self-normalizing neural networks. J. Chem. Inf. Model. 59, 1172–1181.

Kaalia, R., Srinivasan, A., Kumar, A., Ghosh, I., 2016. ILP-assisted de novo drug design. Mach. Learn. 103, 309–341.

Kabsch, W., Sander, C., 1983. Dictionary of protein secondary structure: pattern recognition of hydrogen-bonded and geometrical features. Biopolymers 22, 2577–2637.

Kadurin, A., Aliper, A., Kazennov, A., Mamoshina, P., Vanhaelen, Q., Khrabrov, K., et al., 2017. The cornucopia of meaningful leads: applying deep adversarial autoencoders for new molecule development in oncology. Oncotarget 8, 10883–10890.

Kahraman, A., Morris, R.J., Laskowski, R.A., Favia, A.D., Thornton, J.M., 2010. On the diversity of physicochemical environments experienced by identical ligands in binding pockets of unrelated proteins. Proteins 78, 1120–1136.

Karanicolas, J., Brooks 3rd, C.L., 2002. The origins of asymmetry in the folding transition states of protein L and protein G. Protein Sci. 11, 2351–2361.

Kasson, P.M., 2020. Infectious disease research in the era of big data. Annu. Rev. Biomed. Data Sci. 3.

Keefer, C.E., Chang, G., Kauffman, G.W., 2011. Extraction of tacit knowledge from large ADME data sets via pairwise analysis. Bioorg Med. Chem. 19, 3739–3749.

Khan, T., Ghosh, I., 2015. Modularity in protein structures: study on all-alpha proteins. J. Biomol. Struct. Dyn. 33, 2667–2681.

Khan, T., Panday, S.K., Ghosh, I., 2018. ProLego: tool for extracting and visualizing topological modules in protein structures. BMC Bioinforma. 19, 167.

Khersonsky, O., Lipsh, R., Avizemer, Z., Ashani, Y., Goldsmith, M., Leader, H., et al., 2018. Automated design of efficient and functionally diverse enzyme repertoires. Mol. Cell 72, 178–186. e5.

Kim, P.M., Sboner, A., Xia, Y., Gerstein, M., 2008. The role of disorder in interaction networks: a structural analysis. Mol. Syst. Biol. 4, 179.

Kim, S., Chen, J., Cheng, T., Gindulyte, A., He, J., He, S., et al., 2019. PubChem 2019 update: improved access to chemical data. Nucleic Acids Res. 47, D1102–D1109.

King, N.P., Lai, Y.T., 2013. Practical approaches to designing novel protein assemblies. Curr. Opin. Struct. Biol. 23, 632–638.

Kmiecik, S., Gront, D., Kouza, M., Kolinski, A., 2012. From coarse-grained to atomic-level characterization of protein dynamics: transition state for the folding of B domain of protein A. J. Phys. Chem. B 116, 7026–7032.

Koehl, P., Levitt, M., 2002. Sequence variations within protein families are linearly related to structural variations. J. Mol. Biol. 323, 551–562.

Kolodny, R., Pereyaslavets, L., Samson, A.O., Levitt, M., 2013. On the universe of protein folds. Annu. Rev. Biophys. 42, 559–582.

Krishna, S.S., Grishin, N.V., 2005. Structural drift: a possible path to protein fold change. Bioinformatics 21, 1308–1310.

Kühlbrandt, W., 2014. The resolution revolution. Science 343, 1443–1444.

Kumar, P., 2019. Design and Validation of novel antimalarials using in silico methods. PhD, Jawaharlal Nehru University.

Kumar, P., Ghosh, I., 2020. Molecular multi-target approach on COVID-19 for designing novel chemicals. In: Roy, K. (Ed.), In Silico Modeling of Drugs Against Coronaviruses - Computational Tools and Protocols. Humana Press.

Kumar, P., Shandilya, A., Jayaram, B., Ghosh, I., 2017. Integrative method for finding antimalarials using in silico approach. In: Kholmurodov, K. (Ed.), Computer Design for New Drugs and Materials: Molecular Dynamics of Nanoscale Phenomena. Nova Science Publishers.

Kumar, P., Kaalia, R., Srinivasan, A., Ghosh, I., 2018. Multiple target-based pharmacophore design from active site structures. SAR. QSAR Env. Res. 29, 1–19.

Lai, Y.T., King, N.P., Yeates, T.O., 2012. Principles for designing ordered protein assemblies. Trends Cell Biol. 22, 653–661.

Lammert, H., Schug, A., Onuchic, J.N., 2009. Robustness and generalization of structure-based models for protein folding and function. Proteins 77, 881–891.

Lapenta, F., Jerala, R., 2020. Design of novel protein building modules and modular architectures. Curr. Opin. Struct. Biol. 63, 90–96.

Leach, A.G., Jones, H.D., Cosgrove, D.A., Kenny, P.W., Ruston, L., Macfaul, P., et al., 2006. Matched molecular pairs as a guide in the optimization of pharmaceutical properties; a study of aqueous solubility, plasma protein binding and oral exposure. J. Med. Chem. 49, 6672–6682.

Lee, D.A., Rentzsch, R., Orengo, C., 2010. GeMMA: functional subfamily classification within superfamilies of predicted protein structural domains. Nucleic Acids Res. 38, 720–737.

Li, H., Fast, W., Benkovic, S.J., 2009. Structural and functional modularity of proteins in the de novo purine biosynthetic pathway. Protein Sci. 18, 881–892.

Li, Z.R., Han, X., Liu, G.R., 2004. Protein designability analysis in sequence principal component space using 2D lattice model. Comput. Meth. Prog. Biomed. 76, 21–29.

Li, T., Bonkovsky, H.L., Guo, J.T., 2011. Structural analysis of heme proteins: implications for design and prediction. BMC Struct. Biol. 11, 13.

Lindorff-Larsen, K., Rogen, P., Paci, E., Vendruscolo, M., Dobson, C.M., 2005. Protein folding and the organization of the protein topology universe. Trends Biochem. Sci. 30, 13–19.

Liu, Y., Chen, S., Wang, X., Liu, B., 2019. Identification of intrinsically disordered proteins and regions by length-dependent predictors based on conditional random fields. Mol. Ther. Nucleic Acids 17, 396–404.

Ljubetic, A., Lapenta, F., Gradisar, H., Drobnak, I., Aupic, J., Strmsek, Z., et al., 2017. Design of coiled-coil protein-origami cages that self-assemble in vitro and in vivo. Nat. Biotechnol. 35, 1094–1101.

Lorenz, D.M., Jeng, A., Deem, M.W., 2011. The emergence of modularity in biological systems. Phys. Life Rev. 8, 129–160.

Magwanga, R.O., Lu, P., Kirungu, J.N., Lu, H., Wang, X., Cai, X., et al., 2018. Characterization of the late embryogenesis abundant (LEA) proteins family and their role in drought stress tolerance in upland cotton. BMC Genet. 19, 6.

Mamoshina, P., Vieira, A., Putin, E., Zhavoronkov, A., 2016. Applications of deep learning in biomedicine. Mol. Pharm. 13, 1445–1454.

Marchler-Bauer, A., Derbyshire, M.K., Gonzales, N.R., Lu, S., Chitsaz, F., Geer, L.Y., et al., 2015. CDD: NCBI's conserved domain database. Nucleic Acids Res. 43, D222–D226.

Marks, D.S., Hopf, T.A., Sander, C., 2012. Protein structure prediction from sequence variation. Nat. Biotechnol. 30, 1072–1080.

Martin, J., Letellier, G., Marin, A., Taly, J.F., De Brevern, A.G., Gibrat, J.F., 2005. Protein secondary structure assignment revisited: a detailed analysis of different assignment methods. BMC Struct. Biol. 5, 17.

Martinez Cuesta, S., Furnham, N., Rahman, S.A., Sillitoe, I., Thornton, J.M., 2014. The evolution of enzyme function in the isomerases. Curr. Opin. Struct. Biol. 26, 121–130.

Martinez Cuesta, S., Rahman, S.A., Furnham, N., Thornton, J.M., 2015. The classification and evolution of enzyme function. Biophys. J. 109, 1082–1086.

Mccafferty, C.L., Verbeke, E.J., Marcotte, E.M., Taylor, D.W., 2020. Structural biology in the multi-omics era. J. Chem. Inf. Model. 60, 2424–2429.

Mendez, D., Gaulton, A., Bento, A.P., Chambers, J., De Veij, M., Felix, et al., 2019. ChEMBL: towards direct deposition of bioassay data. Nucleic Acids Res. 47, D930–D940.

Meszaros, B., Simon, I., Dosztanyi, Z., 2009. Prediction of protein binding regions in disordered proteins. PLoS Comput. Biol. 5, e1000376.

Meszaros, B., Erdos, G., Dosztanyi, Z., 2018. IUPred2A: context-dependent prediction of protein disorder as a function of redox state and protein binding. Nucleic Acids Res. 46, W329–W337.

Morris, E.R., Searle, M.S., 2012. Overview of protein folding mechanisms: experimental and theoretical approaches to probing energy landscapes. Curr. Protoc. Protein Sci. 28 (2), 1–22. Chapter 28, Unit.

Morrone, A., Mccully, M.E., Bryan, P.N., Brunori, M., Daggett, V., Gianni, S., et al., 2011. The denatured state dictates the topology of two proteins with almost identical sequence but different native structure and function. J. Biol. Chem. 286, 3863–3872.

Moutevelis, E., Woolfson, D.N., 2009. A periodic table of coiled-coil protein structures. J. Mol. Biol. 385, 726–732.

Mukrasch, M.D., Bibow, S., Korukottu, J., Jeganathan, S., Biernat, J., Griesinger, C., et al., 2009. Structural polymorphism of 441-residue tau at single residue resolution. PLoS Biol. 7, e34.

Mura, C., Veretnik, S., Bourne, P.E., 2019. The Urfold: structural similarity just above the superfold level? Protein Sci. 28, 2119–2126.

Nasir, A., Kim, K.M., Caetano-Anolles, G., 2014. A phylogenomic census of molecular functions identifies modern thermophilic archaea as the most ancient form of cellular life. Archaea 2014, 706468.

Noel, J.K., Levi, M., Raghunathan, M., Lammert, H., Hayes, R.L., Onuchic, J.N., et al., 2016. SMOG 2: a versatile software package for generating structure-based models. PLoS Comput. Biol. 12, e1004794.

O'neill, P.M., Barton, V.E., Ward, S.A., 2010. The molecular mechanism of action of artemisinin—the debate continues. Molecules 15, 1705–1721.

Oliveberg, M., Wolynes, P.G., 2005. The experimental survey of protein-folding energy landscapes. Q. Rev. Biophys. 38, 245–288.

Ou-Yang, S.S., Lu, J.Y., Kong, X.Q., Liang, Z.J., Luo, C., Jiang, H., 2012. Computational drug discovery. Acta Pharmacol. Sin. 33, 1131–1140.

Panchenko, A.R., 2003. Finding weak similarities between proteins by sequence profile comparison. Nucleic Acids Res. 31, 683–689.

Panchenko, A.R., Wolf, Y.I., Panchenko, L.A., Madej, T., 2005. Evolutionary plasticity of protein families: coupling between sequence and structure variation. Proteins 61, 535–544.

Panday, S.K., Sturlese, M., Salmaso, V., Ghosh, I., Moro, S., 2019. Coupling supervised molecular dynamics (SuMD) with entropy estimations to shine light on the stability of multiple binding sites. ACS Med. Chem. Lett. 10, 444–449.

Park, J.M., Niestemski, L.R., Deem, M.W., 2015. Quasispecies theory for evolution of modularity. Phys. Rev. E Stat. Nonlin Soft Matter Phys 91, 012714.

Paul, S.M., Mytelka, D.S., Dunwiddie, C.T., Persinger, C.C., Munos, B.H., Lindborg, S.R., et al., 2010. How to improve R&D productivity: the pharmaceutical industry's grand challenge. Nat. Rev. Drug. Discov. 9, 203–214.

Petrey, D., Fischer, M., Honig, B., 2009. Structural relationships among proteins with different global topologies and their implications for function annotation strategies. Proc. Natl Acad. Sci. USA 106, 17377–17382.

Przytycka, T., Srinivasan, R., Rose, G.D., 2002. Recursive domains in proteins. Protein Sci. 11, 409–417.

Putin, E., Asadulaev, A., Vanhaelen, Q., Ivanenkov, Y., Aladinskaya, A.V., Aliper, A., et al., 2018. Adversarial threshold neural computer for molecular de novo design. Mol. Pharm. 15, 4386–4397.

Rackovsky, S., 2015. Nonlinearities in protein space limit the utility of informatics in protein biophysics. Proteins 83, 1923–1928.

Radoux, C.J., Olsson, T.S., Pitt, W.R., Groom, C.R., Blundell, T.L., 2016. Identifying interactions that determine fragment binding at protein hotspots. J. Med. Chem. 59, 4314–4325.

Rahman, S.A., Cuesta, S.M., Furnham, N., Holliday, G.L., Thornton, J.M., 2014. EC-BLAST: a tool to automatically search and compare enzyme reactions. Nat. Meth. 11, 171–174.

Ramakrishnan, V., Srinivasan, S.P., Salem, S.M., Matthews, S.J., Colon, W., Zaki, M., et al., 2012. Geofold: topology-based protein unfolding pathways capture the effects of engineered disulfides on kinetic stability. Proteins 80, 920–934.

Redfern, O.C., Dessailly, B., Orengo, C.A., 2008. Exploring the structure and function paradigm. Curr. Opin. Struct. Biol. 18, 394–402.

Reymond, J.L., 2015. The chemical space project. Acc. Chem. Res. 48, 722–730.

Richard, J.P., 2019. Protein flexibility and stiffness enable efficient enzymatic catalysis. J. Am. Chem. Soc. 141, 3320–3331.

Rorick, M., 2012. Quantifying protein modularity and evolvability: a comparison of different techniques. Biosystems 110, 22–33.

Sabbadin, D., Moro, S., 2014. Supervised molecular dynamics (SuMD) as a helpful tool to depict GPCR-ligand recognition pathway in a nanosecond time scale. J. Chem. Inf. Model. 54, 372–376.

Sadowski, M.I., Taylor, W.R., 2010. On the evolutionary origins of "fold space continuity": a study of topological convergence and divergence in mixed alpha-beta domains. J. Struct. Biol. 172, 244–252.

Sadowski, M.I., Taylor, W.R., 2012. Evolutionary inaccuracy of pairwise structural alignments. Bioinformatics 28, 1209–1215.

Salvatori, G., Luberto, L., Maffei, M., Aurisicchio, L., Roscilli, G., Palombo, F., et al., 2020. SARS-CoV-2 SPIKE PROTEIN: an optimal immunological target for vaccines. J. Transl. Med. 18, 222.

Saylor, K., Gillam, F., Lohneis, T., Zhang, C., 2020. Designs of antigen structure and composition for improved protein-based vaccine efficacy. Front. Immunol. 11, 283.

Schaeffer, R.D., Liao, Y., Cheng, H., Grishin, N.V., 2017. ECOD: new developments in the evolutionary classification of domains. Nucleic Acids Res. 45, D296–D302.

Scheraga, H.A., Khalili, M., Liwo, A., 2007. Protein-folding dynamics: overview of molecular simulation techniques. Annu. Rev. Phys. Chem. 58, 57–83.

Schonherr, H., Cernak, T., 2013. Profound methyl effects in drug discovery and a call for new C-H methylation reactions. Angew. Chem. Int. Ed. Engl. 52, 12256–12267.

Schuler, B., Eaton, W.A., 2008. Protein folding studied by single-molecule FRET. Curr. Opin. Struct. Biol. 18, 16–26.

Segler, M.H.S., Waller, M.P., 2017. Modelling chemical reasoning to predict and invent reactions. Chemistry 23, 6118–6128.

Segler, M.H.S., Preuss, M., Waller, M.P., 2018. Planning chemical syntheses with deep neural networks and symbolic AI. Nature 555, 604–610.

Senior, A.W., Evans, R., Jumper, J., Kirkpatrick, J., Sifre, L., Green, T., et al., 2020. Improved protein structure prediction using potentials from deep learning. Nature 577, 706–710.

Shandilya, A., Chacko, S., Jayaram, B., Ghosh, I., 2013. A plausible mechanism for the antimalarial activity of artemisinin: a computational approach. Sci. Rep. 3, 2513.

Shi, Q., Chen, W., Huang, S., Wang, Y., Xue, Z., 2019. Deep learning for mining protein data. Brief. Bioinform .

Shimizu, H., Nakayama, K.I., 2020. Artificial intelligence in oncology. Cancer Sci. 111, 1452–1460.

Shirai, T., Terada, T., 2020. Overview of the big data bioinformatics symposium (2SCA) at BSJ2019. Biophys. Rev. 12, 277–278.

Sillitoe, I., Dawson, N., Lewis, T.E., Das, S., Lees, J.G., Ashford, P., et al., 2019. CATH: expanding the horizons of structure-based functional annotations for genome sequences. Nucleic Acids Res. 47, D280–D284.

Skolnick, J., Gao, M., 2013. Interplay of physics and evolution in the likely origin of protein biochemical function. Proc. Natl Acad. Sci. USA 110, 9344–9349.

Smith, B.A., Hecht, M.H., 2011. Novel proteins: from fold to function. Curr. Opin. Chem. Biol. 15, 421–426.

Spicer, C.D., Davis, B.G., 2014. Selective chemical protein modification. Nat. Commun. 5, 4740.

Sponer, J., Bussi, G., Krepl, M., Banas, P., Bottaro, S., Cunha, R.A., et al., 2018. RNA structural dynamics as captured by molecular simulations: a comprehensive overview. Chem. Rev. 118, 4177–4338.

Sterling, T., Irwin, J.J., 2015. ZINC 15–ligand discovery for everyone. J. Chem. Inf. Model. 55, 2324–2337.

Stewart, K.L., Rathore, D., Dodds, E.D., Cordes, M.H.J., 2019. Increased sequence hydrophobicity reduces conformational specificity: a mutational case study of the Arc repressor protein. Proteins 87, 23–33.

Taylor, W.R., 2002. A 'periodic table' for protein structures. Nature 416, 657–660.

Taylor, W.R., 2020. Exploring protein fold space. Biomolecules 10.

Teague, S.J., 2003. Implications of protein flexibility for drug discovery. Nat. Rev. Drug. Discov. 2, 527–541.

Togacar, M., Ergen, B., Comert, Z., 2020. COVID-19 detection using deep learning models to exploit social mimic optimization and structured chest X-ray images using fuzzy color and stacking approaches. Comput. Biol. Med. 121, 103805.

Tompa, P., Fuxreiter, M., 2008. Fuzzy complexes: polymorphism and structural disorder in protein-protein interactions. Trends Biochem. Sci. 33, 2–8.

Tseng, Y.Y., Li, W.H., 2012. Classification of protein functional surfaces using structural characteristics. Proc. Natl Acad. Sci. USA 109, 1170–1175.

Tsubaki, M., Tomii, K., Sese, J., 2019. Compound-protein interaction prediction with end-to-end learning of neural networks for graphs and sequences. Bioinformatics 35, 309–318.

Tsuchiya, Y., Taneishi, K., Yonezawa, Y., 2019. Autoencoder-based detection of dynamic allostery triggered by ligand binding based on molecular dynamics. J. Chem. Inf. Model. 59, 4043–4051.

Tsuchiya, Y., Tomii, K., 2020. Neural networks for protein structure and function prediction and dynamic analysis. Biophys. Rev. 12, 569–573.

Tyrchan, C., Evertsson, E., 2017. Matched molecular pair analysis in short: algorithms, applications and limitations. Comput. Struct. Biotechnol. J. 15, 86–90.

Valastyan, J.S., Lindquist, S., 2014. Mechanisms of protein-folding diseases at a glance. Dis. Model. Mech. 7, 9–14.

Vamathevan, J., Rolf Apweiler, Birney, E., 2019. Biomolecular data resources: bioinformatics infrastructure for biomedical data science. Annu. Rev. Biomed. Data Sci. 2, 199. 122.

Van Der Lee, R., Buljan, M., Lang, B., Weatheritt, R.J., Daughdrill, G.W., Dunker, A.K., et al., 2014. Classification of intrinsically disordered regions and proteins. Chem. Rev. 114, 6589–6631.

Verma, R., Pandit, S.B., 2019. Unraveling the structural landscape of intra-chain domain interfaces: Implication in the evolution of domain-domain interactions. PLoS One 14, e0220336.

Wagner, G.P., Pavlicev, M., Cheverud, J.M., 2007. The road to modularity. Nat. Rev. Genet. 8, 921–931.

Wallmann, A., Kesten, C., 2020. Common functions of disordered proteins across evolutionary distant organisms. Int. J. Mol. Sci. 21.

Wang, J., Oliveira, R.J., Chu, X., Whitford, P.C., Chahine, J., Han, W., et al., 2012. Topography of funneled landscapes determines the thermodynamics and kinetics of protein folding. Proc. Natl Acad. Sci. USA 109, 15763–15768.

Warner, D.J., Griffen, E.J., St-Gallay, S.A., 2010. WizePairZ: a novel algorithm to identify, encode, and exploit matched molecular pairs with unspecified cores in medicinal chemistry. J. Chem. Inf. Model. 50, 1350–1357.

Wathen, B., Jia, Z., 2009. Folding by numbers: primary sequence statistics and their use in studying protein folding. Int. J. Mol. Sci. 10, 1567–1589.

Wensley, B.G., Batey, S., Bone, F.A., Chan, Z.M., Tumelty, N.R., Steward, A., et al., 2010. Experimental evidence for a frustrated energy landscape in a three-helix-bundle protein family. Nature 463, 685–688.

Wensley, B.G., Kwa, L.G., Shammas, S.L., Rogers, J.M., Browning, S., Yang, Z., et al., 2012. Separating the effects of internal friction and transition state energy to explain the slow, frustrated folding of spectrin domains. Proc. Natl Acad. Sci. USA 109, 17795–17799.

Wishart, D.S., Knox, C., Guo, A.C., Cheng, D., Shrivastava, S., Tzur, D., et al., 2008. DrugBank: a knowledgebase for drugs, drug actions and drug targets. Nucleic Acids Res. 36, D901–D906.

Wolynes, P.G., Eaton, W.A., Fersht, A.R., 2012. Chemical physics of protein folding. Proc. Natl Acad. Sci. USA 109, 17770–17771.

Wright, P.E., Dyson, H.J., 2015. Intrinsically disordered proteins in cellular signalling and regulation. Nat. Rev. Mol. Cell Biol. 16, 18–29.

Yao, X.Q., Hamelberg, D., 2019. Detecting functional dynamics in proteins with comparative perturbed-ensembles analysis. Acc. Chem. Res. 52, 3455–3464.

Yoo, S.H., Geng, H., Chiu, T.L., Yu, S.K., Cho, D.C., Heo, J., et al., 2020. Deep learning-based decision-tree classifier for COVID-19 diagnosis from chest X-ray imaging. Front. Med. (Lausanne) 7, 427.

Yruela, I., Contreras-Moreira, B., Dunker, A.K., Niklas, K.J., 2018. Evolution of protein ductility in duplicated genes of plants. Front. Plant. Sci. 9, 1216.

Zeng, X., Song, X., Ma, T., Pan, X., Zhou, Y., Hou, Y., et al., 2020. Repurpose open data to discover therapeutics for COVID-19 using deep learning. J. Proteome Res. 19, 4624–4636.

Zhang, L., Zhang, N., Ruan, J.S., Zhang, T., 2011. Studies on the rules of beta-strand alignment in a protein beta-sheet structure. J. Theor. Biol. 285, 69–76.

Zhavoronkov, A., Ivanenkov, Y.A., Aliper, A., Veselov, M.S., Aladinskiy, V.A., Aladinskaya, A.V., et al., 2019. Deep learning enables rapid identification of potent DDR1 kinase inhibitors. Nat. Biotechnol. 37, 1038–1040.

Zhou, Y., Wang, F., Tang, J., Nussinov, R., Cheng, F., 2020. Artificial intelligence in COVID-19 drug repurposing. Lancet Digit. Health 2, e667–e676.

Artificial intelligence, big data and machine learning approaches in genome-wide SNP-based prediction for precision medicine and drug discovery

15

Isha Joshi[1], *Anushka Bhrdwaj*[1,2], *Ravina Khandelwal*[1], *Aditi Pande*[1], *Anshika Agarwal*[1], *Chillamcherla Dhanalakshmi Srija*[1], *Revathy Arya Suresh*[1], *Manju Mohan*[1], *Lima Hazarika*[1], *Garima Thakur*[1], *Tajamul Hussain*[3,4], *Sarah Albogami*[5], *Anuraj Nayarisseri*[1,2,3,6] *and Sanjeev Kumar Singh*[2]

[1]In silico Research Laboratory, Eminent Biosciences, Indore, Madhya Pradesh, India, [2]Department of Bioinformatics, Computer Aided Drug Designing and Molecular Modeling Lab, Alagappa University, Karaikudi, Tamil Nadu, India, [3]Biochemistry Department, College of Science, King Saud University, Riyadh, Saudi Arabia, [4]Center of Excellence in Biotechnology Research, College of Science, King Saud University, Riyadh, Saudi Arabia, [5]Department of Biotechnology, College of Science, Taif University, Taif, Saudi Arabia, [6]Bioinformatics Research Laboratory, LeGene Biosciences Pvt Ltd, Indore, Madhya Pradesh, India

15.1 Introduction

The healthcare industry is evolving toward the development of precision and personalized medicine. "Customized care" can be achieved by designing strategies that represent a combination of genomics, proteomics, transcriptomics, and metabolomics approaches. Artificial intelligence (AI) is the modern-day tool for laying the foundation of these strategies. Furthermore, these approaches have been proven to provide more effective solutions with lesser side effects when compared with other techniques (Gameiro et al., 2018).

The principle of older treatment practices was to be effective with a considerable section of the population. Precision medicine and personalized medication focus on treatments that are more individual beneficiary in nature. The risk reduction is the result of the gene-centric framework in combination with environmental factors and lifestyle. Large-scale identification of various genetic engineering approaches for the disease prognosis has been reported. These reported datasets can support the development of precision medication (Psaty et al., 2018). Moreover, the uncertainty associated with

Big Data Analytics in Chemoinformatics and Bioinformatics. DOI: https://doi.org/10.1016/B978-0-323-85713-0.00021-9

personalized medicine is high and unreliable to establish the foundation of future therapeutics on it. Overcoming the "more complex—less economic" barrier is necessary to reach strategies with higher achievable working output in lesser time duration, to increase the efficacy of personalized medicine (Laksman and Detsky, 2011).

The alteration of the medical regimen to every patient's characteristics segregates individuals into subpopulations that contradict their susceptibility to a specific disease or their response to a particular treatment. As a result, preventative interventions are robust for those who will benefit, while sparing expenditure, and side effects for those who will not. The potential of precision medicine lies in its competence to usher healthcare decisions, approaching an adequate treatment for an ailing patient. Precision medicine is a type of approach where the genetic architecture of an individual becomes the basis of his medicinal prescription. Reference sequence dependence reduction, structural variants, complex area characterization, on-time scheduling of phasing routine, indel calling (indel: insertion/deletion of bases), and high coverage rate are prime parameters of precision medicine. Precision medicine is developed by applying computational biological techniques in biological phenomena to identify the associated molecular marker. The current application of which is highly prevalent in mendelian disease prediction, pharmacogenomics, and precision oncology. However, optimizing the accuracy levels of ML-based predictions is a concerning limitation in precision medicine that needs to be overcome (Ashley, 2016).

ML-based predictions are concerning different algorithms used for data stratification and mining for drawing a pattern within the input data. It uses targeted labeled (supervised) or unlabeled data(unsupervised) data to perform a comparative analysis with the input data giving a desirable output. The labeled dataset is used to classify the input into category data(classification) and continuous data(regression). The unlabeled dataset is used to cluster data, reduce its dimensionality, and perform outlier detection. Thus, the outputs provide the user with a probable, more straightforward solution to the complex problems in link with a given dataset. By using machine learning (ML) and deep learning (DL), AI can usher precision medicine research toward an unaccustomed accuracy in a time-efficient manner (Tarca et al., 2007; Ashley, 2016).

The application of AI, ML, and DL in precision medicine has moved the healthcare industry toward digitalization, point-of-care testing, database creation, and analytical biosensor development. The use of disruptive technologies has paved the way toward personalized treatment and prevention strategies. The experience-based unprecedented rate of self-learning has provided easy refinement of the necessary information from complex datasets used in translational research (Mesko, 2017).

15.2 Role of artificial intelligence and machine learning in medicine

The use of virtual and physical technology in medicine is already well-known for the past decades. Virtual approaches include maintaining an electronic record system for healthcare, accompanied by treatment decisions guided by the neural network. The physical initiatives are concerned with robot and prostheses development for the medical industry. In addition, clinical correlations can be drawn by network building among the preexisting information stored in the form of a database. Primary care physicians can use this to identify and assist patients who require extra attention (Epstein et al., 2012). The prototype of the healthcare industry dataset to be used by AI broadly includes the following features: diagnostic imaging, genetic testing, electro-diagnosis, physiological monitoring, disability evaluation, and mass screening. Natural language processing uses the above dataset to retrieve specific imperative facts and figures (Table 15.1). Recent surveys

Table 15.1 Application of artificial intelligence in disease prediction and diagnosis.

S. No.	Application	Algorithm	References
1	Electroencephalogram imagery	1. Optimal reverse prediction 2. Cartesian K-means resulting in - > semi-supervised Cartesian K-means (SSCK)	Liu et al. (2020)
2	Iron deficiency anemia and beta thalassemia	Neighborhood component analysis feature selection	Liu et al. (2020)
3	Risk prediction of asthma	Artificial neural network Support vector machine Random forest	Ayyıldız and Tuncer (2020)
4	Prediction of perioperative blood loss in orthognathic surgery	Random forest model Statistical analysis through R programming	Ullah et al. (2019)
5	Prediction of diabetes	Logistic regression method K-nearest neighbor classifier Decision tree classification method Random forest classification	Stehrer et al. (2019)
6		Support vector classifier, random forest classifier, decision tree classifier, extra tree classifier, ADA boost algorithm, perceptron, linear discriminant analysis algorithm, logistic regression, k-nearest neighbor, gaussian naïve Bayes, bagging algorithm, gradient boost classifier.	Shankaracharya et al. (2010)
7	Prediction of brain maturity in infants	Rs-fMRI preprocessing ROI selection Support vector machine	Mujumdar and Vaidehi (2019)

(*Continued*)

Table 15.1 (Continued)

S. No.	Application	Algorithm	References
8	Predicting the availability of hematopoietic stem cell donors	Synthetic minority over-sampling technique BDT model Support vector machine model	Smyser et al. (2016)
9	Heart disease identification	Use of apache spark used DT, SVM, RF, and LR to build the offline model.	Li et al. (2020)
10	Breast Cancer Prediction	GRU-SVM, linear regression, multilayer perceptron, KNN, SoftMax	Ahmed et al. (2020), Mukherjee et al. (2022)
11		Regression.	Conrady and Jouffe (2011)
12		SVM, Naïve Bayesian, KNN	Asri et al. (2016)
13		LDA infused with SVM	Omondiagbe et al. (2019)
14		SVM and relevance vector machines	Agarap (2018)
15		Bayesian	Kourou et al. (2015)
16		Decision trees, ANN & SVM decision trees	Gupta and Garg (2020)
17		Association rules AND neural network	Karabatak and Ince (2009)
18		Naïve Bayesian	Kharya and Soni (2016)
19		Ensemble method	Mohebian et al. (2017)
20		Relevance vector machines, SVM, neural network	Gayathri et al. (2013)
21		Radial bias function neural network (RBFNN)	Zarbakhsh and Addeh (2018)
22	Early Stage Symptoms of SARSCov-2 Infected Patients	Decision tree Random forest Gradient boosting machine Extreme gradient boosting Support vector machine	Pedregosa et al. (2011)

(*Continued*)

Table 15.1 (Continued)

S. No.	Application	Algorithm	References
23	Quantification of Glaucomatous Damage in Fundus pictures	Deep neural network (ResNet34) architecture	Ahamad et al. (2020)
24	Diagnosis of T cell mediated kidney rejection	Linear discriminant analysis (LDA), support vector machines (SVM) random forest (RF)	Medeiros et al. (2019)
25	In silico prediction of unbound brain-to-plasma concentration ratio	Support vector machine(SVM) Random forest (RF), Partial least squares (PLS) algorithm	Liu et al. (2019)
26	Diagnosing Hepatocellular Carcinoma	Decision-tree learning Multilinear regression	Chen et al. (2011)
27	Estimate Sarcopenia on Abdominal CT	U-Net neural network model	Hashem et al. (2020)
28	Prediction of fatty liver disease	Random forest (RF), artificial neural network (ANN), Naïve Bayes (NB), and logistic regression (LR)	Burns et al. (2020)
29	Acute Myocardial Infarction	1. Logistic regression (LR) 2. Naïve Bayes (NB) 3. Support vector machines (SVM) 4. Random forests (RF) 5. Gradient boosting (GB) 6. Deep neural networks (DNN)	Wu et al. (2019)
30	Glioma stages prediction	Naive Bayes (CNB), support vector machine (SVM), K-nearest neighbor (KNN), random forest (RF), and artificial neural network (ANN) algorithms	Gupta et al. (2020)
31	Spinal Ependymoma	Decision trees (DTs), support vector machines (SVMs), and artificial neural networks (ANNs)	Niu et al. (2020)

(*Continued*)

Table 15.1 (Continued)

S. No.	Application	Algorithm	References
32	Detecting Dementia	Naïve Bayes Random forest Multiplayer perceptron	Ryu et al. (2019)
33	Diagnosis of osteoporosis	Evolution migrated AAAML	Bansal et al. (2020)
34	Gene expression profiles of lung cancer	Support vector machine	Devikanniga (2020)
35	Detection and analysis of Alzheimer's disease	Logistic regression (LR), decision tree (DT), random forest (RF), Naive Bayes (NB) support vector machines (SVMs)	Yuan et al. (2020)
36	Prediction of early neurological deterioration in acute minor ischemic stroke	Boosted trees, bootstrap decision forest, deep neural network, and logistic regression	Kishore et al. (2020)
37	Early detection of hepatocellular carcinoma	KNN, reglog, gaussNB, LDA, QDA, Random Forest with 1000 trees, MLP1 with 1 hidden layer (maximum 200 neurons in the hidden layer), SVM (C-SVC, called SVC), SVM (nu-SVC, called nuSVM), and Linear SVM (LinSVM)	Sung et al. (2020)

on AI applications in healthcare industries have highlighted cardiology, neurology, and oncology as the most explored branches; support vector machine (SVM) has been the most used algorithm (Jiang et al., 2017). Furthermore, with the help of AI, the healthcare sector has now upgraded with advancements like big data collection, automation of experiments, prediction, and functional annotation of genes, proteins, and protein binding sites. Literature mining of the high-throughput sequenced dataset with the help of AI can cut short the irrelevant data. Robust predictions of the high-throughput ML algorithms have made molecular dynamics simulation an easily achievable strategy (Somashekhar et al., 2018). The experience-based unprecedented rate of self-learning has provided easy refinement of the necessary information from complex datasets used in translational research (Mesko, 2017).

15.3 Genome-wide SNP prediction

Genomic selection uses Single Nucleotide Polymorphisms (SNPs) to envision quantitative phenotypes for reinforcing traits in the breeding populations; it is extensively used to surge plants' and animals' breeding efficiency. Genome-wide association studies (GWASs) have been the primary method for SNP detection and trait associations but lack combinatorial assessment of both of these features. ML can overcome this limitation by using different algorithms which apply a combination of statistical and biological knowledge for data analysis. In elaboration to the above, multifactor dimensionality reduction (MDR) and neural networks can pool the available genotypes comprising multiple SNPs for attribute construction, bypassing the known parametric statistical approaches. ML algorithm Relief act as a statistical filter for the SNPs present in the GWAS dataset (Moore and Williams, 2009).

ML has replaced classical statistical approaches to acumen predictions and statistical association from the biological data with a relatively higher statistical significance. Emerging DL technology serves as a persuasive ML tool to envision quantitative phenotypes without imputation, and discover potentially associated genotype markers. The available partial/whole genotype is considered at once by ML approaches for predictions and correlations. The prediction-based ML algorithm: SVM, random forest (RF), EN, and correlation-based ML algorithm (CFS) belong to the category of linear mixed models (LMMs). LMM are univariate as well as multivariate: they work on the principle of regression. ML approaches like COMBI, the combination of prediction and correlation algorithms, use SVM/RF algorithms for SNP selection. In COMBI, ML individually and in combination with statistical testing is used to discover the mathematical correlations and relevance of a genetic biomarker like a subset of SNP. The shortlisted SNPs are further subjected to different statistical tests to list out interpretations. The combined method has shown higher clinical significance in testing SNP-trait associations than the individual methods (Mieth et al., 2016a).

The use of MLto evaluates the SNP variants for unraveling the genetic basis of disease has become a global trend. RF models, Breiman's RF, GRRF, wsRF, are known for their application in easy characterization of the disease profile linked with the genome-wide SNP profile. The ts-RF approach is a two-stage quality-based sampling method used for SNP subspace selection. The ts-RF model initially creates two subgroups of "highly informative" and "unimportant SNPs," sampling the relevant SNP subspace by taking the p-value as the parameter from them. The accomplished outputs of ts-Rf have relatively improved test accuracy and high error reduction (Nguyen et al., 2015). SNP analysis using different ML-based analytical methods is made possible by utilizing various computational biology analytical tools: DeepPVP, PhenomeNET Variant Predictor (PVP), Genomiser, WS-SNPs & GO, FATHMM, SuSPect, Snpranker, Snat, MutPhred, PhD-SNP, PolyPhen-2, SNPeff, SparSNP, PLANET-SNP (Nayarisseri, 2020a). Table 15.2 illustrates the application of ML algorithms in different disease predictions using SNP variants.

Table 15.2 SNP variants in different diseases identified by machine learning.

S. No	Gene with SNP variant	Disease	ML algorithm	Reference
1.	CX3CR1, TNFAIP1, YNE1, ALDH5A1, ABCA1, DNMT3A, NF2, SZT2, ACADVL, MED12, TSC2, EP400, RYR2, VCL, BBS2, FUS, L1CAM.	Amyotrophic lateral sclerosis (ALS)	SVM, RF, SVC, LR, DNN, CNN,	Vasilopoulou et al. (2020)
2.	APOE, BIN1, CLU, ABCA7, CR1, PICALM, Ms4A6A, CD33, CD2AP	Alzheimer	BN	Sherif et al. (2015)
3.	HLADQB2, ZBTB34, GALNTL4, RIC8B, DOCK9, RASL11A, IMPA2, SIPA1L3, ARL15, NRP1, HAO2, HS3ST3B1, LY86, ETS1	Rheumatoid Arthritis	SVM	Negi et al. (2013)
4.	RELT, CCL18, TNFRSF10B, LILRB2, TNFRS, F10D, LCP2, CST7, TNFRSF21, FLI1, PRKCH, RGS1, HLA-DOB, HLA-DMB, PTAFR, SERPINB9, NFAM1, CSF2RB, FGR, GBP5, CD72	Inflammatory Bowel Disease	RF, svmPoly, xgbTree, EN, glmnet	Isakov et al. (2017)
5.	IRF5, TNPO3, HLA region, ITGAM, IRF5, STAT4	Systemic Lupus Erythe matosus	(BADTrees), (ADTrees)	Guy et al. (2012)

15.4 Artificial intelligence, precision medicine and drug discovery

The use of computational biology to predict accurate drug responses is a salient application of precision medicine. Moreover, SNP can alter the synthesized protein sensitivity which affects the drug pharmaco-dynamics. Thus, along with risk profiling, the pattern and frequency of SNP in genes hold importance in mediating drug responses and therapeutic drug designing.

ML approaches can use data about natural recovery and drug-based recovery associated with genotypic and phenotypic patterns to devise diverse treatment strategies. Supervised and unsupervised ML can perform "feature selection and identification" used in data partitioning and data mining. In addition, specific algorithms can perform specialized functions: A genetic algorithm is used to perform gene selection.

Prescreening of SNPs for drug formulation and target analysis using ML and AI is enacted by studying domain knowledge, pharmaco-dynamics, nature of the disease, molecular pharmacology, and pharmacokinetics (Shah and Kusiak, 2004).

The pharmaco-genomics industry is growing more specific with time. With the application of ML in precision medicine, it is now possible to customize drug dosage for different individuals with varied genetic profiles. Patient heterogeneity plays a significant part in deciding the sensitivity of specific drug treatment on human. SNP act as important clinical biomarkers for disease and related drug response prediction. Gaussian process regression (GPR) can predict the unknown dependent variables and drug response phenotypes from fed inputs, like the SNP collected by literature mining. GPR deals with genetic heterogeneity by categorizing patients as drug responders or nonresponders (Guan et al., 2019a).

In silico SNP analysis by bioinformatics tools has shown promising results in the prognosis of drug-related toxicity and efficacy abetting in cost-cutting to drug development. Along with the drug efficacy on known targets, it is now also possible to predict new drug targets, and device drugs accordingly. SNP analysis act as a qualitative (functionally significant) and quantitative (combination of different SNP present simultaneously) biomarker for target and drug response prediction. Sequence alignment tools such as SIFT, MAPP, PANTHER, Parepro, PhD-SNP & SNP & GO perform multiple sequence alignment and sequence-based SNP prediction. Structure-based approaches have more relevance than sequence-based SNP prediction methodologies in drug designing dynamics; SNP-based three-dimensional structure prediction diagnoses the altered protein phenotype in different genetic polymorphic alleles. However, tools like PolyPhen, SNPs3D, LS-SNP, SNP effect, and SNAP can integrate structure and sequence-based approaches and achieve the highest benefit. SNPs identified in the drug-metabolizing enzyme and drug target sequences can be studied simultaneously for their secondary and tertiary protein confirmation manipulation ability with the help of AI tools (Mah et al., 2011).

SNP-based prediction models developed by ML can find a diverse set of gene networks involved in intensive drug resistance responses of commercially essential drugs. SNP-based prediction models developed by ML can find a diverse set of gene networks involved in intensive drug resistance responses of commercially important drugs. Algorithm: RF, BP, NB, NN, KNN, LR, LDA can be used to locate and categorize nonsynonymous SNP (nsSNP) variants into the drug-sensitive and resistant phenotype. Ns-SNP found in antituberculous medications (Rifampicin, Isoniazid, Pyrazinamide, and Ethambutol) for resistance-associated genes were analyzed and validated for their prediction efficiency (Dafaalla et al., 2019). Not just a single drug, ML approaches have also identified multidrug resistance diagnostic responses in tuberculosis using SNP as the basis of reference. Machine and statistical learning architectures, neural networks, and penalized regression models were trained with preselected variants of genes involved in first and second line drug resistance (Fig. 15.1); the output was recorded as a predictive performance (Chen et al., 2019).

Alzheimer's disease is leading the panel of old age death causative diseases; Immune system abnormality has been the primary reason for Alzheimer's severity. A web ontology language is in use for integrated prediction analysis of existing

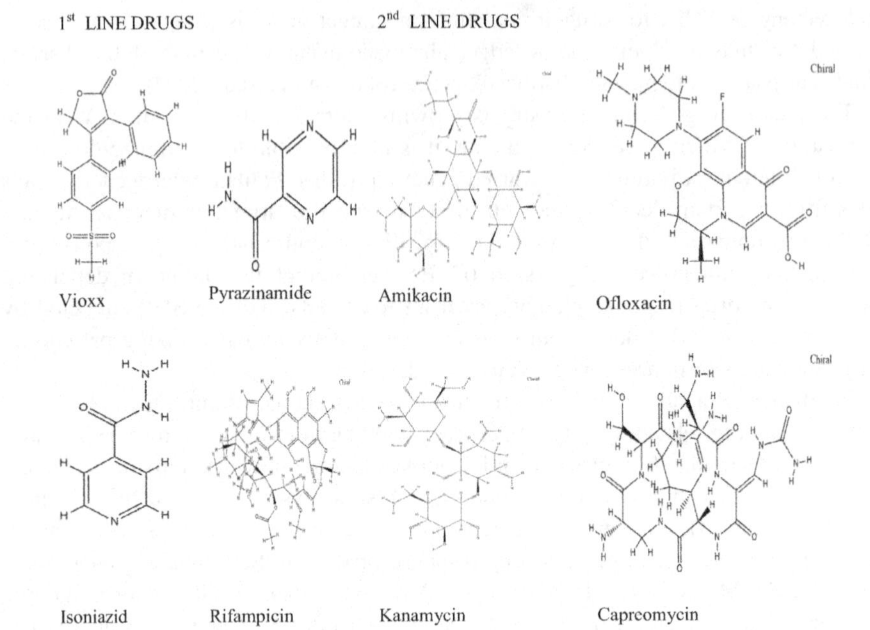

Figure 15.1 Drug architectures used for machine learning training in tuberculosis.

SNP and drug databases. Fig. 15.2 enlists the drugs which got annotated with their direct or indirect target gene variants in Alzheimer's (Han et al., 2018). Azathioprine is a thiopurine drug derivative used in the treatment of inflammatory bowel disease (IBD) because of its immunosuppressant properties. High-throughput targeted sequencing of genes involved in thiopurine metabolism has highlighted many candidate SNPs associated with the same. ML approaches have profiled a number of SNPs in association with IBD (Table. 15.2), some of which belong to the HLA locus. HLA locus had been found to be significantly associated with pancreatitis: a side effect of thiopurine (Park and Jeen, 2019). Hence, indicating the application of ML approaches in IBD detection and drug diagnosis.

Certain unique disruptive gene variants have been discovered in clinical genetic profiles of amyotrophic lateral sclerosis: a neurodegenerative disorder. "Knowledge graph algorithm" can predict and map ALS-related SNPs by conducting functional analysis and phenotypic enrichment profiling. These SNPs can be directly/indirectly linked with the drug Riluzole to lay drug-disease interactions (Bean et al., 2020). Treatment therapy for rheumatoid arthritis (RA) actively involves the use of antitumor necrosis factor drugs: adalimumab, etanercept, infliximab, and methotrexate. The GPR model has used SNP profiles of patients to predict and classify patients into responders and nonresponders, concerning a particular drug-guided treatment selection strategy for precision medicine development (Guan et al., 2019b). Fig. 15.2 is the brief structural depiction of the drugs used to treat the above disease profiles.

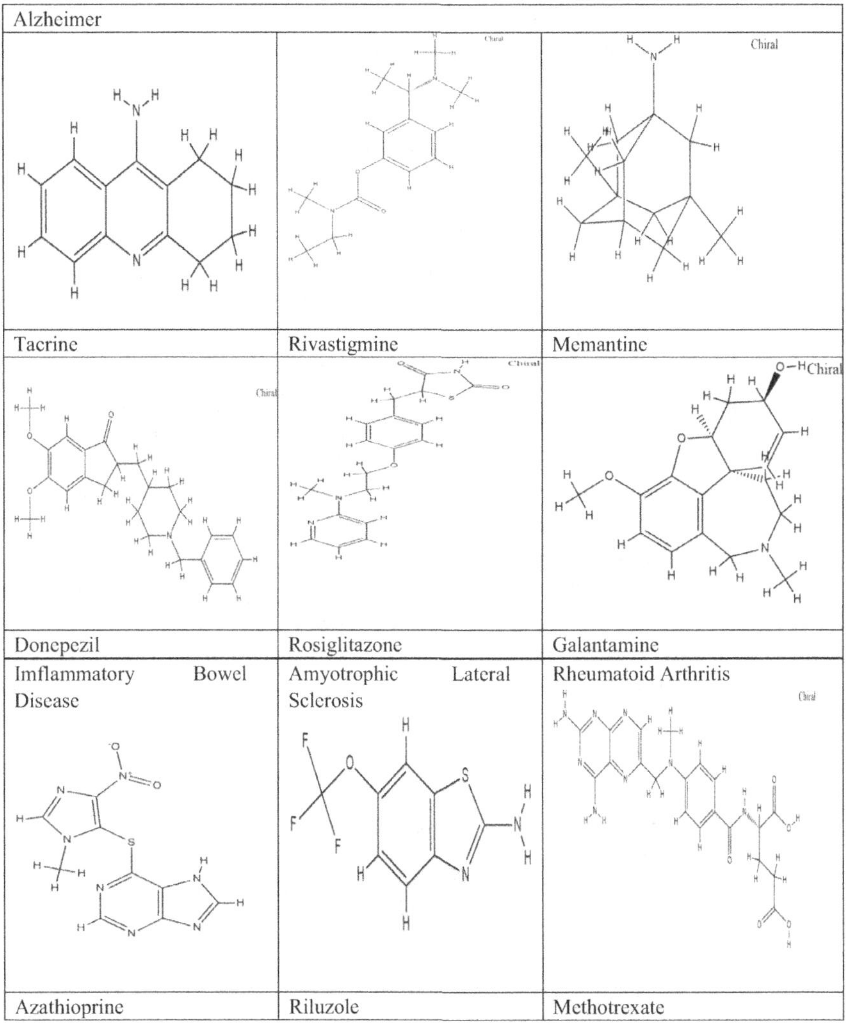

Figure 15.2 Machine learning-based drug-sensitivity prediction in different diseases.

15.5 Applications of artificial intelligence in disease prediction and analysis oncology

AI is the modern-day method for comprehensive data integration and multi-omics analysis of high-throughput complex data types in transcriptomics and genomics oncology. ML algorithms: Bayesian networks, Feature transformation, Heuristic networks, Clustering, Factorization, Deep networks, and Features extraction are used for performing medicine stratification, pathway analysis, biomarker identification, and drug discovery (Nicora et al., 2020); Supervised ML-like RF classifiers, has overcome this limitation by the distinction of nsSNP from the possible true mutations. Cancer-related known SNP variant datasets can be obtained from COSMIC and dbSNP databases, which are analyzed by computational tools like PolyPhen, CHASM, SIFT, can Predict, and Kinase SVM, to predict driver mutations in the gene of interest (Mieth et al., 2016b). Supervised ML algorithms like HMM, NN, DT, RF, and SVM are also used to study the impact of SNP associated with single amino acids polymorphisms (SAPs). However, new ML models like Phd-SNP, SNAP, SNPs & GO, and MutPred have replaced them now because of better functioning and knowledge-based prediction power invariant analysis. The SPF-cancer predictor when combined with the CNO dataset has given results with better performance, more accuracy, and a high correlation coefficient (Capriotti and Altman, 2011). Fig. 15.3 is a brief illustration of the application of ML in the majority of fatal cancers worldwide: breast (Behravan et al., 2018), glioblastoma (Aljouie et al., 2019), prostate (Lee et al., 2018), colorectal (Dorani et al., 2018; Yadav et al., 2022; Nayarisseri et al., 2021), and lung (Erin and Wei, 2011).

Identification of new drug targets and their inhibitor validation is the need of the hour in cancer biology. ML classifiers like SVM and RVF are used to integrate diverse genomic and systematic datasets to rank drug targets in three different cancers: breast (BrCa), pancreatic (PaCa), and ovarian (OvCa) cancers. Several small molecule drugs and synthetic peptides can be tested for their oncological treatment efficacy relative to the novel predicted target. The use of computational networks to assign new therapeutic indications to the known drugs enlisted in Fig. 15.4 can save

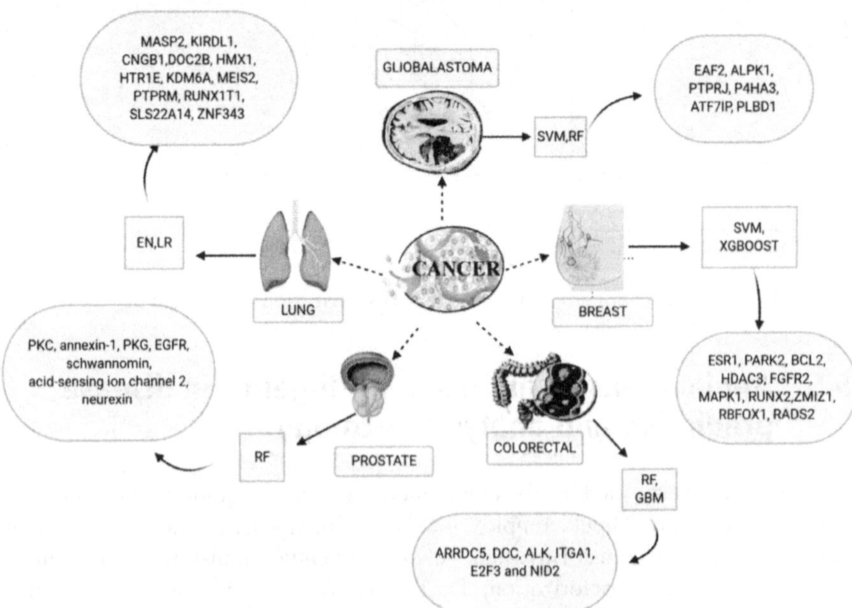

Figure 15.3 Machine learning-based SNP prediction in clinical oncology.

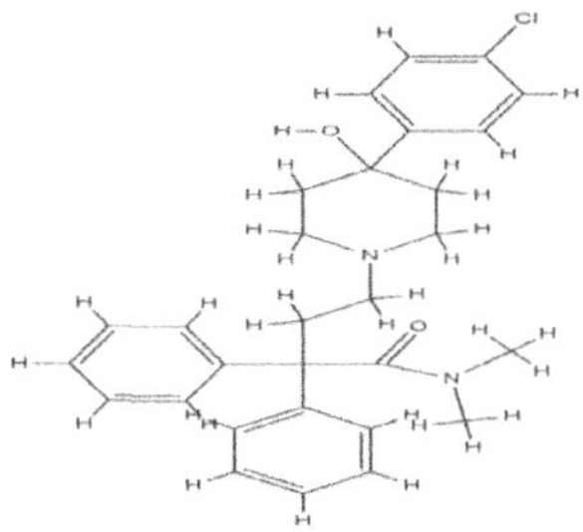

Loperamide

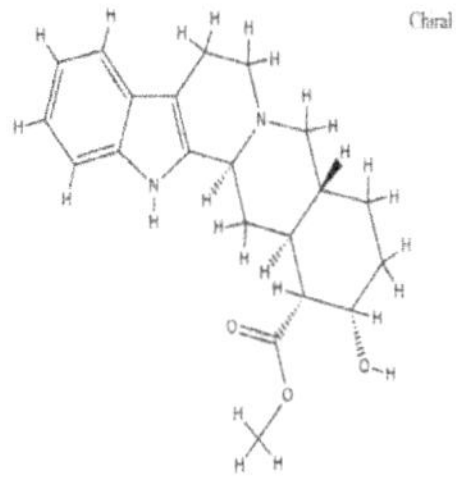

Yohimbine

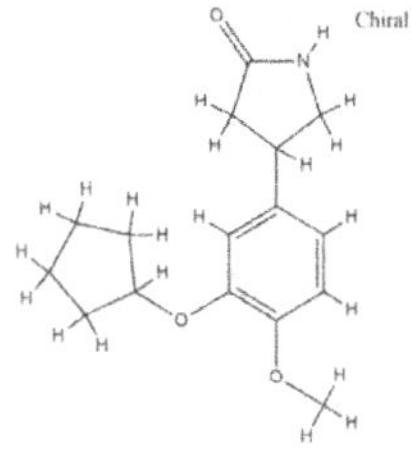

Rolipram

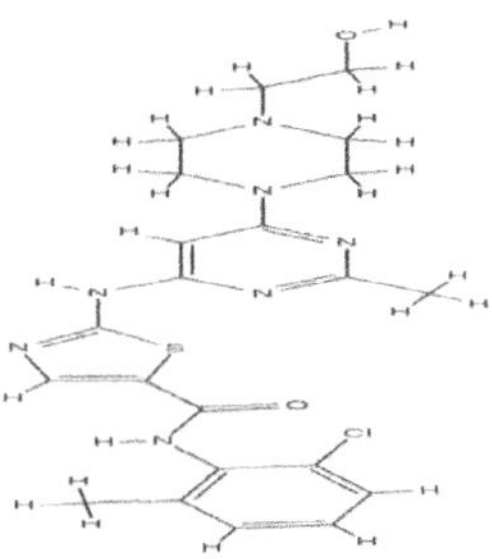

Dasatinib

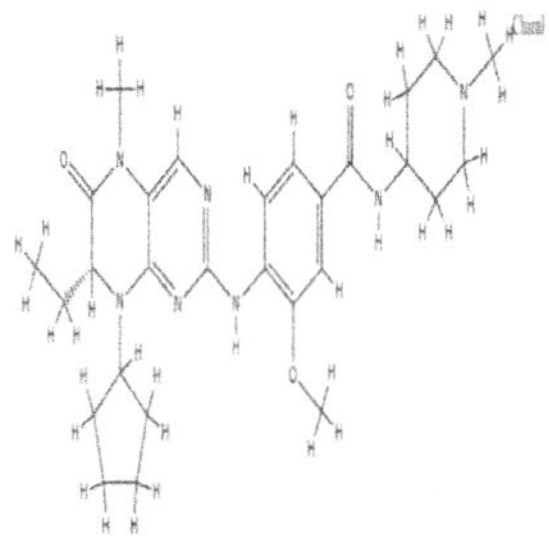

BI-2536

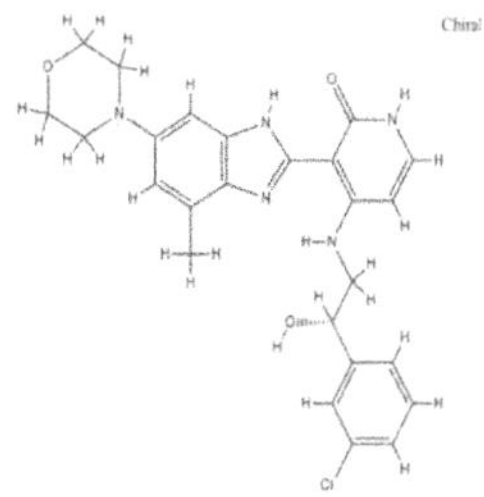

BMS-536924

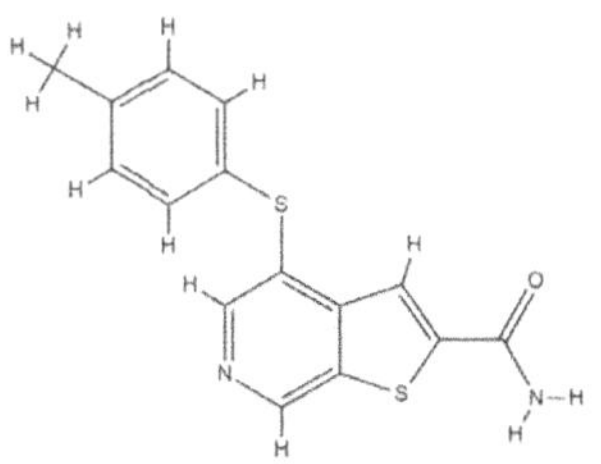

A-205804

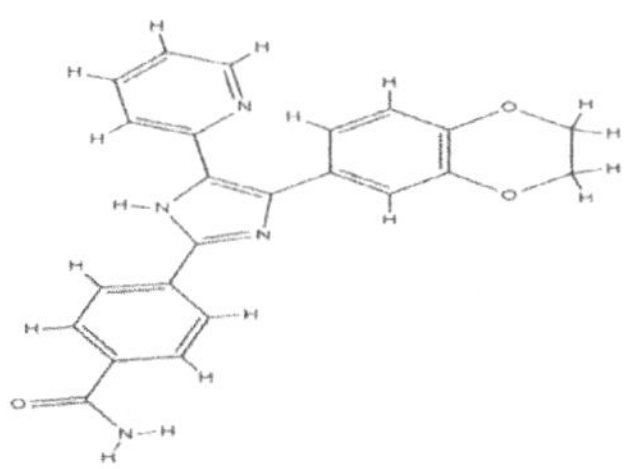

D-4476

Figure 15.4 Inhibitors used in oncological treatment.

time, money, and effort and increase the treatment's usefulness (Jeon et al., 2014; Bandaru et al., 2017; Natchimuthu et al., 2016; Majhi et al., 2018; Khandelwal et al., 2018; Sinha et al., 2018; Patidar et al., 2019; Nayarisseri, 2019; Sharda et al., 2019a; Limaye et al., 2019).

15.6 Cardiology

The incorporation of AI in cardiology increases the operational efficiency of data collection and interpretation. The above has enhanced the ability to make stronger assumptions and implement statistically valid solutions. AI has made pattern grading and risk assessment in patients of different age groups within the domain of cardiology possible. Proteomic measurements by regularized regression have made disease prediction like that of myocardial infarction possible. By transforming linear classifiers into nonlinear, SVM uses a metabolite profile for the prediction of in-stent restenosis. Decision trees can predict the risk associated with a cardiovascular event. The above represents the type of supervised ML. In the case of unsupervised ML, the topological data analysis uses electronic media records. Tensor factorization uses the ejection fraction to subtype diabetes mellitus type II and congestive heart failure. Electronic health records are used as a dataset to deduce DL-based prediction. Precision medicine in cardiology is a significant application of unsupervised ML (Cuocolo et al., 2019a; Johnson et al., 2018; Nayarisseri et al., 2020b, 2020c; Prajapati et al., 2020; Sharda et al., 2019b; Kleandrova et al., 2020; Adhikary et al., 2020; Qureshi et al., 2021).

ML in automatic segmentation of heart and prognostic phenotype differentiation by electrocardiography is a specialized assessment of the cardiovascular system. High accuracy in the deduction of the length of hospital stay by the DL algorithm is noticed compared with classical statistical analysis. Furthermore, neural networks in SNP analysis can provide predominant cross-links between SNPs in genes (ventricular myosin/cardiac myosin binding protein C) and inheritable heart diseases. Speeding up the analysis of large-scale genome-wide association for finding genotypes, further leading to precocious phenotypes, is made conceivable by DL approaches in heart disease diagnosis (Pattarabanjird et al., 2020).

Coronary artery disease (CAD) is the most common and deadly among all heart diseases. ML has been used to predict inferences in CAD by associating SNP and cardiac risk factors. ML models have predicted SNP associated with ID3, FREM1, LDLR, and COMT genes. These variants have the potency to measure vessel disease. Severity in atherosclerosis can be detected years beforehand by the clinical manifestation of cardiac events. The diagnosis of asymptomatic yet severe atherosclerosis is attainable by analyzing the genetic heterogeneity and polymorphism associated with the ID3 gene. Neural networks can predict high and low gensini severity score of novel SNPs, which are associated with CAD severity prediction (coronary artery calcium levels) and risk assessment aiding in clinical decision-making (Hathaway et al., 2019).

Not just diseases with direct damage and misfunctioning of cardiovascular components, ML approaches also contribute to clinical decision support intended for cardiac tissue profiling for diseases with indirect cardiovascular damage like diabetes mellitus. Shapley additive explanations can do binary and multiple classifications; Algorithms: LR, LDA, NB, SVM, and CART have validated the results. SNPs identified in the D-Loop region of mitochondrial DNA hold functional relevance in predicting the frequency of diabetes mellitus outcomes. The classification tree algorithm predicts hypermethylation of the nuclear genome which can act as a future parameter for the clinical diagnosis of disease (Davis et al., 2008).

Side effects associated with drug treatment are very common. Many high-dose drugs used to treat fatal diseases have got linked with increased sensitivity toward cardiovascular diseases. Cardiac myopathy spotted a high expression profile of the COX-2 gene; COX-2 also influences the expression of other indicators of cardiac diseases. Additionally, making it an efficient clinical biomarker and drug target.

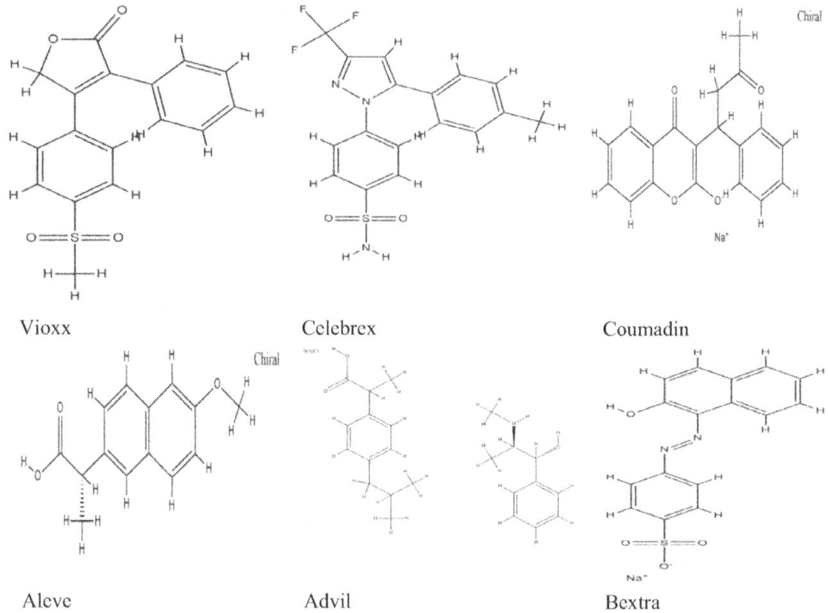

Figure 15.5 COX-2 inhibitory compounds.

ML approaches, SVM and NB, are used to determine the risk profiles of COX-2 inhibitory drugs (Fig. 15.5) based on the SNP variants of the COX-2 gene (O'Callaghan et al., 2009).

15.7 Neurology

DL methods can use effectual algorithms to curate necessary data from a comprehensive and complex dataset. Various SNP candidate genes with neurological disease outcomes have been identified in the ongoing research. With the advancement in computational biology techniques, the influence of genetic makeup on disease progression and aggressiveness can be looked upon easily. Diseases like cerebral palsy (CP), which were previously associated with a lack of oxygen, have now been known to be accountable for different SNP profiles in normal and affected individuals (van Eyk et al., 2019). Drug therapy precision prediction in disorders with varied mental health spectrums is now possible with ML approaches.

The development of a targeted gene panel is a cost-effective way of obtaining a genetic diagnosis with a substantial sequence coverage rate. ML tool: DOMINO has been used to find the association of CP exclusive recurrent genetic variants with the genetically dominant disorder. L1CAM, KIF1A, MAOB, AGAP, TUBA1A, and COL4A1 contribute majorly to the disease burden (Romero et al., 2010). Moreover, DNA variants of IL-6 are found associated with CP in preterm newborns. IPA analysis, TuRF, D′, and MDR algorithms in combination with the SNPper database identified a proportionality between the increased susceptibility to preterm labor/delivery with the occurrence of SNP (A31A) located in IL-6 Receptor in fetal DNA (Oskoui et al., 2015). Furthermore, the role of chromosomal abnormalities like duplication, deletion, and translocation in the form of high copy no. variation (CNV) were identified using the algorithms: iPattern36, PennCNV37, QuantiSNP38, CNVPartition3. The obtained CNV was found to be de novo and rarely inherited (Bahado-Singh et al., 2019).

An epigenetic modification also plays a crucial role in regulating CP dynamics. Hypo-methylation of CpG loci in both coding and noncoding regions are observed in the CP sample and not in the control sample; This pattern was used to predict CP in newborns using ML approaches. Out of all, four identified CpGloci, C6orf27, UFM1, SLC25A36, and RALGDS showed predictive accuracy of AUC $\geq$ 0.90 in CP detection. IPA Analysis can examine metabolic pathways and networks, which includes the differentially methylated genes—using them further to look out for their relevance in brain function and development. The overall ML predictions are supported by multivariate logistic regression validating high predictive accuracy. Illumina Human Methylation 450 K Bead Chip arrays data of the top 25 loci were further confirmed to be unbiased by pyrosequencing (Stern et al., 2018).

The statistics of clinical depression are going high day by day in the modern world. The genetic profiles of various patients have signed posted a link between genetic heterogenicity and depression sensitivity. The mental health spectrum has highlighted various phenotypes common between different mental disorders. Precision medicines in the form of antidepressants and antipsychotics can be designed and implemented based on the occurrence of these common phenotypes and genotypes (Natchimuthu et al., 2022) (Fig. 15.6).

Parallelly, ML strategies: SVM and LASSO regression have formulated procedures to classify and identify variants in the responder and nonresponder population with respect to drug treatment. The response phenotypes of drugs in association with the SNP profiles are looked upon in a no. of studies. BRLMM algorithm identified 40 significant response SNPs out of the 4,30,198 analyzed SNPs in antidepressant citalopram response. Perlegen algorithm has identified SNP hotspots affecting the Antipsychotics responses: olanzapine, quetiapine, risperidone, ziprasidone, and perphenazine (Cuocolo et al., 2019b). Multilayer feed-forward neural networks and LR models have identified SNP-SNP interaction models in antidepressant treatment responses (Lin et al., 2018).

15.8 Conclusion

There is a significant influence of the genetics, surrounding environment, and habits of a person on his health, this has increased the need to use precision medicine as targeted therapeutics. The application of AI in the healthcare sector has made medical decisions more economical and time-efficient. The precision medicine advent

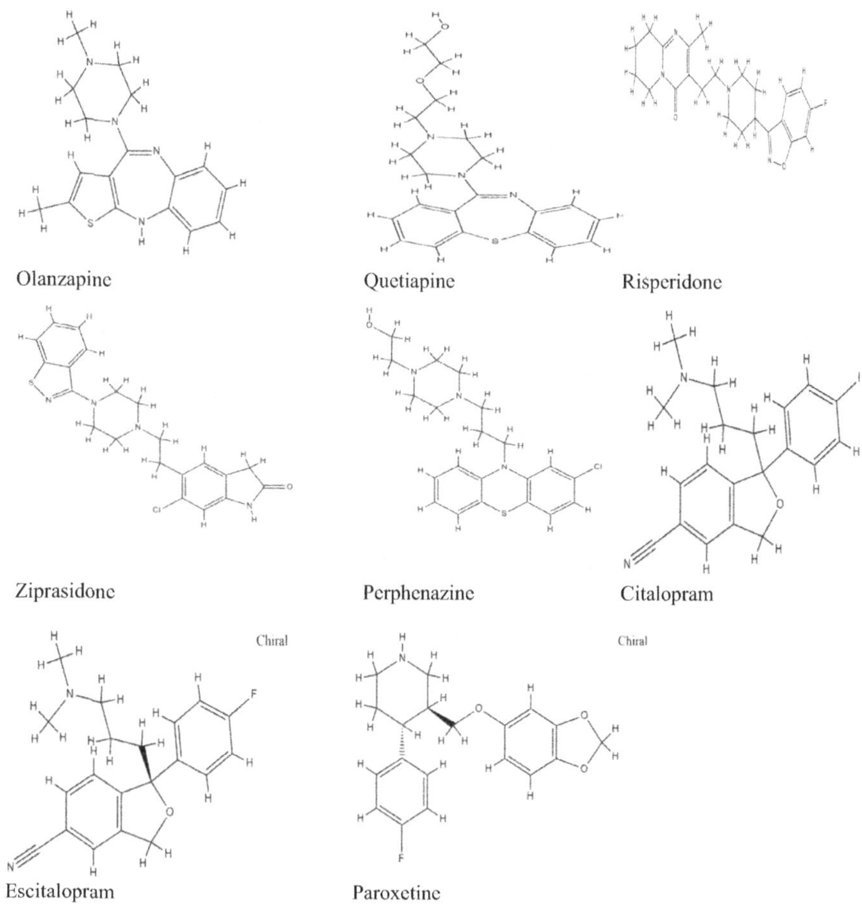

Figure 15.6 Antidepressant and antipsychotics drugs.

leverages a patient's lifestyle, genetic history, environmental factors, location and habits to identify a plan of action to treat ailing disease. AI has successfully classified problems using diversified models and solved precision medicine problems, for example, accurate disease interpretation, disease analysis, and treatment optimization. In addition, the study of multidimensional datasets to capture variations trained using AI algorithms and analyzing cryptic phenotypic structures. In the present study, we explored the role of SNP variants as a biomarker in precision medication by using ML approaches. Personalized medicine is still out of the current technological dimensions but with the application of machine and DL algorithms; it is now possible to group individuals into distinct groups and design treatment strategies based on the most common parameters present. High-throughput biology brought up by AI has given us various parameters to judge the related physiological insights based on technical assumptions. However, imaging techniques can give more of an estimated and less accurate probability than a genetic biomarker. The available research dataset of different human pathologies in diverse diseases and disorders has highlighted the high deposition of SNP variants in clinical data. It is difficult to manually conclude the frequency and type of SNP variants present in such huge and complex datasets. ML can find and link various pestiferous SNP variants with different clinical manifestations via diverse research projects. AI has also made it easy to study the diversity and relevance of the associated polymorphism with molecular dynamic parameters like differential expression regulation, protein structural and sequential variation, and the difference in epigenetic marker enrichment sensitivity. These all together keep the foundation of pharmaco-genomics. We have also shown the use of ML algorithms to estimate the degree of efficacy and associated side effects of various well-known and widely used drugs in numerous clinical pathologies. Novel drug target prediction and drug designing strategies are just another application of this combined approach. The qualitative analysis, the quantitative analysis, and the combinatorial analysis of both by ML can understand the different degrees of clinical manifestation linked with an SNP. The extensive applications of precision medicine in day-to-day life can open doors toward economic health expenses for all sections of society. Multipurpose biomarkers, like SNP, and multidiagnostic tools, provided by AI decline the expenses to generalize precision medicine making it more adaptable.

Abbreviations

ADTrees Alternating Decision Trees
ANN Artificial Neural Network
BADTrees Bagged Alternating Decision Trees
BN Bayesian Network
CART Classification and Regression Tree
CNN Convolutional Neural Network
DCNN Deep Convolutional Neural Network
DT Decision Trees
EN Elastic Net
Glmnet Lasso and Elastic-Net Regularized Generalized Linear Models
HMM Hidden Markov Models
KNN K-Nearest Neighbors
LASSO Least Absolute Shrinkage And Selection Operator
LDA Linear Discriminant Analysis
LR Logistic Regression
MDR Multifactor Dimensionality Reduction
MLP Multilayer Perceptron
NB Naïve Bayes
NN Neural Network
PAM Partition Around Medoids
PLS-DA Partial Least-Squares Discriminant Analysis
PK Polynomial Kernel

QDA	Quadratic Discriminant Analysis
RF	Random Forest
RR	Ridge Regression
SL	Simple Logistic
SVC	Support Vector Classifier
SVM	Support Vector Machine
svmPoly	Support Vector Machine with Polynomial Kernel
TuRF	Tuned ReliefF
xgbTree	Extreme Gradient Boosting

References

Adhikary, R., Khandelwal, R., Hussain, T., Nayarisseri, A., Singh, S.K., 2020. Structural insights into the molecular design of ros1 inhibitor for the treatment of non-small cell lung cancer (NSCLC). Current Computer-aided Drug Design. PubmedID:32364080.

Agarap, A.F.M., 2018. On breast cancer detection: an application of machine learning algorithms on the wisconsin diagnostic dataset. In: Proceedings of the 2nd International Conference on Machine Learning and Soft Computing. pp. 5–9.

Ahamad, M.M., Aktar, S., Rashed-Al-Mahfuz, M., Uddin, S., Liò, P., Xu, H., et al., 2020. A machine learning model to identify early stage symptoms of SARS-Cov-2 infected patients. Expert. Syst. Appl. 160, 113661.

Ahmed, H., Younis, E.M., Hendawi, A., Ali, A.A., 2020. Heart disease identification from patients' social posts, machine learning solution on Spark. Fut. Gener. Comput. Syst. 111, 714–722.

Aljouie, A., Schatz, M., Roshan, U., 2019. Machine learning based prediction of gliomas with germline mutations obtained from whole exome sequences from TCGA and 1000 Genomes Project. In: 2019 Third International Conference on Intelligent Computing in Data Sciences (ICDS). IEEE, pp. 1–8.

Ashley, E.A., 2016. Towards precision medicine. Nat. Rev. Genet. 17 (9), 507.

Asri, H., Mousannif, H., Al Moatassime, H., Noel, T., 2016. Using machine learning algorithms for breast cancer risk prediction and diagnosis. Procedia Comput. Sci. 83, 1064–1069.

Ayyıldız, H., Tuncer, S.A., 2020. Determination of the effect of red blood cell parameters in the discrimination of iron deficiency anemia and beta thalassemia via neighborhood component analysis feature selection-based machine learning. Chemometr. Intell. Lab. Syst. 196, 103886.

Bahado-Singh, R.O., Vishweswaraiah, S., Aydas, B., Mishra, N.K., Guda, C., Radhakrishna, U., 2019. Deep learning/artificial intelligence and blood-based DNA epigenomic prediction of cerebral palsy. Int. J. Mol. Sci. 20 (9), 2075.

Bandaru, S., GangadharanSumithnath, T., Sharda, S., Lakhotia, S., Sharma, A., Jain, A., et al., 2017. Helix-coil transition signatures B-Raf V600E mutation and virtual screening for inhibitors directed against mutant B-Raf. Curr. Drug. Metab. 18 (6), 527–534.

Bansal, D., Khanna, K., Chhikara, R., Dua, R.K., Malhotra, R., 2020. Classification of magnetic resonance images using bag of features for detecting dementia. Procedia Comput. Sci. 167, 131–137.

Bean, D.M., Al-Chalabi, A., Dobson, R.J., Iacoangeli, A., 2020. A knowledge-based machine learning approach to gene prioritisation in amyotrophic lateral sclerosis. Genes 11 (6), 668.

Behravan, H., Hartikainen, J.M., Tengström, M., Pylkäs, K., Winqvist, R., Kosma, V.M., et al., 2018. Machine learning identifies interacting genetic variants contributing to breast cancer risk: a case study in Finnish cases and controls. Sci. Rep. 8 (1), 1–13.

Burns, J.E., Yao, J., Chalhoub, D., Chen, J.J., Summers, R.M., 2020. A machine learning algorithm to estimate sarcopenia on abdominal CT. Academic Radiol. 27 (3), 311–320.

Capriotti, E., Altman, R.B., 2011. A new disease-specific machine learning approach for the prediction of cancer-causing missense variants. Genomics 98 (4), 310–317.

Chen, H., Winiwarter, S., Fridén, M., Antonsson, M., Engkvist, O., 2011. In silico prediction of unbound brain-to-plasma concentration ratio using machine learning algorithms. J. Mol. Graph. Model. 29 (8), 985–995.

Chen, M.L., Doddi, A., Royer, J., Freschi, L., Schito, M., Ezewudo, M., et al., 2019. Beyond multidrug resistance: leveraging rare variants with machine and statistical learning models in *Mycobacterium tuberculosis* resistance prediction. EBioMedicine 43, 356–369.

Conrady, S., Jouffe, L., 2011. Breast cancer diagnostics with Bayesian networks. Conrady Appl. Sci., LLC 5, March.

Cuocolo, R., Perillo, T., De Rosa, E., Ugga, L., Petretta, M., 2019a. Current applications of big data and machine learning in cardiology. J. Geriatric JGC 16 (8), 601.

Cuocolo, R., Perillo, T., De Rosa, E., Ugga, L., Petretta, M., 2019b. Current applications of big data and machine learning in cardiology. J. Geriatric Cardiol.: JGC 16 (8), 601.

Dafaalla, M., Abdullah, M.O.E., Bakhiet, S., Ibrahim, M., 2019. Homology-based prediction of resistance to antituberculous medications using machine learning algorithms.

Davis, J., Lantz, E., Page, D., Struyf, J., Peissig, P., Vidaillet, H., et al., 2008. Machine learning for personalized medicine: Will this drug give me a heart attack. In: Proceedings of International Conference on Machine Learning (ICML).

Devikanniga, D., 2020. Diagnosis of osteoporosis using intelligence of optimized extreme learning machine with improved artificial algae algorithm. Int. J. Intell. Netw. 1, 43–51.

Dorani, F., Hu, T., Woods, M.O., Zhai, G., 2018. Ensemble learning for detecting gene-gene interactions in colorectal cancer. PeerJ 6, e5854.

Epstein, E.A., Schor, M.I., Iyer, B.S., Lally, A., Brown, E.W., Cwiklik, J., 2012. Making watson fast. IBM J. Res. Dev. 56 (3.4), pp. 15-1.

Erin, B., Wei, H., 2011. Identification of a 12-gene signature for lung cancer prognosis through machine learning. J. Cancer Ther. 2011.

Gameiro, G.R., Sinkunas, V., Liguori, G.R., Auler-Júnior, J.O.C., 2018. Precision Medicine: changing the way we think about healthcare. Clinics 73.

Gayathri, B.M., Sumathi, C.P., Santhanam, T., 2013. Breast cancer diagnosis using machine learning algorithms-a survey. Int. J. Distrib. Parallel Syst. 4 (3), 105.

Guan, Y., Zhang, H., Quang, D., Wang, Z., Parker, S.C., Pappas, D.A., et al., 2019a. Machine learning to predict anti–tumor necrosis factor drug responses of rheumatoid arthritis patients by integrating clinical and genetic markers. Arthrit. Rheumatol. 71 (12), 1987–1996.

Guan, Y., Zhang, H., Quang, D., Wang, Z., Parker, S.C., Pappas, D.A., et al., 2019b. Machine learning to predict anti–tumor necrosis factor drug responses of rheumatoid arthritis patients by integrating clinical and genetic markers. Arthrit. Rheumatol. 71 (12), 1987–1996.

Gupta, P., Garg, S., 2020. Breast cancer prediction using varying parameters of machine learning models. Procedia Comput. Sci. 171, 593–601.

Gupta, S., Ko, D.T., Azizi, P., Bouadjenek, M.R., Koh, M., Chong, A., et al., 2020. Evaluation of machine learning algorithms for predicting readmission after acute myocardial infarction using routinely collected clinical data. Can. J. Cardiol. 36 (6), 878–885.

Guy, R.T., Santago, P., Langefeld, C.D., 2012. Bootstrap aggregating of alternating decision trees to detect sets of SNPs that associate with disease. Genet. Epidemiol. 36 (2), 99–106.

Han, Z.J., Xue, W.W., Tao, L., Zhu, F., 2018. Identification of novel immune-relevant drug target genes for Alzheimer's disease by combining ontology inference with network analysis. CNS Neurosci. Therapeut. 24 (12), 1253–1263.

Hashem, S., ElHefnawi, M., Habashy, S., El-Adawy, M., Esmat, G., Elakel, W., et al., 2020. Machine learning prediction models for diagnosing hepatocellular carcinoma with HCV-related chronic liver disease. Comput. Meth. Prog. Biomed. 105551.

Hathaway, Q.A., Roth, S.M., Pinti, M.V., Sprando, D.C., Kunovac, A., Durr, A.J., et al., 2019. Machine-learning to stratify diabetic patients using novel cardiac biomarkers and integrative genomics. Cardiovas. Diabetol. 18 (1), 78.

Isakov, O., Dotan, I., Ben-Shachar, S., 2017. Machine learning–based gene prioritization identifies novel candidate risk genes for inflammatory bowel disease. Inflamm. Bowel Dis. 23 (9), 1516–1523.

Jeon, J., Nim, S., Teyra, J., Datti, A., Wrana, J.L., Sidhu, S.S., et al., 2014. A systematic approach to identify novel cancer drug targets using machine learning, inhibitor design and high-throughput screening. Genome Med. 6 (7), 1–18.

Jiang, F., Jiang, Y., Zhi, H., Dong, Y., Li, H., Ma, S., et al., 2017. Artificial intelligence in healthcare: past, present and future. Stroke Vasc. Neurol. 2 (4), 230–243.

Johnson, K.W., Soto, J.T., Glicksberg, B.S., Shameer, K., Miotto, R., Ali, M., et al., 2018. Artificial intelligence in cardiology. J. Am. Coll. Cardiol. 71 (23), 2668–2679.

Karabatak, M., Ince, M.C., 2009. An expert system for detection of breast cancer based on association rules and neural network. Expert. Syst. Appl. 36 (2), 3465–3469.

Khandelwal, R., Chauhan, A.P., Bilawat, S., Gandhe, A., Hussain, T., Hood, E.A., et al., 2018. Structure-based virtual screening for the identification of high-affinity small molecule towards STAT3 for the clinical treatment of osteosarcoma. Curr. Top. Med. Chem. 18 (29), 2511–2526.

Kharya, S., Soni, S., 2016. Weighted naive bayes classifier: a predictive model for breast cancer detection. Int. J. Comput. Appl. 133 (9), 32–37.

Kishore, P., Kumari, C.U., Kumar, M.N.V.S.S., Pavani, T., 2020. Detection and analysis of Alzheimer's disease using various machine learning algorithms. Mater. Today: Proc.

Kleandrova, V.V., Scotti, M.T., Scotti, L., Nayarisseri, A., Speck-Planche, A., 2020. Cell-based multi-target QSAR model for design of virtual versatile inhibitors of liver cancer cell lines. SAR. QSAR Environ. Res. 31 (11), 815–836.

Kourou, K., Exarchos, T.P., Exarchos, K.P., Karamouzis, M.V., Fotiadis, D.I., 2015. Machine learning applications in cancer prognosis and prediction. Comput. Struct. Biotechnol. J. 13, 8–17.

Laksman, Z., Detsky, A.S., 2011. Personalized medicine: understanding probabilities and managing expectations. J. Gen. Intern. Med. 26 (2), 204–206.

Lee, S., Kerns, S., Ostrer, H., Rosenstein, B., Deasy, J.O., Oh, J.H., 2018. Machine learning on a genome-wide association study to predict late genitourinary toxicity after prostate radiation therapy. Int. J. Radiat. Oncol. Biol. Phys. 101 (1), 128–135.

Li, Y., Masiliune, A., Winstone, D., Gasieniec, L., Wong, P., Lin, H., et al., 2020. Predicting the availability of haematopoietic stem cell donors using machine learning. Biol. Blood Marrow Transplant. .

Limaye, A., Sweta, J., Madhavi, M., Mudgal, U., Mukherjee, S., Sharma, S., et al., 2019. In silico insights on gd2: a potential target for pediatric neuroblastoma. Curr. Top. Med. Chem. 19 (30), 2766–2781.

Lin, E., Kuo, P.H., Liu, Y.L., Yu, Y.W.Y., Yang, A.C., Tsai, S.J., 2018. A deep learning approach for predicting antidepressant response in major depression using clinical and genetic biomarkers. Front. Psych. 9, 290.

Liu, P., Tseng, G., Wang, Z., Huang, Y., Randhawa, P., 2019. Diagnosis of T-cell–mediated kidney rejection in formalin-fixed, paraffin-embedded tissues using RNA-Seq–based machine learning algorithms. Hum. Pathol. 84, 283–290.

Liu, M., Zhou, M., Zhang, T., Xiong, N., 2020. Semi-supervised learning quantization algorithm with deep features for motor imagery EEG recognition in smart healthcare application. Appl. Soft Comput. 89, 106071.

Mah, J.T., Low, E.S., Lee, E., 2011. In silico SNP analysis and bioinformatics tools: a review of the state of the art to aid drug discovery. Drug. Discov. Today 16 (17–18), 800–809.

Majhi, M., Ali, M.A., Limaye, A., Sinha, K., Bairagi, P., Chouksey, M., et al., 2018. An in silico investigation of potential EGFR inhibitors for the clinical treatment of colorectal cancer. Curr. Top. Med. Chem. 18 (27), 2355–2366.

Medeiros, F.A., Jammal, A.A., Thompson, A.C., 2019. From machine to machine: an OCT-trained deep learning algorithm for objective quantification of glaucomatous damage in fundus photographs. Ophthalmology 126 (4), 513–521.

Mesko, B., 2017. The role of artificial intelligence in precision medicine.

Mieth, B., Kloft, M., Rodríguez, J.A., Sonnenburg, S., Vobruba, R., Morcillo-Suárez, C., et al., 2016a. Combining multiple hypothesis testing with machine learning increasees the statistical power of genome-wide association studies. Sci. Rep. 6, 36671. Capriotti, E. and Altman, R.B., 2011. A new disease-specific machine learning approach for the prediction of cancer-causing missense variants. Genomics, 98(4), pp. 310–317.

Mieth, B., Kloft, M., Rodríguez, J.A., Sonnenburg, S., Vobruba, R., Morcillo-Suárez, C., et al., 2016b. Combining multiple hypothesis testing with machine learning increases the statistical power of genome-wide association studies. Sci. Rep. 6, 36671.

Mohebian, M.R., Marateb, H.R., Mansourian, M., Mañanas, M.A., Mokarian, F., 2017. A hybrid computer-aided-diagnosis system for prediction of breast cancer recurrence (HPBCR) using optimized ensemble learning. Comput. Struct. Biotechnol. J. 15, 75–85.

Moore, J.H., Williams, S.M., 2009. Epistasis and its implications for personal genetics. Am. J. Hum. Genet. 85 (3), 309–320.

Mujumdar, A., Vaidehi, V., 2019. Diabetes prediction using machine learning algorithms. Procedia Comput. Sci. 165, 292–299.

Mukherjee, S., Abdalla, M., Yadav, M., Madhavi, M., Bhrdwaj, A., Khandelwal, R., et al., 2022. Structure-based virtual screening, molecular docking, and molecular dynamics simulation of VEGF inhibitors for the clinical treatment of ovarian cancer. J. Mol. Modeling 28 (4), 1–21.

Natchimuthu, V., Bandaru, S., Nayarisseri, A., Ravi, S., 2016. Design, synthesis and computational evaluation of a novel intermediate salt of N-cyclohexyl-N-(cyclohexylcarbamoyl)-4-(trifluoromethyl) benzamide as potential potassium channel blocker in epileptic paroxysmal seizures. Comput. Biol. Chem. 64, 64–73.

Nayarisseri, A., 2019. Prospects of utilizing computational techniques for the treatment of human diseases. Curr. Top. Med. Chem. 19 (13), 1071–1074.

Nayarisseri, A., 2020a. Experimental and computational approaches to improve binding affinity in chemical biology and drug discovery. Curr. Top. Med. Chem. 20 (19), 1651–1660.

Nayarisseri, A., 2020b. Most promising compounds for treating COVID-19 and recent trends in antimicrobial & antifungal agents. Curr. Top. Med. Chem. 20 (24), 2119–2125.

Nayarisseri, A., Khandelwal, R., Madhavi, M., Selvaraj, C., Panwar, U., Sharma, K., et al., 2020c. Shape-based machine learning models for the potential novel COVID-19 protease inhibitors assisted by molecular dynamics simulation. Curr. Top. Med. Chem. 20 (24), 2146–2167.

Nayarisseri, A., Khandelwal, R., Tanwar, P., Madhavi, M., Sharma, D., Thakur, G., et al., 2021. Artificial intelligence, big data and machine learning approaches in precision medicine & drug discovery. Curr. Drug Targets 22 (6), 631–655.

Negi, S., Juyal, G., Senapati, S., Prasad, P., Gupta, A., Singh, S., et al., 2013. A genome-wide association study reveals ARL15, a novel non-HLA susceptibility gene for rheumatoid arthritis in North Indians. Arthritis & Rheumatism 65 (12), 3026–3035.

Nguyen, T.T., Huang, J.Z., Wu, Q., Nguyen, T.T., Li, M.J., 2015. December. genome-wide association data classification and SNPs selection using two-stage quality-based random forests. BMC Genomics 16 (S2), S5.

Nicora, G., Vitali, F., Dagliati, A., Geifman, N., Bellazzi, R., 2020. Integrated multi-omics analyses in oncology: a review of machine learning methods and tools. Front. Oncol. 10, 1030.

Niu, B., Liang, C., Lu, Y., Zhao, M., Chen, Q., Zhang, Y., et al., 2020. Glioma stages prediction based on machine learning algorithm combined with protein-protein interaction networks. Genomics 112 (1), 837–847.

O'Callaghan, M.E., MacLennan, A.H., Haan, E.A., Dekker, G.South Australian Cerebral Palsy Research Group, 2009. The genomic basis of cerebral palsy: aHuGE systematic literature review. Hum. Genet. 126 (1), 149–172.

Omondiagbe, D.A., Veeramani, S., Sidhu, A.S., 2019. Machine learning classification techniques for breast cancer diagnosis. In: IOP Conference Series: Materials Science and Engineering, Vol. 495, No. 1. IOP Publishing, p. 012033.

Oskoui, M., Gazzellone, M.J., Thiruvahindrapuram, B., Zarrei, M., Andersen, J., Wei, J., et al., 2015. Clinically relevant copy number variations detected in cerebral palsy. Nat. Commun. 6, 7949.

Park, S.C., Jeen, Y.T., 2019. Genetic studies of inflammatory bowel disease-focusing on Asian patients. Cells 8 (5), 404.

Patidar, K., Panwar, U., Vuree, S., Sweta, J., Sandhu, M.K., Nayarisseri, A., et al., 2019. An in silico approach to identify high affinity small molecule targeting m-TOR inhibitors for the clinical treatment of breast cancer. Asian Pac. J. Cancer Prevention: APJCP 20 (4), 1229.

Pattarabanjird, T., Cress, C., Nguyen, A., Taylor, A., Bekiranov, S., McNamara, C., 2020. A machine learning model utilizing a novel SNP shows enhanced prediction of coronary artery disease severity. Genes 11 (12), 1446.

Pedregosa, G., Varoquaux, A., Gramfort, V., Michel, B., Thirion, O., Grisel, M., et al., 2011. Scikit-learn: machine learning in python. J. Mach. Learn. Res. 12 (2011), 2825–2830.

Prajapati, L., Khandelwal, R., Yogalakshmi, K.N., Munshi, A., Nayarisseri, A., 2020. Computer-aided structure prediction of bluetongue virus coat protein VP2 assisted by optimized potential for liquid simulations (OPLS). Curr. Top. Med. Chem. 20 (19), 1720–1732.

Psaty, B.M., Dekkers, O.M., Cooper, R.S., 2018. Comparison of 2 treatment models: precision medicine and preventive medicine. JAMA 320 (8), 751–752.

Qureshi, S., Khandelwal, R., Madhavi, M., Khurana, N., Gupta, N., Choudhary, S.K., et al., 2021. A multi-target drug designing for BTK, MMP9, proteasome and TAK1 for the clinical treatment of mantle cell lymphoma. Curr. Top. Med. Chem. .

Romero, R., Edwards, D.R.V., Kusanovic, J.P., Hassan, S.S., Mazaki-Tovi, S., Vaisbuch, E., et al., 2010. Identification of fetal and maternal single nucleotide polymorphisms in candidate genes that predispose to spontaneous preterm labor with intact membranes. Am. J. Obstet. Gynecol. 202 (5), 431–e1.

Ryu, S.M., Lee, S.H., Kim, E.S., Eoh, W., 2019. Predicting survival of patients with spinal ependymoma using machine learning algorithms with the SEER database. World Neurosurg. 124, e331–e339.

Shah, S.C., Kusiak, A., 2004. Data mining and genetic algorithm based gene/SNP selection. Artif. Intell. Med. 31 (3), 183–196.

Shankaracharya, D.O., Samanta, S., Vidyarthi, A.S., 2010. Computational intelligence in early diabetes diagnosis: a review. Rev. Diabet. Studies: RDS 7 (4), 252.

Sharda, S., Khandelwal, R., Adhikary, R., Sharma, D., Majhi, M., Hussain, T., et al., 2019a. A computer-aided drug designing for pharmacological inhibition of mutant ALK for the treatment of non-small cell lung cancer. Curr. Top. Med. Chem. 19 (13), 1129–1144.

Sharda, S., Khandelwal, R., Adhikary, R., Sharma, D., Majhi, M., Hussain, T., et al., 2019b. A computer-aided drug designing for pharmacological inhibition of mutant ALK for the treatment of non-small cell lung cancer. Curr. Top. Med. Chem. 19 (13), 1129–1144.

Sherif, F.F., Zayed, N., Fakhr, M., 2015. Discovering Alzheimer genetic biomarkers using Bayesian networks. Adv. Bioinforma. 2015.

Sinha, K., Majhi, M., Thakur, G., Patidar, K., Sweta, J., Hussain, T., et al., 2018. Computer-aided drug designing for the identification of high-affinity small molecule targeting cd20 for the clinical treatment of chronic lymphocytic leukemia (CLL). Curr. Top. Med. Chem. 18 (29), 2527–2542.

Smyser, C.D., Dosenbach, N.U., Smyser, T.A., Snyder, A.Z., Rogers, C.E., Inder, T.E., et al., 2016. Prediction of brain maturity in infants using machine-learning algorithms. NeuroImage 136, 1–9.

Somashekhar, S.P., Sepúlveda, M.J., Puglielli, S., Norden, A.D., Shortliffe, E.H., Rohit Kumar, C., et al., 2018. Watson for oncology and breast cancer treatment recommendations: agreement with an expert multidisciplinary tumor board. Ann. Oncol. 29 (2), 418–423.

Stehrer, R., Hingsammer, L., Staudigl, C., Hunger, S., Malek, M., Jacob, M., et al., 2019. Machine learning based prediction of perioperative blood loss in orthognathic surgery. J. Cranio-Maxillofacial Surg. 47 (11), 1676–1681.

Stern, S., Linker, S., Vadodaria, K.C., Marchetto, M.C., Gage, F.H., 2018. Prediction of response to drug therapy in psychiatric disorders. Open. Biol. 8 (5), 180031.

Sung, S.M., Kang, Y.J., Cho, H.J., Kim, N.R., Lee, S.M., Choi, B.K., et al., 2020. Prediction of early neurological deterioration in acute minor ischemic stroke by machine learning algorithms. Clin. Neurol. Neurosurg. 105892.

Tarca, A.L., Carey, V.J., Chen, X.W., Romero, R., Drăghici, S., 2007. Machine learning and its applications to biology. PLoSComputBiol 3 (6), e116.

Ullah, R., Khan, S., Ali, H., Chaudhary, I.I., Bilal, M., Ahmad, I., 2019. A comparative study of machine learning classifiers for risk prediction of asthma disease. Photodiag. Photodynm. Ther. 28, 292–296.

Natchimuthu, V., Abdalla, M., Yadav, M., Chopra, I., Bhrdwaj, A., Sharma, K., et al., 2022. Synthesis, crystal structure, hirshfeld surface analysis, molecular docking and molecular dynamics studies of novel olanzapinium 2,5-dihydroxybenzoate as potential and active antipsychotic compound. J. Exp. Nanosci. 17 (1), 247–273.

van Eyk, C.L., Corbett, M.A., Frank, M.S.B., Webber, D.L., Newman, M., Berry, J.G., et al., 2019. Targeted resequencing identifies genes with recurrent variation in cerebral palsy. NPJ Genomic Med. 4 (1), 1–11.

Vasilopoulou, C., Morris, A.P., Giannakopoulos, G., Duguez, S., Duddy, W., 2020. What can machine learning approaches in genomics tell us about the molecular basis of amyotrophic lateral sclerosis? J. Personalized Med. 10 (4), 247.

Wu, C.C., Yeh, W.C., Hsu, W.D., Islam, M.M., Nguyen, P.A.A., Poly, T.N., et al., 2019. Prediction of fatty liver disease using machine learning algorithms. Comput. Meth. Prog. Biomed. 170, 23–29.

Yadav, M., Abdalla, M., Madhavi, M., Chopra, I., Bhrdwaj, A., Soni, L., et al., 2022. Structure-based virtual screening, molecular docking, molecular dynamics simulation and pharmacokinetic modelling of cyclooxygenase-2 (COX-2) inhibitor for the clinical treatment of colorectal cancer. Mol. Simul. 1–21.

Yuan, F., Lu, L., Zou, Q., 2020. Analysis of gene expression profiles of lung cancer subtypes with machine learning algorithms. Biochimica et. BiophysicaActa (BBA)-Molecular Basis Dis. 165822.

Zarbakhsh, P., Addeh, A., 2018. Breast cancer tumor type recognition using graph feature selection technique and radial basis function neural network with optimal structure. J. Cancer Res. Therap. 14 (3), 625.

16 Applications of alignment-free sequence descriptors in the characterization of sequences in the age of big data: a case study with Zika virus, SARS, MERS, and COVID-19

Dwaipayan Sen[1], *Tathagata Dey*[1,2], *Marjan Vračko*[3], *Ashesh Nandy*[1] *and Subhash C. Basak*[4]

[1]Centre for Interdisciplinary Research and Education, Kolkata, West Bengal, India, [2]Department of Computer Science & Engineering, Indian Institute of Technology Bombay, Mumbai, Maharashtra, India, [3]National Institute of Chemistry, Hajdrihova 19, Ljubljana, Slovenia, [4]Department of Chemistry and Biochemistry, University of Minnesota, Duluth, MN, United States

16.1 Introduction

A large pool of data plays a very significant role in modern-day sciences. With time, scientific paradigms have evolved to reach conclusions with experimental verification of large sets of data. From macroscopic astrophysics to microscopic bioinformatics, all forms of scientific fields have their own vision regarding data generation and application. Biological sequences data, in particular, have exploded exponentially in the last few decades and have resulted in complex issues such as how to store, categorize, search and retrieve information of interest.

While in the early days of sequence analyses it was possible to scan the sequence data visually and then use simple computer programs to match and retrieve segments of interest, as instrumentation, computing power, programming tools and sophisticated software became increasingly potent, large amounts of data were generated and new paradigms developed to undertake such analyses and discover hidden patterns and meaningful sequence segments. Software such as Basic Local Alignment Search Tool (BLAST) began to become the most used software to compare and contrast sequences. However, the growth in databases of long, genome length sequences and total basic units of the biological sequences made this approach resource-hungry and time-consuming. Alignment-free software that relied on mapping and mathematical characterization of the sequences began to grow in

Big Data Analytics in Chemoinformatics and Bioinformatics. DOI: https://doi.org/10.1016/B978-0-323-85713-0.00024-4

importance and utility (Nandy, 2015). With such complex and data-intensive search requirements, it is natural to progress toward the big data concept. The new data eco-system seeks to provide more efficient handling of data and predictive data analytics, much required in bioinformatics studies today.

This is especially useful in collating and analyzing the huge amount of sequence data being generated. A single gene sequence of the Zika virus consists of more than ten thousand bases, whereas a sequence of severe acute respiratory syndrome-Coronavirus-2 (SARS-CoV-2) consists of around thirty thousand bases, and more sequences are being added to the databases each day. Besides, these sequences are always accompanied by lots of metadata, so structuring the data also involves cost in terms of time and space. With time, the size of stored data in bioinformatics has grown exponentially, as can be seen from NCBI database statistics.

Fig. 16.1A and B shows how with time the total amount of data has increased. Manipulating these large sets of data needs developed big data architectures. A basic tool in bioinformatics data analysis is traversing the sequence. A traversal of a sequence of length n requires $O(n)$ time and space, that is, an amount of time and space which is of order n. Although there are many alignment-free sequence descriptors, in some cases, it is necessary to compare two sequences, such as finding mutations. These comparisons require their alignment which takes $O(mn)$ time and space, when two sequences of length m and n are aligned (Baichoo and Ouzounis, 2017). So, while comparing two sequences of the same genome, it hits a complexity of square.

Big data has become necessary in almost all fields of science. In bioinformatics, for example, we see the advancements in The Cancer Genome Atlas (TCGA), mRNA expression analysis, protein expressions, ENCODE project, etc. So, it is not about specific regimes, the entire scientific paradigm is shifting toward an era of big data. Big data analytics primarily deal with finding patterns and correlations among large sets of data. These patterns reveal an insightful conclusion to the theories. From cosmic microwave background (CMB) radiation to recording atmospheric changes, these all are examples of continuous and humungous data-generating sources. Dealing with big data includes pattern recognition in these data. For instance, patterns in CMB can reveal the state of the universe at that time and atmospheric changes can

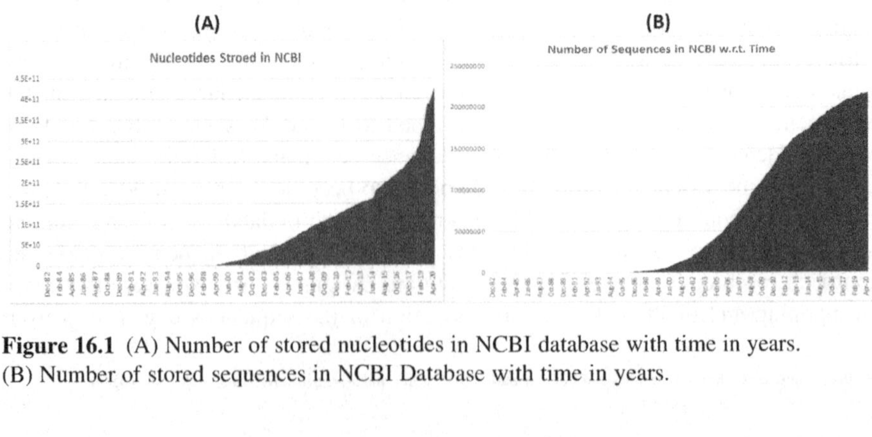

Figure 16.1 (A) Number of stored nucleotides in NCBI database with time in years. (B) Number of stored sequences in NCBI Database with time in years.

predict the weather. Radiation from multiple experiments on a particle in the Large Hadron Collider (LHC) can reveal the quantum information of that particle. In the KamLAND experiment, information on neutrino detection is recorded and performed rigorously. There are many such scientific experiments and data-generative sources which indeed need big data analytics to be handled fruitfully for pattern finding.

So, big data is a large complex data infrastructure that challenges traditional data storing and processing techniques. It is often defined in terms of three properties, which are the 3Vs (volume, velocity, and variety) (Diebold, 2012). Today's digital world contains a massive amount of data. For instance, the daily number of tweets generated is around 750 million, and around 7 billion google searches happen every day. With the advent of internet-of-things (IoT), many sensors are able to capture data automatically. Sources include sensors embedded in phones, wearable devices, video surveillance cameras, MRI scanners, set-top boxes, CCTVs, etc. Structured data are often in the form of excel or csv files or relational databases, whereas unstructured data refer to any raw form of data. As sources of data and their generation methods became widespread, various formats also were incorporated into the world of big data, such as texts, e-mails, messages, tweets, posts, web data, blog data, photos, audios, videos, sensor data, documents, and even sequences. These data are often in the form of structured, semi-structured and unstructured data of various formats.

International Data Corporation (IDC) estimated the size of the digital universe which was around 40 zettabytes (10^{21} bytes/zettabyte). As most modern-day devices contain storage of gigabyte order, we may imagine that, if each gigabyte in a zettabyte were a kilometer, it would be equivalent to 1,300,000 round trips to the moon and back. We see from Fig. 16.2 that the growth of the digital universe has

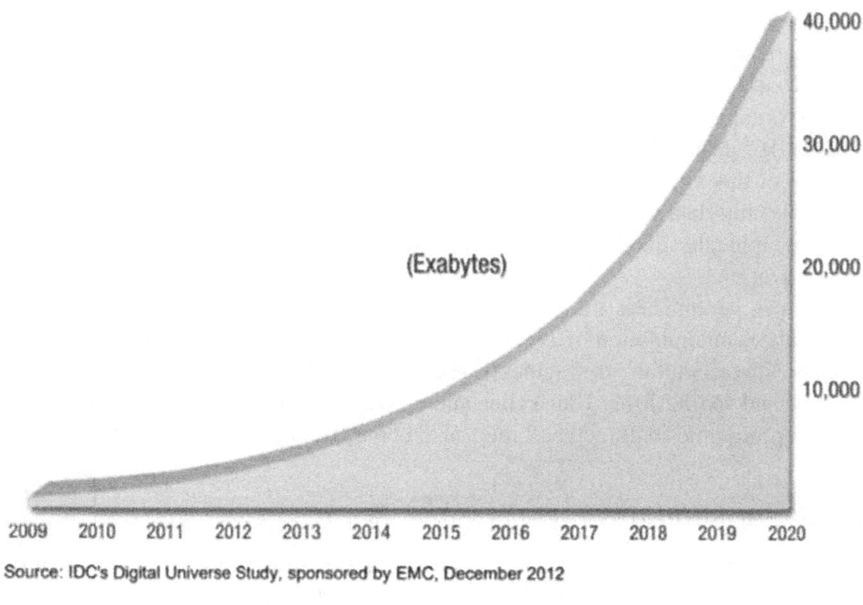

Figure 16.2 Size of the digital universe: International Data Corporation.

been exponential for the past decade and 90% of its growth has been in past few years only. So, it is very evident, that the era of big data is here more than ever. A report by IDC also reveals that this size can be four times today by 2025, making it nearly 175 zettabytes (Turner et al., 2014).

16.2 Section 1—bioinformatics today: problems now

16.2.1 What is bioinformatics and genomics?

Bioinformatics is relatively a new member of the scientific domain. It has a short history of nearly 50 years. It started with the study of inheriting information through biological systems; in simple words, with the study of deoxy-ribonucleic acid or DNA and ribonucleic acid or RNA. During the 1950s, the role of genetic material in transferring information was first discovered (Avery et al., 1944; Griffiths et al., 2000; Hershey and Chase, 1952). Later, Hesper and Hogewag named this field of study as "*Bioinformatics*" (Winkler, 1920).

The period from the 1950s to 1970s saw some revolutionary changes in computer science. From the 1960s, integrated circuit (IC) chips were being used in the development of computers. And 1966 onwards, the so-called "*third-generation*" computers were being used (Denning, 1971). This brought a huge change in the entire computational ecosystem. But this revolution was much more than just a technological explosion. It impacted a lot of previously uncharted territories where the use of computers was inconceivable earlier. Biology was one of these areas and it did not delay its response to this change. Meanwhile, the development of algorithms was incorporated into the use of analyzing biological information, which gave rise to *Modern Bioinformatics* (Gauthier et al., 2018; Hagen, 2000).

Modern bioinformatics deals with studying biological and genomic sequences through computers. Pattern recognition and sequence matching made it easier to identify organisms and track their evolutionary history (Ajith and Nair, 2019; Khan et al., 2016; de Ridder et al., 2013). With the growing complexities in storage and retrieval of this huge load of information, arrived technologies like distributed computing and big data to resolve these problems making technology more available to the geographically distributed scientific community and fueling continual research and innovation.

Whereas genomics is the study of genomes of organisms and interpreting their significance, manipulation of genetic material, that is sequencing DNA, recombination and incorporation of fragments are considered to be parts of genomics too (Bunnik and Roch, 2013; Hauskeller and Calvert, 2004). Multisequence complexities like proteomics and metabolomics also come under the genomics studies group.

16.2.2 Annotations

The information from genomics can be extracted or stored in a variety of different ways. For instance, genome annotations are a part of analyses. It is a "custom" process, often performed before publishing the genome information or storing it in a database. Annotating a genome refers to identifying various separate fragments of the gene sequence and if possible, marking their exact positions. Recognizing a gene in DNA, its RNA section and coded protein can form part of the annotation (Abril and Castellano, 2019; Koonin and Galperin, 2011).

16.2.3 Evolution of sequencing methods

Starting from the 1970s, DNA sequencing has come a long way. It all started at the MRC Center in Cambridge University, when Frederick Sanger first developed his technique to determine the sequence of bases in a DNA. He organized a gel-based method in which the DNA polymerase got mixed with some chain-terminating nucleotides. This method is known as **Sanger sequencing method** (Sanger et al., 1977). This chain-termination method marked the beginning of the **first generation of genetic sequencing**. Sanger sequenced phiX174 of genome 5374 base pairs and bacteriophage λ of length 48501 base pairs (Sanger and Coulson, 1975; Sanger et al., 1980). Later on, the Sanger sequencing method was automated and was used in applied biosystems for commercial purposes. This method was the primary concept used behind the massive Human genome project (Hood and Rowen, 2013). That project took nearly 13 years to complete with an estimated cost of **3 billion dollars** (Yourgenome, 2021). Sanger sequencing method is popular even today for low-throughput sequences.

The first generation of sequencing also contained another significant method, the **Maxam-Gilbert sequencing** (Maxam and Gilbert, 1977). They used the concept of chemical degradation. This concept is primarily based on breaking nucleotides upon chemical reaction. Due to the involvement of radioactive chemical reactions, it is sometimes considered to be dangerous too. Hence, a wide range of scientists prefer Sanger sequencing over this method.

The first-generation sequencing technique was dominant over the market for almost three decades. Sanger sequencing was definitely the most popular among them. However, from around 2000, the cost of sequencing started to make difference and it bothered the scientists' community. Half a decade later, new sequencing technologies started to emerge. These methods were cost-efficient in comparison to the first-generation methods and also completed the tasks in a much shorter time. They used to analyze multiple short strips parallelly and also there was no need for gel electrophoresis to get the final results. These advantages marked the change in generation and the **second generation in sequencing** began. A UK-based company named Solexa, acquired later by Illumina (Balasubramanian, 2015), innovated the "Bridge Amplification" method. This innovation helped sequencing with much-enhanced speed. Along with the Illumina platform, other technologies were also evolving, such as Ion Torrent Platform and Roche.

Illumina, after buying Solexa, rapidly commercialized their sequencing method. According to Kchouk et al., the output data of the last Illumina sequencers is currently higher than 600 Gpb (Kchouk et al., 2017). The error rate in this method is about 1%. Another significant technology was the Ion Torrent method (Rothberg et al., 2011). It used semiconductor concepts to extract base information from a sequence. This also has an error rate of nearly 1%. Roche or 454 Pyrosequencing and ABI or SOLiD sequencing also belonged to the family of second-generation sequencing technologies (Arora, 2019; Singh et al., 2013). The use of short reads made this generation methods much more efficient than the first-generation technologies. But evidently, this was not enough. Even further possibilities of improvement raised the concept of third-generation sequencing (TGS).

The second-generation sequencing technologies ruled the industry quite dominantly until scientists faced challenges further for sequencing. The complicated genomes contained multiple repetitive bases, which were insufficient to extract by the short read method. Also, during the amplification of sequences, polymer chain reaction (PCR) was a compulsory step. PCR amplification is a set of complicated stages which took a huge amount of time and expense. With the advent of newer technologies, scientists came up with better methods. Therefore, the onset of TGS was remarked.

The TGS was characterized by two primary ideas. The single molecular real-time (SMRT) sequencing approach (Ardui et al., 2018; Korlach and Turner, 2013; Shin et al., 2013) and the synthetic approach. SMRT is the most widely used TGS technology. It was developed by Quake Laboratory. Pacific Biosciences was the company that used it first. The sample preparation in this approach was quite fast over the previous ones. However, the error rate in this method is quite large, as much as 13%. On the other hand, Oxford nanopore sequencing (ONT) was developed using the synthetic approach. The MinION technology by ONT is quite cost-efficient and is of small size (Brown et al., 2017; Hughes and Ellington, 2017; Lee et al., 2013). This method can also ensure long reads, as long as 150 kbp. However, along with high throughput, accuracy is compromised here too. The error rate is about approximately 12% with a diverse distribution of 3% mismatches, 4% insertions, and 5% deletions.

Around half a century has passed since the first sequencing technology was launched. Since then, the industry has gone through a lot of advancement and development procedures. The next-generation sequencing (NGS) technologies, starting from 2005 have brought an enormous revolution in the efficiency of sequencing (Pareek et al., 2011). High throughput has been the primary criterion for these NGS methods, even when we had to compromise with accuracy. The articles referred here have a detailed discussion about the evolution of these technologies (Heather and Chain, 2016; de Sá et al., 2018). Table 16.1 shows a summary of sequencing technologies.

Although our current sequencing technologies are pretty evolved, but still there are options for advancement. For now, the various methodologies are incommensurable, that is, they lack common measures to compare (Besser et al., 2018; Buermans and den Dunnen, 2014; Kulski, 2016; Mohamad and Ho, 2011). Automation is another field of improvement. Although, these methods are increasingly being automated but still there is a long way to go. The significance of NGS is quite noticeable in the applicative fields of bioinformatics. Due to high throughput sequencing, information is being generated at a faster rate. This leads to a high

Table 16.1 Table to elaborate the efficiencies of each generation sequencing technologies.

Sl. No.	Name of method	Generation	Reads per traversal	Avg. length of read	Rate of error (%)	Data generated per run (Gb)	Year
1	ABI Sanger	First	96	400–900	0.3	0.00069 to 0.0021	2002
2	Roche	Second	100	700	1	0.07	2014
3	Illumina	Second	6B	150	0.1	1.8 Tb	2014
4	SOLiD	Second	6B	75	~0.1	160	2011
5	Ion Torrent	Second	15M-20M	400	1	03–05	2015
6	PacBio	Third	660	13500	12	0.5–1	2014
7	Oxford nanopore	Third	100	9545	12	1.5	2015

volume of data in the storage banks. The continuous inflow of high volumes of data demands improved computation techniques. Hence, big data architecture comes into the picture. Modern Bioinformatics is thus intensely integrated with big data (Chinmayee et al., 2018; Schmidt and Hildebrandt, 2017; Tripathi et al., 2016).

16.2.4 Alignment-free sequence descriptors

Alignment-based and alignment-free methods leave different impacts in terms of chemical aspects too. Precisely, in case of two structurally similar compounds with dissimilar elements, alignment-based methods cannot work. For example, let us consider two compounds, such as 2,3-dimethyl pyridine and ortho-xylene. Both of them have almost similar structures except for one position where the pyridine derivative has N while ortho-xylene has C. So, in that case, alignment-based methods tend not to work while, if we do not look for alignment and instead calculate molecular similarity separately, it comes in a sound way. This is the hand of advantage with PC analysis of multidimensional feature vectors. Similarly, this observation works good for biological sequences as well.

Alignment-free sequence descriptors (AFSDs) (Zielezinski et al., 2017) are the algorithms which analyze sequences without performing any alignment operations. To be specific in terms of bioinformatics, AFSD algorithms do not compare bases at every position of the sequences before idealizing their properties.

Let us talk about them in a more detailed way. Alignment-based methods often perform base-to-base or protein-to-protein comparisons in order to match the sequences. The basic assumption behind this rigorous step is that most of the real-world biological sequences have more or less conserved stretches and variability occurs only in a small length percentage. However, with rapidly evolving mutations and viral genomes, this notion is certainly quite questionable. The viral genomes are precisely small in length and replicate in a relatively small amount of time, resulting in the occurrence of multiple generations in the span of hours only. This helps them to provide mutations at a faster rate. So, this assumption tends not to work in them more.

On the other hand, aligning sequences may result in ignorance of deletion and addition mutations. For an example let us take two sequences of length 19 (aagctatcgatcctagata) and 18 (agctatcgatcctagata). BLAST analysis of these sequences shows more than 94% match, whereas such a beginning deletion could have changed the complete structure. With alignment-free descriptors the sequences are compared without specific positioning, hence all the addition and deletions result in noticeable value changes. Besides, it does not depend upon evolutionary pathways between the compared sequences.

Speaking in terms of big data, alignment-based approaches are computationally highly expensive and rigorous. Comparing multiple genomes can take up to $O(n^n)$. The possible alignment results of the two sequences increase with the increase in their length. The amount can be given as, $\frac{(2n)!}{(n!)^2}$. Although dynamic programming (Zhang et al., 2000) adds up a reachable real-time solution to this, still the cost doesn't blow away completely. Along with this, there comes a problem of storage in the intermediate stages. The sentential forms can take up to a large amount of storage with increasing global sequencing. On the other hand, alignment-free sequence descriptor algorithms simply traverse the sequence and compute. These minimal operations are often complete in real-time, that is, $O(n)$.

Furthermore, these AFSDs are classified into two major categories. One which works with a constant length subsequence, often regarded as the window length. On the other hand, algorithms use one-to-one correspondence. These alignment-free approaches are also essentially used in linear algebra, information theory, statistical mechanics and even in distance matrices. Our lab-proposed algorithms such as *Graph Radius* (g_r) and *Quotient Radius* (q_R) are of this category only (Dey et al., 2021).

Graphical representation and numerical characterization (GRANCH) bridges the gap between the Cartesian and the graphical space by collating the best of both in return. While in Randic plot (Pan and Chen, 1999) we observe a quadratic, that is, $O(n^2)$ expansion in space and time for the calculation and storage of the information gathered from a sequence of length n, Nandy plot provides a g_R (graphical radius) which constrains the growth till a limit for a given sequence as this g_R in itself becomes the numerical descriptor we need a constant space and a linear time to calculate this. Now a question surfaces, "what does big data have to do with it"? In answering this question, big data must be treated as a technology to store a large amount of fragmented information which is highly available but where the relationships among the data points cannot be qualified by the technology, but the design of the algorithm may lead them to an optimal result. For example, we can break the matrices into submatrices and store them on a large cluster in the case of Randic plot. But what may be a blocker on the way is reconstructing the matrix from this jigsaw puzzle and calculating the leading eigenvalue which is a computation-intensive process (Orman, 2016) anyway.

This makes us more wary about designing some optimized methods first and then focusing on bolstering technologies and not the other way around. Fundamental information theory may help in this area so that we only save the precious characteristic regions with high entropy (as the variability is large) of a single sequence than the entire sequence at large while storing or calculating the same for a given family of sequences. Big data may help us optimize both the storage and retrieval efficiency of these small bits and pieces of information about a sequence rather than the whole sequence alone. That will reduce the infrastructural and temporal expenses at the same time. But to do so, methods must always precede the materials in general.

16.2.5 Metagenomics

Metagenomics (Segre, 2021) is the technique of retrieving microbial genomes directly from the environmental samples regardless of the nature of the sample and abundance of microbial entities (Oulas et al., 2015). In other words, Metagenomics starts where Genomics ends. Genomics, being a more definitive science of extracting meaning from DNA or RNA sequences, one at a time, often needs a disconnect from the very nature where the organisms dwell and seek out seclusion in technology and laboratory for a more mathematical and theoretical churning than an organic one. Though it's very difficult to separate one from the other with their breaking boundaries, Metagenomics is a more organic and empirical science in the first place. This field of study is as concerned with the ecology of and intercommunication between microorganisms as their genomic content (Song et al., 2013). In nature, even a tiny bit of skin will have a host of DNAs clinging to each other (Dash and Das, 2018). A part of metagenomics observes these clusters of DNAs and their mutual impact than studying a single isolated specimen among them, to understand the symbiotic relationship from different and apparently disjoint points of view like biological, biochemical, ecological (Alves et al., 2018), physical and behavioral (Gilbert, 2015) ones. But it does not stop here, this field of study is expanding its breadth to graze more territories to gauge the limits of life itself. Thus, we need more statistical (Calle, 2019; Thomas et al., 2012) and computational power (Kobus et al., 2020; Mende et al., 2012) to deal with and understand this array of features present in such great variety and volume in the ecological symbiosis that we all belong to as cauldrons of microbes.

16.2.6 Software development: scenario and challenges

Programming has always been an integral part of Bioinformatics. Data analysis of bioinformatic information incorporates an active involvement of the silicon industry. Starting from the analysis of sequential patterns to ending with chemical structure analysis, everything involves computational tasks. From the 1950s, the field of computer science and bioinformatics evolved together.

Many software are available these days to compute through the datasets of NGS. These software use data analytical tools to visualize the datasets. Most of them are now available to use on local devices. Although, using a web-based server is recommended. With the advancement of NGS and other data sources, we are more and more discovering the core of big data world. Hence, computing these huge datasets in the hosted kernel is much healthy for local devices. Some of these software are listed in Table 16.2.

With advent of new tools, the data processing is becoming simpler every day. New descriptors are being defined regularly. Hence, with innovation of new descriptors, new tools have to be made to make the computation easier. As datasets are being large, online web-based software are being more popular every day (Ahmad, 2013; Duck et al., 2016).

There are also programming languages modified which make the work more efficient and convenient. Such languages are BioPython (Cock et al., 2009), BioJava (Holland et al., 2008) and BioRuby (Goto et al., 2010).

16.2.7 Data formats

An explosion is taking place at the heart of modern genomic studies; and data formats there have a key role to play both as a catalyst and an inhibitor (Masseroli et al., 2016) at the same time. As for any complex system, options are indicator of

Table 16.2 Important software in the use of bioinformatic study.

Sl. No.	Name	Usage
1	CLCBio (Smith, 2014)	Analyses and visualizes genomic data. Also able to transcript epigenomic information.
2	DNASTAR (Burland, 2000)	Analysis of Sanger Sequencing capabilities.
3	BaseSpace	Solexa sequencing.
4	Geneious Basic (Kearse et al., 2012)	Analyses biological data to show interactive visualizations.
5	Galaxy (Giardine, 2005; Goecks et al., 2010)	Scientific workflow system to show gene expression, proteomics and genome assembly.
6	Globus Genomics (Madduri et al., 2014)	Rapid analysis of large NGS data.
7	PATRIC (Wattam et al., 2013)	Computes protein-protein interaction, structure and other genomic informations.
8	UGENE (Golosova et al., 2014; Okonechnikov et al., 2012)	Annotation of molecular sequences, alignment and ORF finding.
9	CiteAB (Helsby et al., 2014)	Antibody database that ranks them.
10	NCBI (Wheeler, 2000)	Database of genomic sequences and research articles.
11	Protein Data Bank (Berman, 2000)	3D shapes of protein, nucleic acids.
12	Snap Gene	Deals with molecular cloning.
13	DNAMAN	Deals with sequence analysis and data mining.
14	IEDB (Vita et al., 2014)	Epitope database and prediction software
15	BLAST (Syngai et al., 2013)	Alignment search tool.
16	PatchDock (Duhovny et al., 2002; Schneidman-Duhovny et al., 2005)	Performs molecular docking.
17	iMutant (Capriotti et al., 2005)	Performs DNA polymorphism, protein folding and stability check.
18	PyDock (Grosdidier et al., 2007)	Docking using electrostatics and desolvation energy.
19	Chimera (Pettersen et al., 2004)	Used for visualization of protein structures.
20	Cn3D (Porter et al., 2007)	3D structure of sequences in NCBI database.
21	Mega X (Kumar et al., 2018)	Alignment of genomic sequences and phylogenetic relations.
22	SABLE (Adamczak, 2021)	Predicts surface exposure of amino acids in a protein.
23	Phyre 2 (Kelley et al., 2015)	Analyses protein structure function and mutation.
24	Raptor X (Källberg et al., 2012)	Deals with distance based protein folding.
25	PROSPER (Song et al., 2018)	Proteolytic specificity prediction.

progress. But if the options get too many, the process gets a little confused and draws itself a bit back in time. That is what has happened to the data format spectrum in this field. FASTA, FASTQ, SAM/BAM, BED, GFF/GTF, VCF, mmCIF are just a famous few to name among them with widely varying characteristics. There is also a rising debate in case of using full-genome data vs whole-exome sequences (Pabinger et al., 2013) for that matter which diversifies the formats even more. This all leads to the problem of plenty, because the quality assessors or the processing engines should be supportive of all these formats, at least prepared for a fair number of conversions (Yu et al., 2018) from one format to the other, to stay ahead and wide awake with the waking of many more formats with the passage of time.

16.2.8 Storage and exchange

Proliferation of varieties in data formats of biological sequences is a sibling to the explosion in volume of digital sequence data along with the rise of cheap, effective and fast sequencing technologies. This voluminous tide of information has been quite a problem (Marx, 2013) from both storage and analysis points of view. Even if large data centers are devised for larger organizations, exchange of information plays the key role for all of the scientific fraternity let alone the processing of such. But this transmission of information requires both a common space for storage and a network bandwidth to support for local synthesis. Handling the transmission leads to storage of this humongous data on the local infrastructure of smaller organizations which is again a hefty burden. Regular data clean-ups for all such transitional repositories (Baker, 2010) are needed which demand for the analyses to be faster, so that same data does not sit for too long coping with the rapid generation of newer sequences. This leads to a vicious cycle of a systematic crunch in computational resources and mechanisms starting from shared storage, through network connectivity, to local repositories and lastly the computational power of all the systems combined to handle this flood of information.

16.3 Section 2—bioinformatics today and tomorrow: sustainable solutions

The problem is almost settled; and that is concerned more with the infrastructural limitations than intellectual ones, if speaking of the current scenario in Bioinformatics. This ever-increasing astronomical demand for storage and processing power is our biggest impediment toward the progress of Biology and Information systems at the same time; but so is our brightest path to glory. If possible, this can get us more information, rather wisdom than we ever could conceive of, but if not—let us not speak of it for a while when it is certainly possible to do so and that is mostly for the bliss that Big Data is!

16.3.1 The need for big data

"The definition of big data?" "Who cares? It's what you're doing with it," once said Franks (2021), an ex-teradata analytics thought leader and much more. Just like his profile, big data itself demands a lot of space, not just to store but to process it, inside our own heads. But let's pay Bill more attention, it is not "what it is" but "what is does," that matters the most. Among a host of other definitions, a more actionable definition sounds like, "Big data refers to a process that is used when traditional data mining and handling techniques cannot uncover the insights and meaning of the underlying data. Data that is unstructured or time sensitive or simply very large cannot be processed by relational database engines. This type of data requires a different processing approach called big data, which uses massive parallelism on readily-available hardware."

Necessity is the key, just like in all the paradigm-shifts that have ever happened as in the discovery of fire; and just like fire, big data too spread fast and more it spread, more ubiquitous it became, because of the amount of value it could bring to an increasing array of problems on the table. But why was there a necessity? Because traditional systems were great, but they were not prepared for the current problems in hand. We needed something new and big at the same time. Kuhn's concept citing the revolution of disciplinary matrices applies quite literally here, as we move from the dense matrices of micro-proportion (Relational Database Management System or RDBMS) to macro-dimensional matrices consisting of a galaxy of loosely coupled and apparently nuclear submatrices, that is, the Big Data system in general (Orman, 2016).

Looking closer into the problem, there are three increasing dimensions that make big data so very different from the traditional systems. Not only that, these dimensions may help to recognize whether a problem can be solved by big data or not (Konkel, 2013); and they all start with Vs!

Fig. 16.3 already has already shown a lot more about the things we are about to talk about. So, as it has got evident by now, the emergence and sustenance of big data very much relies on these three Vs, namely volume, velocity, and variety to say the least. There are many other factors too, which help us to identify a big data problem, and more so to test a problem's need for having a big data solution in general.

Let us get going then, what and why these three are:

16.3.1.1 Volume

This is the most obvious compared to the other two, but quite not so! Yes, big data should be big, that is for sure, but how big is big data? Or how big the data should be to be eligible for a big data solution (whatever that be, for now)? This answer is relevant because human languages are subjective and English is not an exception. These adjectives are very subjective and personal, but computing systems are not so much. So, when we are speaking of a scale of 100 TB of data, most of the traditional database systems would be happy to handle that load, provided the database design is solid. It is also applicable to any volume less than this limit too. Now

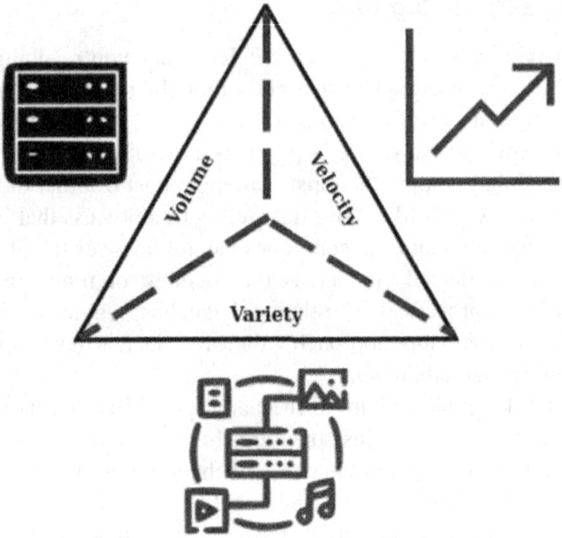

Figure 16.3 3Vs of big data.

coming to a higher order volume, like 1000 TB or 1 PB, things will start to get a bit too interesting to make the traditional systems stumble. Here big data will be our choice by default. So, speaking from the volume perspective, big data starts off when traditional database systems retire, that's pretty much it.

Velocity: This is more gripping, and subtle as well. Let us suppose you have built a computing system which will analyze 10 randomly sampled sequences added by NCBI in every 24 hours. RDBMS or any no-Structured Query Language (SQL) d like Mongo DB would be quite perfect for this. But after a week or so, being unsuccessful in finding a pattern, you start to analyze every other sequence that gets added in the NCBI database to specialize on a problem with SARS-CoV-2. You may not need to move to big data for some weeks may be, but with a constant increase in your data volume (i.e., velocity) will certainly shoot past the traditional ceiling in a month or two. That means your problem is big enough to ask for a big data solution down the line. So why not now?

16.3.1.2 Variety

Traditional databases like MySQL or Oracle are not very well acquainted to deal with the variety of unstructured data ranging from text files to video files. This is because traditional databases are founded on strict schema principles, which support stricter datatypes and columns for examples. Today's era of information (or misinformation) is not so strict both in their type and their quality. To store these, we need some kind of an accommodative indifference, that is, you will be served like a guest in a guesthouse, not like in home. There are not so many restrictions until you are paying for it. So are big data systems, they treat everything as a collection of blocks of information. That's how it can support variety.

16.3.2 Software and development

Some of the attributes of big data are strikingly different from that of the conventional infrastructure we had. With the change of time, demands change and a shift in the demand space, is followed by an inevitable change in the technological space, because technologies are the proprietors of the supply chains, these days. Some of the demands are as follows:

- Support for huge volume;
- Optimal efficiency in storage;
- Good data recovery solution;
- Horizontal scaling;
- Cost effective;
- Ease of access and understanding.

It seems very complex (almost impossible) at the first glance with the traditional toolkit, complex not because it's complicated but as it is new. We need to unlearn a lot of conventions that we took for granted to embrace newness. But once we are free from the conventional bias, a very simple pattern starts to emerge. All these properties start to seem very evident. Why didn't we think of these in the first place?

16.3.2.1 Support for huge volume

If the data is big, that is voluminous, it should require more storage space. Quite obvious.

16.3.2.2 Optimal efficiency in storage

Although we will need more storage, yet there's no point in being wasteful, because storage needs more physical space and electronic devices, which are run by power. All of the three (space, device, and electricity) cost money, so no way! Another important point here is that if the data are not arranged optimally, then read and write operations will become suboptimal too, which will reduce the performance and reliability of such a system.

16.3.2.3 Good data recovery solution

When a system gets big, it becomes less maintainable just like a mansion takes a lot more caretakers than a studio apartment does. So, a large chunk of anything is vulnerable. Like the saying goes, we should not keep all our eggs in the same basket. But here's a catch! Here eggs are not all the same, they're blocks of information and a block is significantly different from another block, unlike eggs. So, if a block of data is in basket A (i.e., node A), it will become unavailable if the basket goes missing (i.e., the node crashes). But our data is our primary property for which this castle of big data is made. We'll do a simple thing, but an effective one. We will keep a copy of block 1 in both node A and node B, so that even if node A (or B) meets a crash, we still will be able to get back our precious block. This is the recovery strategy, a fallback method just as valid as a plan A-B system.

16.3.2.4 Horizontal scaling

What does it mean? Let's take an example of a house. Suppose you have a little sweet cottage in the countryside with a huge plot. You are young and alone. The home needs someone and you find your partner to start a family. Lonely one becomes a couple. A year passes by. Now two becomes three. A little space crunches. But still manageable. After another year or two, three becomes four. You build another floor. Good! Five years pass by. Your brother goes broke (just suppose); his family comes to your place. Don't have enough space? Build another floor (you are a rich fellow, and kind too). Another ten years pass by happily. Your son falls in love with a girl, they want to marry, space crunch again! You start to build another floor, but Government officials come knocking at your door, "enough of floors," they say. "Government can't grant you provision for another floor, as your foundation can't withstand that load." Fair enough (you're a reasonable guy)! What next? Your plot is huge, do you forget? Then build a separate cottage next to your house, for your son's family! This is horizontal scaling. All our systems, that is, bare-metal servers have a capacity, what to do if you outrun that? That means it is not vertically scalable anymore, meaning you cannot build another floor. We borrow another server. Like that, you are actually building a resilient distributed network of computers in place of a single node, thereby reducing the dependency on any single node for that matter, which actually resonates the last principle that we discussed, better recovery. It's all about reducing the risk without losing a bit.

16.3.2.5 Cost effective

This thing we have conceptually covered by this time as this is the underlying contract for anything we do. But there's something more to it. As we discussed recovery and horizontal scaling at the same time, that too for a huge volume of data, this very principle sounds counter intuitive, doesn't it? When we are keeping multiple copies of the same block of data, we're increasing redundancy, how redundancy can be cost effective? It can, because your data is your money, you're storing this huge volume just to ensure that you don't lose it. Right? So, if you get rid of this redundancy, your data security goes down along with it, and once you lose your data, you lose money. So, it is cost-effective! Now comes the horizontal scaling stuff. For vertical scaling we used to need more specialized hardware with specialized capacity, but horizontal scaling does not impose any certain capacity for that matter. A row of commodity hardware on a distributed computing network can be as useful and a lot inexpensive than a single node of hardware with the same sum of their capacities. And ideally it is impossible too. So horizontal scaling makes it cheaper, secure, feasible and flexible at the same time. Nothing is ideal and there is a downside to it anyway, and that is that all these nodes need to have a connection in between, meaning a network. But that's pretty insignificant compared to all these gains.

16.3.2.6 Ease of access and understanding

Traditionally data were consumed and processed by engineers and statisticians or economists used to hopelessly count on these otherwise fine folks. Playing with data needed an engineering skill set. Let's consider that traditional system as a restaurant. The ingredient was data, chefs were engineers and connoisseurs were statisticians and economists. But in today's world, with the advent of Data Science, all these stakeholders need to have the same level of access to get and cook their data as per their own recipes. That means the cooking should be easier but as effective as before so that all these chefs get equality, justice and importance at the same time.

Okay, enough details about the design principles for big data! Let's get down to some implementations. A brief sneak peek into the history of these.

Doug Cutting and Mike Cafarella (Bag, 2020) were working on Apache Nutch for a while since 2002. Apache Nutch was conceived to be an open-source search engine, but the cost of maintaining such with the traditional technology stack seemed to be quite intimidating for a nonprofit open-source project. Redesign started (necessity)! In 2003, Google released a whitepaper on their very own Google File System (Ghemawat et al., 2003) providing a high-level design of a distributed file system on multiple nodes. But that quite solved the entire problem, because though the storage solution sounded pretty neat, yet the mechanics of processing these data on a distributed system remained as challenging as it was. That also didn't take that long; with the publication of another Google paper on MapReduce technique (Dean and Ghemawat, 2004), in 2004, an entirely new path of study emerged on its own.

These were all on paper, like a design framework without any real-world implementation whatsoever. Cutting decided to make the implementation open-source to reach a larger mass of people and collaborated with Mike Cafarella on this purpose to include the same into the ongoing Nutch project. But scale soon stood in the way for these two people. They now needed some organization with a fund and interest to invest in these areas. Yahoo matched this expectation. The year 2006 saw Doug Cutting joining Yahoo and an army of like-minded soldiers eager to operate on the same mission as his.

In January 2008, Hadoop (Agarwal, 2019; Apache Hadoop, 2021) was released as an open-source project under the umbrella of Apache Software Foundation by Yahoo. It revolutionized the entire ecosystem ever since.

16.3.2.6.1 Why "Hadoop"?

Hadoop (Bappalige, 2014) was the name of a stuffed yellow toy elephant belonging to Cutting's son. It was easy, sweet and who doesn't love a soft stuffed elephant on their lap?

16.3.2.6.2 What is Hadoop?

It is very difficult to frame the purpose and function of Hadoop in a couple of words. Let's look into the Wikipedia definition for the same:

> *"Apache Hadoop (/hə'du:p/) is a collection of open-source software utilities that facilitate using a network of many computers to solve problems involving massive amounts of data and computation. It provides a software framework for distributed storage and processing of big data using the MapReduce programming model. Hadoop was originally designed for computer clusters built from commodity hardware, which is still the common use. It has since also found use on clusters of higher-end hardware. All the modules in Hadoop are designed with a fundamental assumption that hardware failures are common occurrences and should be automatically handled by the framework."*

Main components of Hadoop (Narula, 2019) are:

- **Hadoop common:** The common utilities which are used across the product.
- **Hadoop distributed file system (HDFS):** The distributed file system that holds the entire Hadoop architecture in place on a decentralized model. We'll come back to it.
- **Hadoop YARN:** Hadoop YARN stands for Yet Another Resource Negotiator, which has evolved into an almost large-scale distributed operating system focused on big data processing.
- **Hadoop MapReduce:** This is a Hadoop implementation based on the MapReduce model which controls the processing principle of any system using Hadoop as a distributed storage system.

16.3.2.7 Overview of Hadoop distributed file system

HDFS is a distributed file system. All the computers we use, like laptops, desktops or mobile phones, they all have file systems. So, what's so special about HDFS? All the other file systems we have on the aforementioned devices, are local in nature, that is, you can share file from your mobile phone to another through Bluetooth or email or any other messaging service for that matter, but you cannot combine both devices' file systems and think that you only have one. You can transfer stuff between then, but you're unable to access them as two different directories on the same system. HDFS can give you that very power. In a single cluster of Hadoops nodes, you can store blocks of data seamlessly without bothering about their physical locations and that too in a fault tolerant way.

16.3.2.8 Overview of MapReduce

Let's suppose you're a bachelor guy who stays in a rented apartment with three other flatmates with the same status. They're working, professionals, that is, they are almost always focused on their own business for the entire week except for—it is coming. Okay, sufficient introduction, I suppose. Then weekend comes and no; weekend means not only fun; it means shopping too. And a bachelor's weekend-shopping means getting groceries, veggies, meat and dairy. These guys don't even party! I see that you're already feeling sorry about yourself (you're one of these guys, remember). But please don't, more pities are on the way. Now, after shopping you guys need to meet at your apartment and start cooking. This is what MapReduce (Mr) basically is! (Looks like you are fortunate that you thought you were) These four guys are what we call **mappers**, who do a similar job on different premises or nodes, like while you played the grocery shopper, your friend was buying some meat, that is, doing the same thing but on a different instance. Now you guys are lucky enough to have a cook, so each one of you will submit their purchases before this cook and this step in the Mr jargon is called **shuffling**, where each mapper transfers their own data (groceries in your case) to the necessary reducers. Here, there's only one reducer, that is, your cook. But in a more real-life scenario, there should also be other **reducers** for that matter, like a cleaning maid may be, who will consume liquid dishwashers and floor cleaners. Basically, reducers are the ones who consolidate things after they get processed, but in this case your cook will also prepare the food by consuming these data, that is, process it afterwards. To solve a big computational problem, where subroutines can be parallelized, we use Mr. The programming principle backing this is called divide and conquer. This is a pretty old concept in computer science. But Mr has become the de-facto style of programming on any distributed computing system like Hadoop. Working sequentially with big data can cost us a month or so which can be done in fifteen minutes with fifty separate nodes using Mr.

16.3.2.9 Some problems with MapReduce

Mr is a great concept indeed, but there lie some problems with the implementation of MapReduce in Hadoop. They are as follows:

- Lack of support for conceptual visualization.
- Strict programming language support only for Java, which demands a steeper learning curve.
- Requirement of extensive coding even for simple and more common routines like a simple join.
- As a result of these, a lot of time and effort goes into writing a simple Mr job from a development point-of-view.

Looking at all these problems, it seems much like a usability problem rather than a design or a functional one. Fortunately, most of the time we see a common problem template when writing an Mr job which is shown in Fig. 16.4.

So, won't it be useful, if we can write simple codes reusing some of these blocks? That's exactly what Apache Pig does!

16.3.2.10 Apache Pig

Soon after Hadoop was released, nonprogrammers like Data Scientists, Testers or Database Developers observed an increasing need of using Hadoop. But programming being not their niche, writing Mr jobs felt like a painful process. In order to simplify the usage without missing out on the efficiency of an Mr job, Apache Pig was born in 2008, initially released by Yahoo just like Hadoop. Apache Pig is a client

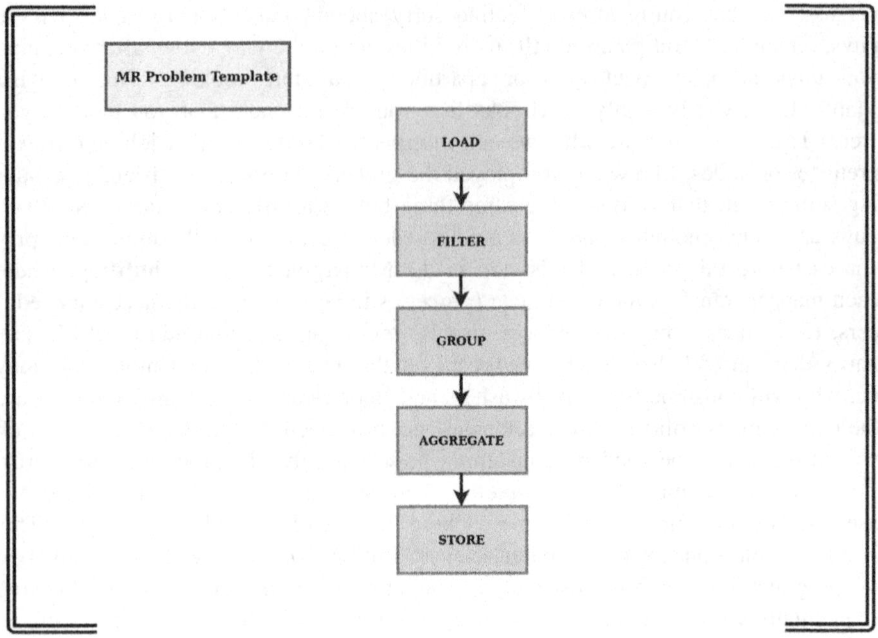

Figure 16.4 A foundational MapReduce problem template.

tool which need not be deployed on each node of a Hadoop cluster. It comes with an in-built data-flow language called Pig Latin, which helps Mr job developers with a rich set of instructions to write Mr jobs which in turn get fed into an engine to translate the instructions into raw Java code for execution as a Hadoop Mr job on your Hadoop cluster. So, Pig can be well off staying only one node in a cluster with an access to the entire cluster, it'll need nothing more. The Pig Latin instructions corresponding to the common stages that we discussed earlier are shown in Fig. 16.5.

Pig also performs some optimizations on the Pig Latin instructions that you'll write for a more efficient execution of the actual resultant Mr jobs. Pig is also based on a set of philosophies:

Pigs eat anything: Pig doesn't care for a strict schema or structure in the data that you feed it.

Pigs fly: Keeping the big data needs in mind, Pig comes with an optimizer which make a less performant Pig Latin script into a well-designed MapReduce job to provide better performance.

Pigs are domestic animals: Pigs can be domesticated and so does Apache Pig. They stay on one node and they don't misbehave with their masters (or users). Pigs allow writing user-defined methods in Java and provide a seamless integration experience. In short, it provides a way of overcoming any limitation on a developer's way while making his life easier.

Pigs live anywhere: Pig Latin is meant to be a language for parallel data processing. Creators of Pig envisage a future where Pig can work with any distributed computing frameworks along with Hadoop.

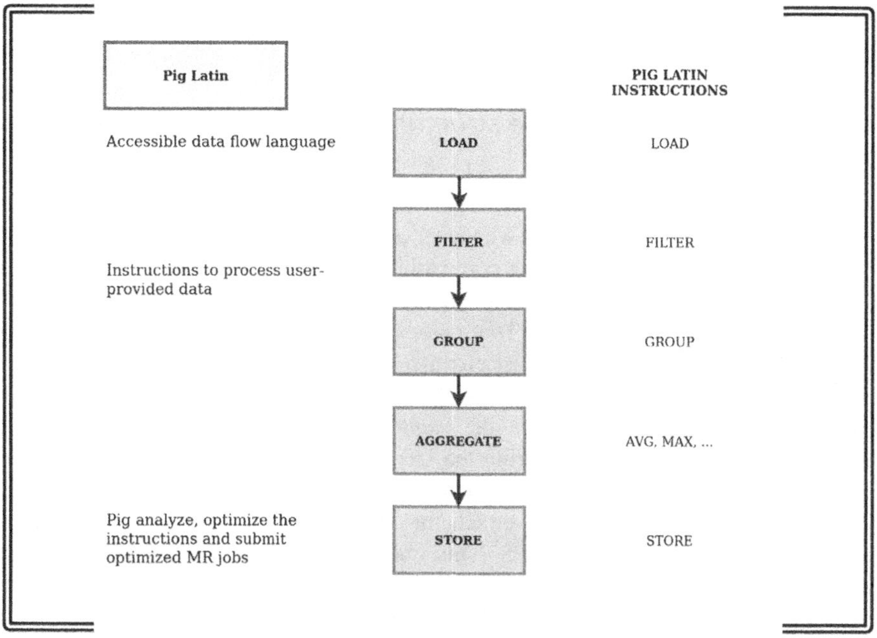

Figure 16.5 Pig Latin workflow: the series of operations that get performed before execution of a MapReduce job.

16.3.2.11 Data formats

We have already discussed about the problem with the proliferation of data formats for the sake of propriety and even the ease of accesses to some extent. But this focus on the specificity of use-cases lead to a mess while creating something flexible as well as less complex to handle all the data that are there along with their processing and storage for historical and analytical reasons. Let's keep the storage part aside for now. But how about the processing?

Processing is almost as critical a process as sequencing itself. It requires a certain way of reading and parsing every other document (Cappelli et al., 2020), that is, sequence in this case, being wary of its format and notations yet decomposing it to a very similar structure in accordance with the other formats. Genomic Data Model has evolved to be a major breakthrough in this area with another tool of GenoMetric Query Language (GMQL) (Masseroli et al., 2015) to create a queryable abstraction to extract and perform tertiary data analysis over almost all of NGS datasets. This technology employs and exploits (Masseroli et al., 2018) a diverse technical stack, though without any particular bias toward any, such as Spark, HDFS, Python, R, Flink, REST API, docker, etc.

Speaking of these, it becomes quite necessary to mention the role of SQL in the entire world of data, just to pull out some miniscule incidents from an ocean of facts and dimensions. GMQL in here very much follows the same principles and looking into big data in general, we find the following more general query system than language.

16.3.2.12 May I have some structured query language please?

This question is often heard in the world of data. SQL is loved and used by so many people from varying disciplines like data science, database engineering, backend development or data analysis, that whenever a technology emerges to handle data, an expectation hangs in the air seeking compatibility with SQL. Hadoop is not an exception too. Everything is file while using Hadoop MapReduce because it is built on top of HDFS, which is a file system. That causes a lack of understanding of the data at large as querying the data is not that effective. To mitigate this demand, **Hive** comes into the picture. It is very similar to Pig and there are certain intersections that both have because both want to solve the same problem, improving the usability of MapReduce. While Pig solves it from the scripting point-of-view, Hive focuses more on the querying front. Hive was originally built by Facebook, and it is a client-side tool too without any restrictions over schemas. It just provides a querying interface to the dataset sitting on top of HDFS. This not only makes the structure clearer and visualization effective but may also lead to writing better Mr or Pig jobs just by having a more holistic understanding of the underlying data.

16.3.2.13 Storage and exchange

Space is limited and expensive, which makes storage a daunting problem in itself. An approach to the solution for efficient storage of data supporting optimal retrieval gave birth to a variety of Database Management Systems (DBMSs). These were relational databases which are now deemed inefficient for today's bioinformatic volume and variety at large. Different strategies like Not Only SQL (NoSQL) (Wercelens et al., 2019) and big data are being tried to solve this problem depending on the shapes and sizes of such. International Nucleotide Sequence Database Collaboration (INSDC) works toward solving the problems of heterogeneity of formats and databases for sequence data, by introducing a consortium of databases namely DNA Data Bank of Japan (DDBJ, Japan) (Tateno and Gojobori, 1997), GenBank (USA) (Benson et al., 1994) and the European Nucleotide Archive (ENA, UK) (Rice et al., 1993). The almost universal syntax is called INSDSeq in which the data get stored on these databases for quick and easy accession throughout the globe.

Big data provides an elegant solution to this problem of many by its nondiscriminating and uniform treatment to all its residents. But it is not always the de-facto standard to adhere to all the big data principles, because it may enlarge the scope of the problem rather than provide a specific solution to the most impending crisis. We have already elucidated the core concepts of MapReduce. In conjunction with Hbase, MapReduce is being used to mine huge XML biological data collections for faster execution and neat storage (Liu et al., 2020; Alnasir and Shanahan, 2018).

Graph databases are also quite in use and exploration to make most of these available sequences that are there. These are databases designed on the concept of graph-like structures with nodes and edges to store and connect a collection of data points to others. Some of these implementations claim to reduce the query-time by a whopping 93% (Fabregat et al., 2018), whereas some are created supporting compatibility with but not limited to Systems Biology Markup Language (SBML) and JSON (Balaur et al., 2016) data formats at the same time. Languages like Cypher (Green et al., 2019; Johnpaul and Mathew, 2017; Kronmueller et al., 2018) also hold high hopes in the community.

So, there are many solutions to similar problems concerning storage and exchange for existing and forthcoming sequence data.

16.3.2.14 Visualization

> *"The ability to visualize data is an essential piece to the puzzle of data exploration and use ... the fear data providers have of having their data trivialized, misunderstood or misrepresented also apply to the process of making visualizations, even simple ones ... Visualizing data changes how data are understood and increases interest in data generally, which will encourage more and better data to be developed."*
>
> *—Robert Kosara*

Visualization is a must for anyone, be it a professional scientist or a newbie in a certain field. Pictures make things clear and often they can detect an anomaly faster than crunching the numbers on our own. But as the volume of the data peaks, visualizations become absolutely essential and difficult at the same time. Essential because the stakes become higher and the data points thicker, and more difficult just because of the same reasons.

But as technologies develop, science progresses—let's not get into the vicious loop for a second. ngs.plot is such a technological advancement (Loh and Shen, 2016) which has made the production of print-ready figures almost seamless with the right set of tools to tweak and tap into the potential of its intelligent projections of data points. Packages like ngsReports (Ward et al., 2019) take it to the next level by making scientific research reports even more appealing with report generation features which support interactive viewability on modern webapps and HTML pages. Apart from being one of the pioneering tools in genomic visualization word, Integrative Genomics Viewer (IGV) (Robinson et al., 2011) belongs to quite a different breed of software which can be used to cite any specific events or structures in a single genome.

However, interactivity is still a huge challenge in terms of effective zooming on any web-browser with an existing package for a really long sequence. But software created for the IT industries are now coming very handy for NGS data visualizations too, such as Tibco Spotfire or SAS (Wexler et al., 2013) packages supporting different graphs, charts and transformations at the same time at pretty megre expenses.

While the majority of GRANCH techniques rely on a few descriptors and visualization of their original graphical representations, they essentially describe the base composition in the case of nucleotide sequences. The information contained in codon structures of the sequences are seldom taken into account in these methods, for instance, the differences in viral sequences as they become endemic to different geographical regions, for example, in our approach to represent the RNA sequences we count the triplets of bases (Vračko et al., 2021a,b). In this approach, we count 64 different triples of bases and thus the dimension of representation space lies at 64. In this way, we encoded sequences of the Zika virus collected from different areas of the globe. In a further study sequences of SARS, MERS, and COVID-19 viruses. For clustering, we used the principal component analysis (PCA) and self-organizing map (SOM). Both methods are often used for the classification and clustering of data (Johnson et al., 1988; Vračko and Zupan, 2015). Fig. 16.6 show the score plot. One can see that the African samples are separated from Asia and South America ones. The SOM is shown in Fig. 16.7. The figure shows that the sequences from three continents occupy separate regions in the map. Inside the regions, one recognizes clusters of objects, which belong to particular countries or sub-regions (Vračko et al., 2021b).

A study was carried out for samples of SARS, MERS, and COVID-19 viruses (Vračko et al., 2021a). Fig. 16.8 shows the PCA score plot separating different virus types.

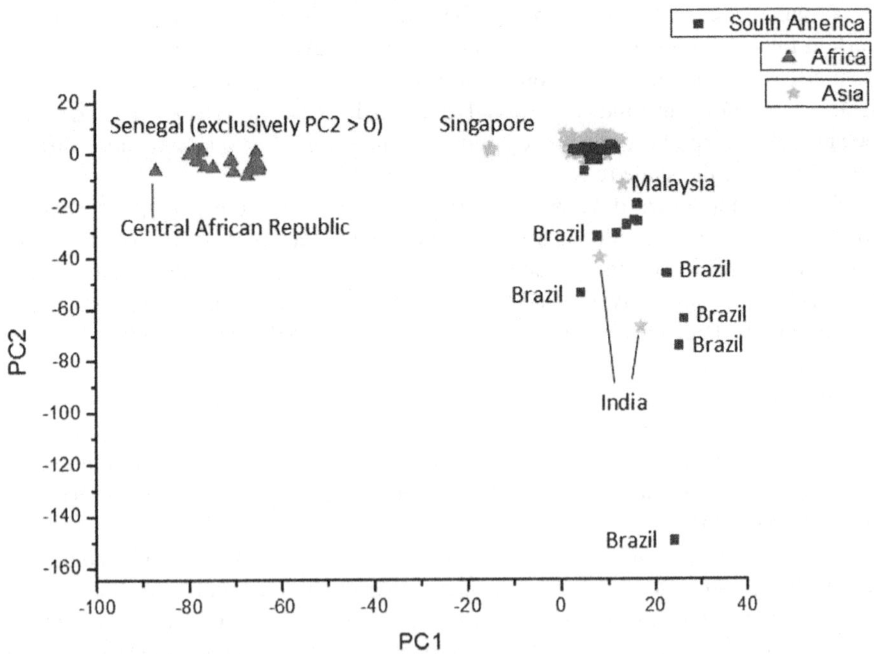

Figure 16.6 Score plots for Zika virus sequences described by the two PCs derived from 64 descriptors (triplets of bases).
Source: From Vračko, M., Basak, S.C., Sen, D., Nandy, A., 2021b. Clustering of Zika viruses originating from different geographical regions using computational sequence descriptors. Curr. Comput. Drug. Des. 17, 314–322.

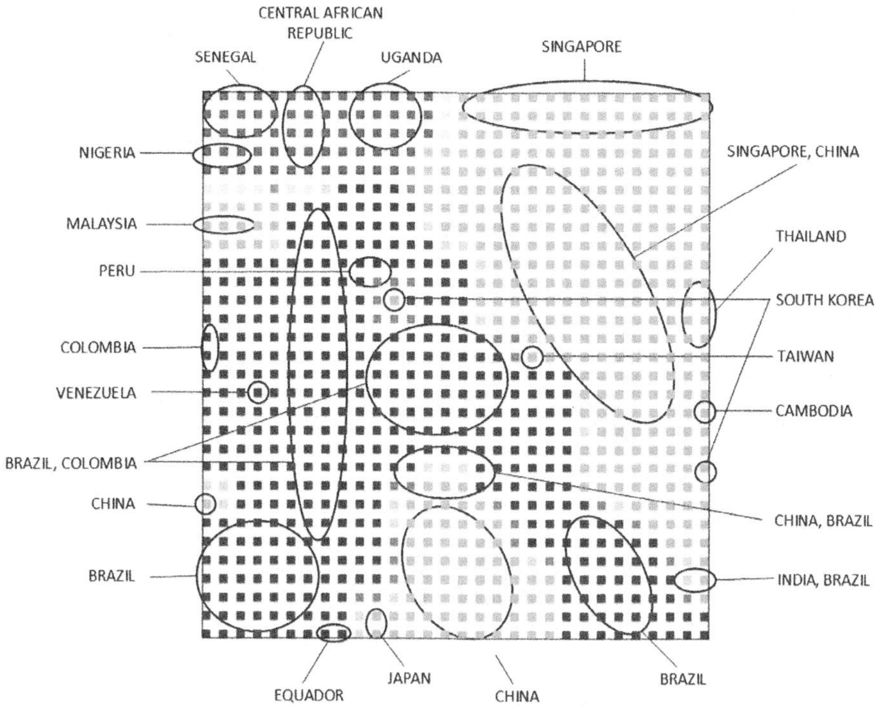

Figure 16.7 Distribution of Zika virus sequences in the self-organizing map (SOM) derived from 64 descriptors (triplets of bases). Red, green, and blue areas represent Africa, Asia, and South America, respectively. Magenta represents the overlap between Asia and South America and orange between Africa and Asia.
Source: From Vračko, M., Basak, S.C., Sen, D., Nandy, A., 2021b. Clustering of Zika viruses originating from different geographical regions using computational sequence descriptors. Curr. Comput. Drug. Des. 17, 314–322.

16.4 Summary

In summary, the growth of biological data has given rise to an extensive collaboration of Bioinformatics and Computer Science with respect to Alignment-free sequence descriptors (AFSDs), calculated by the GRANCH approach.

To interpret patterns in sequences, characterization is crucial but these rigorous algorithms for characterizing genomic sequences are computationally expensive. Comparisons within the family of AFSDs (Sen et al., 2016) have also become essential as some are deemed more complex than others in terms of asymptotic computational complexity (Cormen et al., 2001). Provided we can parallelize operations for these resource-intensive methods, with the likes of MapReduce in specific and big data technologies in general, it might be possible to revolutionize the availability of these descriptors, which in turn can boost the reliability of such an open new avenues up for

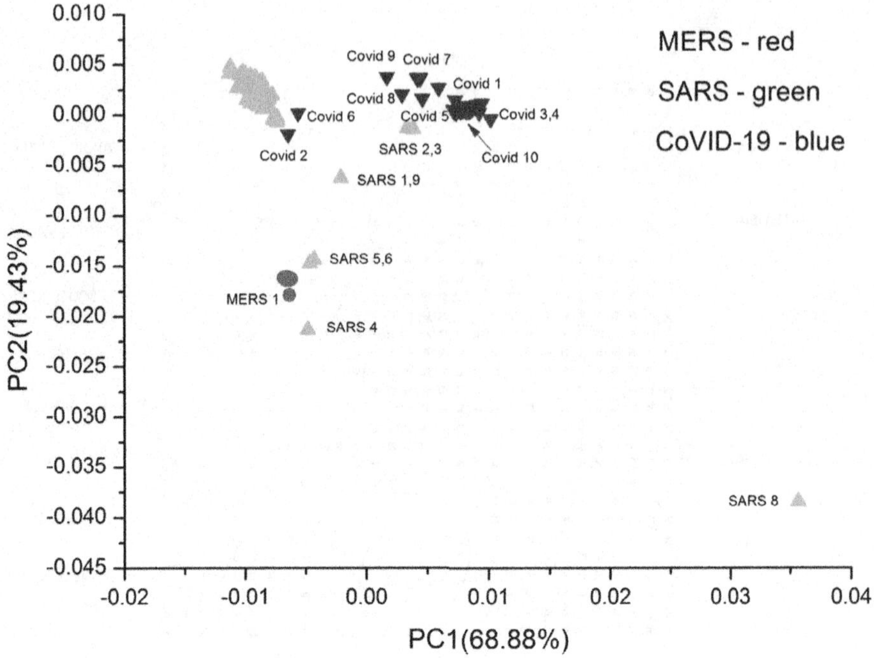

Figure 16.8 The score plot (PC1, PC2) of 573 sequences of SARS, MERS and COVID-19 viruses derived from 65 independent variables: 64 codon and gR.
Source: From Vračko, M., Basak, S.C., Dey, T., Nandy, A., 2021a. Cluster analysis of coronavirus sequences using computational sequence descriptors: with applications to SARS, MERS and SARS-CoV-2 (CoVID-19). Curr. Comput. Drug. Des. 17.

precise applications. That is why this chapter shows the problems that arise due to this and the possible solutions in the context of big data architecture.

This chapter is designed as a stepping stone for people and/or scientists who can either be newbies in the field of big data or are trying to migrate to the same from a traditional relational database system. This chapter highlights the metrics and signals to identify whether a problem really falls under the big data umbrella and paves a very elementary path to initiate the researchers just in case, with a specific focus on the branch of GRANCH under the broad field of Bioinformatics.

References

Abril, J.F., Castellano, S., 2019. Genome annotation. Encyclopedia of Bioinformatics and Computational Biology. Elsevier, pp. 195–209.

Adamczak, R.A.P., 2021. Sable [WWW Document]. SABLE protein structure prediction server. <http://sable.cchmc.org/> (accessed 11.25.21).

Agarwal, A., 2019. Hadoop: history or evolution [WWW Document]. GeeksforGeeks. <https://www.geeksforgeeks.org/hadoop-history-or-evolution/> (accessed 11.22.21).

Ahmad, T., 2013. Software tools in bioinformatics: a survey on the importance and Issues faced in implementation. Glob. Eng. Technol. Rev. 3 (3).

Ajith, V., Nair, A.S., 2019. Pattern recognition in bioinformatics. CSI Commun. 42 (2), 13.

Alnasir, J.J., Shanahan, H.P., 2018. The application of Hadoop in structural bioinformatics. Brief. Bioinform. .

Alves, L., de, F., Westmann, C.A., Lovate, G.L., de Siqueira, G.M.V., Borelli, T.C., et al., 2018. Metagenomic approaches for understanding new concepts in microbial science. Int. J. Genomics 2018, 1–15.

Apache Hadoop [WWW Document], 2021. Wikipedia. <https://en.wikipedia.org/wiki/Apache_Hadoop> (accessed 11.22.21).

Ardui, S., Ameur, A., Vermeesch, J.R., Hestand, M.S., 2018. Single molecule real-time (SMRT) sequencing comes of age: applications and utilities for medical diagnostics. Nucleic Acids Res. 46, 2159–2168.

Arora, P.K., 2019. Next-generation sequencing and its application: empowering in public health beyond reality. Microb. Technol. Welf. Soc. 17, 313–341.

Avery, O.T., MacLeod, C.M., McCarty, M., 1944. Studies on the chemical nature of the substance inducing transformation of pneumococcal types. J. Exp. Med. 79, 137–158.

Bag, S., 2020. Meet Hadoop [WWW Document]. Medium. <https://medium.com/@shrad-dhabag7583/meet-hadoop-d85795c2d587> (accessed 11.22.21).

Baichoo, S., Ouzounis, C.A., 2017. Computational complexity of algorithms for sequence comparison, short-read assembly and genome alignment. Biosystems 156–157, 72–85.

Baker, M., 2010. Next-generation sequencing: adjusting to data overload. Nat. Meth. 7, 495–499.

Balasubramanian, S., 2015. Solexa sequencing: decoding genomes on a population scale. Clin. Chem.

Balaur, I., Mazein, A., Saqi, M., Lysenko, A., Rawlings, C.J., Auffray, C., 2016. RECON2NEO4J: applying graph database technologies for managing comprehensive genome-scale networks. Bioinformatics .

Bappalige, S., 2014. An introduction to apache Hadoop for big data [WWW Document]. Opensource. com. <https://opensource.com/life/14/8/intro-apache-hadoop-big-data/> (accessed 11.22.21).

Benson, D.A., Boguski, M., Lipman, D.J., Ostell, J., 1994. GenBank. Nucl. Acids Res. 22, 3441–3444.

Berman, H.M., 2000. The Protein Data Bank. Nucleic Acids Res. 28, 235–242.

Besser, J., Carleton, H.A., Gerner-Smidt, P., Lindsey, R.L., Trees, E., 2018. Next-generation sequencing technologies and their application to the study and control of bacterial infections. Clin. Microbiology Infect. 24, 335–341.

Brown, B.L., Watson, M., Minot, S.S., Rivera, M.C., Franklin, R.B., 2017. MinION™ nanopore sequencing of environmental metagenomes: a synthetic approach. GigaScience 6.

Buermans, H.P.J., den Dunnen, J.T., 2014. Next generation sequencing technology: advances and applications. Biochimica et. Biophysica Acta (BBA) - Mol. Basis Dis. 1842, 1932–1941.

Bunnik, E.M., Le Roch, K.G., 2013. An introduction to functional genomics and systems biology. Adv. Wound Care 2, 490–498.

Burland, T.G., 2000. DNASTAR's lasergene sequence analysis software. In: Misener, S., Krawetz, S.A. (Eds.), Bioinformatics Methods and Protocols. Humana Press, Totowa, NJ, pp. 71–91.

Calle, M.L., 2019. Statistical analysis of metagenomics data. Genomics Inf. 17, e6.

Cappelli, E., Cumbo, F., Bernasconi, A., Canakoglu, A., Ceri, S., Masseroli, M., et al., 2020. OpenGDC: unifying, modeling, integrating cancer genomic data and clinical metadata. Appl. Sci. 10, 6367.

Capriotti, E., Fariselli, P., Casadio, R., 2005. I-Mutant2.0: predicting stability changes upon mutation from the protein sequence or structure. Nucleic Acids Res. 33.

Chinmayee, C., Nischal, A., Manjunath, C.R., Soumya, K.N., 2018. Next generation sequencing in big data. IJTSRD 2, 379–389.

Cock, P.J., Antao, T., Chang, J.T., Chapman, B.A., Cox, C.J., Dalke, A., et al., 2009. Biopython: freely available Python tools for computational molecular biology and bioinformatics. Bioinformatics 25 (11), 1422–1423.

Cormen, T.H., Leiserson, C.E., Rivest, R.L., Stein, C., 2001. Introduction to Algorithms. The MIT Press, Cambridge, Massachusett.

Dash, H.R., Das, S., 2018. Molecular methods for studying microorganisms from atypical environments. Methods in Microbiology. Elsevier, pp. 89–122.

de Ridder, D., de Ridder, J., Reinders, M.J.T., 2013. Pattern recognition in bioinformatics. Brief. Bioinforma. 14, 633–647.

de Sá, P.H.C.G., Guimarães, L.C., Das Graças, D.A., de Oliveira Veras, A.A., Barh, D., Azevedo, V., et al., 2018. Next-generation sequencing and data analysis: strategies, tools, pipelines and protocols. In: Barh, D., Azevedo, V. (Eds.), Omics Technologies and Bio-Engineering. Elsevier, pp. 191–207.

Dean, J., Ghemawat, S., 2004. MapReduce: simplified data processing on large clusters. In: OSDI'04: Sixth Symposium on Operating System Design and Implementation, San Francisco, CA, pp. 137–150.

Denning, P.J., 1971. Third generation computer systems. ACM Comput. Surv. 3, 175–216.

Dey, T., Chatterjee, S., Manna, S., Nandy, A., Basak, S.C., 2021. Identification and computational analysis of mutations in SARS-CoV-2. Computers Biol. Med. 129, 104166.

Diebold, F.X., 2012. On the origin(s) and development of the term "big data.". SSRN J. .

Duck, G., Nenadic, G., Filannino, M., Brass, A., Robertson, D.L., Stevens, R., 2016. A survey of bioinformatics database and software usage through mining the literature. PLoS ONE 11, e0157989.

Duhovny, D., Nussinov, R., Wolfson, H.J., 2002. Efficient unbound docking of rigid molecules. In Gusfield et al. (Eds.), Proceedings of the 2'nd Workshop on Algorithms in Bioinformatics(WABI) Rome, Italy, Lecture Notes in Computer Science 2452, Springer Verlag, pp. 185–200.

Fabregat, A., Korninger, F., Viteri, G., Sidiropoulos, K., Marin-Garcia, P., Ping, P., et al., 2018. Reactome Graph Database: efficient access to complex pathway data. PLOS Computational Biol. 14.

Franks, B., 2021. Bill Franks, thought leader, speaker, executive, and author [WWW Document]. <https://bill-franks.com/index.html> (accessed 11.22.21).

Gauthier, J., Vincent, A.T., Charette, S.J., Derome, N., 2018. A brief history of bioinformatics. Brief. Bioinforma. 20, 1981–1996.

Ghemawat, S., Gobioff, H., Leung, S.-T., 2003. The google file system. ACM SIGOPS Operating Syst. Rev. 37, 29–43.

Giardine, B., 2005. Galaxy: a platform for interactive large-scale genome analysis. Genome Res. 15, 1451–1455.

Gilbert, J.A., 2015. Social behavior and the microbiome. eLife 4.

Goecks, J., Nekrutenko, A., Taylor, J., Galaxy Team, T., 2010. Galaxy: a comprehensive approach for supporting accessible, reproducible, and transparent computational research in the life sciences. Genome Biol. 11, R86.

Golosova, O., Henderson, R., Vaskin, Y., Gabrielian, A., Grekhov, G., Nagarajan, V., et al., 2014. Unipro UGENE NGS pipelines and components for variant calling, RNA-seq and ChIP-seq data analyses. PeerJ 2, e644.

Goto, N., Prins, P., Nakao, M., Bonnal, R., Aerts, J., Katayama, T., 2010. BioRuby: bioinformatics software for the ruby programming language. Bioinformatics 26, 2617–2619.

Green, A., Guagliardo, P., Libkin, L., Lindaaker, T., Marsault, V., Plantikow, S., et al., 2019. Updating graph databases with cypher. Proc. VLDB Endow. 12, 2242–2254.

Griffiths, A.J.F., Miller, J.H., Suzuki, D.T., Lewontin, R.C., Gelbart, M.L., 2000. An introduction to genetic analysis. W. H. Freeman, Holtzbrinck, p. 860.

Grosdidier, S., Pons, C., Solernou, A., Fernández-Recio, J., 2007. Prediction and scoring of docking poses with pyDock. Proteins 69, 852–858.

Hagen, J.B., 2000. The origins of bioinformatics. Nat. Rev. Genet. 1, 231–236.

Hauskeller, C., Calvert, J., 2004. The meanings of genomics: introduction. N. Genet. Soc. 23 (3), 251–254.

Heather, J.M., Chain, B., 2016. The sequence of sequencers: the history of sequencing DNA. Genomics 107, 1–8.

Helsby, M.A., Leader, P.M., Fenn, J.R., Gulsen, T., Bryant, C., Doughton, G., et al., 2014. CiteAb: a searchable antibody database that ranks antibodies by the number of times they have been cited. BMC Cell Biol. 15, 6.

Hershey, A.D., Chase, M., 1952. Independent functions of viral protein and nucleic acid in growth of bacteriophage. J. Gen. Physiol. 36, 39–56.

Holland, R.C., Down, T.A., Pocock, M., Prlic, A., Huen, D., James, K., et al., 2008. Biojava: An open-source framework for bioinformatics. Bioinformatics 24, 2096–2097.

Hood, L., Rowen, L., 2013. The human genome project: big science transforms biology and medicine. Genome Med. 5, 79.

Hughes, R.A., Ellington, A.D., 2017. Synthetic DNA Synthesis and Assembly: Putting the Synthetic in Synthetic Biology. Cold Spring Harb. Perspect. Biol. 9, a023812.

Johnpaul, C.I., Mathew, T., 2017. A Cypher query based NoSQL data mining on protein datasets using Neo4j graph database. In: 2017 4th International Conference on Advanced Computing and Communication Systems (ICACCS). Presented at the 2017 4th International Conference on Advanced Computing and Communication Systems (ICACCS), IEEE.

Johnson, M., Basak, S., Maggiora, G., 1988. A characterization of molecular similarity methods for property prediction. Math. Comput. Model. 11, 630–634.

Källberg, M., Wang, H., Wang, S., Peng, J., Wang, Z., Lu, H., et al., 2012. Template-based protein structure modeling using the RaptorX web server. Nat. Protoc. 7, 1511–1522.

Kchouk, M., Gibrat, J.F., Elloumi, M., 2017. Generations of sequencing technologies: from first to next generation. Biol. Med. (Aligarh) 09.

Kearse, M., Moir, R., Wilson, A., Stones-Havas, S., Cheung, M., Sturrock, S., et al., 2012. Geneious basic: an integrated and extendable desktop software platform for the organization and analysis of sequence data. Bioinformatics 28, 1647–1649.

Kelley, L.A., Mezulis, S., Yates, C.M., Wass, M.N., Sternberg, M.J.E., 2015. The Phyre2 web portal for protein modeling, prediction and analysis. Nat. Protoc. 10, 845–858.

Khan, S.A., He, D., Valverde, J.C., 2016. Pattern recognition in bioinformatics. BioMed. Res. Int. 2016, 1–2.

Kobus, R., Abuín, J.M., Müller, A., Hellmann, S.L., Pichel, J.C., Pena, T.F., et al., 2020. A big data approach to metagenomics for all-food-sequencing. BMC Bioinforma. 21.

Konkel, F., 2013. Does your agency need big data? Maybe not. [WWW Document]. FCW. <https://fcw.com/articles/2013/03/06/big-data-not-for-all.aspx> (accessed 11.25.21).

Koonin, E.V., Galperin, M.Y., 2011. Sequence - evolution - function: computational approaches in comparative genomics. Springer, New York.

Korlach, J., Turner, S.W., 2013. Single-molecule sequencing. In: Roberts, G.C.K. (Ed.), Encyclopedia of Biophysics. Springer, Berlin Heidelberg, pp. 2344–2347.

Kronmueller, M., Chang, D., Hu, H., Desoky, A., 2018. A graph database of yelp dataset challenge 2018 and using cypher for basic statistics and graph pattern exploration. In: 2018 IEEE International Symposium on Signal Processing and Information Technology (ISSPIT). Presented at the 2018 IEEE International Symposium on Signal Processing and Information Technology (ISSPIT), IEEE.

Kulski, J.K., 2016. Next-generation sequencing—an overview of the history, tools, and "omic" applications. In: Kulski, J. (Ed.), Next Generation Sequencing—Advances, Applications and Challenges. InTech.

Kumar, S., Stecher, G., Li, M., Knyaz, C., Tamura, K., 2018. MEGA X: molecular evolutionary genetics analysis across computing platforms. Mol. Biol. Evolution 35, 1547–1549.

Lee, B.-R., Cho, S., Song, Y., Kim, S.C., Cho, B.-K., 2013. Emerging tools for synthetic genome design. Mol. Cell 35, 359–370.

Liu, J., Liu, Q., Zhang, L., Su, S., Liu, Y., 2020. Enabling massive XML-based biological data management in HBase. IEEE/ACM Trans. Comput. Biol. Bioinf 17, 1994–2004.

Loh, Y.-H.E., Shen, L., 2016. Analysis and visualization of ChIP-Seq and RNA-Seq sequence alignments using ngs.plot. In: Carugo, O., Eisenhaber, F. (Eds.), Methods in Molecular Biology. Springer, New York, pp. 371–383.

Madduri, R.K., Sulakhe, D., Lacinski, L., Liu, B., Rodriguez, A., Chard, K., et al., 2014. Experiences building Globus genomics: a next-generation sequencing analysis service using Galaxy, Globus, and Amazon web services. Concurrency Comput.: Pract. Exper 26, 2266–2279.

Marx, V., 2013. The big challenges of big data. Nature 498, 255–260.

Masseroli, M., Canakoglu, A., Pinoli, P., Kaitoua, A., Gulino, A., Horlova, O., et al., 2018. Processing of big heterogeneous genomic datasets for tertiary analysis of next generation sequencing data. Bioinformatics 35, 729–736.

Masseroli, M., Pinoli, P., Venco, F., Kaitoua, A., Jalili, V., Palluzzi, F., et al., 2015. GenoMetric query language: a novel approach to large-scale genomic data management. Bioinformatics 31, 1881–1888.

Masseroli, M., Kaitoua, A., Pinoli, P., Ceri, S., 2016. Modeling and interoperability of heterogeneous genomic big data for integrative processing and querying. Methods 111, 3–11.

Maxam, A.M., Gilbert, W., 1977. A new method for sequencing DNA. Proc. Natl Acad. Sci. 74, 560–564.

Mende, D.R., Waller, A.S., Sunagawa, S., Järvelin, A.I., Chan, M.M., Arumugam, M., et al., 2012. Assessment of metagenomic assembly using simulated next generation sequencing data. PLoS ONE 7, e31386.

Mohamad, O., Ho, W., 2011. The next generation sequencing technologies. Bul. Persat. Genetik Malays. 18 (1).

Nandy, A., 2015. The GRANCH techniques for analysis of DNA, RNA and protein sequences. In: Basak, S.C., Restrepo, G., Villaveces, J.L. (Eds.), Advances in Mathematical Chemistry and Applications. pp. 96–124.

Narula, P., 2019. Hadoop yarn architecture [WWW Document]. GeeksforGeeks. <https://www.geeksforgeeks.org/hadoop-yarn-architecture/> (accessed 11.22.21).

Okonechnikov, K., Golosova, O., Fursov, M., 2012. Unipro UGENE: a unified bioinformatics toolkit. Bioinformatics 28, 1166–1167.

Orman, T.F., 2016. Paradigm as a central concept in Thomas Kuhn's thought. Int. J. humanities Soc. Sci. 6 (10).

Oulas, A., Pavloudi, C., Polymenakou, P., Pavlopoulos, G.A., Papanikolaou, N., Kotoulas, G., et al., 2015. Metagenomics: tools and insights for analyzing next-generation sequencing data derived from biodiversity studies. Bioinform Biol. Insights 9, BBI.S12462.

Pabinger, S., Dander, A., Fischer, M., Snajder, R., Sperk, M., Efremova, M., et al., 2013. A survey of tools for variant analysis of next-generation genome sequencing data. Brief. Bioinform. 15, 256–278.

Pan, V.Y., Chen, Z.Q., 1999. The complexity of the matrix eigenproblem. In: Proceedings of the Thirty-First Annual ACM Symposium on Theory of Computing - STOC '99. Presented at the the thirty-first annual ACM symposium, ACM Press, pp. 507–516.

Pareek, C.S., Smoczynski, R., Tretyn, A., 2011. Sequencing technologies and genome sequencing. J. Appl. Genet. 52, 413–435.

Pettersen, E.F., Goddard, T.D., Huang, C.C., Couch, G.S., Greenblatt, D.M., Meng, E.C., et al., 2004. UCSF chimera?a visualization system for exploratory research and analysis. J. Comput. Chem. 25, 1605–1612.

Porter, S.G., Day, J., McCarty, R.E., Shearn, A., Shingles, R., Fletcher, L., et al., 2007. Exploring DNA structure with Cn3D. LSE 6, 65–73.

Rice, C.M., Fuchs, R., Higgins, D.G., Stoehr, P.J., Cameron, G.N., 1993. The EMBL data library. Nucl. Acids Res. 21, 2967–2971.

Robinson, J.T., Thorvaldsdóttir, H., Winckler, W., Guttman, M., Lander, E.S., Getz, G., et al., 2011. Integrative genomics viewer. Nat. Biotechnol. 29, 24–26.

Rothberg, J.M., Hinz, W., Rearick, T.M., Schultz, J., Mileski, W., Davey, M., et al., 2011. An integrated semiconductor device enabling non-optical genome sequencing. Nature 475, 348–352.

Sanger, F., Coulson, A.R., 1975. A rapid method for determining sequences in DNA by primed synthesis with DNA polymerase. J. Mol. Biol. 94, 441–448.

Sanger, F., Nicklen, S., Coulson, A.R., 1977. DNA sequencing with chain-terminating inhibitors. Proc. Natl Acad. Sci. 74, 5463–5467.

Sanger, F., Coulson, A.R., Barrell, B.G., Smith, A.J., Roe, B.A., 1980. Cloning in single stranded bacteriophage as an aid to rapid dna sequencing. J. Mol. Biol. 143, 161–178.

Schmidt, B., Hildebrandt, A., 2017. Next-generation sequencing: big data meets high performance computing. Drug. Discov. Today 22, 712–717.

Schneidman-Duhovny, D., Inbar, Y., Nussinov, R., Wolfson, H.J., 2005. PatchDock and SymmDock: servers for rigid and symmetric docking. Nucl. Acids. Res. 33, W363–W367.

Segre, J.A., 2021. Metagenomics [WWW Document]. Genome.gov. <https://www.genome.gov/genetics-glossary/Metagenomics> (accessed 11.22.21).

Sen, D., Dasgupta, S., Pal, I., Manna, S., Basak, S., Nandy, A., et al., 2016. Intercorrelation of major DNA/RNA sequence descriptors - a preliminary study. Comput. Drug. Des. 12 (3).

Shin, S.C., Ahn, D.H., Kim, S.J., Lee, H., Oh, T.-J., Lee, J.E., et al., 2013. Advantages of single-molecule real-time sequencing in high-GC content genomes. PLoS ONE 8, e68824.

Singh, N., Chikara, S., Sundar, S., 2013. SOLiD™ sequencing of genomes of clinical isolates of *Leishmania donovani* from india confirm leptomonas co-infection and raise some key questions. PLoS ONE 8, e55738.

Smith, D.R., 2014. Buying in to bioinformatics: an introduction to commercial sequence analysis software. Brief. Bioinform 16, 700–709.

Song, S., Jarvie, T., Hattori, M., 2013. Our second genome—human metagenome. Advances in Microbial Physiology. Elsevier, pp. 119–144.

Song, J., Wang, Y., Li, F., Akutsu, T., Rawlings, N.D., Webb, G.I., et al., 2018. IProt-sub: a comprehensive package for accurately mapping and predicting protease-specific substrates and cleavage sites. Brief. Bioinform. 20, 638–658.

Syngai, G., Barman, P., et al., 2013. BLAST: an introductory tool for students to bioinformatics applications. Keanean J. Sci. (2), 67–76.

Tateno, Y., Gojobori, T., 1997. DNA Data Bank of Japan in the age of information biology. Nucl. Acids Res. 25, 14–17.

Thomas, T., Gilbert, J., Meyer, F., 2012. Metagenomics—a guide from sampling to data analysis. Microb. Inform. Exp. 2.

Tripathi, R., Sharma, P., Chakraborty, P., Varadwaj, P.K., 2016. Next-generation sequencing revolution through big data analytics. Front. Life Sci. 9, 119–149.

Turner, V., Gantz, J., Reinsel, D., Minton, S., 2014. The digital universe of opportunities: Rich data and the increasing value of the internet of things. In: International Data Corporation, White Paper, IDC 1672.

Vita, R., Overton, J.A., Greenbaum, J.A., Ponomarenko, J., Clark, J.D., Cantrell, J.R., et al., 2014. The immune epitope database (IEDB) 3.0. Nucleic Acids Res. 43, D405–D412.

Vračko, M., Zupan, J., 2015. A non-standard view on artificial neural networks. Chemometr. Intell. Lab. Syst. 149, 140–152.

Vračko, M., Basak, S.C., Dey, T., Nandy, A., 2021a. Cluster analysis of coronavirus sequences using computational sequence descriptors: with applications to SARS, MERS and SARS-CoV-2 (CoVID-19). Curr. Comput. Drug. Des. 17.

Vračko, M., Basak, S.C., Sen, D., Nandy, A., 2021b. Clustering of Zika viruses originating from different geographical regions using computational sequence descriptors. Curr. Comput. Drug. Des. 17, 314–322.

Ward, C.M., To, T.-H., Pederson, S.M., 2019. ngsReports: a bioconductor package for managing FastQC reports and other NGS related log files. Bioinformatics 36, 2587–2588.

Wattam, A.R., Abraham, D., Dalay, O., Disz, T.L., Driscoll, T., Gabbard, J.L., et al., 2013. PATRIC, the bacterial bioinformatics database and analysis resource. Nucl. Acids Res. 42, D581–D591.

Wercelens, P., da Silva, W., Hondo, F., Castro, K., Walter, M.E., Araújo, A., et al., 2019. Bioinformatics workflows with NoSQL database in cloud computing. Evol. Bioinform. Online 15, 117693431988997.

Wexler, J., Thompson, W., Aponte, K., 2013. Time is precious, so are your models. SAS provides solutions to streamline deployment. In: SAS Global Forum 2013.

Wheeler, D.L., 2000. Database resources of the national center for biotechnology information, Nucl. Acids Res., 28. pp. 10–14.

Winkler, H., 1920. Verbreitung und Ursache der Parthenogenesis im Pflanzen—und Tierreiche. Verlag Fischer, Jena.

Yourgenome, 2021. Who was involved in the human genome project? [WWW Document]. <https://www.yourgenome.org/stories/who-was-involved-in-the-human-genome-project> (accessed 11.25.21).

Yu, C., Wu, W., Wang, J., Lin, Y., Yang, Y., Chen, J., et al., 2018. NGS-FC: a next-generation sequencing data format converter. IEEE/ACM Trans. Comput. Biol. Bioinform. 15, 1683–1691.

Zhang, Z., Schwartz, S., Wagner, L., Miller, W., 2000. A greedy algorithm for aligning DNA sequences. J. Comput. Biol. 7, 203–214.

Zielezinski, A., Vinga, S., Almeida, J., Karlowski, W.M., 2017. Alignment-free sequence comparison: benefits, applications, and tools. Genome Biol. 18.

Scalable quantitative structure–activity relationship systems for predictive toxicology

17

Suman K. Chakravarti
MultiCASE Inc., Beachwood, OH, United States

17.1 Background

The modern quantitative structure–activity relationship (QSAR) methodology was introduced almost 60 years ago (Hansch and Fujita, 1964) and evolved very rapidly. As the name suggests, the objective of the QSAR methodology is to find empirical relationships between molecular structures and biological activity to make predictions for new molecules (Hansch and Leo, 1979). Such correlations are established between molecular features or descriptors, also called the predictor variables (*X*) and the biological activity or the response variable (*Y*). When physicochemical properties are modeled, it is called quantitative structure–property relationship (QSPR) modeling. The models are either regression or classification type. Continuous valued response variables are modeled by the regression techniques, whereas categorical responses are often modeled with classification methods. QSARs opened up excellent ways of predicting biological activity using correlative modeling via examples of chemicals instead of computing it directly from the molecular structures. Predictions are rapid, require no or minimal experimental measurements, and therefore save time and cost (Nantasenamat et al., 2010). They have been very valuable for the purpose of predicting biological activity, physicochemical properties, toxicity, chemical reactivity, and metabolism of chemicals.

The process of building QSAR systems can be divided into five main steps: (1) Collection and curation of data, (2) Calculation and selection of molecular descriptors, (3) Building individual models by fitting activity and descriptors, (4) Constructing a network of QSAR models representing individual biological events, and (5) Validation by challenging it with prediction tasks. Not all steps are always essential, however, proper validation is critical to build good QSAR systems (Tropsha, 2010). In this regard, scalability can be considered as an additional validation criterion for building robust QSAR systems, which can survive longer.

The descriptors, or the predictor variables, represent the molecular structures and are crucial for the development of QSAR models (Karelson, 2000). A wide variety of descriptors have been utilized so far, for example, measured or computed physicochemical properties, theoretical quantum chemical descriptors, graph theoretical or mathematical (Basak et al., 1990, 1987; Devillers and Balaban, 1999), properties

Big Data Analytics in Chemoinformatics and Bioinformatics. DOI: https://doi.org/10.1016/B978-0-323-85713-0.00031-1

based on the 3D structure of the molecules, fragments of the molecular structures. However, it is becoming increasingly difficult to use traditional descriptors on very large training sets due to the diverse nature of the datasets and the failure to achieve good predictive performance. Therefore the use of raw molecular structural features or molecular graphs as direct inputs is becoming common in which abstractions of the molecular structure are computed automatically by deep neural networks (Wu et al., 2017). However, traditional descriptors are very valuable for building QSAR models from small and focused datasets and still commonly used.

It is worth discussing the differences between models built from small structurally related (i.e., congeneric) sets of compounds and the ones built from large diverse datasets (Basak and Majumdar, 2016). For the former, a human expert carefully selects the training compounds with the aid of his own understanding of the activity mechanisms. The only task that remains is to find a QSAR model that describes this variation around the fixed core using measurable or computable descriptors. The model is then used for predicting query compounds with the same core structure. Such models usually have limited scope and are only capable of reflecting the reasoning of the expert. On the other hand, QSARs built from large, diverse training sets are fundamentally different. The training sets for such models range from tens of thousands to hundreds of thousands. There is no fixed core scaffold, and the training set is composed of different chemical classes, often with multiple activity mechanisms. The choice of descriptors and the selection of modeling techniques become much more limited. Quantum chemical and other computationally expensive descriptors cannot be used anymore, but topological and graph theoretical descriptors become more suitable. Moreover, the model does not reflect the reasonings of the expert anymore and therefore provides many opportunities and challenges as well. It can now potentially uncover new mechanistic insights and relationships, at the cost of finding spurious correlations that have no mechanistic basis. Both types of models are very useful and are applied to different needs.

A variety of techniques can be used for the QSAR model building. Ordinary/partial least square regression, kernel-based support vector machines (Cristianini and Shawe-Taylor, 2004; Vapnik et al., 1996), decision trees (Quinlan, 1986), ensemble-based methods like random forests (Ho, 1995), simple or deep neural networks (Wu et al., 2017), to name just a few. These are all so-called "supervised" machine learning methods, and their main task is to capture the statistical relationship between the supplied response variable and the predictor variables. However, some of the deep neural network architectures, for example, convolutional neural networks, have the ability to accept the molecular structures directly as a graph and automatically generate successive abstractions of the chemical structures to use them like descriptors (Wu et al., 2017). Therefore these techniques can be used to avoid computing and the selection of descriptors. It is not difficult to imagine that this type of modeling will be commonplace soon.

The nearest neighbor techniques (Itskowitz and Tropsha, 2005) stand apart from the other QSAR methods due to their nonparametric nature. Predictions are produced by searching and retrieving the experimentally observed properties of a few most similar analogs. Structural similarity is used for searching the analogs and is

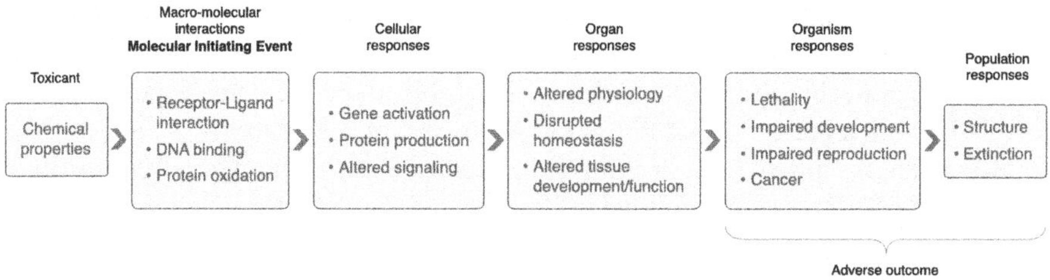

Figure 17.1 Key components of adverse outcome pathways (AOP) (Ankley et al., 2010).

sometimes filtered by structural alerts, physicochemical properties, metabolism, toxicity mechanisms, pharmacokinetics, etc. On the surface they appear to be overly simplistic, but they are very powerful and extremely popular. The "read across" data gap-filling techniques are primarily based on the nearest neighborhood techniques (Ball et al., 2020). They are now commonplace in the regulatory decision-making process for different toxicity endpoints, for example, skin sensitization of cosmetics, repeat-dose toxicity, and reproductive toxicity. In addition, they are valuable in the analysis and interpretation of structural-alert-based QSAR results (i.e., expert review), for example, in the DNA-reactivity of drug impurities. Their simplicity makes them very attractive in terms of scalability.

With the introduction of adverse outcome pathways (AOPs) in predictive toxicology (Ankley et al., 2010), the QSAR modeling approaches can now be applied with a much broader scope. As shown in Fig. 17.1, the AOPs provide a conceptual framework based on the mechanistic knowledge of biological disruption at cellular, organ, organism, and population levels. The AOPs are modular with reusable components and their structure allows effective use of in silico tools for toxicity assessments. They have become an essential tool in the regulatory risk assessment of chemical substances, so much so that the Organization for Economic Cooperation and Development (OECD) launched an international AOP development program in 2012 (Villeneuve et al., 2014; Willett et al., 2014). The modular nature of AOPs allows the incorporation of a variety of resources, for example, databases of compounds, QSAR models, and prediction tools. They work together to provide assessments that are based on both experimental evidence and model predictions and allow continuous evolution and expansion overtime as new mechanistic information and data become available.

17.2 Scalability in quantitative structure–activity relationship modeling

Scalability can be defined as the "ability of a computing process to be used or produced in a range of capabilities" (Oxford dictionary), or in other words, it is the property of a system to keep functioning well when it is expanded in size, volume, scope, workload, etc. (Bondi, 2000). A system can be considered scalable if it does not need significant re-work or complete redesigning when its scope of application is changed. In this regard, all aspects of the scalability of general computing processes are applicable to QSAR modeling. However, there are some aspects specific to QSARs which are described in the following sections. Scalability is vital in terms of effectiveness, quality, applicability, maintenance, and commercial competitiveness of the QSAR systems. Following are some of the common situations when the capability of a QSAR system is challenged:

1. Predicting properties of chemicals with very different or broader structural variation than the training compounds.
2. Applying a QSAR to a different biological system, for example, predicting in vivo endpoints using QSARs built on in vitro assay data, or predicting toxic effects on humans via QSARs based on rodent assays.
3. A significant change in the scope of application of a QSAR model. For example, when a QSAR originally intended for screening in the drug discovery phase is used in regulatory submission processes.
4. Significant increase in a QSAR system's usage due to paradigm shifts in the respective fields or changes in the regulatory landscape. For example, acceptance of QSARs in the assessment of DNA reactive impurities of pharmaceuticals as per the ICH M7 guideline resulted in the widespread use of QSAR models. Similarly, a ban on animal tests caused increased use of QSARs for assessment of the skin sensitization potential of cosmetics ingredients.

17.2.1 Consequences of inability to scale

Quantitative structure–activity relationships that are not designed with scalability features often suffer from the following shortcomings

1. *Failure to take advantage of newly available data:* Some models simply cannot be rebuilt when the training data size is increased considerably. This happens mainly with models that are built with poorly designed custom-made training algorithms or demand too many manual interventions during model building.
2. *Poor performance:* Sometimes prediction performance of a model drops unacceptably or fails to improve when scaled. This is common with models built with greedy descriptor selection algorithms, for example, stepwise selection of structural alerts from a large pool of fragments. They can also display unacceptable behavior during prediction, for example, too many structural alerts in the query molecules.
3. *Failure to incorporate new mechanistic insights:* Modern predictive toxicology is progressing at a rapid pace. New assays and new studies are uncovering previously unknown mechanisms. A good scalable QSAR system should be able to take advantage of such information by incorporating new descriptors (e.g., models of in vitro assays as descriptors to in vivo QSAR models) or adding new nodes in an AOP-based system.

17.2.2 Expandability of the training dataset

A QSAR model faces challenges in predicting properties of chemicals that are very different from its training set. The reason could be the presence of new structural features or novel combinations of features in the query compound. Such chemicals are usually classified as outside of the model's applicability domain and the prediction is labeled as unreliable. As a possible solution, the most common practice is to fill the data gaps and expand the scope of the model by retraining with more data, if available.

When adding a few chemicals to the training set, one can just recalculate descriptors, retrain and revalidate the model. For the predictive systems based on the near-neighbor methodology, new data can simply be appended to the existing database, although fingerprints and physicochemical descriptors need to be recalculated. However, updating a model takes considerably more time if supporting databases with literature references or experimental details are connected.

The nature of the descriptors in the model is an important consideration. It is often difficult to expand the training sets that are based on hand-picked descriptors, because the descriptors often fail to adequately explain the new data. Experimentally measured descriptors can be very problematic because they need to be measured for the newly added training chemicals. It is therefore important to resist the temptation of painfully designing or picking descriptors without considering future expansions, only to achieve high performance for a small-sized dataset. Instead, descriptors that can be computed quickly and suitable for future addition of more data are a better choice. An example is illustrated in Fig. 17.2 of building a group contribution QSPR model for predicting octanol–water partition coefficient (LogP). One set of models was manually built by identifying and coding relevant functional groups that affect

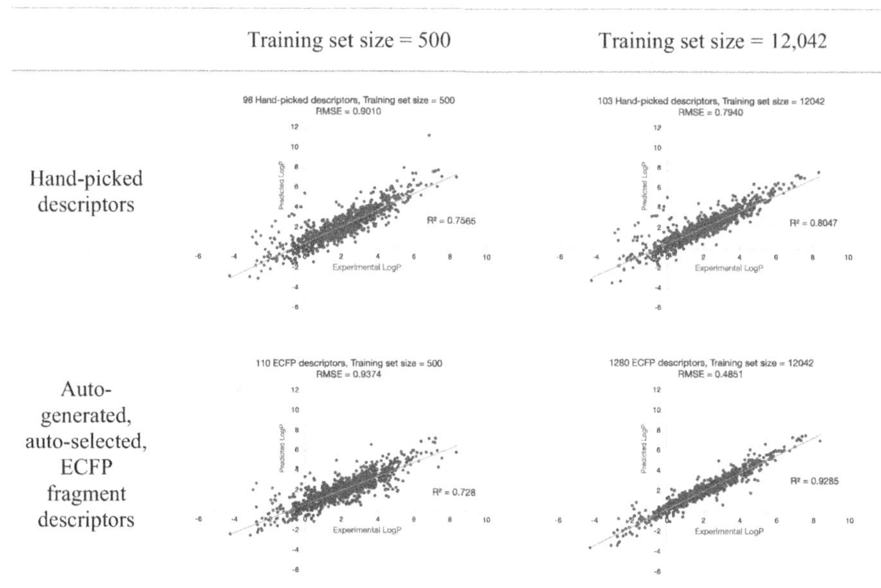

Figure 17.2 Comparison of octanol–water partition coefficient (LogP) prediction for 1000 chemicals between hand-picked group-contribution descriptors and automatically generated ECFP fragment descriptors while expanding the training set from 500 to 12,042 chemicals.

LogP using SMARTS notation. The other set of descriptors were extended-connectivity fingerprint based circular fragments (Rogers and Hahn, 2010), generated automatically from training compounds and employing a variable selection algorithm. The models were evaluated by predicting 1000 compounds that were kept out from the start. Initially, only 500 training compounds were used, and as it is shown in Fig. 17.2, the hand-picked descriptors performed better, in terms of r^2 and root mean squared error (RMSE). However, when the training set was scaled to 12,042 compounds, the ECFP fragments performed substantially better. The r^2 improved from 0.7280 to 0.9285 and *RMSE* decreased from 0.9374 to 0.4851 for the ECFP fragments. Whereas for the hand-picked descriptors, the improvement was not that big, the r^2 improved from 0.7565 to 0.8047 and *RMSE* decreased from 0.9010 to 0.7940. Therefore the hand-picked descriptors offer some marginal advantage for the small training set, which disappears when the training set is expanded. Moreover, it took weeks to create and validate the hand-picked SMARTS descriptors, whereas modeling with ECFP descriptors took less than half an hour. However, the ECFP descriptors will not scale forever if the training set keeps expanding, as the number of unique ECFP fragments will increase and ultimately could become a bottleneck in the descriptor selection step. This can be avoided by using deep neural networks, which can take chemical structures directly as inputs, for example, recurrent neural networks with SMILES input or graph convolution neural networks with molecular graph inputs.

The selection of the model-building algorithm is also key for adding new training data as some of them are better than others in terms of scalability. Commonly used techniques are Naïve Bayes, ordinary regression, logistic regression, k-NN, decision trees, random forests, support vector machines, simple neural networks, neural networks with deeper hidden layers, convolutional neural networks, recurrent neural nets, transformers, and others (Yousefinejad and Hemmateenejad, 2015). Fig. 17.3 is showing the behavior of various modeling techniques with change in training set size. Three modeling methodologies were used for the prediction of two toxicological endpoints. Although the performance of almost all the methods increases with the training data, their relative performance differs. Also, there are some other factors that need to be considered for selecting the most suitable model-building algorithm, for example, is descriptor selection required? What is the cost of retraining? Do multiple hyperparameters need to be optimized? Does the algorithm provide a reasonable balance between performance and interpretability? Naïve Bayes (Hastie, 2001) and random forests are great from a scalability perspective, as they can perform well without requiring descriptor selection. Retraining a Naïve Bayes model is inexpensive because recalculation of conditional probabilities is the only step needed. Random forests do need expensive retraining and optimization of multiple hyperparameters, however, powerful computer hardware and distributed computing over multiple processing cores can help. The majority of deep neural networks can be trained without requiring any variable selection. Instead, they rely on techniques like regularization, dropout, etc., to avoid overfitting. Also, modern neural network frameworks allow fine tuning of existing models by training with new data, avoiding the need for complete retraining of the

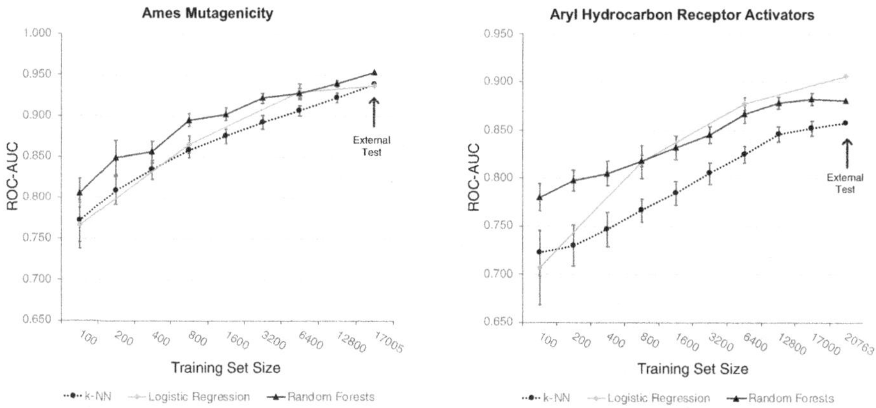

Figure 17.3 Performance of various model-building algorithms for predicting Ames mutagenicity and aryl hydrocarbon receptor (AHR) activation with increasing training set size. The test set size is 2000 compounds for every data point. The external test set (shown by an arrow on right) is 1942 and 2307 for Ames and AHR, respectively.

weights. But as a downside, deep learning techniques often need tuning of several hyperparameters.

Interpretability is very important when QSAR models are used for regulatory decisions. For example, the identification and evaluation of mutagenic structural alerts are essential in the assessment of the DNA reactive potential of drug impurities [covered by the ICH M7 guideline (Barber et al., 2015)]. This limits the available options for descriptors and training algorithms, as fragment-based descriptors are the only option for discovering mutagenic structural alerts. In addition, a good descriptor selection algorithm is required to select the optimal set of fragments to avoid finding too many alerts in the query chemicals. As for the training algorithm, Naïve Bayes or logistic regression are suitable candidates. On the other hand, the accuracy of the prediction is the most important factor for early screening in drug discovery projects, allowing omission of the descriptor selection step and opting for more complex algorithms like random forests or deep neural networks.

17.2.3 Efficiency of data curation

Accuracy and appropriate treatment of the chemical structures and associated information is an essential requirement of any good QSAR project and can be quite expensive in terms of time and effort (Tropsha, 2010). Therefore the efficiency of data curation should be considered an integral factor in the scalability of QSAR systems. There are multiple aspects of chemical data curation, but some steps are typical, for example, the treatment of duplicates and stereoisomers, treatment of salts and mixtures, the correctness of the chemical structures, harmonization of structural representations, and reconciliation of different assay outcomes from different sources. Manual review is feasible for about 10,000 or fewer compounds but becomes increasingly difficult for larger sets, and practically impossible for hundreds of thousands of chemicals. Therefore careful consideration should be given to possible automation or algorithm-based augmentation of some of the manual review steps. Chemical structures can be verified automatically by comparing against different online databases, for example, PubChem (Kim et al., 2019), ChEMBL (Mendez et al., 2019), treatment of inorganic salt and mixtures can also be performed algorithmically.

Chemical structures are often associated with a variety of information, for example, scientific bibliography, experimental assay details, etc. The accuracy of such databases is also important, needs careful curation, therefore, may prove to be expensive during scaling.

17.2.4 Ability to handle stereochemistry

A majority of the available computer software for in silico toxicology tends to ignore the stereochemistry of the compounds. This is not a major issue for toxicity events that are primarily due to the reactivity of small portions of the molecules (e.g., mutagenicity), however, stereochemistry plays important role in some other types of toxicity, for example, teratogenicity. Therefore proper handling of stereochemistry in all aspects of QSAR software is needed to achieve good scalability. This may include reading stereochemical information from SMILES codes, automatic perception of stereochemistry from three-diemsnional structures, the inclusion of stereo flags in molecular fragments, display of stereo information in the molecular displays, and automatic verification of stereochemistry agreement between the existing training set and the new data being added.

17.2.5 Ability to use proprietary training data

In predictive toxicology, a large portion of useful data is generated by commercial entities and therefore proprietary in nature. Using them in public domain QSAR modeling is a major challenge if proprietary information cannot be protected. The choice of descriptors and model building algorithm becomes limited because the chemical structures cannot be disclosed, sometimes even to the model builder. If the model builder gets the privilege to access the chemical structures, structure−activity relationships are extracted, and proprietary structures are removed before public distribution of the model. Alternatively, a list of predefined structural keys can be supplied to the owner of the data and counts of toxic/nontoxic compounds for each key are returned back for modeling purposes. Some associated information can also be confidential in nature, for example, details of the assay results, sources of data, etc. Such information can be a hindrance to the scalability of a model in the absence of proper processes to handle them.

17.2.6 Ability to handle missing data

Relevant data for modeling usually are collected from multiple sources and quite often cover different aspects of chemicals. As a result, gaps appear when such information is compiled by the modeler and is often referred to as missing information. For example, a chemical may not have Ames test results for all the bacterial strains required as per OECD 471 test guidelines, or, for modeling of skin sensitization, very few chemicals may have experimental data for all the assays described in the skin sensitization AOP. The ability to handle such data is vital if we want continuous improvements in a QSAR system and also should be a consideration when selecting a model training algorithm.

17.2.7 Ability to modify the descriptor set

The model descriptors need to be modified when the training set is expanded or rebuilt using a different algorithm. As shown in Fig. 17.2, the hand-picked descriptors usually need to be modified or appended if they fail to explain the new data. This is also a scaling issue, as recalculation of the descriptors and re-training of the model are required. This step can be time-consuming for models with very large training sets. The computational complexity mainly depends on the type of descriptors. Fragment-based descriptors, graph theoretical descriptors, and others that are based on the two-dimensional graph are usually fast to calculate whereas quantum chemical descriptors are expensive.

17.2.8 Scaling expert rule-based systems

Expert rule-based (Q)SAR systems are very common in predictive toxicology, for example, in assessment of mutagenicity (Plošnik et al., 2016), skin sensitization (Barratt and Langowski, 1999; Payne and Walsh, 1994), etc. They are composed of a list of expert rules, usually in the form of hand-coded structural alerts, prepared by experts using their own knowledge or scientific literature. These alerts are commonly annotated with mechanistic details, bibliographic information, etc. Sometimes, a database of reference chemicals is also included.

The addition or modification of existing rules is common and is a slow process requiring significant effort and time. New structural alerts are added when new mechanistic information is discovered, and existing rules are modified to improve the accuracy of the system. Sometimes subrules are added to increase the specificity of existing structural alerts, in the form of deactivating or mitigating features (Fig. 17.4). Subsequently, the rules are validated for predictive ability and interaction with the existing rules. In part, the scalability of such systems depends on the way the toxicity alerts are encoded. SMILES or SMARTS notation is commonly used which can be easily updated.

17.2.9 Scalability of adverse outcome pathway-based quantitative structure–activity relationship systems

AOP-based QSAR systems are composed of multiple models, often hybrid in nature, and may contain a number of different types of QSAR models, structural alert lists, databases etc. The components represent distinct biological events

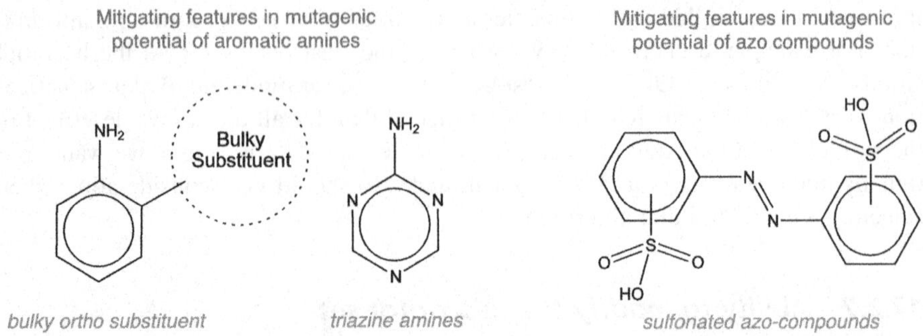

Figure 17.4 Examples of structural features that mitigate the mutagenic potential of aromatic amino and azo compounds.

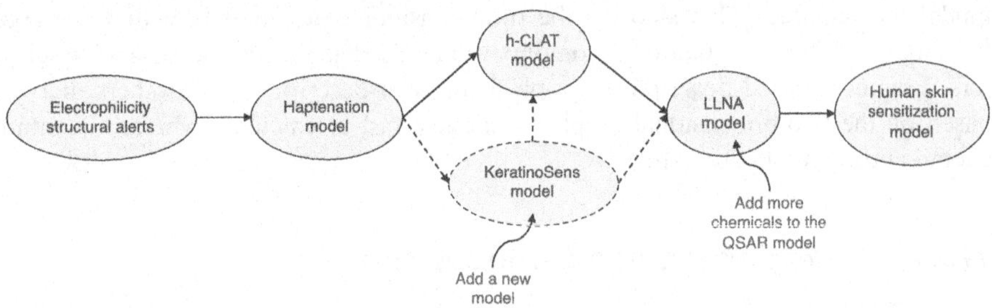

Figure 17.5 Possible modifications on a QSAR system representing the skin sensitization adverse outcome pathways, that is, adding a new node (*dashed lines*) and expanding the training set of an existing model (Local Lymph Node Assay).

connected in a meaningful way. Such networks are usually represented by graphical structures that are composed of nodes and edges (Jaworska et al., 2015). Every node represents a biological event and can be implemented using predictive QSAR models or databases. These nodes are connected by edges that are often *directed* in nature, that is, connecting a cause to its effect. These QSAR systems are built with modularity and scalability in mind. A node can be enhanced using the scalability principles outlined in the preceding sections. In addition, as shown in Fig. 17.5, the network structure itself can be modified, for example, nodes can be added or removed or the connection between the nodes can be modified. If the network is implemented using a statistical method, for example, Bayesian networks, various parameters need to be recalculated in the event of a modification.

17.2.10 Scalability of the supporting resources

State-of-the-art QSAR systems are often connected with a variety of additional components, for example, supporting databases, reporting tools, etc. Supporting databases

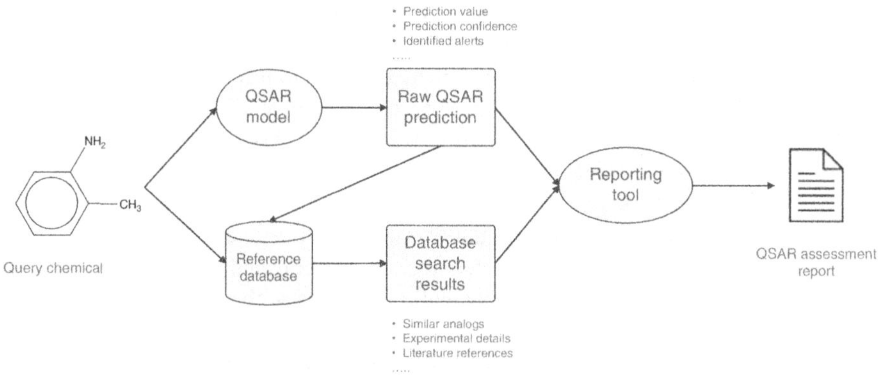

Figure 17.6 Schematic representation of how a reporting tool connects to various components of a quantitative structure–activity relationship system and the flow of information that finally results in a report.

usually contain information from scientific literature, experimental details, etc. A significant amount of manual effort is needed to scale these databases and can be very time-consuming. Therefore, careful consideration should be given to designing these databases and their interaction with the main model.

Many commercial QSAR software is also often connected with software tools for generating reports from raw predictions, as shown in Fig. 17.6. The reports are essential in the interpretation and distribution of the prediction results. Such reports are used in documentation commonly used as part of regulatory submissions. They contain a variety of textual and graphical information, often formatted in many different and nonstandard ways, therefore, pose engineering and design challenges. A report's layout and contents are closely tied to the QSAR algorithm and are usually not interchangeable. For example, a report from random forest predictions will be very different from that of simple regression. If not designed properly, the reporting tools can create difficulties in the scalability, and it is not unusual to spend a lot more time on updating such tools as compared to the actual QSAR system itself. Therefore good planning and design of software interfaces are key elements in this regard.

17.2.11 Scalability of quantitative structure–activity relationships validation protocols

Validation protocols are an integral part of any QSAR system for gauging robustness and predictive ability. A variety of cross-validation methods are generally used, for example, leave-one-out, leave-many-out, bootstrap, Y-scrambling, random split, scaffold-based split, etc. External validations are also commonly employed. However, the validations need to be rethought for large QSARs since some of the cross-validation methods, for example, leave-one-out, are simply not feasible. Cross-validation methods that require model rebuilding become impractical if multiple steps and hyperparameter tuning are part of the model training. In such cases, the model can be fine-tuned instead of full retraining. Validation protocols that involve holding out a certain percentage of data should also be modified, for example, for models with hundreds of thousands of training data, a few cycles of 1% hold-out can be used instead of a 10% hold-out 10 times. The capabilities provided in many of the modern neural network frameworks can be utilized for implementing fully automated validation protocols. More emphasis on external tests is warranted because testing a few thousand chemicals is usually computationally inexpensive. Multiple validation sets with different qualities can be maintained and used to track the effects of any change in the models. A lot of work is needed to develop proper validation methods for hybrid QSAR systems that are comprised of multiple models (as in the AOPs). Currently, they can only be validated by testing external sets of compounds.

17.2.12 Scalability after deployment

A few lucky QSAR systems manage to leave the laboratory where they were originally developed. However, scalability issues still play an important role if the end user wants to make changes to the model, for example, model fine-tuning for in-house compounds or expanding the coverage of the model by adding more training data. Adding in-house compounds to the supporting databases (instead of the training set) is also not uncommon. Users sometimes want to add specific structural alerts to rule-based systems to incorporate knowledge obtained from in-house biological assays. For such changes, all the scalability principles outlined in the preceding sections are applicable. However, such operations are often carried out without the presence of the model developer, therefore proper personnel training and good software design are required.

17.2.13 Ability to use computer hardware resources effectively

The impressive progress in the big data and machine learning field is closely associated with developments in computer hardware and advances in distributed computing software. The majority of the current consumer computing devices, including common laptops, desktops, tablets, and even mobile phones contain multicore GPUs and CPUs. Obviously, QSAR algorithms that can take advantage of such hardware resources are better positioned in terms of scalability. Hardware scalability is mainly important during model building; however, parallel computing is critical for large-scale applications of random forests and deep neural networks, for example, virtual screening of massive compound libraries. Neural network-based model training can easily be scaled to GPUs and multicore CPUs as many deep learning frameworks, for example, Keras (Chollet et al., 2015), TensorFlow (Abadi et al., 2016), are specifically designed with distributed computing in mind and freely available.

17.3 Summary

Based on the above discussion, we can now list some of the main reasons why QSAR systems fail to scale properly and offer some possible solutions:

1. *Use of hand-picked descriptors:* Although the practice of manually choosing molecular descriptors has given us excellent QSAR models for small training datasets, it is nearly impossible to produce good models from tens of thousands of diverse training chemicals in this way. It is therefore necessary to avoid the use of hand-picked descriptors if we need scalable models. Precomputed descriptors that are inexpensive to compute, for example, molecular fragments or fingerprints, are advantageous. Training algorithms that take molecular graphs directly as input are also suitable, for example, graph convolutional neural networks.
2. *Model building algorithms:* Training algorithms that are highly customized for specific domains, or those which use greedy algorithms, or need multiple manual interventions to build a model, are usually very hard to scale. Algorithms in which the expert (the algorithm designer) has inserted his own knowledge to deliver high performance for small datasets, usually fail to scale properly. Instead, commonly available and well-known training algorithms are better suited for large datasets, for example, Naïve Bayes, regression, random forests, etc.
3. *Descriptor selection algorithms:* The majority of the classical descriptor selection algorithms, for example, forward selection, backward elimination (Yousefinejad and Hemmateenejad, 2015), etc., are more suitable for selecting descriptors from a relatively small pool and usually greedy and computationally expensive. Instead, algorithms that do not need any descriptor selection step, for example, random forest, graph convolutional neural networks, or recurrent neural nets are better for large-scale models. If a large set of precomputed descriptor sets is being used, use regularization techniques to prevent overfitting, for example, L1 or L2 regularization, dropout, etc.
4. *Validation procedures:* QSAR models need to be validated for assessing their performance in the wild. Classical protocols, for example, leave-one-out or leave-many-out becomes impractical for very large training sets. Instead, rely on external validations and use a small percentage of holdout data for cross-validations. Also, avoid complete rebuilding of the model during each CV cycle and adopt fine-tuning and re-weighting of the descriptors.
5. *Absence of modularity:* Modularity is the key factor in achieving scalability in QSAR systems. Failure to properly separate logic, data, and computations results into static systems that cannot be properly updated. For example, if an individual QSAR model of a system is modified, then the whole system should be able to be rebuilt with minimal cost. Systems that are based on biological networks or AOPs are usually built with modularity in mind but are very hard to implement in practice. It will be next to impossible to scale them if the removal, addition of a node, or modifying connections between nodes need lots of manual adjustments. Therefore modularity should be considered from the very beginning.
6. *Failure to exploit computing resources:* Algorithms that run on a single core of a CPU or do not use GPUs, perform suboptimally on the vast majority of commonly available modern computers. There is no reason to avoid them, and parallel computing should be an integral part of QSAR-related computations.

References

Abadi, M., et al., 2016. Tensorflow: A system for large-scale machine learning. 12th Symposium on Operating Systems Design and Implementation, pp. 265–283. Software available from tensorflow.org.

Ankley, G.T., Bennett, R.S., Erickson, R.J., Hoff, D.J., Hornung, M.W., Johnson, R.D., et al., 2010. Adverse outcome pathways: a conceptual framework to support ecotoxicology research and risk assessment. Environ. Toxicol. Chem 29 (3), 730–741.

Ball, N., Madden, J., Paini, A., Mathea, M., Palmer, A.D., Sperber, S., et al., 2020. Key read across framework components and biology based improvements. Mutat. Res-Gen. Tox. En 853, 503172.

Barber, C., Amberg, A., Custer, L., Dobo, K.L., Glowienke, S., Van Gompel, J., et al., 2015. Establishing best practise in the application of expert review of mutagenicity under ICH M7. Regul. Toxicol. Pharmacol 73, 367–377.

Barratt, M.D., Langowski, J.J., 1999. Validation and subsequent development of the Derek skin sensitization rulebase by analysis of the BgVV list of contact allergens. J. Chem. Inf. Comput. Sci 39, 294–298.

Basak, S.C., Magnuson, V.R., Niemi, G.J., Regal, R.R., Veith, G.D., 1987. Topological indices: their nature, mutual relatedness, and applications. Math. Model. 8, 300–305.

Basak, S.C., Majumdar, S., 2016. Exploring two QSAR paradigms-congenericity principle versus diversity begets diversity principle analyzed using computed mathematical chemodescriptors of homogeneous and diverse sets of chemical mutagens. Curr. Comput-Aid. Drug. Des. 12, 1–3.

Basak, S.C., Niemi, G.J., Veith, G.D., 1990. A graph-theoretic approach to predicting molecular properties. Math. Comput. Model 14, 511–516.

Bondi, André B., 2000. Characteristics of scalability and their impact on performance. Proc. Second. Int. Workshop Softw. Perform. – WOSP 00, 195–203.

Chollet, F. et al., 2015. Keras <https://keras.io>.

Cristianini, N., Shawe-Taylor, J., 2004. An Introduction to Support Vector Machines and Other Kernel-Based Learning Methods. Cambridge University Press, Cambridge.

Hansch, C., Fujita, T., 1964. p-σ-π analysis. A method for the correlation of biological activity and chemical structure. J. Am. Chem. Soc 86, 1616–1626.

Hansch, C., Leo, A., 1979. Substituent Constants For Correlation Analysis in Chemistry and Biology. John Wiley & Sons, New York.

Hastie, T., 2001. In: Tibshirani, R., Friedman, J.H. (Eds.), The Elements of Statistical Learning: Data Mining, Inference, and Prediction.

Ho, T.K., 1995. Random decision forests. In: Proceedings of the 3rd International Conference on Document Analysis and Recognition. Montreal, QC, pp. 278–282.

Itskowitz, P., Tropsha, A., 2005. k Nearest neighbors QSAR Modeling as a variational problem: theory and applications. J. Chem. Inf. Model 45 (3), 777–785.

Devillers, J., Balaban, L.T. (Eds.), 1999. Topological Indices and Related Descriptors in QSAR and QSPR. Gordon and Breach Science Publishers, Singapore.

Jaworska, J.S., Natsch, A., Ryan, C., Strickland, J., Ashikaga, T., Miyazawa, M., 2015. Bayesian integrated testing strategy (ITS) for skin sensitization potency assessment: a decision support system for quantitative weight of evidence and adaptive testing strategy. Arch. Toxicol 89, 2355–2383.

Karelson, M., 2000. Molecular Descriptors in QSAR/QSPR. John Wiley & Sons.

Kim, S., Chen, J., Cheng, T., Gindulyte, A., He, J., He, S., et al., 2019. PubChem 2019 update: improved access to chemical data. Nucleic Acids Res 47, 1102–1109.

Mendez, D., Gaulton, A., Bento, A.P., Chambers, J., Veij, M.D., Félix, E., et al., 2019. ChEMBL: towards direct deposition of bioassay data. Nucleic Acids Res 47, 930–940.

Nantasenamat, C., Isarankura-Na-Ayudhya, C., Prachayasittikul, V., 2010. Advances in computational methods to predict the biological activity of compounds. Expert. Opin. Drug. Discov 5, 633–654.

Payne, M.P., Walsh, P.T., 1994. Structure-activity relationships for skin sensitization potential: development of structural alerts for use in knowledge-based toxicity prediction systems. J. Chem. Inf. Comput. Sci 34, 154–161.

Plošnik, A., Vračko, M., Dolenc, M.S., 2016. Mutagenic and carcinogenic structural alerts and their mechanisms of action. Arh. Hig. Rada Toksikol 67, 169–182.

Quinlan, J.R., 1986. Induction of decision trees. Mach. Learn 1, 81–106.

Rogers, D., Hahn, M., 2010. Extended-connectivity fingerprints. J. Chem. Inf. Model 50, 742–754.

Tropsha, A., 2010. Best practices for QSAR model development, validation, and exploitation. Mol. Inf 29, 476–488.

Vapnik, V., Golowich, S.E., Smola, A., 1996. Support vector method for function approximation, regression estimation, and signal processing. Adv. Neural Inf. Process. Syst. 9, 281–287.

Villeneuve, D.L., Crump, D., Garcia-Reyero, N., Hecker, M., Hutchinson, T.H., LaLone, C.A., et al., 2014. Adverse outcome pathway (AOP) development I: strategies and principles. Toxicol. Sci 142 (2), 312–320.

Willett, C., Caverly, Rae, J., Goyak, K.O., Landesmann, B., Minsavage, G., et al., 2014. Pathway-based toxicity: history, current approaches and liver fibrosis and steatosis as prototypes. ALTEX 31 (4), 407–421.

Wu, Z., Ramsundar, B., Feinberg, E.N., Gomes, J., Geniesse, C., Pappu, A.S., et al., 2017. MoleculeNet: a benchmark for molecular machine learning. Chem. Sci 9, 513–530.

Yousefinejad, S., Hemmateenejad, B., 2015. Chemometrics tools in QSAR/QSPR studies: a historical perspective. Chemometr. Intell. Lab 149, 177–204. Part B.

From big data to complex network: a navigation through the maze of drug–target interaction

18

Ze Wang[1], *Min Li*[2], *Muyun Tang*[2] *and Guang Hu*[2]
[1]Department of Pharmaceutical Sciences, Zunyi Medical University at Zhuhai Campus, Zhuhai, P.R. China, [2]Department of Bioinformatics, Center for Systems Biology, School of Biology and Basic Medical Sciences, Soochow University, Suzhou, P.R. China

18.1 Introduction

At an average cost of $985 million per drug and at least a decade to reach the market, drug discovery and development are highly expensive, time-consuming, and complex processes (Wouters et al., 2020; Mullard, 2020; Mohs and Greig, 2017). In fact, the attrition rate of drug discovery and the number of clinical trial failures has increased in the last decades (Bolognesi and Cavalli, 2016; Chaudhari, et al., 2017). As pointed out by Hopkins, the fundamental problem may be the core philosophy in drug discovery, which traditionally assumes that the primary goal as designing exquisitely selective "magic bullets" to bind with a single disease target (Hopkins, 2008). With the development of systems biology, scientists realized the one-bullet-one-target assumption is oversimplified and accepted the concept of network pharmacology as a paradigm shift in drug discovery (Hopkins, 2008; Loscalzo and Barabasi, 2011; Yildirim et al., 2007; Liang and Hu, 2016; Yan et al., 2018). Compared to the traditional one-bullet-one-target paradigm, network pharmacology attempts to uncover drug action by considering the interaction between drug molecules and their potential targets through a holistic network, which has great potential to facilitate disease mechanism understanding and drug discovery (Wang et al., 2021).

Identification and discovery of potential therapeutic targets for drugs have largely benefited from high-throughput experimental techniques, which generate numerous biological data (Russell et al., 2013). On the other hand, clarification and characterization of active ingredients from herbal plants also deposited a huge amount of chemical data (French et al., 2018). With the continuous collection and deposition of big data from high-throughput experiments, modern drug discovery and development are moving into the big data era (Zhu, 2020). It is now realized that big data in drug discovery is proposing four challenges to traditional data management and analysis methodologies, including the scale of data, the growth speed of data, the diversity of data source, and the uncertainty of data (Ciallella and Zhu, 2019; Lee and Yoon, 2017).

Big Data Analytics in Chemoinformatics and Bioinformatics. DOI: https://doi.org/10.1016/B978-0-323-85713-0.00016-5

For example, several million compounds were typically investigated in high-throughput experiments in drug development (Santos et al., 2017). More importantly, data uncertainty, especially when considering complex biological mechanisms (e.g., drug responses, side effects), has brought further obstacles to using this data. Therefore, the development of new data analysis tools and computational algorithms to manage and utilize these data is necessary for drug discovery and development.

Modeling the action of drugs through the big data has given birth to the complex network view of drug–target interaction (Hopkins, 2008), which is composed of nodes and lines representing molecular entities (for both drug and target molecules) and their relations, respectively. Network science, which originates from the great mathematician Euler in the Königsberg problem, is growing as a systematic tool for the analysis of complex networks emerging from a wide range of disciplines (Borgatti and Halgin, 2011; Newman, 2003; Parkhe et al., 2006). As shown by Yildirim et al., the application of network science to drug screening data has demonstrated a network map rather than isolated, bipartite nodes for drugs and targets, which revolutionized our understanding of drug–target interactions (Fig. 18.1) (Yildirim et al., 2007). Currently, computational identification and analysis of drug–target interaction are becoming a cutting-edge research areas in drug discovery and development.

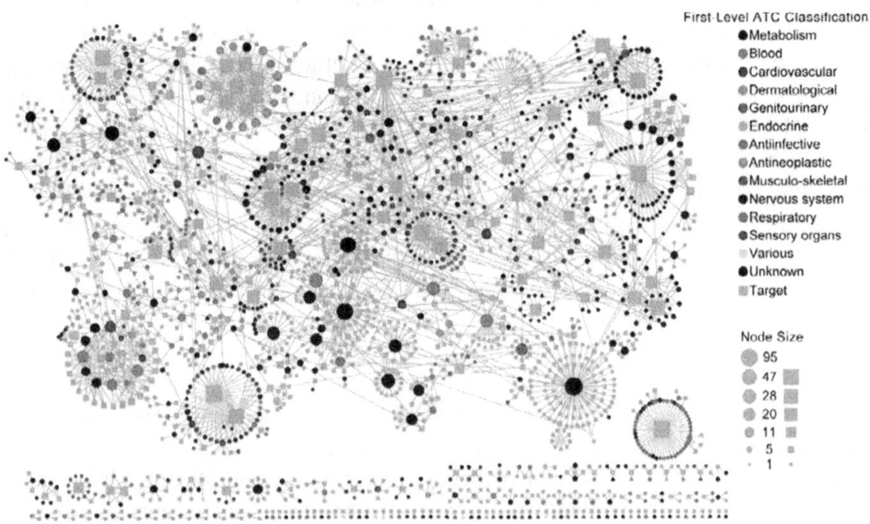

Figure 18.1 Drug–target interaction network constructed from FDA-approved drugs (Yildirim et al., 2007). In the network, drugs and targets are represented as circular and rectangular nodes, respectively. The area of node is proportional to the number of interactions, which are shown as lines. Different colors are used to classify drugs and targets, according to Anatomical Therapeutic Chemical Classification and the Gene Ontology database, respectively.
Source: With permission from Yildirim, M.A., Goh, K.I., Cusick, M.E., Barabasi, A.L., Vidal, M., 2007. Drug-target network. Nat. Biotechnol. 25, 1119–1126.

In this chapter, we reviewed the important data sources in drug discovery and development, including drug screening, active ingredient profiling, and target fishing. These databases are building blocks for the construction and prediction of drug–target interactions. Then, we introduced the algorithms and methodologies in the construction, prediction, and analysis of drug–target interaction. The prediction methods can be roughly divided into structure-based, similarity-based, and machine learning-based. Although structure-based methods showed high accuracy, the application of these methods is often limited by the lack of three-dimensional structures. Therefore, we only focused on the other two methods. In the second part, we also reviewed important computational tools and methods in network construction and analysis. We hope the content of this chapter will highlight the critical role of the network view of drug–target interaction, which is driven by the continuously expanding databases.

18.2 Databases

The construction of drug–target interaction networks relies on databases, which are generally composed of a hierarchical collection of alphabetical, numerical, graphical, and structural data. This section will introduce the most commonly used databases covering small molecules (Table 18.1), biological macromolecules (Table 18.2), and traditional Chinese medicine (TCM) (Table 18.3), as well as their interactions.

18.2.1 Chemical databases

18.2.1.1 DrugBank

Released in 2006, DrugBank (https://go.drugbank.com/) is one of the most used drug-related resources for bioinformatics, chemoinformatics, and medicinal chemistry (Wishart et al., 2017). It is a freely available internet-based database that aims to comprehensively include detailed information on targets, mechanisms, and interactions of both FDA-approved and investigational drugs. The current version contains a total number of 14,460 drug entries, including 2683 FDA-approved small molecule drugs, 2585 biotech drugs such as proteins and peptides, 6643 phase I/II/III drugs, and 131 nutraceuticals (Table 18.1, Fig. 18.2, data collected at the end of April 2021) (Wishart et al., 2017). Besides, 5236 non-redundant protein sequences and annotations related to the drugs were included. Each drug entry is composed of over 200 distinct data fields covering chemical identification, pharmacology, pharmaceutics, clinical trial, target sequence, pathway, and spectra information (Wishart et al., 2017). Data in DrugBank can be accessed and retrieved from a field search engine. Additionally, the database provides alternative format and datasets for data mining and analysis. For example, DrugBank contains a portal for machine-learning algorithms, which require labeled datasets including drug, target, side-effect, and toxicity (Wishart et al., 2017).

Table 18.1 Chemical databases for drugs and small molecules.

Database	Description	Database statistics	Website	Reference
DrugBank	FDA-approved and investigational drug	2683 FDA-approved small molecule drugs, 2585 biotech drugs, 6643 investigational drugs, 131 nutraceuticals, 5236 nonredundant protein sequences, and annotations	https://go.drugbank.com/	Wishart et al. (2017)
PubChem	Resources for chemical compounds	270,998,024 chemical entities (109,891,884 unique chemical structures) and 1,366,263 bioassays	https://pubchem.ncbi.nlm.nih.gov/	Kim et al. (2018)
ChEMBL	Structure, bioassays, affinity data for drug	More than 2 million compounds,17 million activity data, >1600 distinct cell lines, 500 tissues/organs, 3600 organisms, >14,300 targets	https://www.ebi.ac.uk/chembl/	Mendez et al. (2018)
ChemSpider	Pure chemical structure and property	103 million chemical structures and links to original data sources	http://www.chemspider.com/	Pence and Williams (2010)

Table 18.2 Biological databases for targets.

Database	Description	Database statistics	Website	Reference
UniProt	Comprehensive database for protein sequence and annotations	564,638 reviewed protein sequences for over 84 thousand species	https://www.uniprot.org/	Consortium (2018)
PDB	Structural data for biomacromolecules	177,009 structural entities for biological macromolecules	https://www.rcsb.org/	Burley et al. (2018)
STRING	Interaction database	24,584,628 proteins, 3,123,056,667 total interactions from 5090 organisms	https://string-db.org/	Szklarczyk et al. (2018)
BindingDB	Affinity database	2.2 million protein–ligand affinity data, involving 977,487 small molecules and 8516 targets	https://www.bindingdb.org/	Gilson et al. (2015)

Table 18.3 Databases for traditional Chinese medicine.

Database	Description	Database statistics	Website	Reference
TCM database@Taiwan	Currently the most comprehensive and largest noncommercial TCM database available for download	37,170 (32,364 nonduplicate) TCM compounds from 352 TCM ingredients.	http://tcm.cmu.edu.tw/	Chen (2011)
TCMSP	Highlight the role that the systems pharmacology plays across the TCM discipline.	All the 499 herbs registered in Chinese pharmacopoeia (2010), with a total of 12144 chemicals.	https://www.tcmspw.com/tcmsp.php	Ru et al. (2014)
TCMID	Information on all respects of TCM including formulae, herbs, and herbal ingredients, and information for drugs and diseases	8159 herbs, 46,914 TCM formulae, and more than 25,210 herb ingredients.	http://119.3.41.228:8000/tcmid/	Huang et al. (2017)

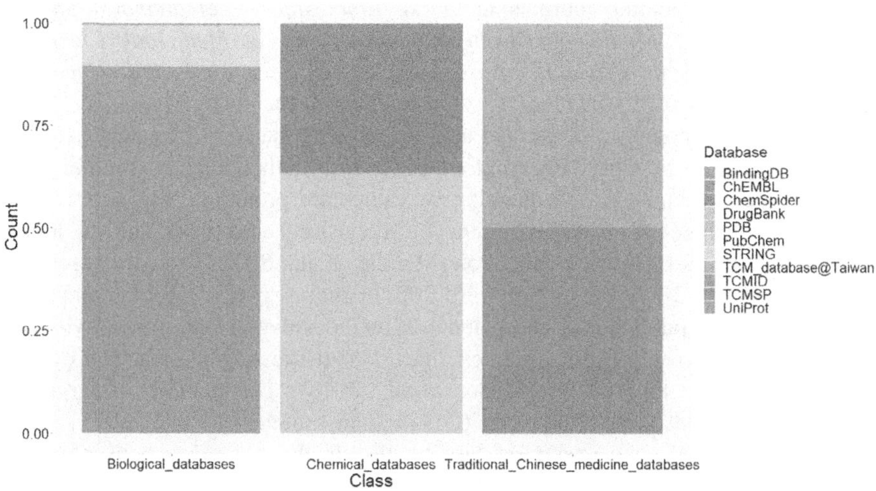

Figure 18.2 Statistics for data size of each database in three different categories, that is, small molecules, biological targets, and traditional Chinese medicine.

18.2.1.2 PubChem

Initiated and maintained by the US National Institutes of Health (NIH), PubChem (https://pubchem.ncbi.nlm.nih.gov/) is an open database that collects chemical information and resources (Kim et al., 2018). PubChem supports bidirectional data transfer between users and the database, allowing contributors to create, upload, and edit data freely. Since its first version in 2004, PubChem has continually become a huge chemical database that contains 270,998,024 chemical entities (109,891,884 unique chemical structures) and 1,366,263 bioassays (carbohydrates, nucleotides, peptides, etc.) contributed by PubChem users (Table 18.1, Fig. 18.2) (Kim et al., 2018). It provides spectral information including ^{1}H NMR, ^{13}C NMR, 2D NMR, FT-IR, Ms, UV-Vis, and Raman data for more than 590,000 compounds. Spectral data in PubChem are linked with external spectral databases such as SpectraBase (http://spectrabase.com) and the MassBank of North America (https://mona.fiehnlab.ucdavis.edu/). By the end of April 2021, the database archived 296,907,771 biological activity data, 90,426 gene data, 96,561 protein data, 4849 taxonomy, and 237,925 pathways involved with chemical entities (Kim et al., 2018). Data in PubChem is organized as three dependent databases, including substance which collects descriptions of substances contributed by users, Compound which enumerates chemical compounds according to unique chemical structure, and Bioassay containing biological assays and experiments related to the compounds.

18.2.1.3 ChEMBL

ChEMBL (https://www.ebi.ac.uk/chembl/) is a manually maintained drug discovery database that deposits medicinal chemistry data from clinical development candidates and academic journals including *Bioorganic & Medicinal Chemistry Letter, Journal of Medicinal Chemistry, Bioorganic & Medicinal Chemistry, Journal of Natural Products, European Journal of Medicinal Chemistry, MedChemComm, ACS Medicinal Chemistry Letters*, etc. (Mendez et al., 2018). Structures of compounds, assays, and activity information were manually extracted from the literature by ChEMBL curators. Since information such as structure connectivity, stereochemistry, and quantitative values are prone to error, it is encouraged to contribute to ChEMBL data by depositing chemical and biological information during scientific publication (Mendez et al., 2018). The current released version ChEMBL 28 (at the end of April 2021) contains over 2 million compounds from over 80,000 publications and patents. It includes over 17 million activity data annotating from over 1600 distinct cell lines, 500 tissues/organs, and 3600 organisms (Table 18.1, Fig. 18.2) (Mendez et al., 2018). The number of targets in ChEMBL has exceeded 14,300, with 6311 human proteins (Mendez et al., 2018). Except for human, mouse, and rat targets, the database also contains plenty of experimental data from other model organisms such as *Staphylococcus aureus*. ChEBML is embracing new data sources from bacteria, viruses, and pathogens, making it an ideal platform for multipurpose drug development (e.g., antimicrobial). Clinical data in ChEMBL is continuing to be incorporated with other public databases such as the ClinicalTrials.gov database (https://clinicaltrials.gov/), FDA Orange Book (https://www.accessdata.fda.gov/scripts/cder/ob/), FDA New Drug Approvals (https://www.fda.gov/Drugs/DevelopmentApprovalProcess/DrugInnovation/default.htm), the British National Formulary (https://bnf.nice.org.uk/), Medicinal Subject Headings (MeSH, https://www.nlm.nih.gov/mesh/). Bioactivity data are also timely exchanged with external databases like PubChem (https://pubchem.ncbi.nlm.nih.gov/) and BindingDB (http://www.bindingdb.org/). Other properties of deposited compounds were calculated by RDKit (https://www.rdkit.org/). For data accessibility, ChEMBL supports text search through its webpage and download from FTP site (ftp://ftp.ebi.ac.uk/pub/databases/chembl/ChEMBLdb/latest/) with a variety of data formats including SD file and FASTA file (Mendez et al., 2018).

18.2.1.4 ChemSpider

From the perspective of pure chemical structure and property, researchers hope to obtain a variety of information about a compound, including molecular structure, systematic nomenclature, physical properties, spectral data, reactions and synthetic methods, and safety information. The information is typically distributed in different literatures, libraries, and databases. ChemSpider (http://www.chemspider.com/) was born to collectively integrate chemical structure-related information from different data sources (Table 18.1, Fig. 18.2) (Pence and Williams, 2010). In 2009, ChemSpider was purchased by the Royal Society of Chemistry (RSC), allowing the accessibility of a wealth of information from RSC, that is, scientific publications and databases. ChemSpider has also been connected with other databases such as Wikipedia (https://en.jinzhao.wiki/wiki/Main_Page), PubChem, and Kyoto Encyclopedia of Genes and Genomes (KEGG) (https://www.kegg.jp/). To avoid errors in the data input process, ChemSpider is curated by only registered users. The data in ChemSpider can be accessed from text search, structure searches as well as substructure search. With over 103 million chemical structures and links to original data sources, ChemSpider is becoming a portal to the property, annotation, synthesis, spectral information of the expanding chemical universe (Table 18.1) (Pence and Williams, 2010).

18.2.2 Databases for targets

18.2.2.1 UniProt

The Universal Protein Resource (UniProt, https://www.uniprot.org/) is aimed to provide a comprehensive and high-quality data source of protein sequences and annotations (Table 18.2) (Consortium, 2018). The behavior and physiology of cells are defined by proteins that respond to environmental signals. Understanding the time-dependent protein expression at a whole proteome level is crucial to interpret life in a quantitative way. With the improvements of experimental techniques, the information on protein sequence, structure and function is increasing broadly and deeply. It is therefore challenging to manage the information and make it conveniently accessible to users. UniProt data are managed by more than 100 experts hosted by the collaboration of the European Bioinformatics Institute (EMBL-EBI), the Swiss Institute of Bioinformatics (SIB), and the Protein Information Resource (PIR). UniProt (release 2020_05) now provides 564,638 reviewed entries for over 84 thousand species including humans, rice, *Arabidopsis thaliana*, mouse, zebrafish, etc. UniProt entry is composed of the core data field (protein sequence, protein name, description, taxonomy, citation) and peripheral field including as much annotation information (Consortium, 2018). Although the database can provide rich information by simple text query and search, it actively supports in-depth data mining through various online training such as webinars (https://www.ebi.ac.uk/training/online/), YouTube videos (https://www.youtube.com/user/uniprotvideos/), Facebook (https://www.facebook.com/uniprot.org/), and Twitter (@uniprot).

18.2.2.2 Protein Data Bank

Structural biology has witnessed frequent advances in the structural determination of proteins, RNA, DNA, and their complexes with small molecules. Since 1971, the Protein Data Bank (PDB, https://www.rcsb.org/) established an open-access database in structural biology by depositing only seven protein structures at the beginning (Table 18.2) (Burley et al., 2018). With continuing development, PDB has grown up to a comprehensive database consisting of 177,009 structural entities for biological macromolecules (Fig. 18.2) (Burley et al., 2018). PDB data entry is originated from experimental sources including X-ray diffraction, nuclear magnetic spectroscopy (NMR), and three-dimensional electron microscopy (Table 18.2). Structural data are validated and biocurated by a global expert team to ensure the accurate representation of the structural data and the underlying annotation information. Data exploration service in PDB allows convenient accessibility to every structural entry via any popular web browser (e.g., Chrome, Firefox, Microsoft Edge). The website rcsb.org supports the keywords and unstructured text search, whilst the obtained data are sorted and tabulated to include atomic coordinates, experimental methods, sequence, description, citation, specific chemical components, taxonomy, and enzyme classification. Additionally, PDB data can be explored by multiple online tools for data manipulation and visualization. For example, the PDB website enables metabolic pathway mapping for user-interested structures, drug, and ligand discovery through external links such as DrugBank and BindingDB, as well as the fast and interactive three-dimensional display through NGL Viewer (Burley et al., 2018).

18.2.2.3 String

With impressive advances in elucidating the interaction between individual proteins, it is realized cellular machinery depends on the global network of physical (direct) and functional (indirect) protein–protein interactions. The information space of protein–protein interactions is far more complicated than the intrinsic properties and annotations of individual proteins. STRING (https://string-db.org/) is a knowledgebase of known and computationally predicted protein–protein interactions (Szklarczyk et al., 2018). It collects and stores protein–protein interaction data from a variety of publicly available data sources: genomic predictions, high-throughput experiments, co-expression, automated text-mining, and online databases such as Database of Interacting Proteins (DIP, http://dip.doe-mbi.ucla.edu/), Biomolecular Interaction Network Database (BIND, http://bind.ca/), Molecular Interaction Database (MINT, http://mint.bio.uniroma2.it/mint/), KEGG (http://www.kegg.jp/), and Reactome (http://www.reactome.org/). STRING v11.0 contains 24,584,628 proteins and 3,123,056,667 total interactions from 5090 organisms including Homo sapiens, Mus musculus, *A. thaliana*, and so on (Table 18.2, Fig. 18.2) (Szklarczyk et al., 2018). STRING defines a functional association unit as the basic building blocks, which is an edge between two proteins both having functional contributions to a specific biological process. By the definition, protein–protein interaction does not necessarily require physical contact between proteins. STRING website provides user-friendly access to the interaction network for single protein and multiple proteins, which can be enquired either by name or sequence. Also, through the STRING online server, users can compute functional enrichment for a set of proteins involving the interaction network (Szklarczyk et al., 2018).

18.2.2.4 BindingDB

BindingDB (https://www.bindingdb.org/) is an open database of experimental affinity data of protein–ligand interaction (Gilson et al., 2015). With steady growth since 2000, BindingDB now contains about 2.2 million protein–ligand affinity data, involving 977,487 small molecules and 8516 protein targets (Table 18.2,

Fig. 18.2) (Gilson et al., 2015). The data source for BindingDB includes scientific publications and patents. Affinity data of at least one protein–ligand complex is supplied in the database along with information on publication source and experimental conditions (e.g., temperature, pH, buffer composition). BindingDB supports interactive connection to several public databases including PDB, UniProt, DrugBank, ChEMBL, PubChem, Reactome, MarinLit (http://pubs.rsc.org/marinlit), and ZINC (http://zinc.docking.org/). Data in BindingDB is organized as hyperlinks listed in a table format and can be accessed through flexible web tools for query, browsing, download, visualization, and analysis (Gilson et al., 2015).

18.2.3 Databases for traditional Chinese medicine

TCM often comprises over thousands of chemical compounds from different botanical species, hitting multiple biological targets (Cheung, 2011). The herbal compounds and corresponding targets form a complex network that involves various nodes and edges (Li et al., 2011; Tao et al., 2013). To comprehensively characterize and analyze the network, the wet experiment is time-consuming and expensive due to the dozens of chemical entities and biological targets involved. Systems pharmacology is a big data-driven strategy that deals with prior experimental data of herbal compounds as well as biological assays (Li et al., 2011; Ru et al., 2014). With increasing attention towards discovering novel lead compounds from TCM, a database for TCM is necessary. Besides, the prediction power of systems pharmacology is enhanced by online target prediction algorithms. This section briefly reviews some typical TCM databases (Table 18.3).

18.2.3.1 Traditional Chinese medicine Database@Taiwan

TCM Database@Taiwan includes more than 20000 chemical compounds from 453 herbs, animals, and minerals in TCM (Chen, 2011). The database is evolving to cover more compound data from folk herbs. In TCM Database@Taiwan, drug molecules were classified into 22 different categories according to clinical applications (Chen, 2011). The classification model is based on the theories of TCM involving the Yin-yang and the Five Elements theory. TCM ingredients were collected from publications on Medline and ISI Web of Knowledge. Through simple and advanced search, TCM Database@Taiwan provides both two-dimensional and three-dimensional structures of each TCM constituent, as well as physical properties such as ALogP, polar surface area, rotatable bonds, and so on (Table 18.3, Fig. 18.2) (Chen, 2011).

18.2.3.2 Traditional Chinese medicine systems pharmacology

The TCM systems pharmacology (TCMSP) database and analysis platform is built for this purpose (Ru et al., 2014). TCMSP contains 499 Chinese herbs collected in Chinese Pharmacopeia (Ru et al., 2014). Through deep data mining and analysis, 29,384 chemical compounds, 3311 targets, and 837 associated diseases were manually curated in the database (Table 18.3, Fig. 18.2) (Ru et al., 2014). ADME-related properties were computed in TCMSP, including oral bioavailability, half-life, drug-likeness, Caco-2 permeability, blood–brain barrier, and Lipinski's rule of five (Ru et al., 2014). For drug targets, TCMSP includes all experimentally validated targets and SysDT model predicted targets. The strengths of the TCMSP platform allow the analytical decomposition of TCM through data and network methodology (Ru et al., 2014).

18.2.3.3 Traditional Chinese medicine integrated database

TCM integrated database (TCMID) is aiming to provide convenient online information on TCM for pharmacologists and scholars (Huang et al., 2017). Established in 2013, TCMID integrated online databases including TCM Database@Taiwan (Chen, 2011), HIT (Ye et al., 2010) to collect over 49,000 prescriptions, 8159 herbs, 25,210 ingredients, 3791 diseases, 6828 drugs and 17,521 targets (Table 18.3, Fig. 18.2). Since most publications on TCM, especially separation and pharmacological research, were written in Chinese, TCMID manually collects original data from the Chinese national knowledge infrastructure (CNKI) and translated the related information into English. Users can easily retrieve detailed descriptions and information from external databases such as Drugbank, OMIM, and STITCH. Additionally, TCMID has documented mass spectra (Ms) of Chinese herbs through CNKI. TCMID collects the original place of the herb, Ms spectrum, chromatography spectrum, as well as compound information (Huang et al., 2017). With new features added, TCMID is growing as an important hub for the modernization of TCM.

18.3 Prediction, construction, and analysis of drug–target network

The drug–target network is mathematically described as a bipartite network graph $G(D, T, P)$. In the network, the drug set D and target set T is defined as

$$D = D(d_1, \quad d_2, \ldots, d_n)$$

$$T = T(t_1, \quad t_2, \ldots, t_m)$$

And the interaction set P is defined as a Kronecker matrix:

$$P = \begin{bmatrix} p_{11} & p_{12} & \cdots & p_{1n} \\ & & \cdots & \\ p_{m1} & p_{m2} & \cdots & p_{nm} \end{bmatrix}$$

where $p_{kl} = 1$ when drug d_k binds with target t_l, otherwise $p_{kl} = 0$. Practically, a binding affinity threshold is used to obtain the interaction p_{kl}. The purpose of the prediction, construction, and analysis of the drug–target network is to identify drug targets from the whole target pool, formulate the interactome configuration, and characterize the property and module both globally and locally. A landscape on the drug–target interaction network is crucial to the understanding of therapeutic mechanisms and side effects. In this section, we briefly review advances in algorithms, computational tools, and network analysis methods in drug–target interaction.

18.3.1 Algorithms to predict drug–target interaction network

Prediction of biological networks containing thousands of compounds and targets is still challenging to the traditional experimental approach, such as high-throughput screening and biological assays (Haggarty et al., 2003; Kuruvilla et al., 2002; Wang et al., 2015; Whitebread et al., 2005). Therefore, the computational prediction method is important for biological network analysis. Although the virtual screening method for three-dimensional compounds and targets is well developed, the lack of three-dimensional structural data and time-consuming algorithms for most biological molecules still makes this approach limited in real application (Cheng et al., 2007; Morris et al., 2009). Alternatively, several knowledge-based computational methods have been developed to efficiently address the drug–target prediction problem (Table 18.4). In this section, we will briefly review the typical algorithms and methods for drug–target prediction.

Table 18.4 Algorithms to predict drug–target interaction.

Algorithms	Classification	Description	Reference
Bipartite graph algorithm	Supervised machine learning	A supervised machine learning algorithm for a BG model, mapping drugs in chemical space and targets in genomic space	Yamanishi et al. (2008)
Advanced BG algorithm	Supervised machine learning	Advanced version of BG algorithm with pharmacological data involved	Yamanishi et al. (2010)
BLM	Supervised machine learning	A BLM incorporating the concepts of local models to predict drug–target interaction	Bleakley and Yamanishi (2009)
BLM-NII	Supervised machine learning	An updated version of BLM by introducing neighbor-based interaction-profile inferring	Mei et al. (2013)
RBM	Supervised machine learning	A RBM with a two-layer graphic model effectively capture the features of drug–target interaction and predict different types of drug–target interaction	Wang and Zeng (2013)

(*Continued*)

Table 18.4 (Continued)

Algorithms	Classification	Description	Reference
Random forest algorithm	Supervised machine learning	Combines the information from chemical, biological, and network features to predict drug–target interaction with high accuracy	Cao et al. (2014)
Negative dataset selection method	Supervised machine learning	Two methods to assist the selection of negative dataset in the machine learning-based algorithms	Wang et al. (2014)
NetLapRLS	Semisupervised machine learning	Adopts both labeled and unlabeled data in machine learning	Xia et al. (2010)
Chemical similarity	Chemical similarity	Chemical similarity method based on the assumption that similar drug structures are more likely to interact with similar targets	Keiser et al. (2009)
Two-step similarity	Chemical similarity	A similarity score was obtained by graph representation and chemical functional group representation in two steps	Chen and Zeng (2013)
Phenotypic side-effect similarity	Network similarity	An algorithm to determine if two drugs will interact with the same target	Campillos et al. (2008)
NRWRH	Network similarity	Based on the framework of random walk and the assumption that similar drugs often corresponding to similar targets	Xia et al. (2010)
DBSI	Network similarity	Drug-based similarity inference based on complex network theory	Cheng et al. (2012)
TBSI	Network similarity	Target-based similarity inference based on complex network theory	Cheng et al. (2012)
NBI	Network similarity	Network-based inference based on complex network theory	Cheng et al. (2012)
Within scores and between scores	Network similarity	Considering features from target similarity and drug similarity	Shi et al. (2015)
MTOI	Network similarity	Multiple target optimal intervention finding algorithm to identify potential drug targets and their optimal combinations restoring to a normal state	Yang et al. (2008)

18.3.1.1 Machine learning-based methods

Yamanishi et al. have developed a bipartite graph (BG) algorithm to probe drug–target interaction for four target classes including enzymes, ion channels, GPCRs, and nuclear receptors (Table 18.4) (Yamanishi et al., 2008). By introducing in-prior knowledge of chemical structures and genomic sequence information, they have built a supervised machine learning algorithm for a BG model, mapping drugs in chemical space and targets in genomic space (Fig. 18.3). The machine learning models f_c and f_g were defined based on a modified kernel regression function:

$$f: \quad X \times X \to \mathbf{R}^q$$

$$f(x, x_i) = \sum_{i=1}^{n} s(x, x_i)\mathbf{w}_i + \epsilon$$

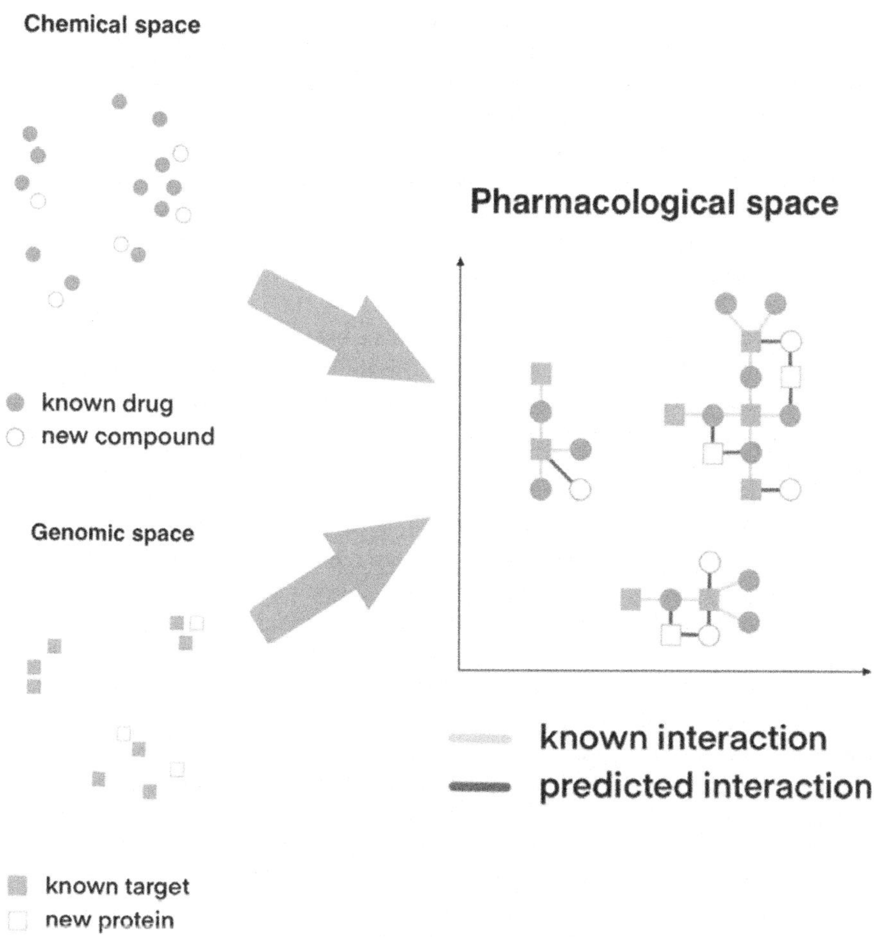

Figure 18.3 An illustration of bipartite graph algorithm.

where **w** represents a weighing vector, and *s* stands for similarity score for chemical structures or sequence (Yamanishi et al., 2008). In this BG pharmacological space algorithm, the structural and sequence similarity were considered, and the drug–target interactions were predicted by the closeness between drugs and targets.

Regarding the side effect of a drug, it is assumed that drugs with similar side effects are more likely to interact with similar targets. Taking pharmacological data into consideration may further improve the performance of a machine learning-based algorithm. Yamanishi et al. have further improved the BG method by involving pharmacological knowledge (Yamanishi et al., 2010). The pharmacological effect similarity, computed from the chemical structures of drugs, was introduced into the BG model to identify drug–target interactions (Table 18.4) (Yamanishi et al., 2010).

Based on the BG method, Bleakley et al. proposed a bipartite local model (BLM) incorporating the concepts of local models to predict drug–target interaction (Table 18.4) (Bleakley and Yamanishi, 2009). By involving local models, the edge-prediction problem was transformed into the binary classification of labeled points (Bleakley and Yamanishi, 2009). Targets are predicted by comparing sequence similarities, and drug–target interactions are predicted based on structural similarities. Finally, independent drug–target interactions were obtained putatively. Since it combines the strengths of the BG model and the local model, the BLM algorithm showed an excellent computational performance to predict drug–target interaction (Bleakley and Yamanishi, 2009).

Despite the computational speed, BLM is not able to predict drug–target interaction without training data. Therefore, the prediction of drug–target interaction for new drug molecules is not possible by using BLM. Mei et al. have proposed an updated version of BLM by introducing neighbor-based interaction-profile inferring (BLM-NII, Table 18.4) (Mei et al., 2013). The BLM-NII method derived the initial weighted interactions for the new drug from its neighbor interaction profile and then labeled this interaction to train the BLM model (Mei et al., 2013). For nuclear receptors, BLM-NII enhances the BLM method, especially for the dataset that contains drug–target with no prior interaction information.

Zeng et al. have developed a restricted Boltzmann machine (RBM) method that can predict drug–target interactions and the types of interaction (Table 18.4) (Wang and Zeng, 2013). In the RBM method, a contrastive divergence algorithm was applied to a two-layer graphic model which represents drug–target interaction. Zeng et al. has tested the RBM method on MATADOR and STITCH database (Günther et al., 2008; Szklarczyk et al., 2015; Wang and Zeng, 2013). It has shown the RBM method can effectively capture the features of drug–target interaction and predict different types of drug–target interaction.

Cao et al. proposed a random forest algorithm to predict drug–target interaction (Table 18.4) (Cao et al., 2014). The novelty of the algorithm was the combination with the information from chemical, biological, and network features. The accuracy of the algorithm was evaluated as 93.52%, 94.84%, 89.68%, and 84.72% for enzymes, ion channels, GPCRs, and nuclear receptors, respectively (Cao et al., 2014). The performance of the algorithm showed the importance of network topology as training information for the prediction of drug–target interaction.

In the prediction of drug–target interaction, a common problem for the machine learning-based method is the lack of a negative dataset. Wang et al. have proposed two methods to assist the selection of negative datasets in the machine learning-based algorithms (Table 18.4) (Wang et al., 2014). In the first method, a drug–protein deviation function is defined as:

$$\xi(\mathbf{X}_i) = \sum_j \left| \frac{\langle x_j \rangle (x_{ij} - \langle x_j \rangle)}{\mathrm{var}(x_j) \sum (\langle x_j \rangle^2 / \mathrm{var}(x_j))} \right|$$

vector $\mathbf{X}_i$ ($i = 1,2,\ldots, m$) is an m-dimension vector representing m properties of the ith target, x stands for the jth value for the property of the ith targets. Wang et al. used $\xi > 0.42$ as a threshold value to select a negative dataset (Wang et al., 2014). In the second method, a probability function for the ith unknown target to be a negative sample was defined as

$$P(\mathbf{X}_i) = \frac{(\xi(\mathbf{X}_i) - \langle \xi_{\text{positive}} \rangle)^2}{\sum (\xi(\mathbf{X}_i) - \langle \xi_{\text{positive}} \rangle)^2}$$

$P = 0.5$ was used to consider the negative dataset. Wang et al. have improved the prediction accuracy and identified 1797 and 227 drug–target interactions by using these two methods, respectively (Wang et al., 2014).

As discussed above, labeling the positive or negative dataset is often a challenging problem in the development of algorithms based on supervised machine learning methods. The problem can be addressed by introducing a semisupervised method, which adopts both labeled and unlabeled data in machine learning. Wong et al. developed a manifold regularization semisupervised machine learning method (Table 18.4) (NetLapRLS) to predict drug–target interaction, which generates a biological space by combining information of chemical space, sequence space, and drug–target interaction network (Xia et al., 2010). In the drug domain, classification functions are defined as:

$$\mathbf{F}_d^* = \min_{\mathbf{F}_d} J(\mathbf{F}_d) = \mathbf{Y} - \mathbf{F}_{df}^{\,2} + \beta_d \mathrm{Trace}(\mathbf{F}_d^T \mathbf{L}_d \mathbf{F}_d \big)$$

$$\mathbf{F}_d^* = \mathbf{W}_d \alpha_d^*$$

$$\alpha_d^* = \arg \min_{\alpha_d \in R^{n_d \times n_p}} \left\{ \mathbf{Y} - \mathbf{W}_d \alpha_{df}^{\,2} + \beta_d \mathrm{Trace}(\alpha_d^T \mathbf{W}_d \mathbf{L}_d \mathbf{W}_d \alpha_d) \right\}$$

where $\mathbf{F}_d$ is the prediction function on drug domain, α_d is a cost function, and is Frobenius norm, β_d is the trade-off in the drug domain, Trace is the matrix trace, $\mathbf{Y}$ is the adjacent matrix of the known drug–target interaction network (Xia et al., 2010). The similar function $\mathbf{F}_t$ was defined in the target domain. Applying representer theorem and optimization, the prediction function in drug and protein domain were derived as:

$$\mathbf{F}_d^* = \mathbf{W}_d(\mathbf{W}_d + \beta_d \mathbf{L}_d \mathbf{W}_d)^{-1}\mathbf{Y}$$

$$\mathbf{F}_t^* = \mathbf{W}_t(\mathbf{W}_t + \beta_t \mathbf{L}_t \mathbf{W}_t)^{-1}\mathbf{Y}$$

The predictions are then obtained by combining drug and target domain as (Xia et al., 2010)

$$\mathbf{F}^* = \frac{\mathbf{F}_d^* + (\mathbf{F}_t^*)^T}{2}$$

18.3.1.2 Similarity-based methods

Side effects and efficacy of a drug could be explained by the multiple physiological targets of a drug. It is reasonable to assume that similar drug structures are more likely to interact with similar targets. But one should keep in mind that molecular similarity should be well defined first, since the concept of similarity is subjective and the similarity space is complex (Basak et al., 2002; Basak et al., 2006). By using two-dimensional chemical similarity method, Keiser et al. predicted thousands of drug–target interactions (Table 18.4) (Keiser et al., 2009). Among them, 23 associations were experimentally confirmed, including the inhibition of the 5-hydroxytryptamine transporter by the ion channel drug Vadilex, and antagonism of the histamine H4 receptor by the enzyme inhibitor Rescriptor (Keiser et al., 2009).

Chen et al. have developed a two-step similarity-based method to predict the target group of drugs (Table 18.4) (Chen and Zeng, 2013). In this method, drugs were encoded as their graph representations. Then the target group *T*(*d*) for drug *d* was defined as a vector containing five elements, which are Boolean values representing whether a drug target belongs to the five target groups, that is, G-protein-coupled receptors (GPCRs), cytokine receptors, nuclear receptors, ion channels and enzymes (Chen and Zeng, 2013). A similarity score was obtained by graph representation and chemical functional group representation in two steps, respectively. The method provided more than one target group for each drug, and the prediction accuracy was 79.01% and 76.43% for the training and test set, respectively (Chen and Zeng, 2013).

Using phenotypic side-effect similarity, which describes the similarity of in vitro target binding profiles of drugs, Kuhn et al. proposed an algorithm to determine if two drugs will interact with the same target (Table 18.4) (Campillos et al., 2008). Two-dimensional Tanimoto similarity coefficient of a chemical structure and a linear function P_{SE} describing the probability of sharing the same considering the side-effect similarity were used in the algorithm. Combining these functions, Kuhn et al. defined a sigmoid function P_{2D} to characterize the probability of sharing the same target from chemical structures (Campillos et al., 2008):

$$P_{2D} = \left(1 + e^{\frac{B-y}{A}}\right)^{-1}$$

where A and B are function parameters. Kuhn et al. used the method to analyze 746 approved drugs, and build a side-effect network with 1018 drug–drug relations, which contains 261 with no chemical similarity (Campillos et al., 2008).

Based on the framework of random walk and the assumption that similar drugs often correspond to similar targets, Yan et al. have developed a network-based random walk with restart on the heterogeneous network (NRWRH) algorithm to predict drug–target interaction (Table 18.4) (Xia et al., 2010). Different from machine learning approaches, the NRWRH algorithm utilized network analysis techniques by introducing random walk on the heterogeneous network. With information on known drug–target interactions, Yan et al. have integrated three different networks into a heterogeneous network, including target–target similarity network, drug–drug similarity network, and drug–target interaction network (Xia et al., 2010). To implement a random walk, a transition matrix **M** was calculated as:

$$\mathbf{M} = \begin{bmatrix} \mathbf{M}_{TT} & \mathbf{M}_{TD} \\ \mathbf{M}_{DT} & \mathbf{M}_{TT} \end{bmatrix}$$

where $\mathbf{M}_{TT}$ and $\mathbf{M}_{DD}$ are the probability for target-to-target and drug-to-drug transition in the random walk, respectively; $\mathbf{M}_{TD}$ and $\mathbf{M}_{DT}$ are the transition probability for target-to-drug and drug-to-target, respectively (Xia et al., 2010). Then the random walk was implemented by the following iteration equation:

$$p_{t+1} = (1 - r)M^T p_t + rp_0$$

where the probability p is iteratively calculated with the restart probability r and transition matrix **M**. Yan et al. has shown that NRWRH has improved prediction performance in four classes of drug–target interactions, that is, enzymes, ion channels, GPCRs, and nuclear receptors (Xia et al., 2010).

Based on complex network theory, Tang et al. proposed three supervised inference methods to predict drug–target interaction (Table 18.4), namely drug-based similarity inference (DBSI), target-based similarity inference (TBSI), and network-based inference (NBI) (Cheng et al., 2012). For the three inference methods, different similarity score functions were defined based on chemical structure similarity, sequence similarity, or network similarity. For example, the final score $f(i)$ of drug d_i in the NBI method is obtained from:

$$f(i) = \sum_{l=1}^{m} \frac{a_{il}}{k(t_l)} \sum_{o=1}^{n} \frac{a_{ol} f_o(o)}{k(d_o)}$$

where $k(d_o)$ represents the number of targets interacting with drug d_o, and $k(t_l)$ denotes the number of drugs interacting with target t_l. NBI method showed the best performance despite it neglects chemical structure similarity (Cheng et al., 2012).

Shi et al. introduced the drug target pair as a vector of within-scores and between-scores, which utilizes features from target similarity and drug similarity

(Table 18.4) (Shi et al., 2015). By doing this, Shi et al. has created a global classifier and a uniform vector of all different types of drug–target pair (Shi et al., 2015). Besides, the unknown drug–target pair can be analyzed in the same visualization space.

Tang et al. have developed a multiple target optimal intervention (MTOI) finding algorithm which aims to identify potential drug targets and their optimal combinations restoring to a normal state (Table 18.4) (Yang et al., 2008). To implement the algorithm, ODEs and parameters for the network were obtained from the Michaelis–Menten equation and experimental data. Monte Carlo simulation was performed to achieve the desired state by optimizing an objective function F_{obj} (Yang et al., 2008). Tang et al. applied the MTOI method to understand the side-effects of traditional nonsteroidal antiinflammatory drugs in an inflammation-related network (Yang et al., 2008).

18.3.2 Tools for network construction

18.3.2.1 Cytoscape

Modeling complex biological network from a set of experimental data is crucial to understand various layers in systems biology, including biochemical reactions, gene transcription kinetics, cellular physiology, and metabolic control. Researchers have developed different computer-aided software to facilitate the management and visualization of big data from lab experiments and mathematical predictions. Cytoscape is an important tool to build a unified biological framework from high-throughput expression data and bio-molecular states (Table 18.5) (Shannon et al., 2003). The network graph is the core concept in Cytoscape, which represents molecular species and their interactions as nodes and edges, respectively. The basic functionality of Cytoscape generates a graph representation of imported biological data. By defining attributes, nodes are paired according to their names and values. Hierarchical

Table 18.5 Computational tools for network construction.

Tools	Description	Website	Reference
Cytoscape	Build biological framework from high-throughput expression data and bio-molecular states	https://go.drugbank.com/	Shannon et al. (2003)
Pajek	Efficiently analyze large network structures by storing sparse networks	https://pubchem.ncbi.nlm.nih.gov/	Batagelj et al. (2003)
Gephi	Utilizes 3D graphics engine to explore and manipulate large networks in-time	https://www.ebi.ac.uk/chembl/	Bastian et al. (2009)
NetworkX	Python package aiming to create, explore and analyze network structures	http://stitch.embl.de/	Hagberg et al. (2008)

classification is allowed by using graph annotation. Users can customize graph layout, attribute-to-visual mapping, and complete graph selection and graph-filtering with plugin functions (Shannon et al., 2003). With the help of external databases of drug–protein interaction, protein–protein interaction, protein-nucleic acid interaction, and genetic interaction, Cytoscape is powerful in modeling, analyzing, and visualizing biological networks for humans and other organisms.

18.3.2.2 Pajek

Pajek is a program to analyze large network structures efficiently (Table 18.5) (Batagelj et al., 2003). Networks in biological systems are usually large, and contains thousands of nodes and edges. The common network analysis tool is mathematically based on a matrix, which is inefficient when dealing with large graphs. Since modern computers have enough memory for storing sparse networks, Pajek proposed an alternative approach to efficiently analyze large graphs by compensating for space complexity (Batagelj et al., 2003). Data structures in Pajek are implemented as six layers, namely network, permutation, vector, cluster, partition, and hierarchy. Also, different transition methods were defined to allow data structure transformation. Theoretically, most of the algorithms in Pajek have subquadratic time complexities (Batagelj et al., 2003).

18.3.2.3 Gephi

To obtain high-quality visualization and data processing experience, a network exploration tool should develop to incorporate high flexible and scalable interactive functions. Gephi is a freely available program that uses a three-dimensional graphics engine to explore and manipulate large networks in time (Table 18.5) (Bastian et al., 2009). The three-dimensional rendering technique is based on a computer graphic card. Due to its multitask nature, Gephi can deal with large graphs with over 2000 nodes (Bastian et al., 2009). Gephi loaded network data into the workspace where each network can be managed separately. And, the function can be extended with external plugin programming. The manipulated networks can be exported as SVG or PDF files.

18.3.2.4 NetworkX

NetworkX is a Python package aiming to create, explore and analyze network structures (Table 18.5) (Hagberg et al., 2008). NetworkX can deal with arbitrary graph objects including simple graphs, directed graphs, graphs with self-loops, and parallel edges based on its basic data structure. The standard data structure in NetworkX contains edge lists, adjacency matrices, and adjacency lists (Hagberg et al., 2008). Since the computation storage and speed depends on the choice of data structure, NetworkX uses adjacent lists for real-world networks with sparse nature. Search and update algorithms for adjacent lists can be achieved through dictionary data structure in Python (Hagberg et al., 2008). Once a graph object is created in NetworkX, users can analyze the network through standard algorithms, such as degree distribution, clustering coefficient, shortest path computing, and spectral measures. NetworkX allows graph visualization through its hooks into Matplotlib. In application, NetworkX has been used to perform spectral analysis of network dynamics and to investigate the synchronization of oscillators. The installation of NetworkX is easy, which requires NumPy, SciPy and Matplotlib installed in prior (Hagberg et al., 2008).

18.3.3 Network topological analysis

18.3.3.1 Degree distribution

The degree of a node is the number of edges linking to the node. It has shown in many biological networks are scale-free, which means the degree distribution of a network follows a power-law $k^{-\lambda}$, where λ is the degree exponent. In a scale-free network, the distribution of degrees is not evenly distributed. Cohen et al. showed a scale-free network is very robust to random attacks (Cohen et al., 2000). Therefore, the proteins with a high degree, also named hubs, evolve slowly and are crucial for the cell's survival (Cheng et al., 2014; Eisenberg and Levanon, 2003; Hahn and Kern, 2004; He and Zhang, 2006; Jeong et al., 2001).

18.3.3.2 Path and distance

The shortest path for a pair of nodes is defined as the shortest length linking the two nodes out of all possible path lengths. Analysis of the shortest path is important to investigate regulatory pathways in protein–protein interaction networks through direction assignment (Blokh et al., 2013; Silverbush and Sharan, 2014). By using the concept of path and distance, it is possible to evaluate the proximity in the drug–target network. Guney et al. have proposed different distance measurement methods (including the closest, shortest, kernel, center, and separation distances) to analyze the therapeutic effect of drugs (Guney et al., 2016). They have investigated 238 drugs used in 78 diseases and found that the therapeutic effect is localized in the neighborhood of a small network (Guney et al., 2016). Guney et al. have shown the network-based distance analysis are useful in drug repurposing and adverse effect detection (Guney et al., 2016).

Another important measure based on the shortest path is efficiency, which is defined as

$$E(G) = \frac{1}{N(N-1)} \sum_{i \neq j \in G} \frac{1}{d_{ij}}$$

where N is the number of nodes, and d_{ij} is the shortest path for nodes i and j (Latora and Marchiori, 2001). Efficiency measures the traffic capacity of a network and how efficiently it exchanges information (Latora and Marchiori, 2001). Csermely et al. found it is possible to efficiently inhibit targets through a small number of inhibitors instead of a complete inhibition of a single target (Csermely et al., 2005). In addition, the concept of network efficiency rationalizes multitarget strategy in drug design, which is useful in the development of drug combinations (Cheng et al., 2019; Csermely et al., 2005; Vazquez, 2009).

Based on the network shortest path, it is possible to measure the importance of a node, which is characterized through betweenness in the following equation:

$$B(v) = \sum_{i \neq j} \frac{\delta_{ij}(v)}{\delta_{ij}}$$

where δ_{ij} is the number of the shortest paths from i to j, $\delta_{ij}(v)$ is the number of the shortest paths that travel through node v. It should be noted the degree and betweenness of a node is not correlated, which means nodes with a small degree could have large betweenness (Guimerà et al., 2005; Joy et al., 2005; Yu et al., 2007). As mentioned, the betweenness characterizes the importance of a node. A node with high betweenness is known as a bottleneck in a network. Bottlenecks control the flow of information in a network and improve network efficiency. It has shown that proteins with high betweenness are essential and tend to be highly pleiotropic (Ahmed et al., 2018; Estrada and Ross, 2018; Zou et al., 2008).

18.3.3.3 Module and motifs

In complex networks, a dense subgraph or subnetwork is referred to as a module. The modularity of a network is defined as (Clauset et al., 2004; Newman and Girvan, 2004; Newman, 2012):

$$M = \frac{1}{2E} \sum [A_{ij} - P_{ij}] \delta_{C_i, C_j}$$

where E is the number of network edges, A is an adjacent matrix, P is the expected number of edges from node i to j, δ_{ij} is a Kronecker function which equals 1 only if node i and j belongs to the same community. A number of methods have been developed to identify modules and communities in a network (Ahn et al., 2010; Palla et al., 2005; Palla et al., 2007; Rosvall and Bergstrom, 2007; Rosvall and Bergstrom, 2008). Module analysis provided an effective approach to investigating complex networks by identifying specific modules instead of unfolding the entire network. Increasing results showed that modules are important in uncovering new drug targets and promoting drug development (Derry et al., 2012).

Motifs in a network are defined as connection patterns with a high occurring number in a network than in randomized networks (Alon, 2007; Milo et al., 2002; Shen-Orr et al., 2002). Universal classes of networks can be defined through motifs. Some of the basic motifs, including 13 3-node directed motifs and 30 undirected motifs (also named graphlets) with node numbers ranging from 2 to 5, are shown in Fig. 18.4 (Milo et al., 2002; Pržulj, 2007). Analysis of network motifs is useful to identify druggable targets. Tan et al. used network motifs analysis to uncover basic principles of cellular target druggability, which describes the capacity of a target

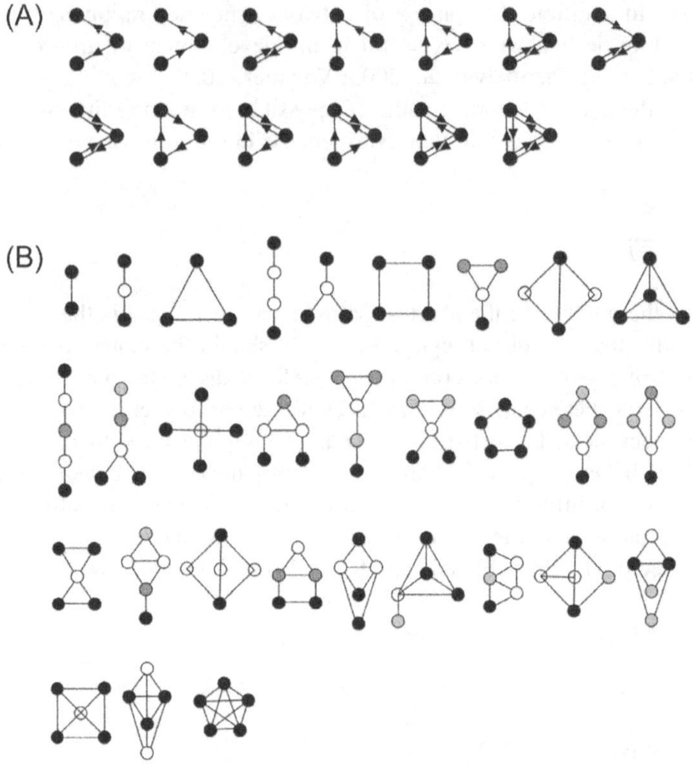

Figure 18.4 An illustration of (A) 13 3-node directed motifs and (B) 30 graphlets with node numbers ranging from 2 to 5.

modulated by a drug (Wu et al., 2016). They found that highly druggable motifs share a consensus topology of a negative feedback loop without any positive feedback loops (Wu et al., 2016). On the opposite, the motifs of low druggability consist of multiple positive direct regulations and positive feedback loops. In addition, Tan et al. showed druggability can be reduced by adding direct regulations to a drug–target network (Wu et al., 2016).

18.4 Conclusion and perspectives

The paradigm has shifted from one-bullet-one-target to a network view in drug discovery and development. As demonstrated by Yildirim et al., the complex network nature of drug–target interaction imposes a holistic philosophy in drug discovery and drug repurposing in the next decades (Yildirim et al., 2007). As high-throughput experimental data is expanding rapidly and dramatically, novel databases and data management methodologies are emerging, especially when considering chemicals derived from complicated herbal plants and their related targets. A uniform data format or data transformation platform will facilitate data utilization more efficiently, which is also fundamental to the construction, prediction, and analysis of drug–target interaction networks.

Various computational algorithms have been proposed for drug–target interaction prediction and analysis. Through the framework of network science, it is possible to reconstruct drug–target interaction networks without concerning the three-dimensional structures of drugs and targets. These methods showed high performance in accuracy and speed, which are important in real applications including target prediction and mechanism elucidation. However, these methods still need to be further completed, including drug–target interaction prediction method without a prior knowledge of ligand, as well as the development of quantitative methods for the analysis of drug–target interaction. Nevertheless, the coupling of big data and network science in drug–target interaction has opened a new era in drug discovery and development.

Acknowledgments

This research was funded by the National Natural Science Foundation of China (31872723), the Priority Academic Program Development (PAPD) of Jiangsu Higher Education Institutions, Zunyi Science and Technology Project (2018(21)), and Guizhou Provincial Natural Science Foundation (QKH-J[2020]1Y045).

References

Ahmed, H., Howton, T.C., Sun, Y., Weinberger, N., Belkhadir, Y., Mukhtar, M.S., 2018. Network biology discovers pathogen contact points in host protein-protein interactomes. Nat. Commun. 9, 2312.

Ahn, Y.-Y., Bagrow, J.P., Lehmann, S., 2010. Link communities reveal multiscale complexity in networks. Nature 466, 761–764.

Alon, U., 2007. Network motifs: theory and experimental approaches. Nat. Rev. Genet. 8, 450–461.

Basak, S.C., Gute, B.D., Mills, D., Hawkins, D.M., 2002. Quantitative molecular similarity methods in the property/toxicity estimation of chemicals: a comparison of arbitary versus tailored similarity spaces. J. Mol. Struct.: Theochem 622, 127–145.

Basak, S.C., Gute, B.D., Mills, D., 2006. Similarity methods in analog selection, property estimation and clustering of diverse chemicals. Arch. Org. Chem. 9, 157–210.

Bastian, M., Heymann, S., Jacomy, M. 2009. Gephi: an open source software for exploring and manipulating networks. In: International AAAI Conference on Weblogs and Social Media.

Batagelj, V., Andrej, M., Jünger, M.M.P., 2003. Pajek - analysis and visualization of large networks. Graph drawing software. Springer, Berlin.

Bleakley, K., Yamanishi, Y., 2009. Supervised prediction of drug–target interactions using bipartite local models. Bioinformatics 25, 2397–2403.

Blokh, D., Segev, D., Sharan, R., 2013. The approximability of shortest path-based graph orientations of protein–protein interaction networks. J. Comput. Biol. 20, 945–957.

Bolognesi, M.L., Cavalli, A., 2016. Multitarget drug discovery and polypharmacology. ChemMedChem 11, 1190–1192.

Borgatti, S.P., Halgin, D.S., 2011. On network theory. Organ. Sci. 22, 1168–1181.

Burley, S.K., Berman, H.M., Bhikadiya, C., Bi, C., Chen, L., Di Costanzo, L., et al., 2018. RCSB protein data bank: biological macromolecular structures enabling research and education in fundamental biology, biomedicine, biotechnology and energy. Nucleic Acids Res. 47, D464–D474.

Campillos, M., Kuhn, M., Gavin, A.-C., Jensen, L.J., Bork, P., 2008. Drug target identification using side-effect similarity. Science 321, 263.

Cao, D.-S., Zhang, L.-X., Tan, G.-S., Xiang, Z., Zeng, W.-B., Xu, Q.-S., et al., 2014. Computational prediction of drug-target interactions using chemical, biological, and network features. Mol. Inf. 33, 669–681.

Chaudhari, R., Tan, Z., Huang, B., Zhang, S., 2017. Computational polypharmacology: a new paradigm for drug discovery. Expert. Opin. Drug. Dis. 12 (3), 279–291.

Chen, C.Y.-C., 2011. TCM database@taiwan: the world's largest traditional chinese medicine database for drug screening in silico. PLoS One 6, e15939.

Chen, L., Zeng, W.-M., 2013. A two-step similarity-based method for prediction of drug's target group. Protein Pept. Lett. 20, 364–370.

Cheng, A.C., Coleman, R.G., Smyth, K.T., Cao, Q., Soulard, P., Caffrey, D.R., et al., 2007. Structure-based maximal affinity model predicts small-molecule druggability. Nat. Biotechnol. 25, 71–75.

Cheng, F., Liu, C., Jiang, J., Lu, W., Li, W., Liu, G., et al., 2012. Prediction of drug-target interactions and drug repositioning via network-based inference. PLoS Comput. Biol. 8, e1002503.

Cheng, F., Jia, P., Wang, Q., Lin, C.-C., Li, W.-H., Zhao, Z., 2014. Studying tumorigenesis through network evolution and somatic mutational perturbations in the cancer interactome. Mol. Biol. Evol. 31, 2156–2169.

Cheng, F., Kovács, I.A., Barabási, A.-L., 2019. Network-based prediction of drug combinations. Nat. Commun. 10, 1197.

Cheung, F., 2011. TCM: made in china. Nature 480, S82–S83.

Ciallella, H.L., Zhu, H., 2019. Advancing computational toxicology in the big data era by artificial intelligence: data-driven and mechanism-driven modeling for chemical toxicity. Chem. Res. Toxicol. 32, 536–547.

Clauset, A., Newman, M.E.J., Moore, C., 2004. Finding community structure in very large networks. Phys. Rev. E 70, 066111.

Cohen, R., Erez, K., ben-Avraham, D., Havlin, S., 2000. Resilience of the internet to random breakdowns. Phys. Rev. Lett. 85, 4626–4628.

Consortium, T.U., 2018. Uniprot: a worldwide hub of protein knowledge. Nucleic Acids Res. 47, D506–D515.

Csermely, P., Ágoston, V., Pongor, S., 2005. The efficiency of multi-target drugs: the network approach might help drug design. Trends Pharmacol. Sci. 26, 178–182.

Derry, J.M.J., Mangravite, L.M., Suver, C., Furia, M.D., Henderson, D., Schildwachter, X., et al., 2012. Developing predictive molecular maps of human disease through community-based modeling. Nat. Genet. 44, 127–130.

Eisenberg, E., Levanon, E.Y., 2003. Preferential attachment in the protein network evolution. Phys. Rev. Lett. 91, 138701.

Estrada, E., Ross, G.J., 2018. Centralities in simplicial complexes. Applications to protein interaction networks. J. Theor. Biol. 438, 46–60.

French, K.E., Harvey, J., McCullagh, J.S.O., 2018. Targeted and untargeted metabolic profiling of wild grassland plants identifies antibiotic and anthelmintic compounds targeting pathogen physiology, metabolism and reproduction. Sci. Rep. 8, 1695.

Gilson, M.K., Liu, T., Baitaluk, M., Nicola, G., Hwang, L., Chong, J., 2015. Bindingdb in 2015: a public database for medicinal chemistry, computational chemistry and systems pharmacology. Nucleic Acids Res. 44, D1045–D1053.

Guimerà, R., Mossa, S., Turtschi, A., Amaral, L.A.N., 2005. The worldwide air transportation network: anomalous centrality, community structure, and cities' global roles. Proc. Natl. Acad. Sci. 102, 7794.

Guney, E., Menche, J., Vidal, M., Barábasi, A.-L., 2016. Network-based in silico drug efficacy screening. Nat. Commun. 7, 10331.

Günther, S., Kuhn, M., Dunkel, M., Campillos, M., Senger, C., Petsalaki, E., et al., 2008. Supertarget and matador: resources for exploring drug-target relationships. Nucleic Acids Res. 36, D919–D922.

Hagberg, A., Swart, P., Chult, D. 2008. Exploring network structure, dynamics, and function using networkx. In: Proceedings of the 7th Python in Science Conference (SciPy 2008); Pasadena, CA, USA.

Haggarty, S.J., Koeller, K.M., Wong, J.C., Butcher, R.A., Schreiber, S.L., 2003. Multidimensional chemical genetic analysis of diversity-oriented synthesis-derived deacetylase inhibitors using cell-based assays. Chem. Biol. 10, 383–396.

Hahn, M.W., Kern, A.D., 2004. Comparative genomics of centrality and essentiality in three eukaryotic protein-interaction networks. Mol. Biol. Evol. 22, 803–806.

He, X., Zhang, J., 2006. Why do hubs tend to be essential in protein networks? PLoS Genet. 2, e88.

Hopkins, A.L., 2008. Network pharmacology: the next paradigm in drug discovery. Nat. Chem. Biol. 4, 682–690.

Huang, L., Xie, D., Yu, Y., Liu, H., Shi, Y., Shi, T., et al., 2017. Tcmid 2.0: a comprehensive resource for tcm. Nucleic Acids Res. 46, D1117–D1120.

Jeong, H., Mason, S.P., Barabási, A.L., Oltvai, Z.N., 2001. Lethality and centrality in protein networks. Nature 411, 41–42.

Joy, M.P., Brock, A., Ingber, D.E., Huang, S., 2005. High-betweenness proteins in the yeast protein interaction network. J. Biomed. Biotechnol. 2005, 594674.

Keiser, M.J., Setola, V., Irwin, J.J., Laggner, C., Abbas, A.I., Hufeisen, S.J., et al., 2009. Predicting new molecular targets for known drugs. Nature 462, 175–181.

Kim, S., Chen, J., Cheng, T., Gindulyte, A., He, J., He, S., et al., 2018. Pubchem 2019 update: improved access to chemical data. Nucleic Acids Res. 47, D1102–D1109.

Kuruvilla, F.G., Shamji, A.F., Sternson, S.M., Hergenrother, P.J., Schreiber, S.L., 2002. Dissecting glucose signalling with diversity-oriented synthesis and small-molecule microarrays. Nature 416, 653–657.

Latora, V., Marchiori, M., 2001. Efficient behavior of small-world networks. Phys. Rev. Lett. 87, 198701.

Lee, C.H., Yoon, H.-J., 2017. Medical big data: promise and challenges. Kidney Res. Clin. Pract. 36, 3–11.

Li, S., Zhang, B., Zhang, N., 2011. Network target for screening synergistic drug combinations with application to traditional chinese medicine. BMC Syst. Biol. 5, S10.

Liang, Z., Hu, G., 2016. Protein structure network-based drug design. Mini-Rev. Med. Chem. 16, 1330–1343.

Loscalzo, J., Barabasi, A.-L., 2011. Systems biology and the future of medicine. WIRES Syst. Biol. Med. 3, 619–627.

Mei, J.P., Kwoh, C.K., Yang, P., Li, X.L., Zheng, J., 2013. Drug–target interaction prediction by learning from local information and neighbors. Bioinformatics 29, 238–245.

Mendez, D., Gaulton, A., Bento, A.P., Chambers, J., De Veij, M., Félix, E., et al., 2018. Chembl: towards direct deposition of bioassay data. Nucleic Acids Res. 47, D930–D940.

Milo, R., Shen-Orr, S., Itzkovitz, S., Kashtan, N., Chklovskii, D., Alon, U., 2002. Network motifs: Simple building blocks of complex networks. Science 298, 824.

Mohs, R.C., Greig, N.H., 2017. Drug discovery and development: role of basic biological research. Alzheimers Dement. 3 (4), 651–657.

Morris, G.M., Huey, R., Lindstrom, W., Sanner, M.F., Belew, R.K., Goodsell, D.S., et al., 2009. Autodock4 and autodocktools4: automated docking with selective receptor flexibility. J. Comput. Chem. 30, 2785–2791.

Mullard, A., 2020. $1.3 billion per drug? Nat. Rev. Drug. Discov. 19, 226.

Newman, M.E.J., 2003. The structure and function of complex networks. SIAM Rev. 45, 167–256.

Newman, M.E.J., Girvan, M., 2004. Finding and evaluating community structure in networks. Phys. Rev. E 69, 026113.

Newman, M.E.J., 2012. Communities, modules and large-scale structure in networks. Nat. Phys. 8, 25–31.

Palla, G., Derényi, I., Farkas, I., Vicsek, T., 2005. Uncovering the overlapping community structure of complex networks in nature and society. Nature 435, 814–818.

Palla, G., Barabási, A.-L., Vicsek, T., 2007. Quantifying social group evolution. Nature 446, 664–667.

Parkhe, A., Wasserman, S., Ralston, D.A., 2006. New frontiers in network theory development. Acad. Manage. Rev. 31, 560–568.

Pence, H.E., Williams, A., 2010. Chemspider: an online chemical information resource. J. Chem. Edu. 87, 1123–1124.

Pržulj, N., 2007. Biological network comparison using graphlet degree distribution. Bioinformatics 23, e177–e183.

Rosvall, M., Bergstrom, C.T., 2007. An information-theoretic framework for resolving community structure in complex networks. Proc. Natl. Acad. Sci. 104, 7327.

Rosvall, M., Bergstrom, C.T., 2008. Maps of random walks on complex networks reveal community structure. Proc. Natl. Acad. Sci. 105, 1118.

Ru, J., Li, P., Wang, J., Zhou, W., Li, B., Huang, C., et al., 2014. Tcmsp: a database of systems pharmacology for drug discovery from herbal medicines. J. Cheminform 6, 13.

Russell, C., Rahman, A., Mohammed, A.R., 2013. Application of genomics, proteomics and metabolomics in drug discovery, development and clinic. Ther. Deliv. 4, 395–413.

Santos, R., Ursu, O., Gaulton, A., Bento, A.P., Donadi, R.S., Bologa, C.G., et al., 2017. A comprehensive map of molecular drug targets. Nat. Rev. Drug. Discov. 16, 19–34.

Shannon, P., Markiel, A., Ozier, O., Baliga, N.S., Wang, J.T., Ramage, D., et al., 2003. Cytoscape: a software environment for integrated models of biomolecular interaction networks. Genome Res. 13, 2498–2504.

Shen-Orr, S.S., Milo, R., Mangan, S., Alon, U., 2002. Network motifs in the transcriptional regulation network of *Escherichia coli*. Nat. Genet. 31, 64–68.

Shi, J.-Y., Liu, Z., Yu, H., Li, Y.-J., 2015. Predicting drug-target interactions via within-score and between-score. BioMed. Res. Int. 2015, 350983.

Silverbush, D., Sharan, R., 2014. Network orientation via shortest paths. Bioinformatics 30, 1449–1455.

Szklarczyk, D., Santos, A., von Mering, C., Jensen, L.J., Bork, P., Kuhn, M., 2015. Stitch 5: Augmenting protein–chemical interaction networks with tissue and affinity data. Nucleic Acids Res. 44, D380–D384.

Szklarczyk, D., Gable, A.L., Lyon, D., Junge, A., Wyder, S., Huerta-Cepas, J., et al., 2018. String v11: protein–protein association networks with increased coverage, supporting functional discovery in genome-wide experimental datasets. Nucleic Acids Res. 47, D607–D613.

Tao, W., Xu, X., Wang, X., Li, B., Wang, Y., Li, Y., et al., 2013. Network pharmacology-based prediction of the active ingredients and potential targets of chinese herbal radix curcumae formula for application to cardiovascular disease. J. Ethnopharmacol. 145, 1–10.

Vazquez, A., 2009. Optimal drug combinations and minimal hitting sets. BMC Syst. Biol. 3, 81.

Wang, Y., Zeng, J., 2013. Predicting drug-target interactions using restricted Boltzmann machines. Bioinformatics 29, i126–i134.

Wang, J.T., Liu, W., Tang, H., Xie, H., 2014. Screening drug target proteins based on sequence information. J. Biomed. Inf. 49, 269–274.

Wang, J., Zhang, C.-J., Chia, W.N., Loh, C.C.Y., Li, Z., Lee, Y.M., et al., 2015. Haem-activated promiscuous targeting of artemisinin in *Plasmodium falciparum*. Nat. Commun. 6, 10111.

Wang, F., Han, S., Yang, J., Yan, W., Hu, G., 2021. Knowledge-guided "community network" analysis reveals the functional modules and candidate targets in non-small-cell lung cancer. Cells 2021 (10), 402.

Whitebread, S., Hamon, J., Bojanic, D., Urban, L., 2005. Keynote review: In vitro safety pharmacology profiling: an essential tool for successful drug development. Drug. Discov. Today 10, 1421–1433.

Wishart, D.S., Feunang, Y.D., Guo, A.C., Lo, E.J., Marcu, A., Grant, J.R., et al., 2017. Drugbank 5.0: a major update to the drugbank database for 2018. Nucleic Acids Res. 46, D1074–D1082.

Wouters, O.J., McKee, M., Luyten, J., 2020. Estimated research and development investment needed to bring a new medicine to market, 2009–2018. JAMA 323 (9), 844–853.

Wu, F., Ma, C., Tan, C., 2016. Network motifs modulate druggability of cellular targets. Sci. Rep. 6, 36626.

Xia, Z., Wu, L.-Y., Zhou, X., Wong, S.T.C., 2010. Semi-supervised drug-protein interaction prediction from heterogeneous biological spaces. BMC Syst. Biol. 4, S6.

Yamanishi, Y., Araki, M., Gutteridge, A., Honda, W., Kanehisa, M., 2008. Prediction of drug–target interaction networks from the integration of chemical and genomic spaces. Bioinformatics 24, i232–i240.

Yamanishi, Y., Kotera, M., Kanehisa, M., Goto, S., 2010. Drug-target interaction prediction from chemical, genomic and pharmacological data in an integrated framework. Bioinformatics 26, i246–i254.

Yan, W., Zhang, D., Shen, C., Liang, Z., Hu, G., 2018. Recent advances on the network models in target-based drug discovery. Curr. Top. Med. Chem. 18, 1031–1043.

Yang, K., Bai, H., Ouyang, Q., Lai, L., Tang, C., 2008. Finding multiple target optimal intervention in disease-related molecular network. Mol. Syst. Biol. 4, 228.

Ye, H., Ye, L., Kang, H., Zhang, D., Tao, L., Tang, K., et al., 2010. Hit: linking herbal active ingredients to targets. Nucleic Acids Res. 39, D1055–D1059.

Yildirim, M.A., Goh, K.I., Cusick, M.E., Barabasi, A.L., Vidal, M., 2007. Drug-target network. Nat. Biotechnol. 25, 1119–1126.

Yu, H., Kim, P.M., Sprecher, E., Trifonov, V., Gerstein, M., 2007. The importance of bottlenecks in protein networks: correlation with gene essentiality and expression dynamics. PLoS Comput. Biol. 3, e59.

Zhu, H., 2020. Big data and artificial intelligence modeling for drug discovery. Annu. Rev. Pharmacol. Toxicol. 60, 573–589.

Zou, L., Sriswasdi, S., Ross, B., Missiuro, P.V., Liu, J., Ge, H., 2008. Systematic analysis of pleiotropy in C. elegans early embryogenesis. PLoS Comput. Biol. 4, e1000003.

19 Dissecting big RNA-Seq cancer data using machine learning to find disease-associated genes and the causal mechanism

Dipanka Tanu Sarmah[1,*], *Shivam Kumar*[1,*], *Samrat Chatterjee*[1] *and Nandadulal Bairagi*[2]

[1]Complex Analysis Group, Translational Health Science and Technology Institute, NCR Biotech Science Cluster, Faridabad, Haryana, India, [2]Department of Mathematics, Centre for Mathematical Biology and Ecology, Jadavpur University, Kolkata, West Bengal, India

19.1 Introduction

The credit for coining the now-ubiquitous term "big data" goes to John R. Mashey (Diebold, 2012). It refers to a field that systematically collects, analyses, and extracts information from data sets that are too broad and complicated to manage with conventional application tools for data processing. In oncology, big data narrates the rapid acquisition, processing, and analysis of considerable amounts of information from various sources such as population cancer registries, large-scale genetic sequencing studies, etc. Decades of research to delineate the underlying mechanism of diseases in different cells, tissues, or organisms have resulted in a considerable accumulation of experimental data that have been stored across multiple databases and repositories. Based on the importance that big data carries, it is not imprudent to term the current era as the era of big data.

Recently, the act of harnessing big data to extract the disease-associated gene has been dominated by machine-learning approaches (Sarmah et al., 2020; Tang et al., 2019). These approaches have been applied in systems biology, genomics, proteomics, and various other domains. Arthur Samuel, a computer scientist, working at IBM, first coined the phrase "machine learning" (ML) (Samuel, 1988). ML approaches intend to build predictive models relying on an underlying algorithm and a given dataset. The input data to the algorithm contains "features" and "labels" across a set of samples where features are the measurable properties across the samples, while labels are what the model aims to predict, that is, the model's output. However, ML algorithms can also deal with datasets without labels. These algorithms can further be classified into supervised and unsupervised methods. The former deals with the known categorization of samples, while the labels remain

* These authors are equally contributed to this chapter.

Big Data Analytics in Chemoinformatics and Bioinformatics. DOI: https://doi.org/10.1016/B978-0-323-85713-0.00015-3

unknown in the latter. The supervised algorithms determine the associations between the features and the labels, thereby building a classification model to precisely predict new sample class labels. On the other hand, unsupervised learning can cluster similar samples based on their distance or similarity without relying on the sample class labels. The supervised method possesses an exact idea about the classes in the training data. In supervised methods, the classifier can be trained in such a way that the individual classes remain entirely distinct. These methods, however, are not free of shortcomings, such as they are guided and fail to capture the natural patterns. Again, unlike unsupervised learning, it cannot group or classify data by discovering its features solitarily. Unsupervised methods excel in detecting hidden patterns in the data. On the other hand, the major downside of unsupervised learning is that the output is less accurate as the input data is not prelabeled. The choice of selection of methods is sometimes limited. If the data contains no labels, then the unsupervised method becomes the spontaneous choice. However, if the data include labels, then a supervised method may not always be the best approach because the supervised method assumes that the distribution used for generating the training data set is the same as the distribution that generates the test data set (Libbrecht and Noble, 2015). This assertion is considerable if a single labeled data is randomly subdivided into testing and training sets. However, in general, the distribution for generating train and test data sets is different. In the case of the feasibility of supervised learning and easy accessibility of additional unlabeled data points, a supervised or hybrid approach can be considered. The hybrid methods combine supervised and unsupervised learning methods to solve the problem. A hybrid method starts with unsupervised learning and then proceeds with supervised learning. However, like the two methods mentioned above, hybrid methods also struggle. For example, most of the available hybrid models first split the training set into many clusters. However, if a not well-prepared training set is given to analyze, many samples may be located near the boundaries of the clusters, and such samples are often very hard to classify.

Apart from ML, various other methods also exist which can bear the responsibility of identifying disease-associated genes. One among them is the protein–protein interaction (PPI) network analysis. Such networks are very easy to construct as the number of PPI databases is proliferating. Various methods to analyze the PPI network have been used to identify culprit genes in multiple diseases that include the likes of nonalcoholic steatohepatitis, cancer, type 2 diabetes mellitus, Alzheimer's disease, etc. Each interaction of a PPI network is given a unique confidence score by the database from which it is curated. This score is given based on the data curation method, and hence, the score of a particular interaction may vary from database to database. The higher the confidence score, the more is the accuracy of the interaction. Although the scores help pick the most accurate interactions in the database, it is also the Achilles heel of a PPI network. This is because, while constructing a network by choosing the most confident interactions from a particular database, a lot of useful interactions might get overlooked, which may lead to an incorrect prediction of disease genes. Thus it is safe to say that ML methods can produce more accurate results than conventional PPI network analysis methods.

However, conventional PPI network analysis methods help to capture the causal mechanism of disease. These methods have been very helpful in providing insight into the underlying mechanism of disease. Integrating with the ML approach in a study, the network analysis methods can be used to check whether the disease-associated genes identified using ML capture the causal mechanism of the disease or not.

The continuous evolution of in silico approaches has improved the understanding of complex diseases like cancer. The cancer spectrum consists of more than 200 distinct subtypes. This large set of diseases can grow in any tissue or organ of the body. Cancer initiates from the aberration of multiple steps of gene expression programs. According to the World Health Organization, cancer is currently the second leading cause of death globally. With the assistance of technological advances in transcriptomic profiling, it has become possible to identify the different cancer stages, indicating that several gene mutations are involved in cancer pathogenesis. Transcriptome analysis is a quintessential tool for identifying and characterizing genes and pathways responsible for the progression of the disease. Two of the most used transcriptional methods include microarray and RNA-Seq. RNA-Seq is just a decade old, but it has overshadowed the popular DNA microarray technology (Fig. 19.1). It is because full sequencing of the whole transcriptome can be done with RNA-Seq, while only profiles of predefined transcripts through hybridization can be done with microarrays. The expression of over 70,000 noncoding RNAs is measured with RNA-Seq that are generally not measured with microarrays (Zhao et al., 2016). RNA-Seq also improves the accuracy for low-abundance transcripts (Ge et al., 2009) and is capable of distinguishing between the expression of various splice variants (Richard et al., 2010), which can have well-defined biological functions (Kelemen et al., 2013) and interaction partners (Yang et al., 2016). This implies that RNA-Seq can help identify more differentially modulated transcripts, splice variants, and noncoding transcripts and provide more insight into many

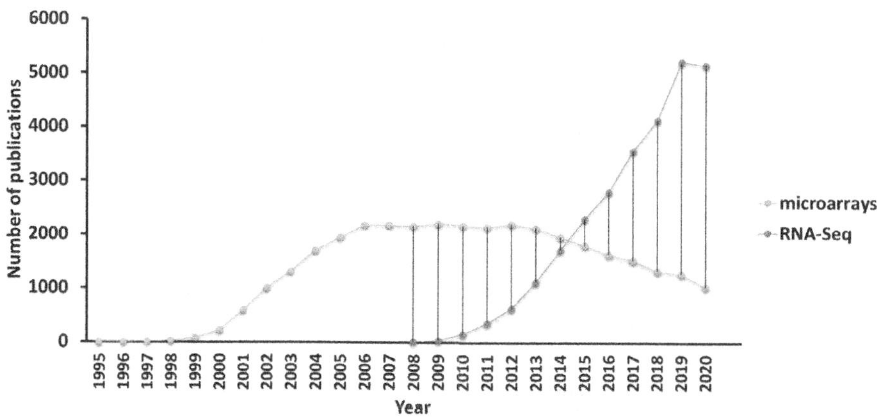

Figure 19.1 RNA-Seq leading the race to decipher cancer. We searched PubMed using the keywords RNA-Seq', "microarrays," and "cancer" in the title/abstract category.

biomedical and biological questions. Due to these advantages and the general advancement of the field, RNA-Seq has expanded well beyond the genomics community and has become a regular feature of the science toolkit.

In this chapter, we have discussed a relatively advanced and successful hybrid machine learning workflow that may help to unravel causative agents of disease from high throughput RNA-Seq datasets. The method is then applied to a breast cancer dataset taken from the Gene Expression Omnibus (GEO) (Barrett et al., 2013) repository, and disease genes associated with breast cancer are identified. Finally, using the PPI network analysis approach, we observed the significance that the detected disease genes possess a role in the causal mechanism of disease. This method discussed here is universal and can be applied to any RNA-Seq data independent of disease.

19.2 Bird's eye view of the analysis of cancer RNA-Seq data using machine learning

This section has summarized some of the recent works done in cancer RNA-Seq data analysis using machine-learning approaches. This literature review helps to understand how ML methods are used to decrypt RNA-Seq data.

Zhang et al. (2020) used a machine learning approach to diagnose early-stage pancreatic ductal adenocarcinoma (PDAC), where the computational model used was based on within-sample relative expression orderings (REOs). The optimal REOs were selected by using the minimum redundancy maximum relevance feature selection method. Support vector machines were found to be the best classifier method for early PDAC biomarkers. The study found nine gene pair signatures that could identify PDAC with an accuracy of up to 97.53% on the training set in fivefold cross-validation. Zhao et al. (2020) developed an RNA-based classifier named CUP-AI-Dx, which is capable of highly accurate prediction of tumor primary site and molecular subtype. This classifier uses a 1D inception convolutional neural network model to deduce the primary tissue of origin of a tumor. Machine learning methods have also been used to predict the prognosis of various cancers (Huang et al., 2020; Tang et al., 2020). For example, Chen et al. (2020) used data repositories like GEO, ArrayExpress, and the cancer genome atlas (TCGA) to obtain single-cell RNA sequencing (scRNA-seq) and gene expression data of lung adenocarcinoma patients and identified differentially expressed ligand-receptor interactions. Based on these interactions, they further built a prognostic machine learning model intending to predict the prognosis of lung adenocarcinoma patients.

In a nutshell, using RNA-Seq data, ML methods have been very widely used with RNA-Seq data for prognosis, diagnosis, and even tailoring personalized medicine in cancer. To get a clear insight into the methods applied, it is of utmost importance to understand the basic principles of RNA-Seq data analysis. In the next section, we have provided a brief introduction to these principles.

19.3 Materials and methods

The basic principles of RNA-Seq data entail different stages. A more accurate prediction of results depends upon the successful pass-through of these stages. These stages are explained below.

19.3.1 Preprocessing of the data

Data preprocessing is when the data is transformed or encoded to make it easy for the machine to parse. The RNA-Seq data obtained from GEO needs to go through the data quality assessment phase as the data obtained may contain errors due to limitations of measuring devices or errors in the data collection process. Also, the possibility of human error cannot be neglected. These errors result in missing values, duplicate values, and outliers in the data. The missing values and outliers can either be eliminated or estimated. Like, mean, median, mode, nearest neighbor, etc., various methods are used to calculate these values. The duplicate values are often removed from the data so that a particular data object does not benefit. We recommend that the readers read the following articles for better insight into missing values treatment and outlier detection methodologies (Batista et al., 2003; Hodge and Austin, 2004).

Data normalization is a crucial step in RNA-Seq study. Here, raw data are adjusted to account for factors like length (Oshlack and Wakefield, 2009), GC-content (Risso et al., 2011), and sequencing depth (Robinson and Oshlack, 2010). These factors influence the number of reads mapped to a gene. The data normalization error causes a substantial impact in the downstream analysis; inflated false positive is to be named among few. Detailed information about the normalization method can be found in the study done by Evans et al. (2018).

19.3.2 Feature selection

Feature selection, also known as attribute selection or variable selection, is a pivotal stage in RNA-Seq data analysis using ML. The high dimensionality of features often characterizes ML; however, not all are necessarily pertinent and advantageous. When the input is high-dimensional, many algorithms that function well in low dimensions become unsolvable, a phenomenon that was coined as the "curse of dimensionality" by Richard Bellman in 1961. The feature selection step identifies the features that will contribute more to the prediction variable or the model's output and, thus, reduces the model's input variables. Again, the selection of irrelevant features degrades the model's accuracy. The feature selection thus reduces the model overfitting, improves the model's accuracy, and reduces computational time. The feature selection process can be divided into three approaches—"filters," "wrappers," and "embedded." The details about these approaches can be found in Mladenić (2006).

The feature selection process can also be accomplished by using dimensionality reduction methods. Two of the most used algorithms for dimensionality reduction are principal component analysis (PCA) and linear discriminant analysis (LDA). LDA is a supervised ML algorithm and uses the information of classes to discover new features to maximize its separability. On the other hand, PCA is an unsupervised ML algorithm and transforms the data into different components extracted from the original data. The feature selection method is applied to the training set, and the component with the most variance is selected.

19.3.3 Classification learning

When faced with data analysis with ML, one of the most crucial and tricky stages is selecting a classification algorithm. Classification alludes to a predictive modeling problem where a class label is predicted from input features. Various classification models are available in the literature that includes artificial neural network (ANN) (Mohandes et al., 2019), random forest (Chen and Ishwaran, 2012), J48 (Bhargava et al., 2017), Naive Bayes (Dou et al., 2015) etc. In classification models, training errors referring to misclassification errors and generalization errors referring to expected errors in testing results can be generated. A good classification model refers to a classification that properly fits the training set as well as precisely classifies all the instances. It is, therefore necessary to approximate the efficiency of the classifier once a classification model is acquired using one or more ML techniques. The model's efficiency is measured in terms of accuracy, sensitivity, specificity, and area under the curve (AUC). Accuracy measures the total number of correct predictions by the model. Sensitivity and specificity are the measures of the proportion of true positives and true negatives that the classifier correctly observed. On the other hand, the AUCs measure the performance of the model based on the receiver operating characteristic (ROC) curve. ML methods, boosted by a proper classification method, can extract genes responsible for the propagation of disease.

19.3.4 Extraction of disease-associated genes

Biomolecules often work in a synergy that drives the system they represent. To unveil the underlying mechanism driving a disease, it is of utmost importance to decipher the uncanny synergy flowing through the biomolecules. Irrespective of the system's size, there always exist a core set of biomolecules that govern the entire system. Experimental detection of such biomolecules is a challenging task because such approaches are often laborious, expensive, and time-consuming. To optimize the search, various algorithms to identify such an optimal set of biomolecules are available in the literature. Many of these methods endorse the network analysis perspective (Anand et al., 2018; Vinayagam et al., 2016), while many take a machine-learning approach to catch the culprit genes or proteins (Asif et al., 2018; Barman et al., 2019). The identified list of proteins or genes may differ in different algorithms, and hence a proper validation of the identified set is required.

19.3.5 Validation

The final stage of RNA-Seq data analysis is the validation stage to assess the classifier's performance. A critical need exists to understand the functionality of the identified genes (or their protein products) and their association with the disease. If the method chosen results in clusters, it is of utmost importance to check whether the cluster's proteins work in tandem. The validation method can be *in silico*, *in vivo*, or *in vitro*. The *in silico* validation can be done in various ways. The first being the extensive mining of the literature where the goal is to find the association of identified disease genes with the disease. Gene ontology (GO) analysis is another way of validation. The GO project furnishes a controlled vocabulary to enable high-quality functional gene annotation for all species (Lomax, 2005). Here, the genes are mapped to their linked biological process, pathways, and cellular compartment. Identification of the process or pathways helps to determine whether the identified set of genes paves the way for disease or not.

Another ingenious way of finding the significance of the identified disease-associated genes is to use the PPI network. The PPI network can easily be constructed by using various PPI databases like STRING (Szklarczyk et al., 2019), SIGNOR (Licata et al., 2020), RegNetwork (Liu et al., 2015), etc. Irrespective of the size, such networks are always driven by a core set of proteins, which can easily be identified by investigating the network topology parameters or using novel approaches such as identifying the minimum set of nodes, removing which will disconnect the disease network (Anand et al., 2018). The disease-associated genes, identified using the ML approach, can then be checked with the list of proteins obtained using the network analysis approach. The overlapping proteins can then be assumed to play a role in the causal mechanism of disease.

19.4 Hand-in-hand walk with RNA-Seq data

In this section, we have used a class of ML algorithms, that is, a deep learning-based approach, for extracting important gene sets that work in synergy to govern the disease dynamics. The aim here is to achieve an accurate, optimal set of genes, which, along with a successful classification of genes, also play a role in the causal mechanism of breast cancer progression. Here, we have used the basic principles of machine learning techniques, as described in the materials and methods section and the workflow shown in Fig. 19.2.

19.4.1 Dataset selection

There are many publically accessible repositories/databases that contain high throughput data like RNA-Seq and many others. GEO, TCGA, and Arrayexpress are among the most popular public repositories. These databases can be queried for requirements like various disease conditions, experiment protocols, etc. To build a workflow, we queried the GEO database with the keywords "breast cancer" and

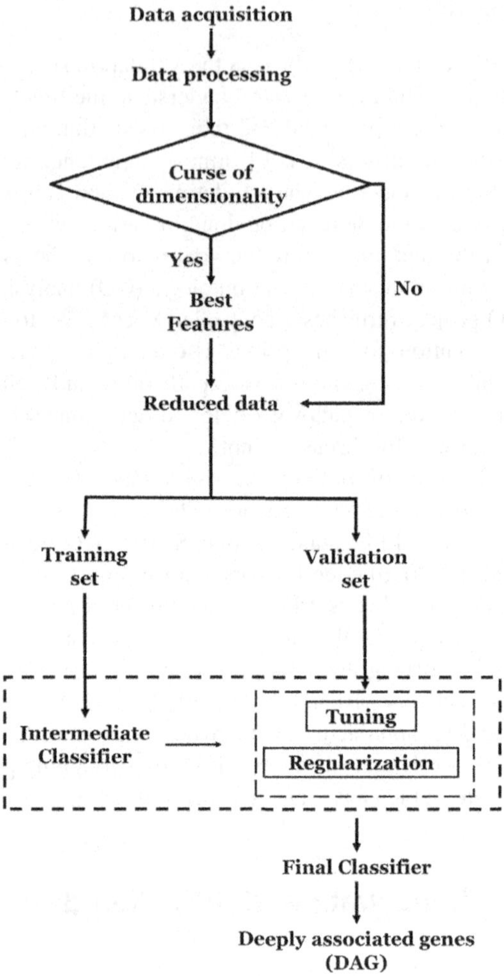

Figure 19.2 The machine learning pipeline. The process starts with data collection from various sources and eventually identifies disease-associated genes in the system. The significance of these genes is then further investigated using PPI network analysis.

"tumor grading." We have selected a dataset identified as GSE96058, which contains expression values of 3k individuals. These individuals are placed in one of three histological grading G1, G2, and G3. Further details can be found in the data source (Brueffer et al., 2018).

19.4.2 Data preprocessing

The example dataset GSE96058 was scrutinized for the preprocessing steps. Missing values are identified as black spaces, and zeros and outliers are identified by defining upper and lower bounds. The lower bound begins at Q1−1.5*IQR, and the upper bound ends at Q3 + 1.5*IQR, where Q1 is the first quartile, Q3 is the third quartile of the data, and IQR is the interquartile range. Data was clean and well preprocessed already, so it passed our scrutinization, and no additional work was needed for the preprocessing step. For the next steps, we randomly separated the dataset in a ratio of 8:1:1 in the train set, validation set, and test set.

19.4.3 Feature selection

The gene expression data contained approximately 30k transcripts for sample 3069. To deal with the curse of dimensionality, we build a PCA model on our train set to extract the important features from the dataset. Later we selected those principal components out of 30k , which explains the 95% variance of our dataset. To evaluate the significance of each feature of the original sample, we have created a score that will rank the feature in order of its importance. Each feature of the original samples has a heterogeneous effect on feature extraction results, that is, on the eigenvectors or principal components of the covariance matrix of the dataset. To estimate the score, we have averaged each feature's absolute contribution to filtered principal components (Song et al., 2010). Later, to decide the cut-off for selecting the best feature, we have plotted the distribution of scores and observed the first inflection in score distribution. We have chosen the inflection point as the cut-off value (1.5), that is, we have selected features with scores greater than 1.5, resulting in a list of 1249 transcripts.

Our hypothesis is that these 1249 transcripts can describe most of our dataset. To verify this further, we used the expression of selected genes as an important feature for classification models built using ANN (Fig. 19.3).

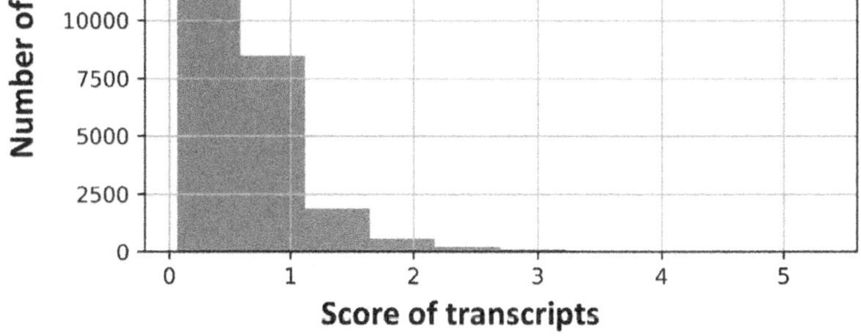

Figure 19.3 Distribution of feature score.

19.4.4 Classification model

The predictive accuracy of this classification model will assess the validity of the selected features. We have applied a supervised learning algorithm known as ANN. The reason for selecting ANN is that it captures nonlinearity very well. It also creates abstract features from original features, which might help us identify strongly connected gene clusters working as functional modules to drive pathophysiology. We used a training set for building the classification model with 1249 top-ranked transcripts from our feature selection algorithm. We have evaluated the model performance in terms of classification accuracy. To fine-tune the model's hyperparameters, we have applied systematic grid searching for selecting the optimal hyperparameters. This optimization is done using a validation dataset so that no unintended bias is incorporated into our model. After tuning, we have observed that optimal values of the learning rate, regularization parameter, hidden neuron, and hidden layers are 0.001, 0.0001, 50, and 1, respectively.

To rely on the observations of a model, it must be evaluated in many dimensions so that we do not get any poor or unnecessarily optimistic results. So, we intend to check that the model is neither underfitted nor overfitted. Hence, we have plotted our model's learning curves, which shows model performance on training and test dataset for various epochs of the model. We have observed 90% accuracy on the train set and 70% on the test set (Fig. 19.4). We trained the model up to 1000 epochs and observed that the learning curve for training and test data was approaching saturation. This implies that the model was neither underfitted nor overfitted and hence fit for further use.

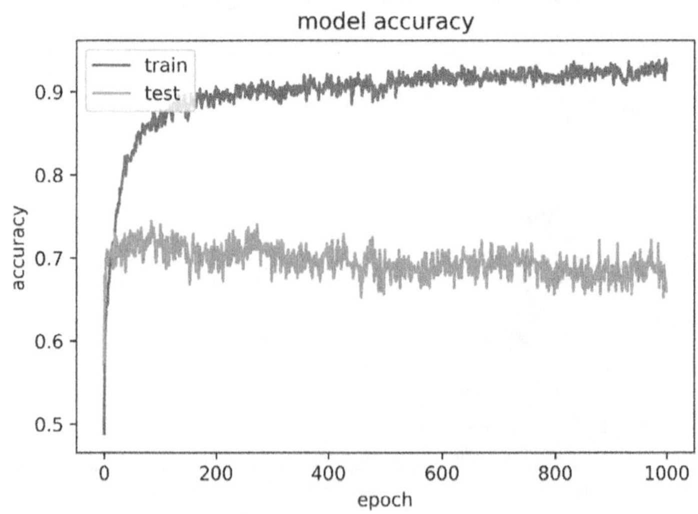

Figure 19.4 The learning curve of the model, which shows the accuracy of the classification algorithm on training and test datasets with increasing epoch size.

19.4.5 Identification of the genes involved in disease progression

Now we want to leverage our trained ANN model to answer questions about biological systems. As we know, biomolecules do not work in isolation. They maintain synergetic communication and perform tasks in a symphony. So with the help of ANN, we wanted to extract these communication patterns unbiasedly from the data. We achieved this by leveraging the beauty of the hidden layer of our ANN model. The hidden layer considers each gene's contribution in some proportion, that is, high from some genes and low from some genes. To identify this, we regularize and drop out some nodes and edges in our model. If an edge is not contributing much towards the next layer, regularization will try to make that edge weight zero or closer to zero through penalization. On the other hand, dropout will randomly reduce the weight of hidden neurons to zero or will check if it affects the model performance or not. So after applying dropout and regularization, we have eliminated neurons and edges having weight close to zeros or exactly zero. In turn, we get 11 deeply associated genes (DAG) modules, which might play a vital role in the causal mechanism of breast cancer progression.

19.4.6 Significance of the identified deeply associated genes

To investigate whether the DAGs are significant and might play a role in the causal mechanism of breast cancer progression, we opt for a PPI network analysis approach. We first export the human interactome data from the STRING database. The data contained 19,356 proteins with 11,759,455 interactions between them. We took a confidence score cut off of 900 to consider a highly confident network for our study. This step has reduced the number of proteins to 12,394 with 6,48,276 interactions between them (including the self-loops) and constructed a network of these proteins.

We found that 334 genes out of the 1249 transcriptomes obtained from PCA are mapped to the network. The reduction in numbers is because we have considered a very high-confidence network for our study. The constructed network contained many disconnected components. To study the progression of breast cancer, the disconnected components should be converted into a giant component. We next looked for a minimal number of proteins to connect these disconnected proteins and found that an additional 146 proteins are sufficient to construct a single giant component consisting of 423 nodes and 3590 edges. The network is shown in Fig. 19.5, where the larger the node's size, the greater its degree. The average degree of the nodes in the network was 15.39, with a maximum degree of 124 and the lowest degree of 2. Among the 158 DAG, we found 43 mapped to the network we created, and their topological properties are studied. The degree of these nodes is shown in Fig. 19.6A. We have considered those proteins as hubs whose degree is twice the average degree. The figure shows that six of the DAGs are hub proteins in the network. Clustering is an important property of a PPI network. A score ranging from 0 to 1 measure how the proteins in the network tend to cluster. We have measured

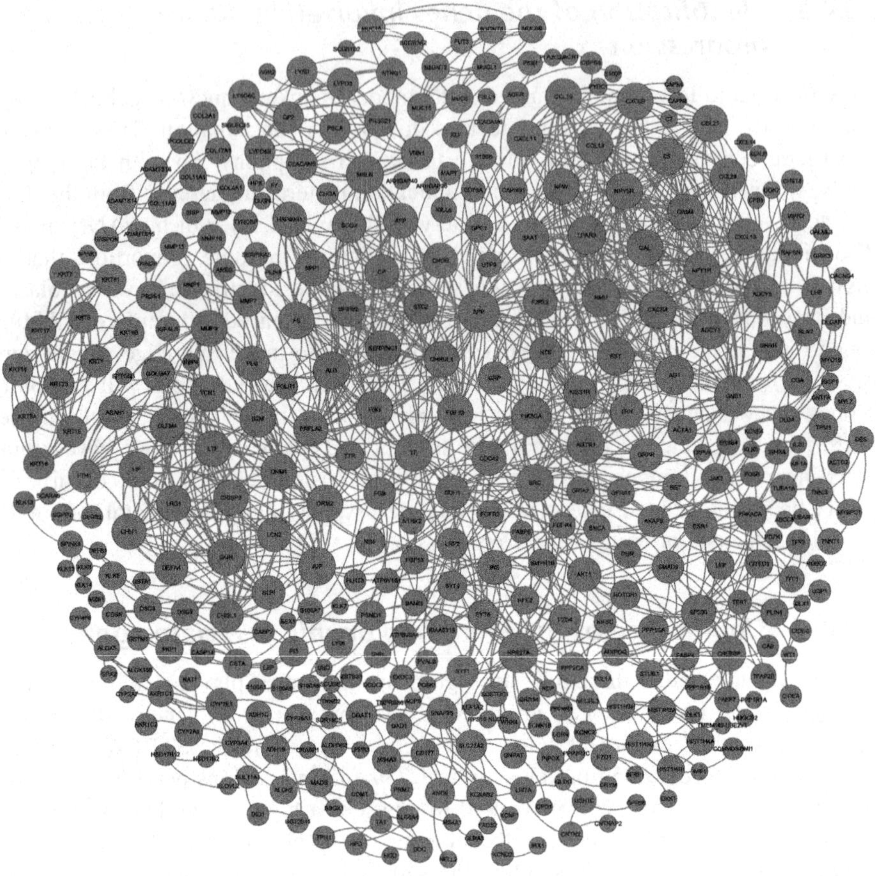

Figure 19.5 The network of the genes obtained from the feature selection method. Here nodes represent the proteins, and the edge represents the interaction between them, which can be both functional and physical. In the network, the node size denotes the degree of that particular node. Hence, the larger nodes denote the hub proteins of the network.

the clustering coefficient of the DAG proteins in the network and found that most of them have a strong tendency to form clusters. The result is shown in Fig. 19.6B.

The topologically strong proteins in the network, that is, hubs, nodes with higher clustering coefficients, etc., are crucial in the network. The removal of hubs increases the network diameter, that is, the number of nonreachable proteins in the network increases. Therefore these proteins are vital for maintaining the global network structure. On the other hand, the clustering coefficient measures the abundance of connected triangles in the network, it portrays the well-connectedness of the neighbors of a protein. Our results found that the DAGs are among the topologically strong proteins in the network and might play a role in breast cancer's causal mechanism.

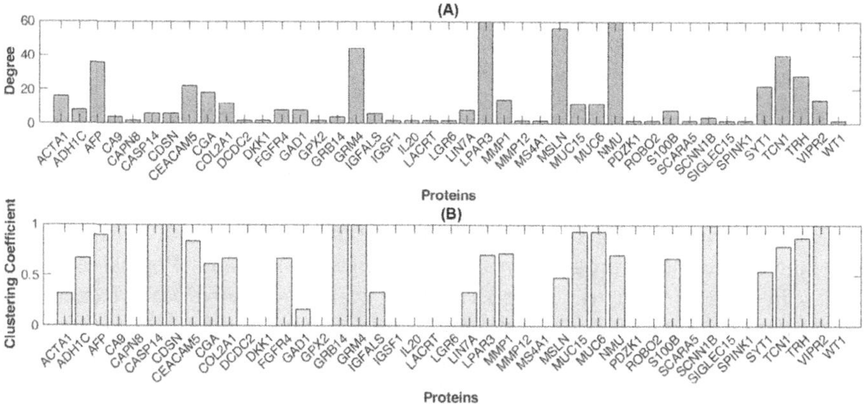

Figure 19.6 Topological properties of DAGs. (A) Degree of the 43 DAGs. This figure shows that six of the DAGs, AFP, GRM4, LPAR3, MSLN, NMU, and TCN1 are hubs. (B) The clustering coefficient of the DAGs. It shows that some proteins do not tend to form clusters, but most of the proteins have a strong clustering affinity.

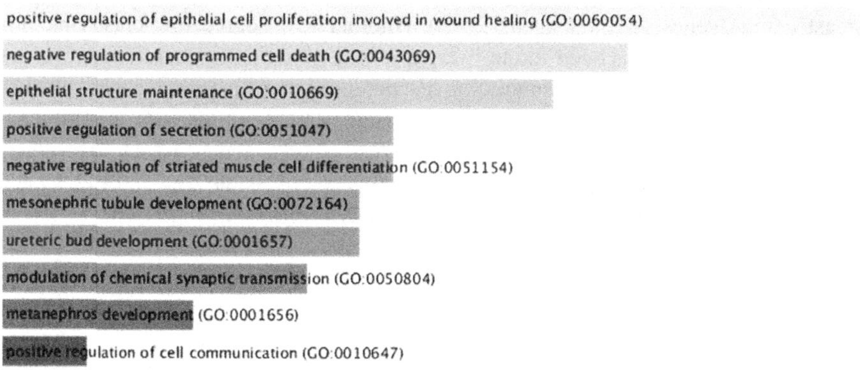

Figure 19.7 The results of the gene set enrichment analysis from EnrichR. Here the figure shows the top 10 biological processes associated with the DAGs obtained in our study. The biological processes are ranked according to the *p*-value.

We next performed the GO analysis using the EnrichR (Chen et al., 2013) and looked for the biological process in which the 43 DAGs play a part. The results are shown in Fig. 19.7. The most enriched biological process was the positive regulation of epithelial cell proliferation in wound healing, followed by the negative regulation of programmed cell death. Wound healing and cancer are the two faces of the same coin. A literature study has revealed that in breast cancer, wound healing fluid that originates from surgical sites increases the aggressiveness of cancer cells that last even after the surgery (Agresti et al., 2019). The negative regulation of programmed cell death is directly associated with cancer as it has been shown in the literature that tumor cell growth results not only from abnormal cell proliferation but also from reduced apoptosis (Sant et al., 2018).

We identified five hubs among these DAGs, which are already known to play a role in breast cancer. For example, AFP is an oncofetal antigen and is found in many, if not all, types of cancer, including breast cancer (Moro et al., 2012). GRM4, a member of Glutamate metabotropic receptors, plays a role in inhibiting cell proliferation, invasion, and migration in breast cancer cells (Xiao et al., 2019). The overexpression of LPAR3 in epithelial cells has been reported to be associated with aggressive tumor progression in human breast carcinoma (Popnikolov et al., 2012). In patients with triple-negative breast cancer, elevated MSLN expression was found to correlate with poor prognosis (Bera and Pastan, 2000; Li et al., 2014), while overexpression of NMU and TCN1 has been reported in breast cancer (Garczyk et al., 2017; Lee et al., 2017). CASP14 was found to possess a higher clustering coefficient. The investigation of its potential role in breast cancer has revealed that it is highly overexpressed in breast cancer (Handa et al., 2017). These results show that the DAGs are not only topologically strong but are also capable of capturing the disease genes of breast cancer.

A low-degree low-clustering protein is SCARA5. The literature survey has revealed that the downregulation of SCARA5 expression levels is associated with breast carcinogenesis (Ulker et al., 2018). Another low-degree low-clustering protein is GPX2, which has been found overexpressed in mammary carcinogenesis and cell proliferation in humans, suggesting its potential role in breast cancer therapy (Naiki-Ito et al., 2007). These results again suggest that, along with topologically strong proteins, the DAG set also includes lower-degree nodes, which also play a role in the causal mechanism of disease, omitting any possible degree of clustering biasedness. The list of DAGs obtained in the study is shown in Fig. 19.8.

Module1	Module2	Module3	Module4	Module5	Module6	Module7	Module8	Module9	Module10	Module11
JN111512	MIR7111	U73752	TRNA_Pseudo	MIR4459	DQ578783	CEACAM5	FDCSP	FDCSP	AJ236552	JN111512
C2orf54	CEACAM5	MMP1	SNORD10	JN576489	MIR8078	SNORA3	CGA	PRAME	MIR6834	HV975509
SLC1A1	TCN1	CA9	COL2A1	FDCSP	JC221905	MSMB	SCGB3A1	MMP1	MSLN	MUC6
TFPI2	SLITRK6	MIR3135A	SNORD117	CBLN2	MIR4523	CASP14	DCD	SNORA81	TRNA_Pro	AJ236552
MMP12	DL492076	DQ573287	CASP14	FAM3B	SNORA27	SCARNA4	SNORD21	MIR3135B	CAPN8	LACRT
CAPN8	DQ573287	SCARNA16	CA9	SNORA64	MIR4442	CA9	SNORA38B	FAM3B	STMND1	MIR6834
MIR6847	RTBDN	FAM196A	ELF5	HW120475	SNORD15B	SYT1	SLC5A8	TPSD1	IGFALS	IGHD
TRNA_Arg	HRASLS5	U4	AK310094	FAM196A	DQ571333	BC011773	SNORA81	SCARNA16	DQ341457	ODZ4
IGFALS	SPINK4	SPINK4	GRB14	ROBO2	PDZK1	ANKRD30B	MAPT-AS1	FAM196A	DKK1	ZNF385B
SLC44A5	AK125684	AK125684	ANKRD30B	GLYATL1	TRH	SNORA64	FAM196A	ROBO2	FGF10-AS1	TFPI2
BC010924	CNTD2	MT1H	SNORA64	CDSN	MIR6883	TMPRSS4	ROBO2	DQ597539	MIR21	MMP12
MIR21	CDSN	JQ778268	IL20	JQ778268	PEBP4	AK125684	UGT2B28	ABCA12	VIPR2	NCCRP1
SLC6A14	JA783658	FGFR4	HRASLS5	VSTM2L	FAM196A	AFP	DQ597539	MIR95	SCARA5	OBP2A
SIGLEC15	JQ778268	KLHDC7B	MAPK15	EF101778	MIR4489	MT1H	MIR106B	SNORA55	GAD1	WT1
DQ589204	TMEM145	TCRBV22S1A2N1	IGSF1	C1orf116	TRNA_Glu	MUC15	MIR8066	ACTA1	HV555659	SLC26A3
LIN7A	FGFR4	IL8	IGHE		MIR106B	CDSN	MIR95	JQ778268	LPAR3	IGFALS
S100B	RIIAD1	SCNN1B	NMU		SNORD116-1	NMU	AF086346	SNORD75	DCDC2	FGF10-AS1
	KLHDC7B	ADH1C	GPX2		SCUBE1	AK310634	ST8SIA6	RIIAD1	MS4A1	BC010924
	LOC101927722		M34430			JQ778268	VSTM2L	LGR6	LIN7A	MYEOV
	T-Cell Receptor V-alpha region					DQ601422	SPINK1	EF101778	DL490867	MIR21
						TMEM145	CYP4F11	C1orf116		SCARA5
								GRM4		LPAR3
								SCNN1B		SIGLEC15
										AX746649
										DCDC2
										LIN7A
										DQ584254

Figure 19.8 The list of DAGs obtained in the study. It contains both the topologically strong proteins having a higher degree and high clustering coefficient as well as the weak proteins having a low degree and low clustering coefficient.

19.5 Conclusion

In this chapter, we have recapitulated a ML-based approach to analyze RNA-Seq data in cancer. We started with the basic principles of ML-based analysis and later enacted the understating on a breast cancer dataset. Among the rabble of available methods for identifying disease-associated genes, we opted for the ANN method, which helped us capture 158 deeply associated genes that are possibly playing a role in breast cancer's pathophysiology. Here, the precedence of the selected method is that it is free from various invariable conditionality like threshold selection, which does not hold any concrete definition and might incorporate user's bias in the analysis. This process is also sensitive for less expressed genes, which is a major shortcoming of an abundance-based method to identify disease-associated genes. Our choice of neural network-based method for identifying DAG has provided an advantage of extracting condition-specific associations, which is not possible in molecular interaction network-based approaches. To signify the importance of DAGs, we opted for a PPI network analysis approach. We extracted the human interactome information from the STRING database and constructed a network of the genes obtained from the feature selection method. The interaction of this network was both physical and functional. We carried out a topological analysis in the network where we studied the degree of centrality and clustering affinity. The topological analysis demonstrates the potential behavior of the network. Such an analysis helps to identify the essential set of nodes in the network. The analysis reveals that the DAGs are among the topologically strong proteins in the network and thus very important in disease progression and possess high spreading power in the network. In other words, the identified DAGs have the potential to regulate the breast cancer network. This elucidation was further boosted by literature evidence. A significant association was observed between the DAGs and breast cancer, hypothesizing that these genes are possibly responsible for the disease's pathology.

The reliability of RNA-Seq technology to decipher the complexity of gene expression data is unequivocal. In this era of big data, the amalgamation of RNA-Seq with ML approaches has proven to be very promising in identifying the delinquent genes in various diseases. In this chapter, we have studied such an approach and provide a hybrid methodology that can be applied to study any RNA-Seq data. We believe that our method, which combines both PPI and ML, will provide the researchers a valuable perspective and will help to untangle a few knots of the inherent complex nature of diseases.

References

Agresti, R., Triulzi, T., Sasso, M., Ghirelli, C., Aiello, P., Rybinska, I., et al., 2019. Wound healing fluid reflects the inflammatory nature and aggressiveness of breast tumors. Cells 8.

Anand, R., Sarmah, D.T., Chatterjee, S., 2018. Extracting proteins involved in disease progression using temporally connected networks. BMC Syst. Biol. 12, 78.

Asif, M., Martiniano, H.F.M.C.M., Vicente, A.M., Couto, F.M., 2018. Identifying disease genes using machine learning and gene functional similarities, assessed through gene ontology. PLoS One 13, e0208626.

Barman, R.K., Mukhopadhyay, A., Maulik, U., Das, S., 2019. Identification of infectious disease-associated host genes using machine learning techniques. BMC Bioinforma. 20, 736.

Barrett, T., Wilhite, S.E., Ledoux, P., Evangelista, C., Kim, I.F., Tomashevsky, M., et al., 2013. NCBI GEO: archive for functional genomics data sets—update. Nucleic Acids Res. 41, D991–D995.

Batista, G.E.A.P.A., Gustavo, E.A.P., Monard, M.C., 2003. An analysis of four missing data treatment methods for supervised learning. Appl. Artif. Intell. 17, 519–533.

Bera, T.K., Pastan, I., 2000. Mesothelin is not required for normal mouse development or reproduction. Mol. Cell. Biol. 20, 2902–2906.

Bhargava, N., Sharma, S., Purohit, R., Rathore, P.S., 2017. Prediction of recurrence cancer using J48 algorithm. In: 2017 2nd International Conference on Communication and Electronics Systems (ICCES).

Brueffer, C., Vallon-Christersson, J., Grabau, D., Ehinger, A., Häkkinen, J., Hegardt, C., et al., 2018. Clinical value of RNA sequencing-based classifiers for prediction of the five conventional breast cancer biomarkers: a report from the population-based multicenter sweden cancerome analysis network-breast initiative. JCO Precis. Oncol. 2.

Chen, X., Ishwaran, H., 2012. Random forests for genomic data analysis. Genomics 99, 323–329.

Chen, E.Y., Tan, C.M., Kou, Y., Duan, Q., Wang, Z., Meirelles, G.V., et al., 2013. Enrichr: interactive and collaborative HTML5 gene list enrichment analysis tool. BMC Bioinforma. 14, 128.

Chen, Z., Yang, X., Bi, G., Liang, J., Hu, Z., Zhao, M., et al., 2020. Ligand-receptor interaction atlas within and between tumor cells and T cells in lung adenocarcinoma. Int. J. Biol. Sci. 16, 2205–2219.

Diebold, F.X., 2012. On the origin(s) and development of the term "big data." SSRN Electronic Journal.

Dou, Y., Guo, X., Yuan, L., Holding, D.R., Zhang, C., 2015. Differential expression analysis in RNA-Seq by a Naive Bayes classifier with local normalization. Biomed. Res. Int. 789516. 2015.

Evans, C., Hardin, J., Stoebel, D.M., 2018. Selecting between-sample RNA-Seq normalization methods from the perspective of their assumptions. Brief. Bioinform. 19, 776–792.

Garczyk, S., Klotz, N., Szczepanski, S., Denecke, B., Antonopoulos, W., von Stillfried, S., et al., 2017. Oncogenic features of neuromedin U in breast cancer are associated with NMUR2 expression involving crosstalk with members of the WNT signaling pathway. Oncotarget 8, 36246–36265.

Ge, W., Ma, X., Li, X., Wang, Y., Li, C., Meng, H., et al., 2009. B7-H1 up-regulation on dendritic-like leukemia cells suppresses T cell immune function through modulation of IL-10/IL-12 production and generation of Treg cells. Leuk. Res. 33, 948–957.

Handa, T., Katayama, A., Yokobori, T., Yamane, A., Horiguchi, J., Kawabata-Iwakawa, R., et al., 2017. Caspase14 expression is associated with triple negative phenotypes and cancer stem cell marker expression in breast cancer patients. J. Surg. Oncol. 116, 706–715.

Hodge, V., Austin, J., 2004. A survey of outlier detection methodologies. Artif. Intell. Rev. 22, 85–126.

Huang, Z., Johnson, T.S., Han, Z., Helm, B., Cao, S., Zhang, C., et al., 2020. Deep learning-based cancer survival prognosis from RNA-seq data: approaches and evaluations. BMC Med. Genomics 13, 41.

Kelemen, O., Convertini, P., Zhang, Z., Wen, Y., Shen, M., Falaleeva, M., et al., 2013. Function of alternative splicing. Gene 514, 1–30.

Lee, Y.-Y., Wei, Y.-C., Tian, Y.-F., Sun, D.-P., Sheu, M.-J., Yang, C.-C., et al., 2017. Overexpression of transcobalamin 1 is an independent negative prognosticator in rectal cancers receiving concurrent chemoradiotherapy. J. Cancer 8, 1330–1337.

Li, Y.R., Xian, R.R., Ziober, A., Conejo-Garcia, J., Perales-Puchalt, A., June, C.H., et al., 2014. Mesothelin expression is associated with poor outcomes in breast cancer. Breast Cancer Res. Treat. 147, 675–684.

Libbrecht, M.W., Noble, W.S., 2015. Machine learning applications in genetics and genomics. Nat. Rev. Genet. 16, 321–332.

Licata, L., Lo Surdo, P., Iannuccelli, M., Palma, A., Micarelli, E., Perfetto, L., et al., 2020. SIGNOR 2.0, the signaling network open resource 2.0: 2019 update. Nucleic Acids Res. 48, D504–D510.

Liu, Z.-P., Wu, C., Miao, H., Wu, H., 2015. RegNetwork: an integrated database of transcriptional and post-transcriptional regulatory networks in human and mouse. Database 2015.

Lomax, J., 2005. Get ready to GO! A biologist's guide to the gene ontology. Brief. Bioinform. 6, 298–304.

Mladenić, D., 2006. Feature selection for dimensionality reduction. Subspace, Latent Structure and Feature Selection. pp. 84–102.

Mohandes, S.R., Zhang, X., Mahdiyar, A., 2019. A comprehensive review on the application of artificial neural networks in building energy analysis. Neurocomputing. .

Moro, R., Gulyaeva-Tcherkassova, J., Stieber, P., 2012. Increased alpha-fetoprotein receptor in the serum of patients with early-stage breast cancer. Curr. Oncol. 19, e1–e8.

Naiki-Ito, A., Asamoto, M., Hokaiwado, N., Takahashi, S., Yamashita, H., Tsuda, H., et al., 2007. Gpx2 is an overexpressed gene in rat breast cancers induced by three different chemical carcinogens. Cancer Res. 67, 11353–11358.

Oshlack, A., Wakefield, M.J., 2009. Transcript length bias in RNA-seq data confounds systems biology. Biol. Direct 4, 14.

Popnikolov, N.K., Dalwadi, B.H., Thomas, J.D., Johannes, G.J., Imagawa, W.T., 2012. Association of autotaxin and lysophosphatidic acid receptor 3 with aggressiveness of human breast carcinoma. Tumour Biol. 33, 2237–2243.

Richard, H., Schulz, M.H., Sultan, M., Nürnberger, A., Schrinner, S., Balzereit, D., et al., 2010. Prediction of alternative isoforms from exon expression levels in RNA-Seq experiments. Nucleic Acids Res. 38, e112.

Risso, D., Schwartz, K., Sherlock, G., Dudoit, S., 2011. GC-content normalization for RNA-Seq data. BMC Bioinforma. 12, 480.

Robinson, M.D., Oshlack, A., 2010. A scaling normalization method for differential expression analysis of RNA-seq data. Genome Biol. 11, R25.

Samuel, A.L., 1988. Some studies in machine learning using the game of checkers. II—recent progress. Computer Games I.

Sant, D.W., Mustafi, S., Gustafson, C.B., Chen, J., Slingerland, J.M., Wang, G., 2018. Vitamin C promotes apoptosis in breast cancer cells by increasing TRAIL expression. Sci. Rep. 8, 5306.

Sarmah, D.T., Bairagi, N., Chatterjee, S., 2020. Tracing the footsteps of autophagy in computational biology. Brief. Bioinform. .

Song, F., Guo, Z., Mei, D., 2010. Feature selection using principal component analysis. In: 2010 International Conference on System Science, Engineering Design and Manufacturing Informatization.

Szklarczyk, D., Gable, A.L., Lyon, D., Junge, A., Wyder, S., Huerta-Cepas, J., et al., 2019. STRING v11: protein-protein association networks with increased coverage, supporting functional discovery in genome-wide experimental datasets. Nucleic Acids Res. 47, D607–D613.

Tang, B., Pan, Z., Yin, K., Khateeb, A., 2019. Recent advances of deep learning in bioinformatics and computational biology. Front. Genet. 10, 214.

Tang, W., Cao, Y., Ma, X., 2020. Novel prognostic prediction model constructed through machine learning on the basis of methylation-driven genes in kidney renal clear cell carcinoma. Biosci. Rep. 40.

Ulker, D., Ersoy, Y.E., Gucin, Z., Muslumanoglu, M., Buyru, N., 2018. Downregulation of SCARA5 may contribute to breast cancer via promoter hypermethylation. Gene 673, 102–106.

Vinayagam, A., Gibson, T.E., Lee, H.-J., Yilmazel, B., Roesel, C., Hu, Y., et al., 2016. Controllability analysis of the directed human protein interaction network identifies disease genes and drug targets. Proc. Natl. Acad. Sci. U. S. A 113, 4976–4981.

Xiao, B., Chen, D., Zhou, Q., Hang, J., Zhang, W., Kuang, Z., et al., 2019. Glutamate metabotropic receptor 4 (GRM4) inhibits cell proliferation, migration and invasion in breast cancer and is regulated by miR-328-3p and miR-370-3p. BMC Cancer 19, 891.

Yang, X., Coulombe-Huntington, J., Kang, S., Sheynkman, G.M., Hao, T., Richardson, A., et al., 2016. Widespread expansion of protein interaction capabilities by alternative splicing. Cell 164, 805–817.

Zhang, Z.-M., Wang, J.-S., Zulfiqar, H., Lv, H., Dao, F.-Y., Lin, H., 2020. Early diagnosis of pancreatic ductal adenocarcinoma by combining relative expression orderings with machine-learning method. Front. Cell Dev. Biol. 8, 582864.

Zhao, Y., Li, H., Fang, S., Kang, Y., Wu, W., Hao, Y., et al., 2016. NONCODE 2016: an informative and valuable data source of long non-coding RNAs. Nucleic Acids Res. 44, D203–D208.

Zhao, Y., Pan, Z., Namburi, S., Pattison, A., Posner, A., Balachander, S., et al., 2020. CUP-AI-Dx: a tool for inferring cancer tissue of origin and molecular subtype using RNA gene-expression data and artificial intelligence. EBioMedicine 61, 103030.

Index

Note: Page numbers followed by "*f*" and "*t*" refer to figures and tables, respectively.

B

D

Q

S

Made in the USA
Las Vegas, NV
10 October 2025

32144924R00275